Heterogeneous Nanocatalysis for Energy and Environmental Sustainability

Heterogeneous Nanocatalysis for Energy and Environmental Sustainability

Volume 1 - Energy Applications

Edited by

Putla Sudarsanam
Department of Chemistry
Indian Institute of Technology Hyderabad
Kandi, Telangana, India

Yusuke Yamauchi
Australian Institute for Bioengineering and Nanotechnology (AIBN)
School of Chemical Engineering
The University of Queensland
Brisbane, Queensland, Australia

Pankaj Bharali
Department of Chemical Sciences
Tezpur University
Nappam, Assam, India

WILEY

Registered Offices
John Wiley & Sons, Inc., 111 River Street, Hoboken, NJ 07030, USA
John Wiley & Sons Ltd, The Atrium, Southern Gate, Chichester, West Sussex, PO19 8SQ, UK

Editorial Office
The Atrium, Southern Gate, Chichester, West Sussex, PO19 8SQ, UK

For details of our global editorial offices, customer services, and more information about Wiley products visit us at www.wiley.com.

Wiley also publishes its books in a variety of electronic formats and by print-on-demand. Some content that appears in standard print versions of this book may not be available in other formats.

Library of Congress Cataloging-in-Publication Data
Names: Sudarsanam, Putla, editor. | Yamauchi, Yusuke, editor. | Bharali, Pankaj, editor.
Title: Heterogeneous nanocatalysis for energy and environmental sustainability / Putla Sudarsanam, Yusuke Yamauchi, Pankaj Bharali.
Description: First edition. | Chichester, West Sussex, UK ; Hoboken, NJ, USA : Wiley, 2023. | Includes bibliographical references and index. | Contents: volume 1. Energy applications – volume 2. Environmental applications.
Identifiers: LCCN 2022032470 (print) | LCCN 2022032471 (ebook) | ISBN 9781119771999 (cloth ; volume 1) | ISBN 9781119772026 (cloth ; volume 2) | ISBN 9781119772002 (adobe pdf ; volume 1) | ISBN 9781119772019 (epub ; volume 1) | ISBN 9781119772033 (adobe pdf ; volume 2) | ISBN 9781119772040 (epub ; volume 2)
Subjects: LCSH: Catalysts. | Heterogeneous catalysis. | Nanochemistry–Industrial applications. | Green chemistry. | Electric power production from chemical action. | Biomass energy. | Environmental protection. | Pollution prevention.
Classification: LCC TP159.C3 H48 2023 (print) | LCC TP159.C3 (ebook) | DDC 660/.2995–dc23/eng/20220720
LC record available at https://lccn.loc.gov/2022032470
LC ebook record available at https://lccn.loc.gov/2022032471

Cover Design: Wiley
Cover Images: © Intographics/Pixabay, Ractapopulous/Pixabay

Set in 9.5/12.5pt STIXTwoText by Straive, Pondicherry, India
Printed and bound by CPI Group (UK) Ltd, Croydon, CR0 4YY

C9781119771999_281022

Contents

Preface

Heterogeneous Nanocatalysis for Energy and Environmental Sustainability, Volume 1: Energy Applications

Indubitably, fossil fuels (coal, petroleum, and natural gas) are the primary sources of energy for modern society. Concurrently, there are several concerns associated with fossil fuels, including price volatility, long-term availability, and environmental detrimental effects. Thus, utilizing alternative energy and technology (e.g. biomass/solar, clean hydrogen, fuel cells) has received tremendous attention to meet the increasing energy demand while controlling environmental pollution, thus moving toward a sustainable society.

Heterogeneous catalysis is a key technology for various energy applications, for instance, renewable energy (solar/biomass) valorization, clean hydrogen production, and photo/electro/chemical energy conversion processes. Over the last two decades, a variety of heterogeneous solid catalysts (bare/mixed oxides, supported metals, zeolites, carbons, perovskites, metal-organic frameworks, MXenes, hybrid materials, etc.) have been developed with desirable active sites and specific functionalities. The particle size of a solid catalyst is a key determining factor in improving the catalytic properties, such as specific surface area, active site density, and active site–reactant interactions. This breakthrough discovery has paved the way for the development of a new type of catalysis, i.e. nanocatalysis that plays a key role in bridging the gap between homogeneous and heterogeneous catalysis. Thus, the key focus of current catalysis research is to develop advanced nanocatalysts with well-controlled particle sizes (1–100 nm in at least one dimension) and morphologies because they exhibit promising catalytic activities compared with bulk solid catalysts.

This book focuses on the synthesis, characterization, and applications of a variety of nanosized catalysts for the conversion of renewable energy (biomass/solar) into fuels and chemicals, clean hydrogen production, photoelectrochemical energy conversion processes, etc. Chapter 1 elucidates various factors intervening in oxide and oxide-composite supports of nanocatalysts in energy conversion. Chapters 2 and 3 summarize the role of nanocatalysis for renewable aromatics from biomass and biodiesel production, respectively. Chapter 4 discusses the role of hybrid electrocatalysts for oxygen electrode reactions. The applications of porous graphitic carbon nitride nanostructures and 2D transition metal carbides (MXenes) for photocatalytic hydrogen evolution reaction and electrocatalysis are summarized in Chapters 5 and 6, respectively. Chapter 7 provides the advances and challenges in Pt-free Pd-based catalysts for oxygen electro-reduction in alkaline media. Chapter 8 reports the applications of the morphology- and size-selective Pd-based electrocatalysts for fuel cell reactions. The significance of Prussian blue analogues and nanostructured graphene oxide-based catalysts for H_2O_2 production and Fischer–Tropsch synthesis are highlighted in Chapters 9 and 10, respectively. Chapter 11 addresses the production of light olefin via catalytic oxidative cracking. The magic of heterogeneous nanocatalysis in VLPC for sustainable organic synthesis is discussed in Chapter 12. Hydrogen generation from ammonia borane using various metal nanoparticle-based catalysts is summarized in Chapter 13.

The book covers various aspects of nanocatalysis and nanosized catalysts, including synthesis, characterization, and their role in achieving energy sustainability. Thus, the book will be of great interest to researchers/students and industrial chemists familiar with the fields of heterogeneous catalysis and nanocatalysis. We hope that catalysis researchers will find this book as a key reference for the fundamental understanding of surface science, rational catalyst design, and the applications of various nanocatalysts for energy applications. We would like to thank the authors for accepting the invitation and submitting the chapters to this book. We also thank the production team at Wiley for their invaluable assistance and support.

Dr. Putla Sudarsanam
Department of Chemistry
Indian Institute of Technology Hyderabad
Kandi, Telangana, India

Professor Yusuke Yamauchi
Australian Institute for Bioengineering and Nanotechnology (AIBN)
School of Chemical Engineering
The University of Queensland
Brisbane, Queensland, Australia

Dr. Pankaj Bharali
Department of Chemical Sciences
Tezpur University
Napaam, Assam, India

List of Contributors

Nicolas Alonso-Vante
IC2MP, UMR-CNRS 7285
University of Poitiers
Poitiers Cedex, France

Harshitha N. Anchan
Department of Chemistry
National Institute of Technology
Karnataka (NITK)
Mangalore, Karnataka, India

Bindu Antil
Department of Chemistry
University of Delhi
Delhi, India

Pankaj Bharali
Department of Chemical Sciences
Tezpur University
Nappam, Assam, India

Navya Subray Bhat
Department of Chemistry
National Institute of Technology
Karnataka (NITK)
Mangalore, Karnataka, India

Biraj Jyoti Borah
Department of Chemical Sciences
Tezpur University
Nappam, Assam, India

Rashmi Chetry
Department of Chemical Sciences
Tezpur University
Nappam, Assam, India

Bhugendra Chutia
Department of Chemical Sciences
Tezpur University
Nappam, Assam, India

Sasanka Deka
Department of Chemistry
University of Delhi
Delhi, India

Saikat Dutta
Department of Chemistry
National Institute of Technology
Karnataka (NITK)
Mangalore, Karnataka, India

Luis Alberto Estudillo-Wong
IC2MP, UMR-CNRS 7285
University of Poitiers
Poitiers Cedex, France

and

Departamento de Biociencias e Ingeniería
CIIEMAD
Instituto Politécnico Nacional
CDMX, Mexico

Xianfeng Fan
School of Engineering
The University of Edinburgh
Edinburgh, United Kingdom

Sarah Farrukh
School of Chemical and Materials Engineering
National University of Science and Technology
Islamabad, Pakistan

Ujjal K. Gautam
Department of Chemical Sciences
Indian Institute of Science Education and Research
Mohali, Punjab
India

Chiranjita Goswami
Department of Chemical Sciences
Tezpur University
Nappam, Assam, India

Harikrishna
Sustainable Materials and Catalysis Research Laboratory
Department of Chemistry
Indian Institute of Technology Jodhpur
Karwar, Rajasthan, India

Kumar Kashyap Hazarika
Department of Chemical Sciences
Tezpur University
Nappam, Assam, India

Nadeem Iqbal
Centre for Undergraduate Studies
University of the Punjab
Lahore, Pakistan

Suresh Iyer
CSIR-National Chemical Laboratory
Pune, Maharashtra, India

Huda Sharbini Kamaluddin
Department of Chemistry
Faculty of Science
King Abdulaziz University
Jeddah, Saudi Arabia

Mala Khan
Bangladesh Reference Institute for Chemical Measurements (BRiCM)
Dhaka, Bangladesh

R. Krishnapriya
Sustainable Materials and Catalysis Research Laboratory
Department of Chemistry
Indian Institute of Technology Jodhpur
Karwar, Rajasthan, India

Divya Kumar
Sustainable Materials and Catalysis Research Laboratory
Department of Chemistry
Indian Institute of Technology Jodhpur
Karwar, Rajasthan, India

Devika Laishram
Sustainable Materials and Catalysis Research Laboratory
Department of Chemistry
Indian Institute of Technology Jodhpur
Karwar, Rajasthan, India

Xiugang Li
Key Laboratory of Functional Small Molecules for Ministry of Education
College of Chemistry and Chemical Engineering
Jiangxi Normal University
Nanchang, China

and

College of Material and Chemical Engineering
Tongren University
Tongren, China

Zhang-Hui Lu
Key Laboratory of Functional Small Molecules for Ministry of Education
College of Chemistry and Chemical Engineering
Jiangxi Normal University
Nanchang, China

Sara Musaddiq
Department of Chemistry
The Women University Multan
Multan, Pakistan

Kiran Mustafa
Department of Chemistry
The Women University Multan
Multan, Pakistan

Katabathini Narasimharao
Department of Chemistry
Faculty of Science
King Abdulaziz University
Jeddah, Saudi Arabia

Suranjana Patowary
Department of Chemical Sciences
Tezpur University
Nappam, Assam, India

Alwar Ramani
School of Energy, Geoscience, Infrastructure and Society
Heriot-Watt University
Edinburg, Scotland

Lipipuspa Sahoo
Department of Chemical Sciences
Indian Institute of Science Education and Research
Mohali, Punjab, India

Bhagirath Saini
Sustainable Materials and Catalysis Research Laboratory
Department of Chemistry
Indian Institute of Technology Jodhpur
Karwar, Rajasthan, India

Swagotom Sarkar
Bangladesh Reference Institute for Chemical Measurements (BRiCM)
Dhaka, Bangladesh

Rakesh Kumar Sharma
Sustainable Materials and Catalysis Research Laboratory
Department of Chemistry
Indian Institute of Technology Jodhpur
Karwar, Rajasthan, India

Kiran P. Shejale
Sustainable Materials and Catalysis Research Laboratory
Department of Chemistry
Indian Institute of Technology Jodhpur
Karwar, Rajasthan, India

Nittan Singh
Catalysis and Inorganic Chemistry Division
CSIR-National Chemical Laboratory
Pune, Maharashtra, India

and

Academy of Scientific and Innovative Research (AcSIR)
Ghaziabad, Uttar Pradesh, India

Arumugam Sudalai
CSIR-National Chemical Laboratory
Pune, Maharashtra, India

Putla Sudarsanam
Department of Chemistry
Indian Institute of Technology Hyderabad
Kandi, Telangana, India

Hiroyasu Tabe
Department of Chemistry and Bioengineering
Graduate School of Engineering
Osaka Metropolitan University
Osaka, Japan

and

Research Center for Artificial Photosynthesis (ReCAP)
Osaka Metropolitan University
Osaka, Japan

Yusuke Yamada
Department of Chemistry and Bioengineering
Graduate School of Engineering
Osaka Metropolitan University
Osaka, Japan

and

Research Center for Artificial Photosynthesis (ReCAP)
Osaka Metropolitan University
Osaka, Japan

1

Factors Intervening in Oxide and Oxide-Composite Supports on Nanocatalysts in the Energy Conversion

Luis Alberto Estudillo-Wong[1,2] *and Nicolas Alonso-Vante*[1]

[1] *IC2MP, UMR-CNRS 7285, University of Poitiers, Poitiers Cedex, France*
[2] *Departamento de Biociencias e Ingeniería, CIIEMAD, Instituto Politécnico Nacional, CDMX, México*

Acronyms

AES	Auger Electron Spectroscopy
AST	Accelerated Stability Test
BET	Bruanuer-Emmett–Teller
CCR	Carbonyl Chemical Route
DFA	Debye Function Analysis
DFT	Density Functional Theory
DPR	Derivate Peak fitting of diffuse Reflectance
EC	Electrocatalyst
ECSA	Electrochemical Surface-active Area
EIS	Electrochemical Impedance Spectroscopy
FC	Fuel Cell
FFT	Fast Fourier Transform
FWHM	Full-Width at Half Maximum
HER	Hydrogen Evolution Reaction
HOR	Hydrogen Oxidation Reaction
HR-TEM	High Resolution - Transmission Electron Microscopy
MA	Mass Surface Area
NPs	Nanoparticles
OBGE	Optical Band-Gap Energy
OER	Oxygen Evolution Reaction
ORR	Oxygen Reduction Reaction

Heterogeneous Nanocatalysis for Energy and Environmental Sustainability: Energy Applications, First Edition. Edited by Putla Sudarsanam, Yusuke Yamauchi, and Pankaj Bharali.

PDEIS	Potentiodynamic Electrochemical Impedance Spectroscopy
PDM	Photo-assisted Deposition Method
PDOS	Projected Electronic Density of States
PXRD	Powder X-ray Diffraction
RDS	Rate-Determining Step
RF	Rietveld Refinement
SA	Specific Activity
SEM	Scanning Electron Microscopy
SMCI	Strong Metal–Carbon Interaction
SMSI	Strong Metal–Support Interaction
TIP	Titanium Isopropoxide
TEM	Transmission Electron Microscopy
TOF	Turnover Frequency
W-H	Williamson–Hall
WMCI	Weak Metal–Carbon Interaction
WMSI	Weak Metal–Support Interaction
WPPF	Whole Powder Pattern Fitting
XAS	X-ray Absorption Spectroscopy
XPS	X-ray Photoelectron Spectroscopy

1.1 Overview in Materials Used as Supports in Electrocatalysis

The stability of the catalytic center has always been the biggest challenge in any reaction of technological interest. For example, precious materials, such as platinum (Pt), are used as electrocatalysts (ECs) and widely studied in fuel cell (FC) systems [1]. The nature of its degradation impacts the overall performance of the power generation system [2]. This degradation is due to the strength between the metal and the support, which compromises the stability of the catalytic center and consequently decreases the lifetime of the electrochemical system. The materials used as support and their interaction with the catalytic center play a substantial role in the degradation mechanisms.

Carbon-based materials are widely used as support due to their high electrical conductivity and surface area [3]. The corrosion of this material occurs at potentials higher than 0.97 V (RHE) [4]. For this purpose, different carbon allotropies have been explored as supports, which are those with low or high graphitization levels, such as amorphous carbon (XC-72) [5], carbon nanotube [1], and graphene [6]. On the other hand, oxide-based materials have gained interest due to their stability under operating conditions [7]. However, they present low electrical conductivity and surface area compared to carbon-based materials [8]. The oxide-based materials that have been studied as supports are TiO_2 [9], SnO_2 [10], CeO_2 [11], WO_3 [12], TiO_2, or SnO_2 doped with Nb [13], Mo [14], Ru [15], Sb [16, 17], or mixed oxides such as CeO_2-TiO_2 [18].

The oxides can have different shapes, some of which, more typical, are spheres (0D) [19], nanowires, or nanotubes (1D) [20, 21].

For the supports presented in this chapter, we can classify them into two domains, which are (i) carbon-based and (ii) oxide-based materials. The combination of these materials has allowed designing hybrid materials (composites), which combine properties of oxide and carbon materials [22–27]. Here, the purpose is to show how the synthesis route used in the synthesis of these materials is based on the synthesis of carbon-based materials.

We also intend to show how the synthesis route, used for the catalytic center, generates different interactions with the support. For this reason, the physicochemical characterization between the support and the catalytic center plays an important role. The latter aims to determine the structural parameters, which are modified by the hetero-bonding between the catalytic center and the support. Additionally, the use of density functional theory (DFT) simulation on these catalysts, previously characterized physical-chemically, allows the prediction and design of new materials applied to a model reaction.

The electrochemical characterization gives us the material fingerprint, and the photoelectrochemical characterization gives the degree of recombination of the electron–hole pairs (e^-/h^+) of the oxide materials used. Electrochemical parameters such as electrochemically active surface area (ECSA), mass surface area (MA), and specific activity (SA) as a function of structural and optoelectronic properties give clues to the performance applied to a reaction of interest.

Under this context, the case studies confirm how different metal–support interactions allow to modify the catalytic center being a precious metal or its substitution by a nonprecious material.

1.2 Chemical Synthesis of Materials

Synthesis and design of materials are the most crucial part to undertake for different applications, such as medicine [28, 29], electronics [30], energy conversion [31], remediation [32], and pharmaceutical [33]. Here, we review three support and catalytic center synthesis routes, the physicochemical characterization, the support interaction with the catalytic center, and the theoretical approach that has been done so far.

1.2.1 Routes of Synthesis

1.2.1.1 Sol–Gel

This technique focuses on the hydrolysis of different precursors and their conversion into an inorganic solid induced by inorganic polymerization in water to the synthesis of oxide supports. The sol–gel process generally follows four or five steps [34], namely: (i) dissolution of the chemical precursor in an organic solvent, (ii) conversion of the mixture solution into a sol by adding water/acid–base solution, (iii) aging/shaping, and (iv) thermal treatment.

A sol–gel reaction starts with hydrolysis and continues with a condensation of the precursors. According to the precursor, the sol–gel method can be classified into (i) inorganic salt and (ii) metal-organic compound [34]. Different metal oxide materials can be synthesized via the sol–gel method, such as TiO_2 [35], SnO_2 [36], and ZnO [37]. Also, the sol–gel route's main advantage is modifying steps or combining methods to tailor novel materials [38–40]. In this latter case, a composite is a combination of properties from two (or more) individual materials incorporated into one material. For example, electrical conductivity is the most critical issue on catalysts support for electrochemical applications. Therefore, the oxide-carbon composite commits to increase its electrical conductivity and avoid carbon corrosion [23, 26, 41]. As a result, the central oxide can also be doped by a second element in the composite material [42, 43]. Figure 1.1 shows a modified sol–gel method for the preparation of M:TiO_2-C composites.

Figure 1.1 is a setup where a modified sol–gel route to synthesize oxide-carbon composites is shown. In the beginning, the MCl_x precursor is mixed with isopropanol solvent in an ice bath. Thereafter, the titanium isopropoxide (TIP) is added and mixed for 16 hours. The steps followed are (i) increase of temperature to 80 °C under nitrogen atmosphere, (ii) addition of XC-72 carbon support, and (iii) addition of 2 ml of a solution of 0.7 M H_2SO_4.

The synthesis conditions are continued, and after 12 hours, the solvent is evaporated at 98 °C. The as-prepared material is thereafter heat treated at 400 °C in an air atmosphere. Lastly, the support is washed with deionized water and dried at 60 °C for 12 hours [42]. This synthesis chemical route leads to a change in the optoelectronic properties of the oxide.

1.2.1.2 Carbonyl Chemical Route

This method has been applied to prepare a wide range of electrocatalytic materials, namely, noble [25, 26, 44], non-noble metals [45], alloys [46–48], and semiconductors [49]. The Pt cluster complex was initiated by Longoni et al. [50], and the reduction of several Pt precursors and various reducing agents to produce various dianions $\left[Pt_3(CO)_6\right]_n^{2-}$ $(n = 10, 5, 4, 3)$ was subsequently reported using catalytic materials supports [51]. Figure 1.2 shows this route to obtain carbon supported Pt or Pt-nanoalloys materials. Here, three essential aspects

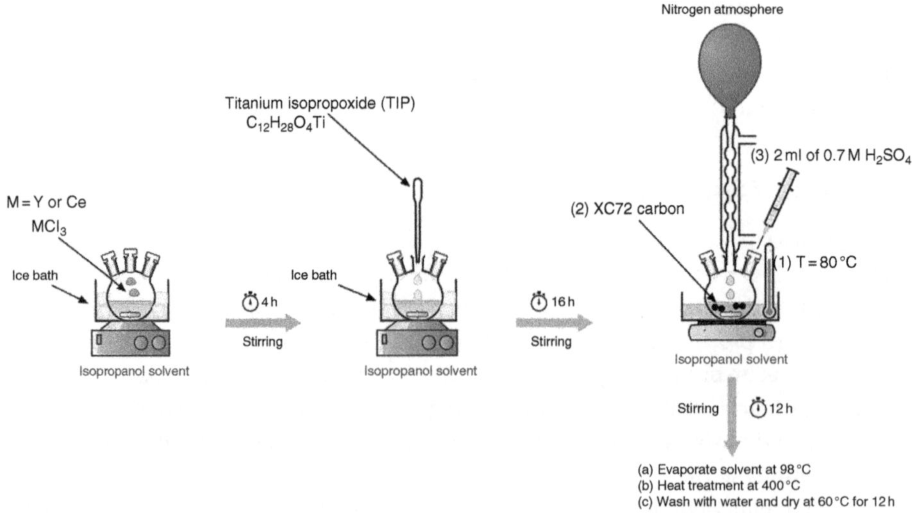

Figure 1.1 Modified sol–gel route for the synthesis of M:TiO_2/C composites.

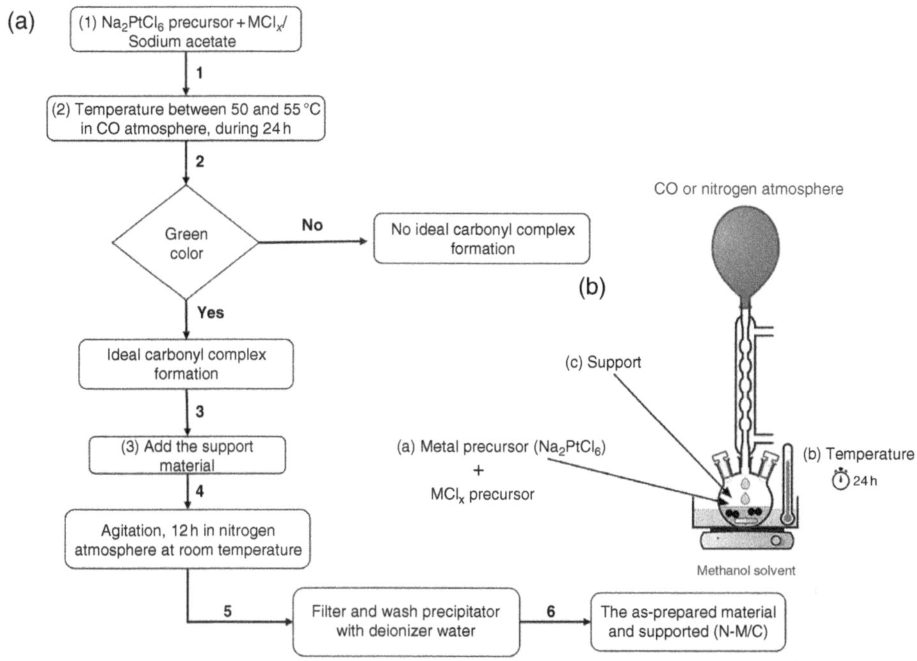

Figure 1.2 (a) Flowchart of Pt-based material low-temperature synthesis under CO atmosphere. M refers to Cu, Ni, or Cr elements. (b) Simple setup scheme.

can be identified [51]: (i) the ideal carbonyl complex formation, (ii) the optimal temperature, and (iii) the support material.

Concerning the first aspect, the carbonyl chemical route starts with the reaction (1.1), i.e. the preparation of Pt carbonyl cluster complexes in methanol solvent, step 2, Figure 1.2 [50].

$$\left[\mathrm{Pt}_{3x}M_y(\mathrm{CO})_6\right]_n^{2-} \rightarrow \mathrm{Pt}_xM_y(\mathrm{CO})_z \tag{1.1}$$

The easiest carbonyl complex consists of $n = 5$, step 2; when the solution turned green. This property depends on the molar ratio between $CH_3COONa + 3H_2O$ and $Na_2PtCl_6 + 6\ H_2O$. In our group, this molar ratio was fixed to 6 [48, 50].

For Pt-alloys, reaction (1.1) is modified by the addition of the metallic salt, MCl_m, Eq. (1.2) [48, 51].

$$\left[\mathrm{Pt}_{3x}M_y(\mathrm{CO})_6\right]_n^{2-} \rightarrow \mathrm{Pt}_xM_y(\mathrm{CO})_z \tag{1.2}$$

The molar ratio, Pt: M (M = Cu, Ni, Cr), is linked to x and y [48]. Therefore, the $\left[\mathrm{Pt}_{3x}M_y(CO)_6\right]_n^{2-}$ carbonyl complex is converted to $Pt_xM_y(CO)_z$, Eq. (1.3) [51].

$$\left[\mathrm{Pt}_{3x}M_y(\mathrm{CO})_6\right]_n^{2-} \rightarrow \mathrm{Pt}_xM_y(\mathrm{CO})_z \tag{1.3}$$

In the second aspect, the optimal temperature corresponds to the formation or decomposition of the carbonyl complex. For reactions (1.1) or (1.2), the optimal synthesis of Pt-based carbonyl complex temperature lies between 50 and 55 °C, Figure 1.2.

This carbonyl complex can decompose at room temperature in a nitrogen atmosphere, to generate supported nanoparticles (NPs), step 4, Figure 1.2. For non-metals or semiconductors, the optimal temperature is the solvent's boiling point ($T = 140\,°C$ or $180\,°C$) [52]. The carbonyl complex and another precursor, e.g. selenium, can react to yield chalcogenide nanoalloys in a nitrogen atmosphere.

Carbon black is the standard material used for support because of its electrical conductivity and high surface area [46, 53, 54]. The catalytic center's dispersion and stability are also a design criterion [2, 42]. Thus, the carbonyl chemical route has the importance of providing a good dispersion on the carbon XC-72 support, step 3, Figure 1.2. The metal–support interaction is discussed in Section 1.2.3. Finally, the synthesis of the as-prepared carbon-supported Pt-alloys NPs are heat treated at 400 °C in H_2-N_2 atmosphere, for 2 hours.

1.2.1.3 Photo-assisted Deposition Method

In 1965, Clark et al. [55] reported a black end product (i.e. metallic silver) due to a photoreaction between TiO_2 and Ag^+ ions. Whereas Kraeutler and Bard [56] revealed the synthesis of platinized TiO_2 material in 1978. Consequently, the PDM took a crucial role to prepare supported metal catalysts. In general, a large-gap semiconductor under ultraviolet (UV) illumination reduces the metal cations due to photogenerated electron–hole pairs. Therefore, different metals have been synthesized, such as Rh, Pt [9], Pd-alloys [57], Au [58], Ag [59], Ru [60], and Co-based [61] to be used in photocatalysis [62, 63].

For the energy conversion and design of novel materials, the PDM is considered as an economical method. For the sake of comparison, this synthesis method favors the metal–support interaction [27]. Figure 1.3 shows the process of preparing Pt-based materials and supported onto oxide-carbon composite materials.

From Figure 1.3, two critical aspects can be identified in this type of synthesis, such as (i) the support material and (ii) the nature of the hole scavenger.

On the semiconducting material used as support, the first reaction consists of electron–hole (e^-/h^+) pairs generation under UV illumination, Eq. (1.4) step 2, Figure 1.3 [64].

$$\text{Oxide or Carbon Support} + hv \rightarrow h^+ + e^- \tag{1.4}$$

The photogenerated electrons, transferred from the conduction band, reduce the metal ion precursor at the support's surface, reactions (1.5)–(1.6).

$$\left[\text{PtCl}_6\right]^{2-}_{(\text{abs})} + 2e^- \rightarrow \left[\text{PtCl}_4\right]^{2-}_{(\text{abs})} + 2\text{Cl}^- \tag{1.5}$$

$$\left[\text{PtCl}_4\right]^{2-}_{(\text{abs})} + 2e^- \rightarrow \text{Pt} + 4\text{Cl}^- \tag{1.6}$$

Semiconductors (e.g. TiO_2) and carbon materials (C-sp^2 domains) have been proposed as supports. In this connection, oxide-carbon composite materials have been considered due to the increase of conductivity [65]. The investigated composite materials are TiO_2/C [22,

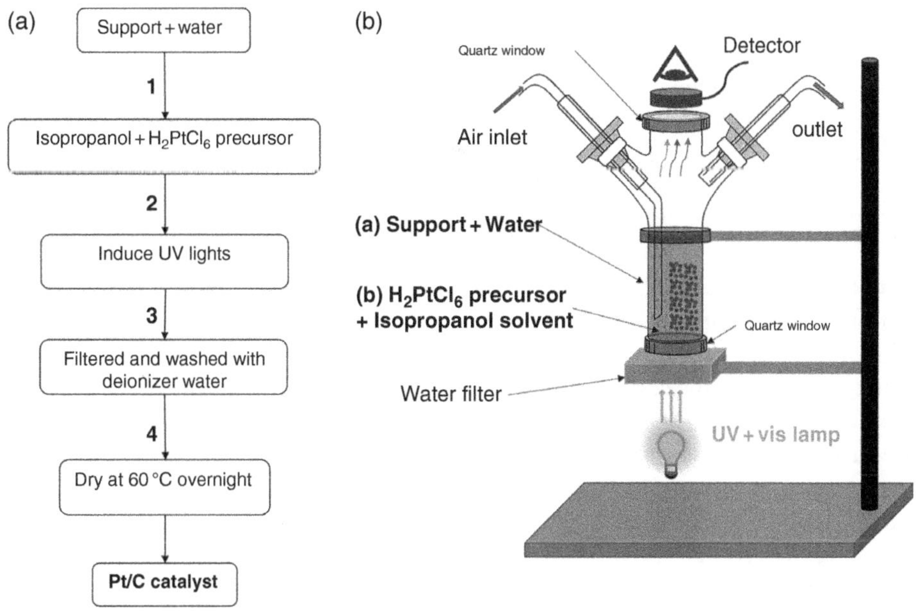

Figure 1.3 (a) Flowchart showing the Pt-based material synthesis via the photo-assisted deposition method. (b) Diagram of the experimental setup.

26, 43], SnO_2/C [23, 41], and $M:TiO_2/C$ [42]. It is important to note that the photoreduction of $[PtCl_6]^{2-}$ can, in addition, be activated by photolysis, in the absence of semiconducting material. Indeed, in the bulk solution, the hole scavenger and the $[PtCl_6]^{2-}$ precursor can produce $[PtCl_6]^{3-}$, in the presence of UV illumination, Eq. (1.7).

$$[PtCl_6]^{2-} + RCH_2OH + h\nu \rightarrow [PtCl_6]^{3-} + RCH_2O^* + H^+ \tag{1.7}$$

In turn, $[PtCl_6]^{3-}$ in equilibrium can produce $[PtCl_5]^{2-}$, Eq. (1.8).

$$[PtCl_6]^{3-} \rightleftharpoons [PtCl_5]^{2-} + Cl^- \tag{1.8}$$

Whereas $[PtCl_5]^{2-}$ in the presence of RCH_2O^* transforms to $[PtCl_4]^{2-}$, Eq. (1.9).

$$[PtCl_5]^{2-} + RCH_2O^* \rightarrow [PtCl_4]^{2-} + RCH_2O + HCl \tag{1.9}$$

The disproportionation reaction takes place to form $[PtCl_6]^{2-}$, Eq. (1.10), and the cycle of photoreduction proceeds via Eq. (1.9).

$$2[PtCl_5]^{2-} \rightleftharpoons [PtCl_6]^{2-} + [PtCl_4]^{2-} \tag{1.10}$$

This mechanism has been proposed by Einaga et al. [66, 67], and the formation of Pt metal particles takes place only after more than 90% of $[PtCl_4]^{2-}$ ions (i.e. Pt^{2+} ions) is formed, Eq. (1.6) [66].

Different alcohols have been used as hole scavenger for the PDMs, as shown in Figure 1.3 [68]. These small organic compounds, serving as hole scavengers, inhibit the semiconductor's recombination process, Eq. (1.4), thus, favoring the electrons to be transferred to $[PtCl_4]^{2-}$ ions at the conduction band, Eq. (1.6). The obtained product is further heat treated at 400 °C in H_2-N_2 atmosphere, for 2 hours [22].

1.2.2 Physical–chemical characterization of supports

Different techniques and methods of analysis of materials have been employed to generate the so-called metal–support interaction. Here, we can consider a classification shown in Figure 1.4 [31, 69].

Figure 1.4 considers the compositional, structural, and optical or magnetic characterization as a first step. The principal evaluation in the physicochemical characterization is the surface and/or morphology analyses. This stage involves the analysis of the technique and the method as a second step. In general, transmission electron microscopy (TEM) and powder X-ray diffraction (PXRD) data are used to obtain the microstructural properties of the material. For further calculation details, the following references are suggested [31, 70, 71].

1.2.2.1 Transmission Electron Microscopy Analysis

The electron microscope aids in analyzing the morphology and the microstructural properties of the material in the nano-sized domain. Since the advent of electron microscopes in the 1930s [70], various improvements in the analysis have been made in this domain. So,

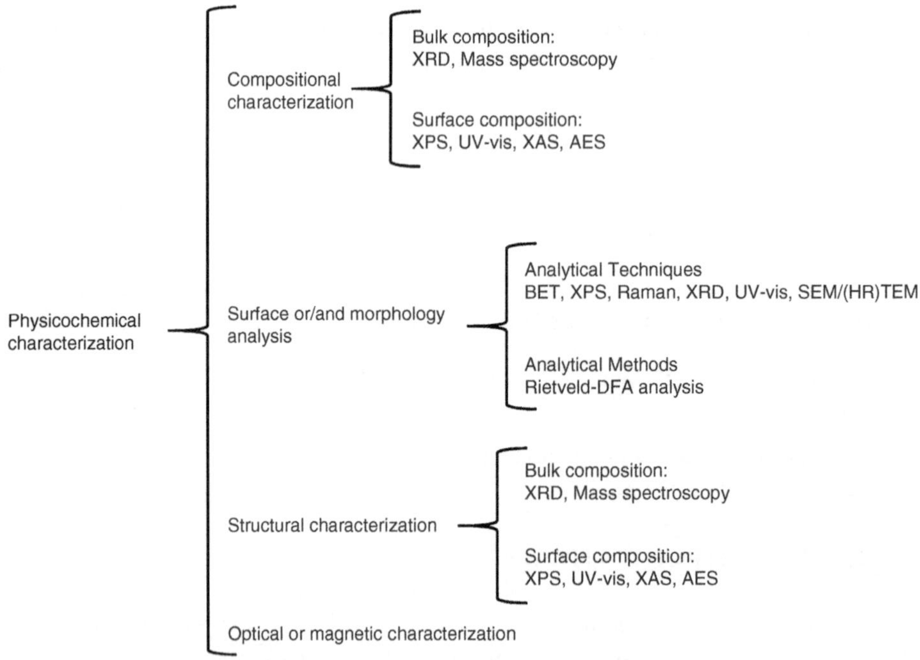

Figure 1.4 Physicochemical classification overview for the characterization of materials.

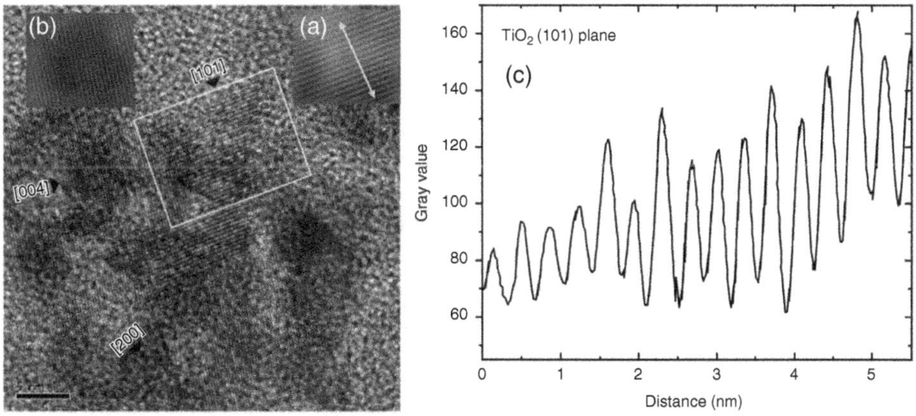

Figure 1.5 TEM image of a TiO_2 sample. (a) (101) and (b) (004) plane reconstructed from FFT from square region, and (c) profile on the (101) plane, see double sided arrow, in (a). *Source:* From [74] / with permission of Elsevier.

high-resolution transmission electron microscopy (HRTEM) is used to determine the planes and the electron diffraction pattern [72]. Here, the fast Fourier transform (FFT) of the image indexes the planes according to the X-ray diffraction (XRD) analysis [73]. For example, Figure 1.5 shows the anatase phase from the TiO_2 sample with three directions.

According to Figure 1.5, it is possible to identify the TiO_2 (101) plane with an interplanar distance of 3.5460 ± 0.0091 Å (measured from Figure 1.5c. This is slightly in contrast to 3.52 Å, from the Joint Committee on Powder Diffraction Standards (JCPDS) database [74]. Besides, HRTEM images can elucidate the morphology and the phases present. Figure 1.6 shows the morphology of Pt NPs supported on Y-doped TiO_2 support. As shown in the figure, Pt and anatase phases are both identified in the sample. The dispersion of Pt suits the heterogeneous dispersion of the oxide. This observation assesses the fact that the photodeposited Pt is selective on the oxide sites of the support. The HRTEM analysis shows that the particles' size is within the interval from 7 to 19 nm, for Pt in agreement with XRD data analysis [43]. Thus, HRTEM and XRD analyses are complementary techniques to characterize materials. However, a robust interpretation can be delivered by an XRD refinement to complement HRTEM analysis. In some cases, however, it is also possible to simulate HRTEM images using a computer program [75–77].

1.2.2.2 X-ray Powder Diffraction Analysis

The diffraction pattern can be analyzed and indexed in three different ways [78], which are: (i) the reported patterns used to compare with the experimental profile; (ii) a diffraction standard of a sample, given in JCPDS or crystallographic open database (COD) [79], used to index the peaks; and (iii) a calculated and background spectrum used to extract the microstructural properties (whole powder pattern fitting, WPPF). The conditions (i) and (ii) can help to implement the integral range method (Williamson–Hall [W-H]) [80]. For the condition (iii), Rietveld-like [80] and Debye Function Analysis (DFA) method [81] are used with a robust refinement analysis. The W-H, Rietveld Refinement (RF), and DFA methods are described next.

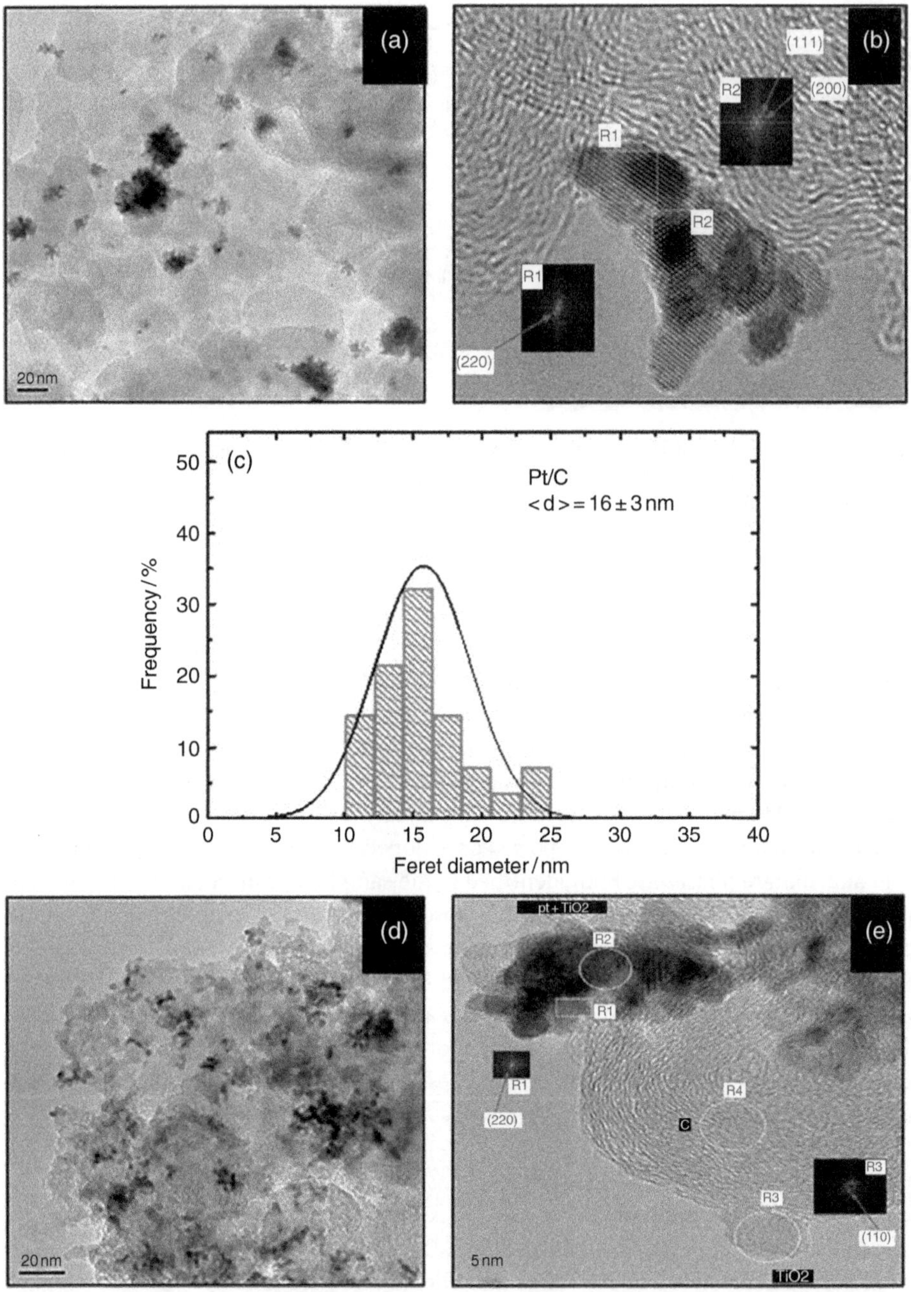

Figure 1.6 HRTEM images and particle size histograms deduced from TEM images for Pt/C (a-c), Pt/TiO_2-C (d-f), and Pt/Y: TiO_2-C (g-**i**). Insets in b, e, and h are the fast Fourier transform (FFT) on high-resolution images. *Source:* From [43] / with permission of Elsevier.

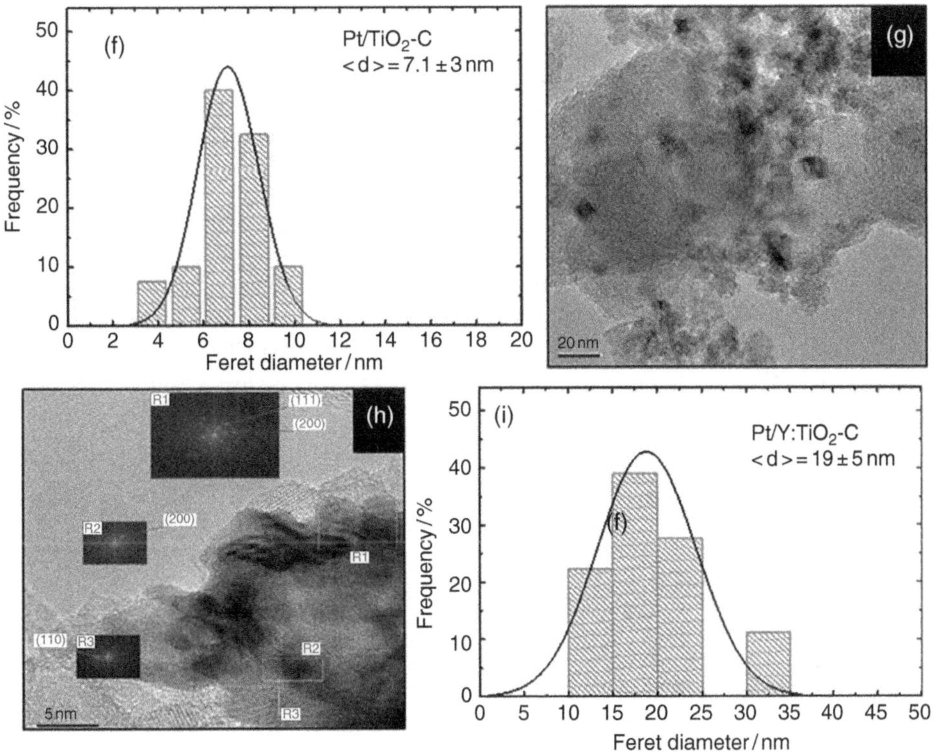

Figure 1.6 (Continued)

1.2.2.3 Williamson–Hall Method

The W-H method is the first approach to obtain microstructural properties in XRD analysis, which considers the line broadening and is applied as an integral amplitude method. First, this analysis finds that the peak width is a function of 2θ. Each Bragg diffraction peak is deconvoluted to extract the size induced and strain-induced effects. Eq. (1.11) defines the line broadening parameter (*db*).

$$\mathrm{d}b(\theta_{hkl}) = \frac{K}{\langle d \rangle} + 2\varepsilon b(\theta_{hkl}) \tag{1.11}$$

$$\mathrm{d}b(\theta_{hkl}) = \beta_{f,hkl}\frac{\cos\theta_{hkl}}{\lambda} \tag{1.12}$$

$$b(\theta_{hkl}) = 2\frac{\sin\theta_{hkl}}{\lambda} \tag{1.13}$$

where λ is the wavelength, K is a constant that depends on the shape [82], <d> is the crystallite size, ε is the micro-strain in the Bragg diffraction angle at the *hkl*-plane (θ_{hkl}). The full-width at half maximum (FWHM) $\beta_{f,hkl}$ is to correct the instrumental broadening, Eq. (1.14).

$$\beta_{f,hkl} = \sqrt{\beta_{f,hkl}^{M\ 2} - \beta_{f,hkl}^{Inst\ 2}} \tag{1.14}$$

$\beta_{f,hkl}^{M}$ is the experimental broadening for each Bragg diffraction peak, whose value is obtained to be used as a peak function [43]. Besides, $\beta_{f,hkl}^{Inst}$ is measured with a standard sample, and used with a Caglioti pseudo-Voigt (PV) function [83]. The θ_{hkl}, the lattice constant (a), and $\beta_{f,hkl}^{M}$ parameters are obtained with the weighted sum of squared residuals (WSSR) [84], Eq. (1.15) [43].

$$\chi^2 = \sum_{i}^{n} \frac{1}{\sigma_i^2}\left[I_i^{\exp} - I_i^{\text{calc}}\right]^2 \tag{1.15}$$

Then, the microstructural properties are applied in Eq. (1.11), and the obtained graph is named the W-H plot.

1.2.2.4 Rietveld Refinement Method

This method has been widely applied as a standard analysis method to obtain microstructural properties from PXRD data. This refinement requires the modeling of the entire dispersion spectrum, i.e. WPPF. This weighted sum of squares (χ^2) is the first quantity to minimize with the Levenberg–Marquardt algorithm, Eq. (1.16).

$$\chi^2 = \sum_{i}^{n}\left[w_i\left(I_i^{\exp} - I_i^{\text{calc}}\right)\right]^2 \tag{1.16}$$

Where the parameter w_i is defined in Eq. (1.17).

$$w_i = \frac{1}{\sqrt{I_i^{\exp}}} \tag{1.17}$$

The theoretical intensity is evaluated with Eqs. (1.18) and (1.19).

$$I_i^{\text{calc}} = S_j \sum_{k=1}^{N_{\text{peaks}}} L_k \left|F_{k,j}\right|^2 S\left(2\theta_i - 2\theta_{k,j}\right) + bkg_i \tag{1.18}$$

$$S_j = S_F \sum_{j=1}^{N_{\text{peaks}}} \frac{f_j}{V_j^2} \tag{1.19}$$

Here, S_F, f_j, V_j, L_k, $F_{k,j}$, $S(2\theta_i - 2\theta_{k,j})$ parameters are the beam intensity, the phase volume fraction, the phase cell volume, the Lorentz-polarization factor, the structure factor, and the profile shape function, at the k-peak, and bkg_i is the background function, respectively. Hence, each phase's mass weight fraction can be calculated with the phase volume fraction according to Eqs. (1.20) and (1.21).

$$wt_j = \frac{f_j \rho_j}{\sum_j^N f_j \rho_j} \tag{1.20}$$

$$\rho_j = \frac{n_{A,j} A_j}{V_j N_A} \tag{1.21}$$

Here, ρ_j, $n_{A,j}$, A_j, and N_A parameters are: (i) the theoretical phase density, (ii) the number of atoms associated with the unit cell, (iii) the atomic weight for each phase, and (iv) the Avogadro number.

Finally, the R-indices, calculated with Eqs. (1.22)–(1.24), are used to obtain a good refinement.

$$R_{\text{wp}} = \sqrt{\frac{\chi^2}{\sum_{i=i}^n \left[w_i I_i^{\text{exp}}\right]^2}} \tag{1.22}$$

$$R_{\text{exp}} = \sqrt{\frac{n-P}{\sum_{i=i}^n \left[w_i I_i^{\text{exp}}\right]^2}} \tag{1.23}$$

$$GofF = \frac{R_{\text{wp}}}{R_{\text{exp}}} \tag{1.24}$$

Therefore, n and P are the numbers of points and parameters, respectively. Some free computer software [80], such as Maud [85], Profex [86], and FullProf [87] programs, perform the RF method.

1.2.2.5 Debye Function Analysis Method

Debye first proposed his Debye Function in 1915 [88, 89], implemented as DFA. The DFA method has been employed by different groups. For example, Vogel et al. applied the DFA method on surface oxidation–reduction of small Pt clusters in 1989 [90]. In the same year, Hall et al. reported DFA results in experimental electron diffraction patterns of unsupported silver NPs [91]. However, Vogel provided hints on how to apply the DFA method on noble metal groups in 1998 [92]. Hence, the Debye Function is shown in Eq. (1.25).

$$I_{\chi,n}(q) = I_0 \sum_i \sum_j f_i(q) f_j(q) \frac{\sin(qr_{ij})}{(qr_{ij})} \tag{1.25}$$

I_0 is the incident beam intensity for i and j atoms, $f_i(q)$, $f_j(q)$, and r_{ij} are the scattering factors, and the distance between i and j. Therefore, Eq. (1.26) defines the length of the dispersion factor.

$$q = |q| = 4\pi \sin\theta / \lambda \tag{1.26}$$

Where θ and λ are the Bragg diffraction angle and the wavelength. The DFA simulation spectrum is done by Debye scattering equation (DSE), Eq. (1.27).

$$I_i^{\text{calc}}(q) = \sum_{\chi} x_{\chi} \sum_{n=1}^{S_{\chi}} D_{\chi}(n) I_{\chi,n}(q) + bkg_i \tag{1.27}$$

The cluster size distribution n up to a size $S\chi$ is $D\chi(n)$, $x\chi$ is the fraction number of each phase (χ), whose total sum is equal to 1. The mass fraction distribution is calculated from Eq. (1.28).

$$\%x_{\chi} = \frac{M_{\chi}}{\sum_{\chi} M_{\chi}} 100 \tag{1.28}$$

Here, Eq. (1.29) defines $M\chi$, and $N\chi(n)$ represents the number of atoms in the cluster.

$$M_{\chi} = \sum_{n=1}^{S_{\chi}} D_{\chi}(n) N_{\chi}(n) \tag{1.29}$$

The R_{wp} and $GofF$ are calculated from Eqs. (1.30) and (1.31).

$$R_{\text{wp}} = \sqrt{\frac{\chi^2}{\sum_{i=1}^{M} w_i \left[I^{\text{exp}}(q_i) \right]^2}} \tag{1.30}$$

$$GofF(r) = \sqrt{\frac{1}{N} \sum_{i=1}^{N} \left(\frac{I^{\text{exp}}(q_i) - I^{\text{calc},r}(q_i)}{\sigma_i} \right)^2} \tag{1.31}$$

The chi-square (χ^2) has the same definition in Eq. (1.16). In addition, since there are not many computer programs that apply the DFA or DSE method, the Debussy software is the only program, developed by Cervellino et al. [89, 93, 94], that allows it.

1.2.2.6 Microstructural Properties from X-ray diffraction Analysis

The main microstructural properties are the lattice constant, a; the crystallite size, $< d >$; the micro-strain, ε; and stacking fault, α. RF and DFA analysis can help obtain all these parameters. As an example, Figure 1.7 shows the RF and DFA analysis performed on the TiO_2 sample. The anatase phase (PDF#84–1286) is the main phase that was only present in the powder material.

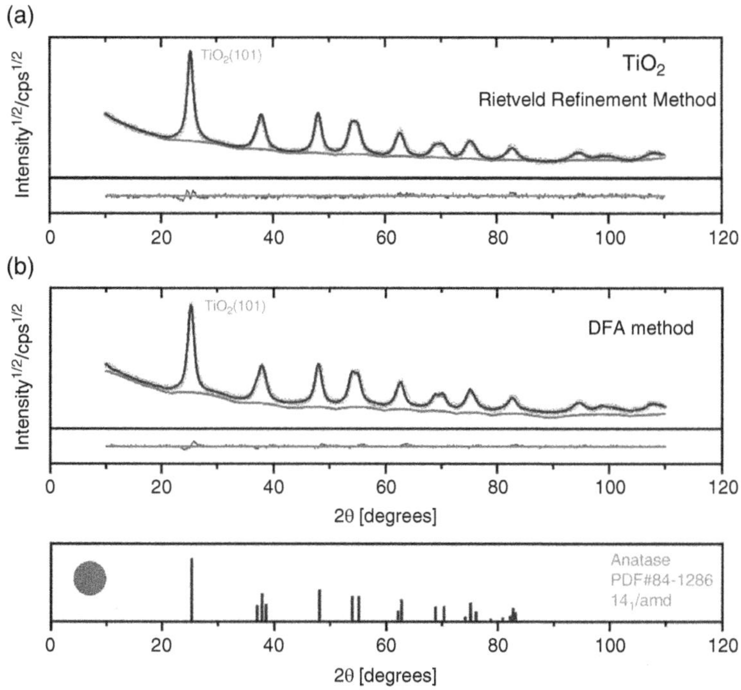

Figure 1.7 (a) RF and (b) DFA method applied on TiO_2 sample. The spheric morphology was considered in both refinement methods.

From this figure, we can observe that WPPF fits the experimental diffraction pattern in both methods. For the analysis with RF, the $S(2\theta_i - 2\theta_{k,j})$ parameter considers the profile shape function at the k-peak, where $2\theta_{k,j}$ depends on the Bravais' lattice (crystal systems). For example, Eq. (1.32) represents a tetragonal system (e.g. the anatase phase) with the provided crystallographic information file (CIF), as a first approximation.

$$2\theta_{k,j} = 2\sin^{-1}\left[\frac{\lambda}{2}\sqrt{\frac{h^2+k^2}{a^2}+\frac{l^2}{c^2}}\right] \tag{1.32}$$

Here, λ and a and c parameters are the wavelength and lattice parameters. Besides, the $S(2\theta_i)$ function describes the peak broadening, which is related to the finite size of crystallites, $\langle d \rangle$, and the micro-strain, ε. The broadening of the peak associates these two components to extract FWHM, using Lorentzian (L) and Gaussian (G) functions. The crystallite size and micro-strain are obtained with FWHM implemented in Eq. (1.33) and (1.34), like that of the W-H method, Eq. (1.11).

$$\varepsilon = \frac{\beta_{f,hkl,G}}{4\tan\theta_{hkl}} \tag{1.33}$$

$$\langle d \rangle = \frac{\lambda}{\beta_{f,hkl,L}\cos\theta_{hkl}} \tag{1.34}$$

With the help of the CIF file, the Debussy software builds the population for the morphology with different crystallite sizes for the DFA analysis and implements six lattice expansion cases for each morphology [94]. Hence, Eq. (1.34) considers a spheric morphology, see Eq. (1.11), and the same spherical morphology is performed in the DFA analysis. Table 1.1 summarizes the microstructural properties for the TiO_2 sample (cf. Figure 1.7).

One can conclude that similar results from both RF and DFA analysis can be obtained. Moreover, the DFA analysis is the most versatile method, where different morphologies can be explored.

1.2.3 Effect of supports on catalytic centers

The understanding of the interaction between the support and the catalytic center is fundamental in electrocatalysis. Such interactions depend on the synthesis method used. Therefore, it is possible to find out the three significant effects, namely, electronic [95, 96], geometric [96, 97], and bifunctional mechanism [96]. Figure 1.8 displays the different impacts that support materials can have on the catalytic center.

The two strategies of inducing interaction, Figure 1.8, which greatly affect the electrocatalytic phenomenon, in view of developing a highly stable and long-lasting catalyst [2], consider (i) the particle size, (ii) inter-particle distance, (iii) the nature of the support, (iv) the thermal treatment, and (v) the stability.

Here, we focus on chemical and physical syntheses, namely, CCR and PDM, where the support (carbon and oxide) has a higher impact on the microstructure and on the properties of the optoelectronic materials.

1.2.3.1 Carbon Domains

Carbon is the most popular catalysts' support materials. Carbon can be present as low or highly graphitized, having high electrical conductivity and high surface area [98]. However, the catalytic center's support interaction is associated with the carbon hybridization, known as C-sp^3 and C-sp^2. First, for the results of the electrostatic force between C-sp^3 domains with nanoparticulated metals or alloys, no modification of electronic properties is observed, i.e. there is no electronic effect. This effect is generally known as weak metal–carbon interaction (WMCI), where amorphous carbon black (e.g. Carbon Vulcan or Carbon Ketjen) has been used as support for Pt-based NPs with good performances [97]. Different chemical synthesis routes show this phenomenon, in which the metal base material must be stabilized to form

Table 1.1 Microstructural properties calculated from RF and DFA analysis for the TiO_2 sample.

Method	Lattice constant	Crystallite size/nm	Micro-strain/ ε
RF analysis	$a = b = 3.7888$ $c = 9.4904$	11.5 ± 0.09	0.0047 ± 0.0002
DFA analysis	$a = b = 3.7888$ $c = 9.4904$	8.16 ± 3.3	n.a.

n.a. = It was not considered in the lattice expansion.

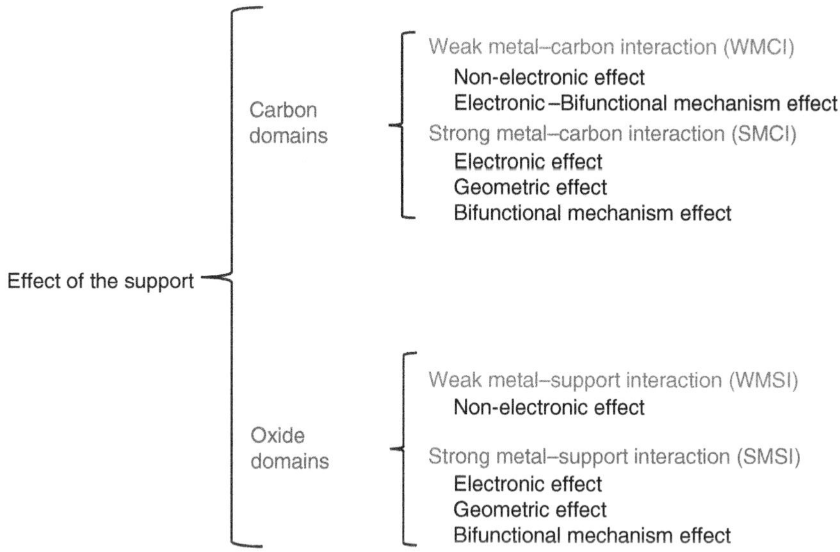

Figure 1.8 Different effects of the supporting material made of carbon and oxide.

nanoalloyed materials to optimize their particle size, i.e. the electronic effect [99]. In this context, Wang et al. [100] reviewed the Pt-based alloy materials and different amorphous carbon used as support. On the other hand, the electronic properties of Pt_3Ni NPs can be modified by using Mo atoms to form $MoPt_3Ni/C$. Thus, this material showed higher stability (10.3 mA/cm^2 and 7 A $mg^{-1}{}_{Pt}$) with respect to the Pt/C catalyst [101]. Therefore, the electronic properties of Pt_3Ni NPs are associated with the Mo atoms located at the edges of the particle.

The electronic-bifunctional mechanism effect was first demonstrated with PtRu alloys material. Here, Pt sites are associated with the active catalytic sites, whereas ruthenium sites adsorb OH species at lower potential. Thus, this effect is also known as oxophilic effect, which appears during the hydrogen electrode reactions [102].

For C-sp^2 domains, these surfaces are graphitic, hydrophobic, and non-reactive, where metal NPs cannot anchor via the chemical synthesis route. Different methods of synthesis were used to stabilize metal NPs on their planes [103, 104]. These modifications included creating defects, i.e. functionalization to anchor metal particles (Figure 1.9) [105]. As shown in Figure 1.9, such defects can promote the strong metal–carbon interaction (SMCI) electronic and geometric effect. Furthermore, Figure 1.10 shows additional clues as to how the C-sp^2 interaction with metal NPs occurs. Therefore, this type of functionalization can generate an electronic effect between the C-sp^2 surface and metal NPs (Figure 1.10d).

When the metal NPs are encapsulated in the support, this phenomenon is called the geometric effect [106, 107]. This effect is like that in the oxide domains, and in some cases, the metal's surface, partially covered, promotes the formation of interfacial phases (see Figure 1.10c).

Additionally, the C-sp^2 domains possess another attractive property, a photocatalytic behavior [64], a property that facilitates anchoring metal NPs via the PDM. This synthesis route also generates the SMCI effect. Figure 1.11 shows the photo-assisted deposition process in two different materials: amorphous carbon black (XC72) and multi-walled carbon

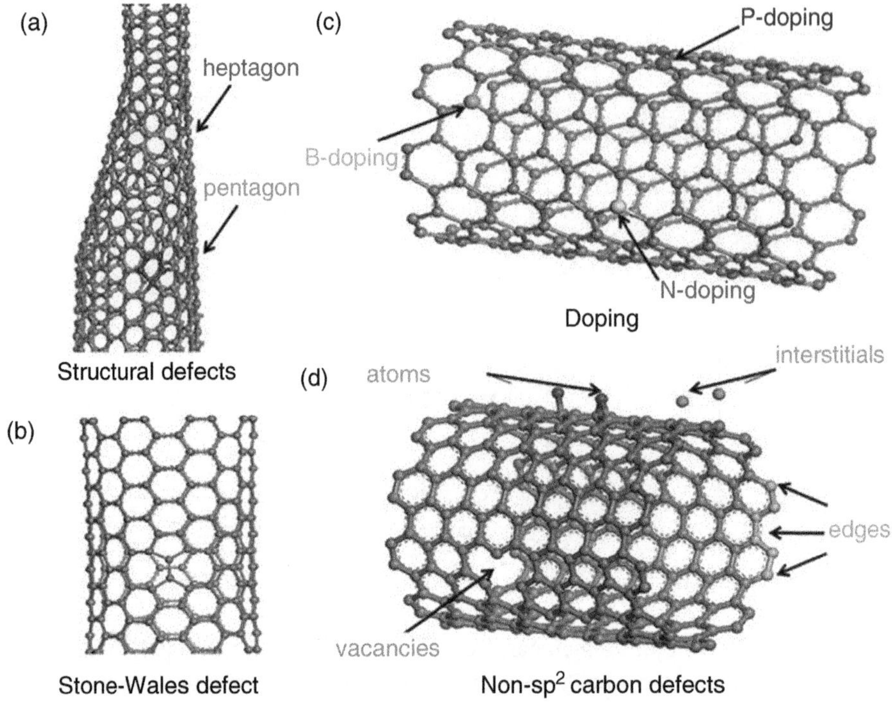

Figure 1.9 Schematic models represented different types of defects and heteroatom doping on carbon surfaces such as structural defect (a), stone-wales defect (b), doping atoms (c) and non-sp^2 carbon defect (d). *Source:* From [97] / with permission of John Wiley & Sons.

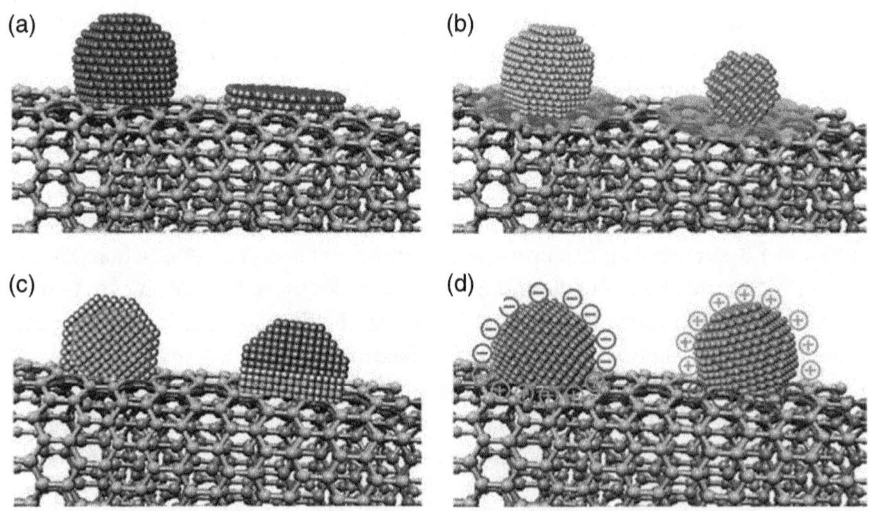

Figure 1.10 The metal–carbon support interaction on surface-modified CNT: (a) weak interaction; **(b)** strong interactions; (c) left – encapsulation of the metal NP by carbon, right – interphase or compounds at the interface (geometric effect); and (d) electron transfer between metal NPs and carbon support (electronic effect). *Source:* From [97] / with permission of John Wiley & Sons.

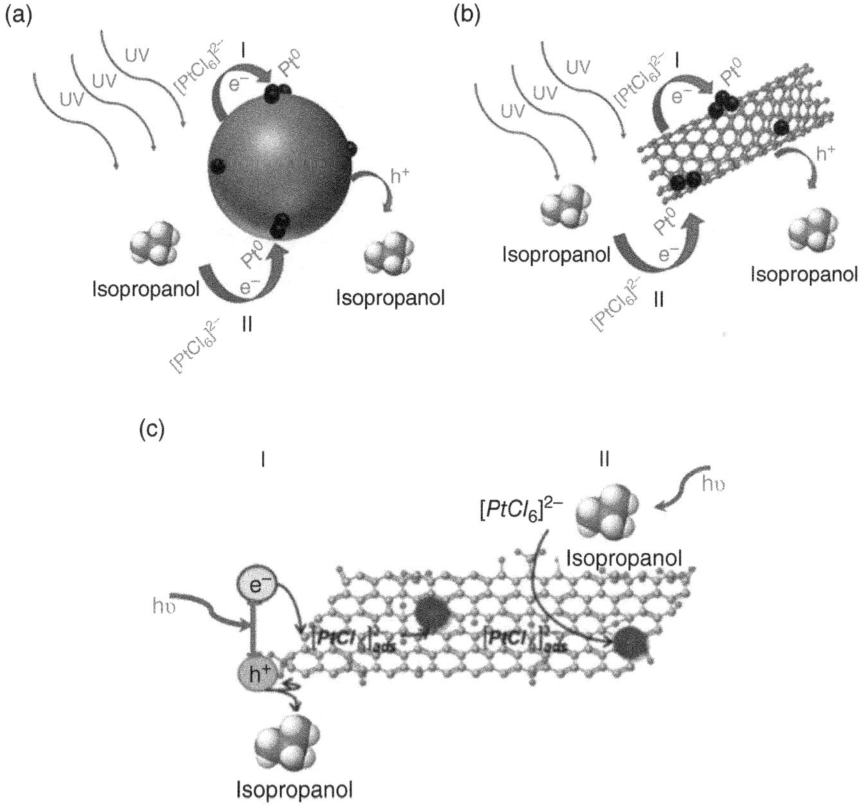

Figure 1.11 UV-light induced deposition of Pt NPs onto carbon-based substrates. (a) Pt NPs formation onto Vulcan XC-72 surface, (b) Pt NPs formation onto MWCNT, and (c) general scheme of Pt NPs formation onto a defect-containing graphitic plane. I and II refer to the photo-assisted deposition process involving electrons from the carbonaceous materials' conduction band and the photochemical process occurring in the bulk solution, respectively. *Source:* From [64] / with permission of Elsevier.

nanotubes (MWCNT). The interaction effect between C-sp^2 domains and metal NPs is a clear electronic effect [64].

Besides, the bifunctional mechanism effect appears when the Pt NPs are anchored on the C-sp^2 surface. This phenomenon is relevant in the hydrogen electrode reaction, precisely in alkaline media [108].

1.2.3.2 Oxide Domains

Tauster and coworkers first used the term SMSI in 1978 as a strong metal–support interaction [109]. However, the interaction between metal and oxide support has an additional effect, namely, weak metal–support interaction (WMSI), like WMCI interaction [42, 96]. Thus, Figure 1.12 indicates that the oxide support and nanoparticulated metal present an identical effect to those discussed on carbon domains.

Pan et al. [96] reviewed the SMSI in oxide domains. It follows that the electronic and geometric effects are present. Also, a bifunctional mechanism is relevant, in which the support acts as a cocatalyst during the reaction of interest, see Figures 1.13 and 1.8.

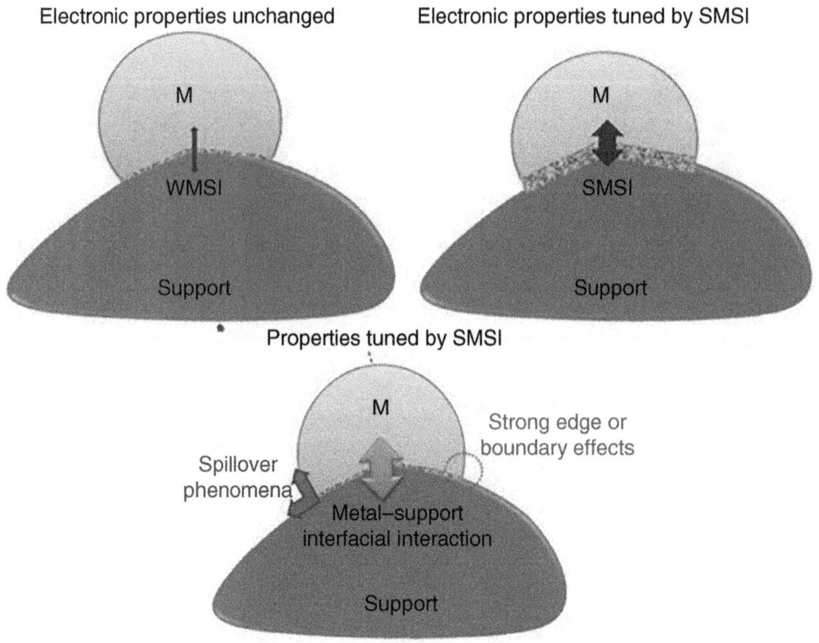

Figure 1.12 Weak-, and strong-metal–support-interaction (WSMI, SMSI), and the SMSI effect on the catalytic center. *Source:* From [96] / with permission of Elsevier.

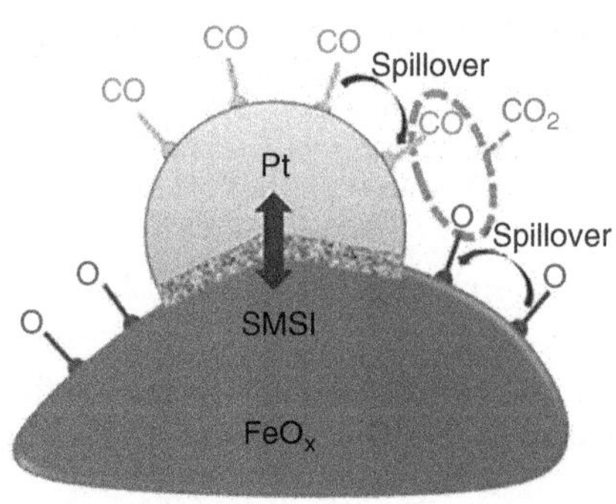

Figure 1.13 Bifunctional mechanism effect on CO electro-oxidation. *Source:* From [96] / with permission of Elsevier.

This term indicates that the metal and the oxide support provide active sites to improve catalytic performance, like the SMCI effect. Thus, Figure 1.13 illustrates this effect observed in the stripping process of an adsorbed monolayer of CO.

Metal NPs have been supported on different oxides such as TiO_2 [9, 110], CeO_2 [111], and SnO_2 [112] materials. However, a new class of support emerges as hybrid materials [113], known as composite material [59]. The corrosion of carbon is a problem in FC reactions electrocatalysis. Consequently, oxide-carbon composite materials that combine both SMSI and SMCI effects, when the catalytic center is photo-deposited, can cope with the corrosion problem [23, 26, 41, 114]. Furthermore, the M-doped oxide/C is another possibility to improve the interaction between support and catalytic center [42, 43].

1.2.4 Theoretical Approach

Designing an efficient and selective catalyst for the electrocatalytic performance of a reaction of interest is always a challenge. The different reactions of interest include oxygen reduction-evolution reaction (ORR-OER) [115–117], hydrogen evolution-oxidation reaction (HER-HOR) [118, 119], nitrogen reduction reaction (NRR) [120], and CO_2 reduction [121]. The Sabatier principle, represented as a volcano plot, has provided a conceptual framework for designing an EC, where the bond strength energy must be optimum, see Figure 1.14.

The intermediates' binding energy can be accurately calculated with the first-principles quantum chemical calculation, using the DFT. This simulation implies the calculation of the projected electronic density of states (PDOS) for the catalytic center, support, interaction between catalytic center-support, and their effect with a probe molecule or intermediates species in a reaction of interest. For example, Greely et al. [118] applied DFT to investigate Pt and PtBi alloys materials. Their calculated data agreed with the experimental results. They concluded that the DFT-based method could aid to develop new metals alloy catalyst [123]. In this context, Giustino published the basis of DFT simulation in materials science [124].

1.2.4.1 CO_2 Reduction

The electro-reduction of CO_2 is a multi-electron charge transfer reaction, as summarized in the reaction, Eq. (1.35).

$$k\mathrm{CO_2} + n\left(\mathrm{H^+} + e^-\right) \rightleftharpoons P + m\mathrm{H_2O} \tag{1.35}$$

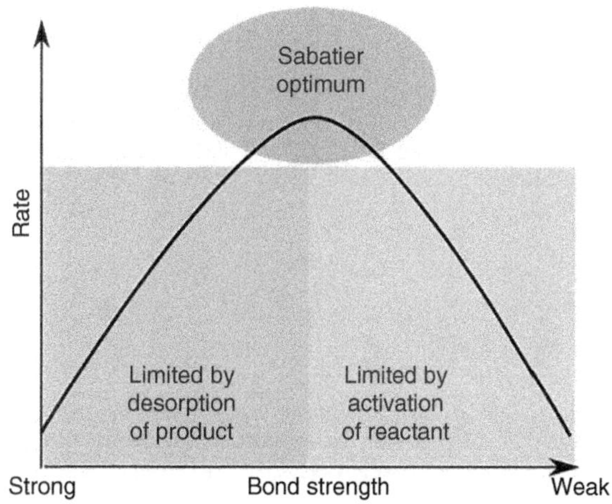

Figure 1.14 Schematic representation of the Sabatier principle. *Source:* From [122] / with permission of Elsevier.

The most typical products, P, in aqueous media, are reported to be carbon monoxide (CO), formic acid (HCOOH), formaldehyde (HCHO), methanol (CH_3OH), methane (CH_4), ethanol (CH_3CH_2OH), ethylene (C_2H_4), and k, n, and m are the coefficients at the equilibrium potential [121]. In this context, Koper and Calle-Vallejo [125] proposed a mechanism explaining the experimental observation on Cu(100) surface, Figure 1.15. Here, the rate-determining step (RDS) consisted of two CO-coupled molecules to form $^*C_2O_2$, and the main products were C_2H_4 and CH_3CH_2OH.

The CO_2 adsorption has been studied on various materials, such as TiO_2 [126], zeolite [127], and CeO_2 [128]. Different possible adsorbed surface structures of CO_2 can be found on the TiO_2 surface (Figure 1.16a).

Figure 1.16b shows the PDOS for the reactivity of the (101) and (001) anatase surface toward CO_2 adsorption and compared to two bonding behaviors of the CO_2 molecule. These findings explained the difference in reactivity on the TiO_2 planes, like that of a bifunctional mechanism.

Mino et al. [126] suggested that the formation of carbonate and bicarbonate species can occur preferentially on the TiO_2 (001) surface since this surface has the strongest basicity of the surface oxygen sites. The TiO_2 (101) surface weakly absorbed the CO_2 molecule. This phenomenon can be exported to understand the interaction between metal NPs and oxide or carbon domains. For example, the interaction of Cu(111)/ZnO was compared against ZnCu(211) plane in the potential energy diagram, Figure 1.17. Here, two main reaction pathways were considered for the calculations: (i) the reverse water-gas shift (RWGS)

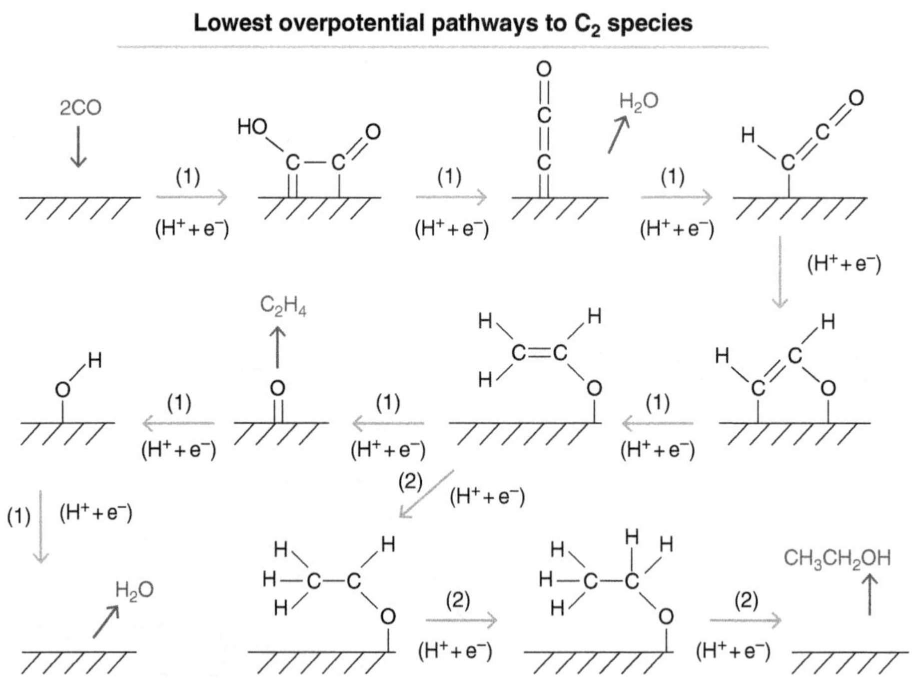

Figure 1.15 Schematic representations of the species involved in the reaction pathway to C_2H_4 (**1**) and CH_3-CHO/CH_3-CH-OH (**2**). *Source:* From [125].

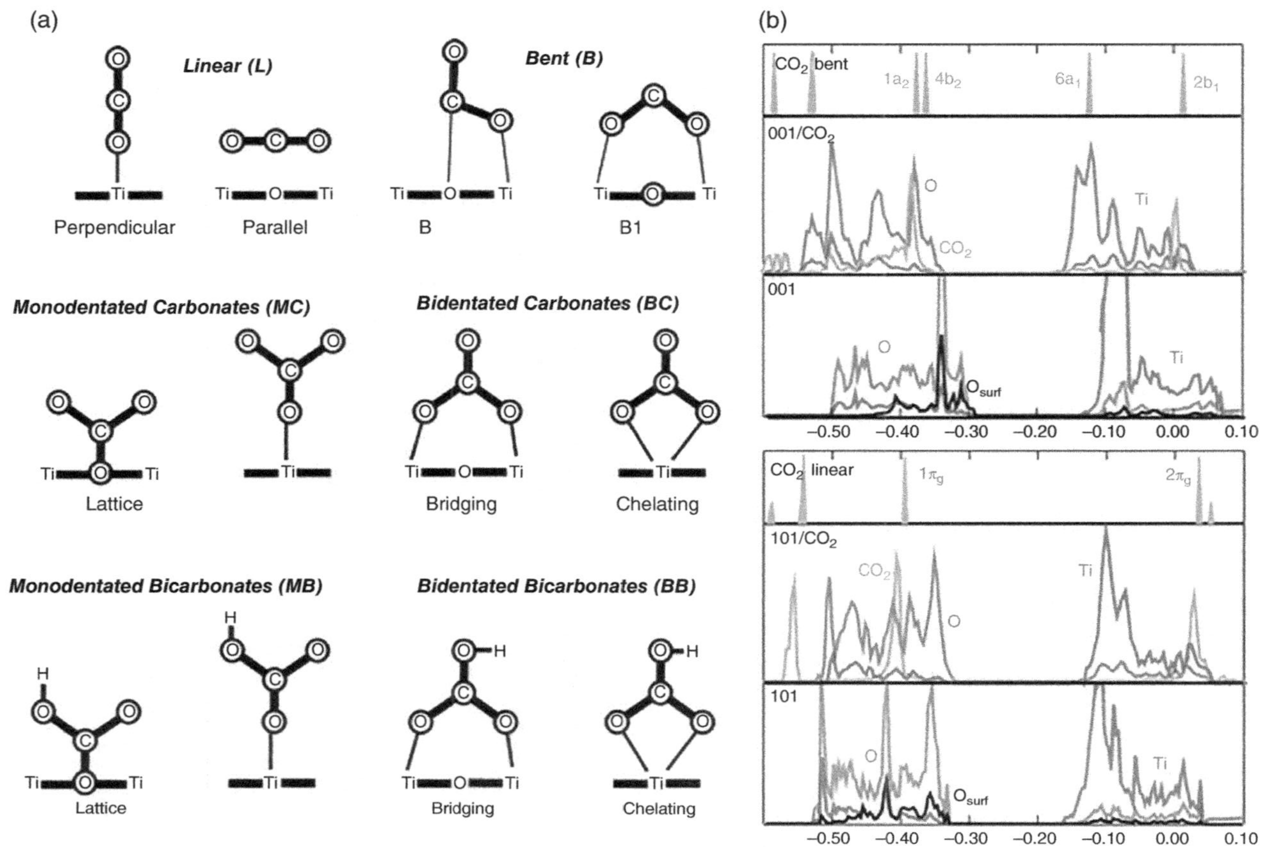

Figure 1.16 (a) Surface structures of CO_2 on TiO_2 surfaces (anatase phase). (b) PDOS of 101-L1 (bottom panels) and 001-MC (top panels) systems compared to the reference bare surfaces and free CO_2. The energy equals zero corresponds to the vacuum level. Energy in Hartree units. *Source:* From [126] / with permission of American Chemical Society.

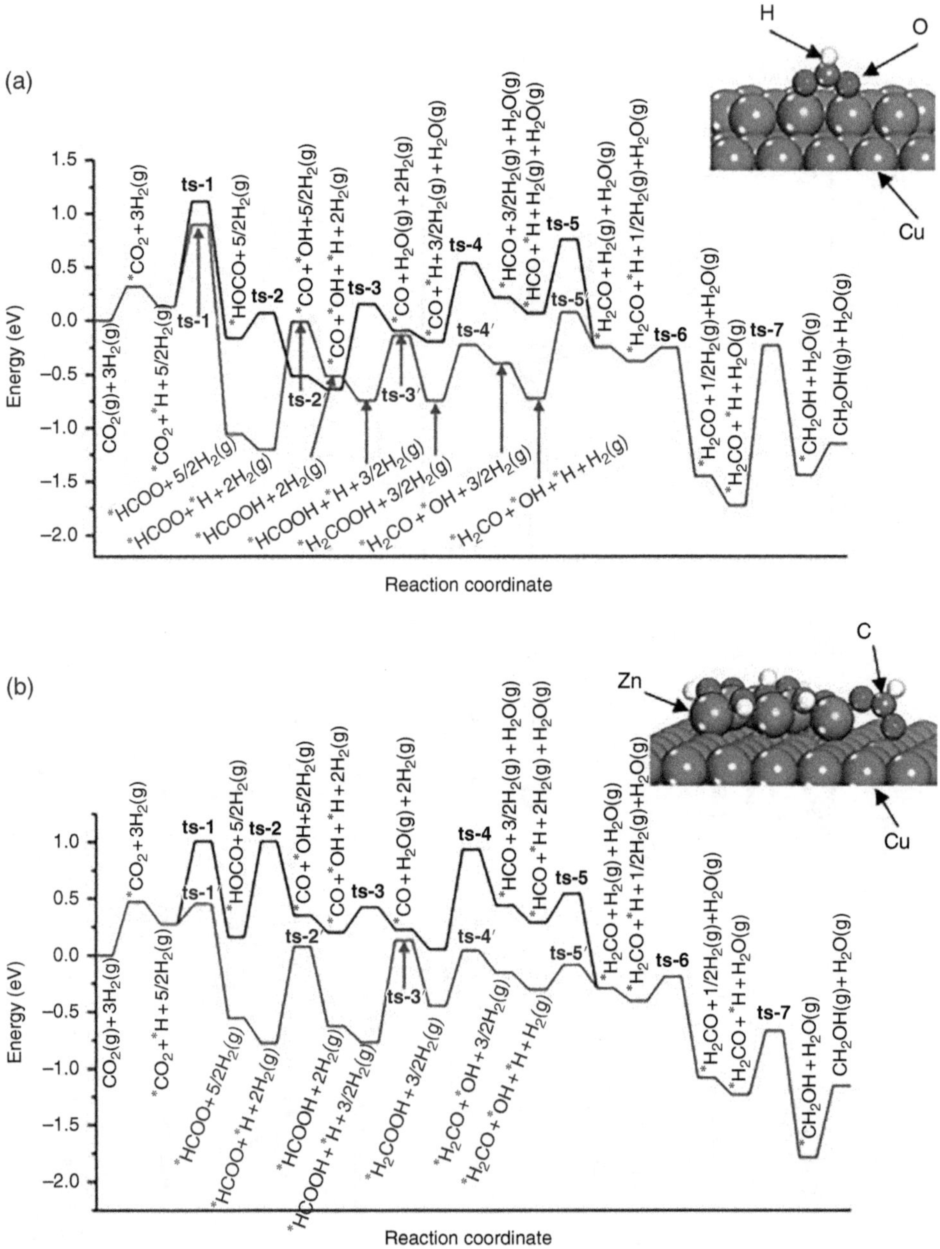

Figure 1.17 Potential energy diagram for the hydrogenation of CO_2(g) to CH_3OH(g) on (a) ZnCu(211) and (b) ZnO/Cu(111) via the RWGS + CO-hydro and formate pathways. Transition state (ts). (Inset) Structures of *HCOO on ZnCu(211) (a) and ZnO/Cu(111) (b). Cu, Zn, O, H, C. *Source:* Republished with permission of AAAS from [129]; permission conveyed through Copyright Clearance Center, Inc.

reaction to produce a CO intermediate followed by its hydrogenation to methanol (RWGS + CO-hydro pathway), and (ii) initial hydrogenation of CO_2 to an *HCOO intermediate followed by its hydrogenation and dissociation to methanol (formate pathway). Kattel et al. [129] attributed the formate mechanism to the lower energy potential, where *HCOO (E_a = 0.85 eV) and *HCOOH (E_a = 0.9 eV) hydrogenation reactions have lower energy barriers in comparison with ZnCu(211) (E_a = 1.49 eV). These calculated results agreed with the experimental data. Higher production of methanol was found on Cu/ZnO material.

Summing up, the DFT methods can help to understand the mechanism of a reaction of interest and develop novel catalyst materials. Furthermore, it allows understanding the modification of the optoelectronic properties generated by the interaction between the catalytic center and the support [105]. However, the interfacial electrochemistry simulation is still a challenge via DFT. Moreover, considerable effort has been put into this topic, but some difficulties still remain to be solved [130].

1.3 Surface Electrochemistry

The surface electrochemical response of oxide-carbon composites, Pt-based, and chalcogenides materials is discussed in this section.

1.3.1 Carbon Composite Materials

The carbon corrosion potential is defined in Eq. (1.36). Moreover, the corrosion product detection of CO_2 was probed, by the mass spectrometry technique, to occur at >0.97 V (RHE) [4]. This material is not stable for electrochemical applications at higher anodic electrode potential [1].

$$C + 2H_2O \rightarrow CO_2 + 4H^+ + 4e^- \; E_0 = 0.207 V_{SHE} \tag{1.36}$$

For this reason, different degradation mechanism can occur using carbon as support in hydrogen FCs. Meier et al. [2] revealed some strategies by comparing carbon black and hollow graphitic spheres. In this, a weakening of particle and support interaction can deliver different degradation mechanisms, such as metal dissolution, Ostwald ripening, agglomeration or coalescence, particle detachment, and primary carbon corrosion; see Figure 1.18.

The SMSI effect between metal particles and oxide has attracted attention, as previously discussed in Section 1.2.3. The interaction of Pt NPs and the oxide-carbon composite (TiO_2-C) for the study of the ORR was reported by Timperman et al. [26], and it turned out that the interaction was favored. Therefore, the modification of the TiO_2 was the next step. For example, the current-potential behavior of carbon (XC-72), TiO_2/C, and Y: TiO_2/C supports are contrasted in Figure 1.19. Before carbon corrosion, E > 1.0 V (RHE), the current-potential characteristics remain non-faradic within the explored potential interval. The determined capacitance trend, in each system, is as follows: Vulcan XC-72 (489 $\mu C/cm^2_{geo}$) > TiO_2/C (401 $\mu C/cm^2_{geo}$) > Y:TiO_2/C (323 $\mu C/cm^2_{geo}$). The absence of a surface redox reaction is shown curves (1) in Figure 1.19.

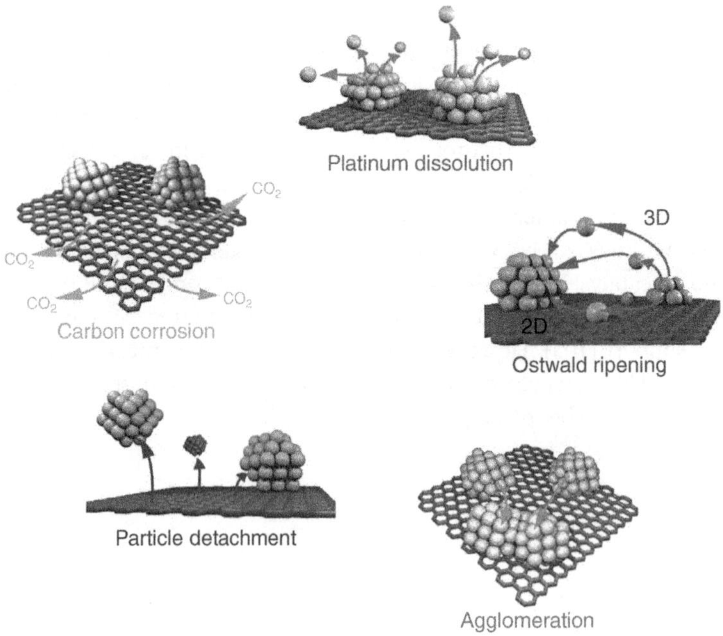

Figure 1.18 Simplified representation of proposed degradation mechanisms for carbon-supported Pt NPs in fuel cells. *Source:* From [2] / Beilstein Journal of Nanotechnology / CC BY 2.0.

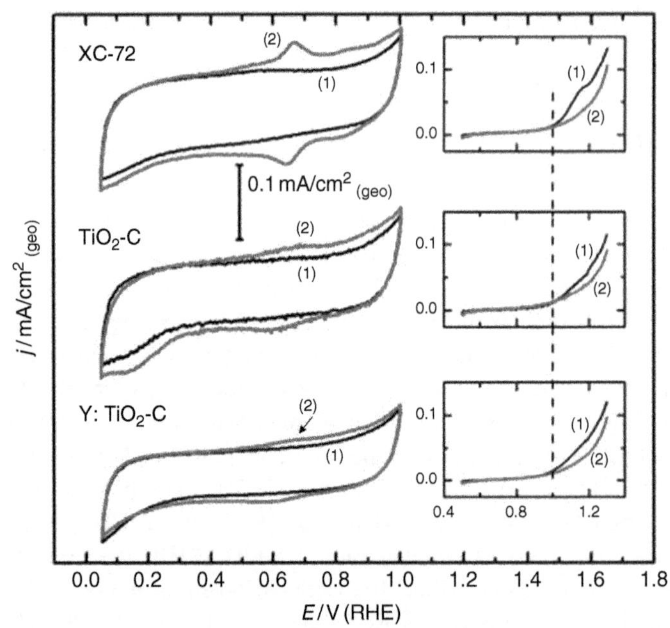

Figure 1.19 Current-potential characteristics of carbon (XC-72), TiO_2-C, and Y: TiO_2-C at 25 °C in 0.5 M H_2SO_4. Before corrosion (**1**) and after corrosion (**2**). The insets show the linear sweep voltammetry for all supports. *Source:* From [43] / with permission of Elsevier.

After submission to corrosion process, i.e. polarized at potential higher than 1.0 V (RHE), a wave emerged, centered at 0.65 V (RHE), which corresponds to the redox-potential of quinone/hydroquinone chemical groups, whose trend, which can be related to the CO_2 production, is Vulcan (36%) > TiO_2/C (24%) > Y: TiO_2/C (14%). Therefore, the chemical (in situ synthesis of oxide in the presence of carbon) versus the physical mixing of oxide with carbon revealed that the interaction of carbon with the oxide, in the in situ composite syntheses, can be attributed to the WMCI effect, as shown in Figure 1.20a. This figure also compares the anodic electrical charge variation of both systems. Electrical conductivity is an important issue for electrochemical science. Indeed, the combination of oxide with carbon suggests a new hybrid material or oxide-carbon composite. After 50 wt.% carbon, the composite material increases the electrical conductivity. This latter parameter, as expected, depends on the semiconducting oxide mass loading, Figures 1.20b and c.

Similar as earlier mentioned, with the Vulcan (XC-72) support, Y: TiO_2/C and Ce: TiO_2/C composites show a similar trend after submission to carbon corrosion potential. The quinone/hydroquinone redox wave decreases as a function of oxide mass loading. This point is important as it is crucial to anchor the catalytic centers to the oxide sites of the composites [42] (Figure 1.21).

In this context, different oxides such as SnO_2 [23, 41, 131], WO_3 [27], V_2O_5 [132, 133], ZnO [41], NiO [134], MnO_2 [135], RuO_2 [136], S-GO/C [137], and Co_3O_4/MWCNT [138] have been developed as oxide-carbon composite for different applications. The surface electrochemistry of Pt-based materials and chalcogenides supported on an oxide-carbon composite is discussed next.

1.3.2 Pt-based Materials

Pt surface electrochemistry is a straightforward approach to reveal the surface nature of Pt-based materials via the underpotential deposition of hydrogen, H_{upd}, and CO stripping. In this context, the strength of H_{upd} on Pt(111), Pt(100), and Pt(110) planes are revisited [139]. The mechanism for the HER follows the Tafel and Heyrovsky or Volmer reactions in acid, Eqs. (1.37)–(1.39) and alkaline media, Eqs. (1.40)–(1.41) for the adsorption–desorption of H_{upd} [140].

In acid medium:

$$\text{Heyrovsky}: H_{ad} + H^+ + e^- \rightleftharpoons H_2 + Pt(hkl) \quad (1.37)$$

$$\text{Volmer}: Pt(hkl) + H^+ + e^- \rightleftharpoons H_{ad} \quad (1.38)$$

$$\text{Tafel}: 2H_{ad} \rightleftharpoons H_2 + 2Pt(hkl) \quad (1.39)$$

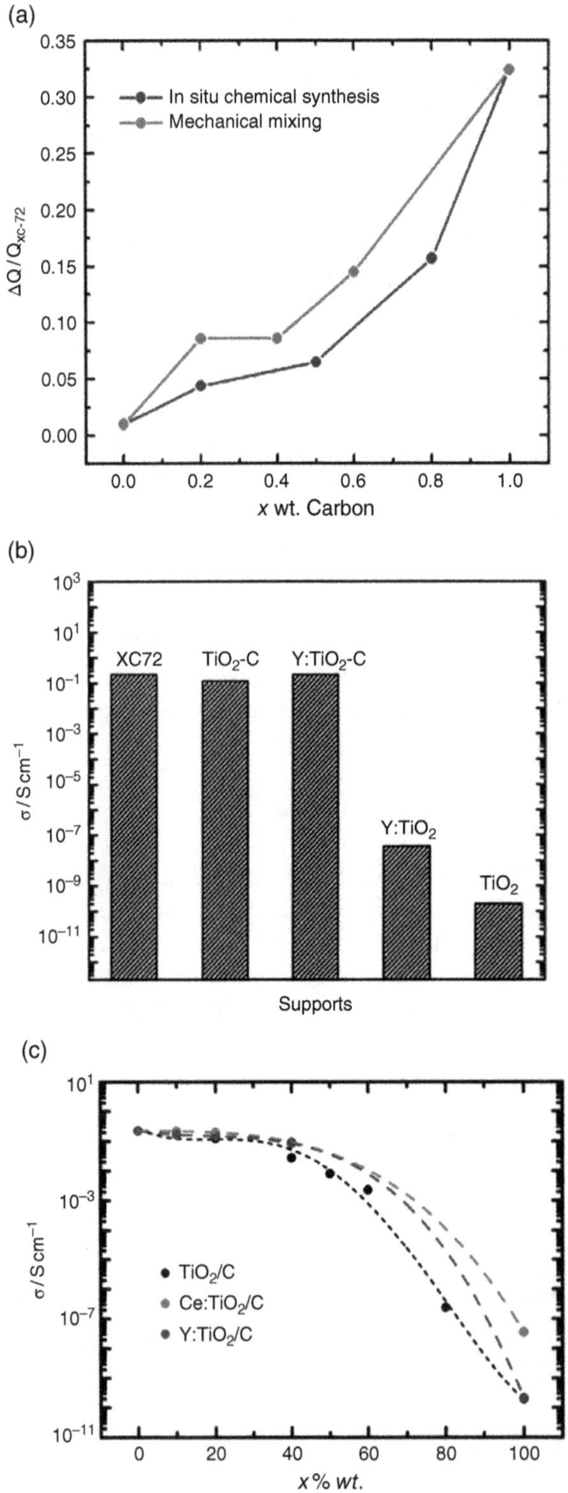

Figure 1.20 (a) Variation of the integrated electrical charge as a function of the mass loading of carbon deduced from cyclic voltammetry (CV)s experiments before and after the corrosion process. Electrical relative conductivity σ/S/cm for (b) Vulcan (XC-72), TiO_2-C, Y: TiO_2-C, Y-doped TiO_2 and TiO_2 supports, and (c) as a function of mass loading and the nature of the oxide. Each powder sample was submitted to a torque of 25 lbf-inch. *Source:* From [43] / with permission of Elsevier.

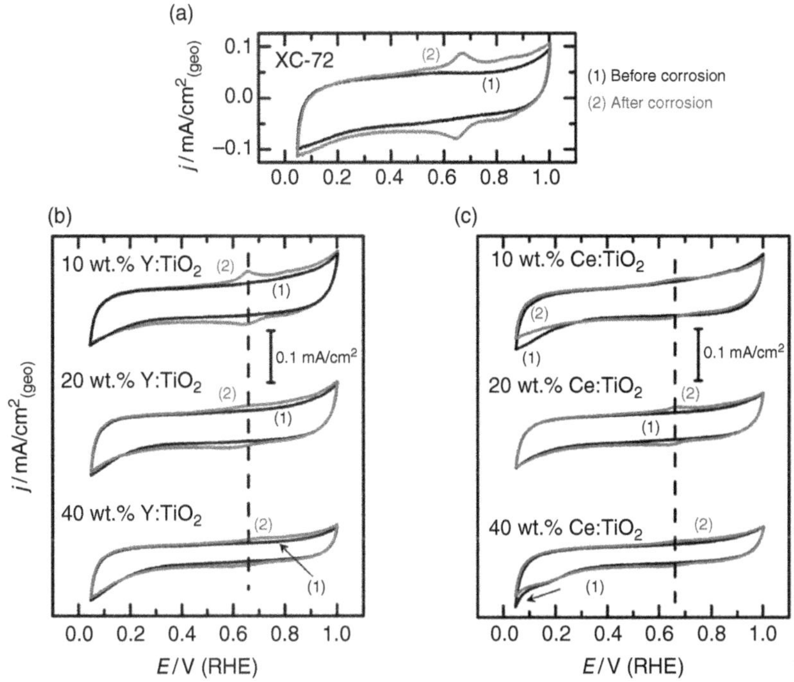

Figure 1.21 Current-potential curves for (a) Carbon Vulcan (XC72), (b-c) Y: TiO_2/C, and Ce: TiO_2/C supports with different oxide mass loading (i.e. 10, 20, and 40 wt.%) in 0.5 M H_2SO_4 solution.

In alkaline medium:

$$\text{Heyrovsky}: H_{ad} + H_2O + e^- \rightleftharpoons H_2 + OH^- + Pt(hkl) \tag{1.40}$$

$$\text{Volmer}: Pt(hkl) + H_2O + e^- \rightleftharpoons H_{ad} + OH^- \tag{1.41}$$

The crystallographic domains of the metal can be characterized as a fingerprint. This includes OH adsorption and the CO monolayer oxidation – CO stripping, Eq. (1.42)–(1.43).

$$H_2O + Pt(hkl) \rightarrow OH_{ads} + H^+ + e^- \tag{1.42}$$

$$CO_{ads} + OH_{ads} \rightarrow CO_2 + H^+ + e^- + 2Pt(hkl) \tag{1.43}$$

The CO-stripping reaction also confirms different crystallographic domains on single crystalline surfaces [141, 142]. However, this reaction is more complex, and the challenge is to separate the contribution of each plane. On polycrystalline Pt, the surface electrochemistry confirms the crystallographic domains involved in the reaction [143]. The voltammetric behavior is illustrated in Figure 1.22, which shows the H_{upd} and CO-stripping region in acid and alkaline media.

Pt NPs with different morphologies show different surface electrochemistry behavior. Vidal-Iglesias et al. [144] synthesized Pt NPs with spherical (Pt NP_{sphe}), cubic (Pt NP_{cubic}), octahedral and tetrahedral (Pt NP_{tetra}), and octahedral and tetrahedral truncated (Pt NP_{trunc}) shapes, cf. Figure 1.23. From this figure, we can deduce four contributions, namely: (i) the peak is related to Pt(110)-type site, at 0.125 V; (ii) the peak contains two contributions, Pt(100) step site on Pt(111) and Pt(100) terraces at 0.27 V; (iii) the signal at 0.35–0.37 V is attributed to Pt(100) bidimensional terraces; and (iv) at 0.5 V, the signal is related to the bidimensional ordered Pt(111) terraces.

On the other hand, in an alkaline medium, the effect of the OH adsorption is important. The authors confirmed that the peaks correspond to the Pt(110) and Pt(100) steps on the Pt(111) terraces [144]. The CO-stripping performed on the various morphologies, in the

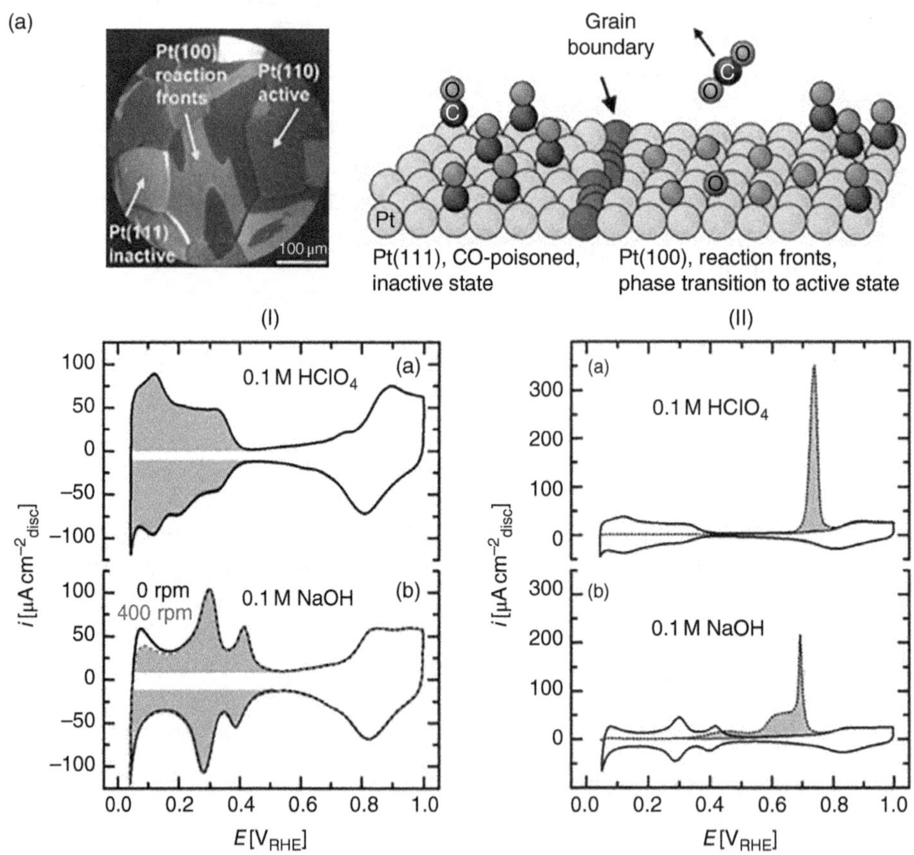

Figure 1.22 (a) Surface electrochemistry on polycrystalline Pt: (I) H_{upd} and (II) CO stripping in $HClO_4$ and NaOH solutions. *Source:* Figure (A) From [145] / with permission of John Wiley & Sons, and Figures (I) and (II) From [139] / IOP Publishing / CC BY 4.0.

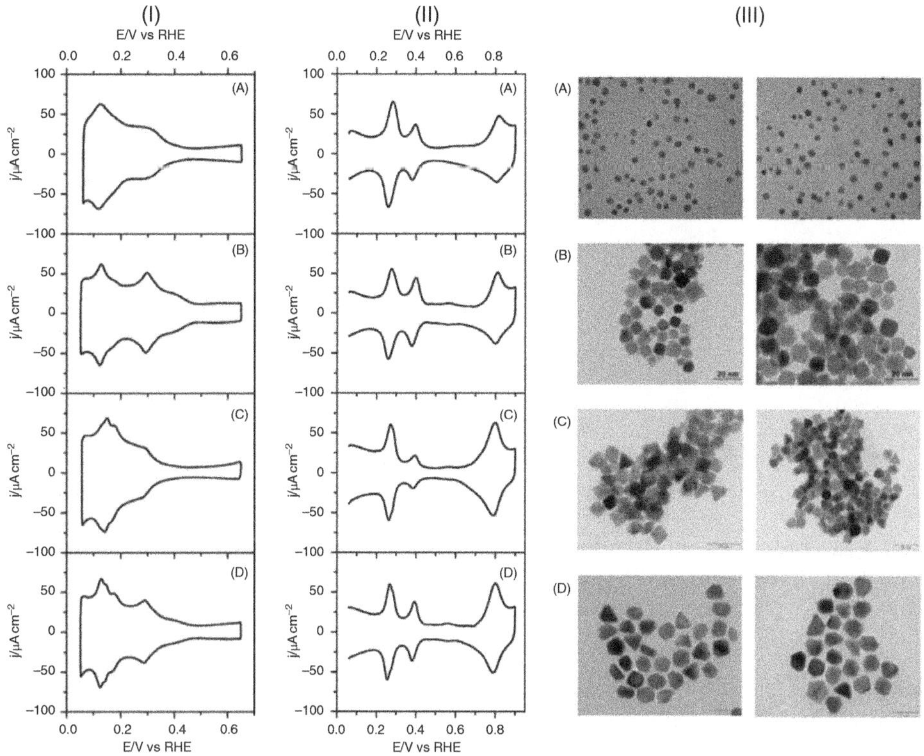

Figure 1.23 (I) Voltammograms corresponding to (A) Pt NP_{sphe}, (B) Pt NP_{cubic}, (C) Pt NP_{tetra}, and (D) Pt NP_{trunc} in 0.1 M $HClO_4$. Scan rate, 50 $mV s^{-1}$, and (II) in 0.1 M NaOH. Scan rate, 50 $mV s^{-1}$ (III). Representative corresponding TEM images of Pt NP. *Source:* From [144] / with permission of American Chemical Society.

alkaline medium, showed no differences for each morphology [144]. Moreover, on carbon domains, Pt NPs supported on C-sp^2 domains showed an SMCI effect according to CO-stripping behavior. Figure 1.24 displays the CO-stripping experiment for Pt/C, Pt/CNT, Pt/p-CNT, and Pt/e-CNT (p and e stand for purified and exfoliated CNTs, respectively) in acid and alkaline, respectively [108].

In an acid medium, the Pt NPs show that the oxidation peak on carbon is shifted to a lower potential with respect to Pt/C (C-sp^3 domains). We can, however, observe a small wave at c. 0.75 V (RHE), which is associated with the interaction between Pt and C-sp^3 domains, in agreement with the CO-stripping behavior on polycrystalline Pt [144], cf. Figure 1.22. This observation confirms that different crystallographic domains interact with CO and OH adsorption. The electronic effect induced between Pt and C-sp^2 surface is observed in the data represented in Figure 1.24a, whereas the chemical effect or the oxophilicity of carbon is clearly observed in alkaline medium, Figure 1.24b. In other words, the bifunctional mechanism effect is activated by a non-precious component – carbon [108].

Pt NPs preferentially deposited onto TiO_2 sites, via the photo-deposition technique, show similar electrochemical behavior, Figure 1.25, to that of spherical-shaped Pt NPs, cf.

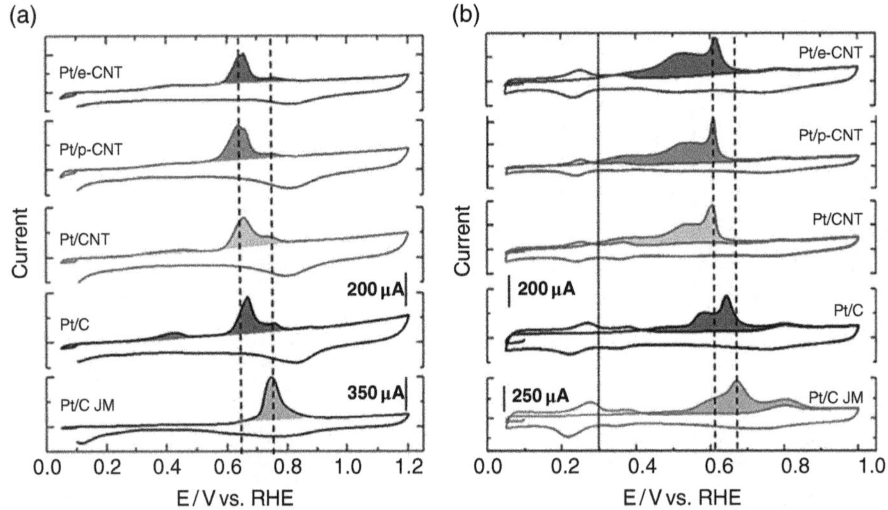

Figure 1.24 CO-stripping profiles (5 mV/s, 25 °C) of PD Pt on C, CNT, p-CNT, and e-CNT. (a) 0.1 M $HClO_4$. (b) In 0.1 M KOH. *Source:* From [108] / with permission of Elsevier.

Figure 1.23(I-IIIA). The H_{upd} region exposes some details regarding the Pt (*hkl*) domains, Figure 1.25a. In the range of potential from 0.05 to 0.35 V (RHE), we can visibly associate the contribution of different planes of Pt (*hkl*) clusters.

According to Vidal-Iglesias' work [145], Pt clusters can expose a family of planes in contact with the electrolyte as spherical morphology, marked as ({*hkl*}, {*hk0*}, and {*h00*}) sites.

In alkaline media, the Figure 1.25b shows that ({*hkl*}, {*hk0*}, and {*h00*}) domains are visible in the H_{upd} region. Similar features were observed on Pt NPs photo-deposited on Y: TiO_2/C and Ce: TiO_2/C support [42]. To account for the electronic state of Pt NPs photo-deposited on oxides, CO-stripping measurements were performed. The results are summarized in Figure 1.26.

In an acid medium, one can notice that peak (1) and (2) are associated with Pt/C-sp^2 and Pt/C-sp^3 domains. The former is the result of the PDM, whereas the latter corresponds to Pt generated by the carbonyl chemical route. The measurements carried out on Pt black attest that Pt/C-sp^2 and Pt/C-sp^3 domains present SMCI and WMCI effects, respectively. In the frame labeled 3, the peak is centered in the same electrode potential as that of Pt/C-sp^2 domains. This result stresses the fact that a similar metal–support interaction (SMSI) phenomenon is present on both supports [42].

In alkaline media, however, the CO-stripping behavior is like polycrystalline particles, revealing that the oxide-supported Pt NPs' properties are electronically modified. Here, it is important to assess further that the shift to a lower potential, also observed on photo-deposited Pt onto carbon (Pt/C), resulted in SMSI as the main effect. The oxophilicity of Pt remains unchanged.

To compare the materials' performance, the ECSA is calculated from H_{upd} and CO-stripping region, Eq. (1.44).

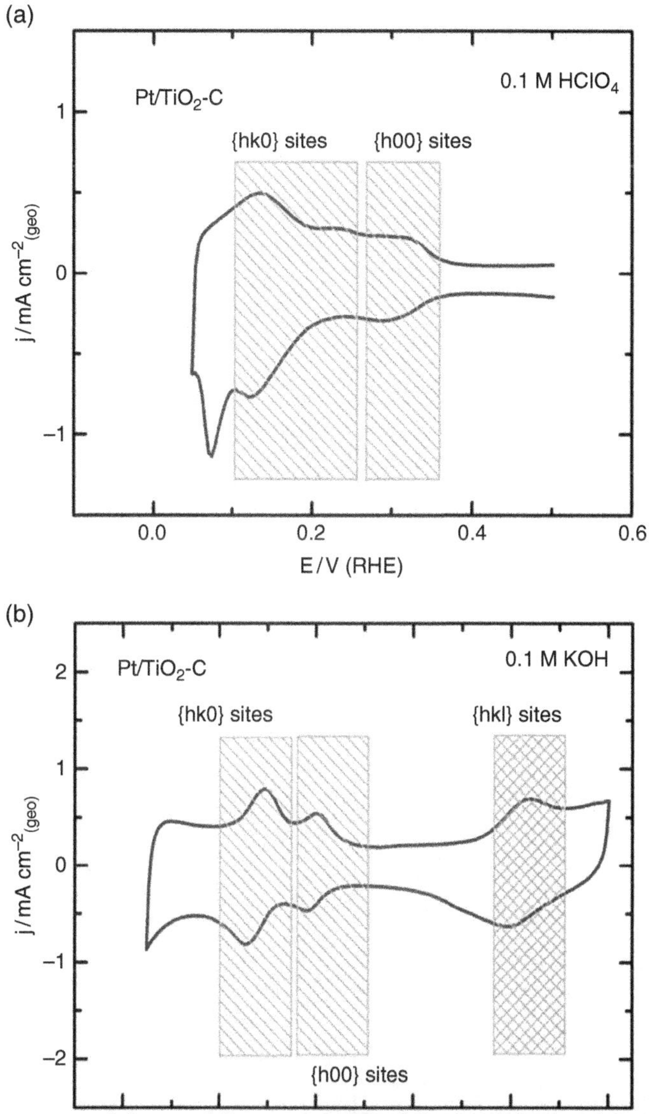

Figure 1.25 Surface electrochemistry characteristic for (a) Pt/TiO_2-C electrocatalyst onto a gold disk in 0.1 M $HClO_4$, and (b) Pt/TiO_2-C on a glassy carbon disk in 0.1 M KOH at 50 mVs^{-1} and 25 °C. *Source:* Adapted from [42].

$$\mathrm{ECSA}_n = \frac{1}{q_{Pt}v}\left[\int_{E_0}^{E_f} j_n(E)\partial E - \int_{E_0}^{E_f} j_n^{dl}(E)\partial E\right] \tag{1.44}$$

Where n refers to the method used to calculate the area (H_{upd} or CO stripping), j is the current (μA), E is the electrode potential (V), q is the charge density for Pt (210 or 420 $\mu C/cm^2$,

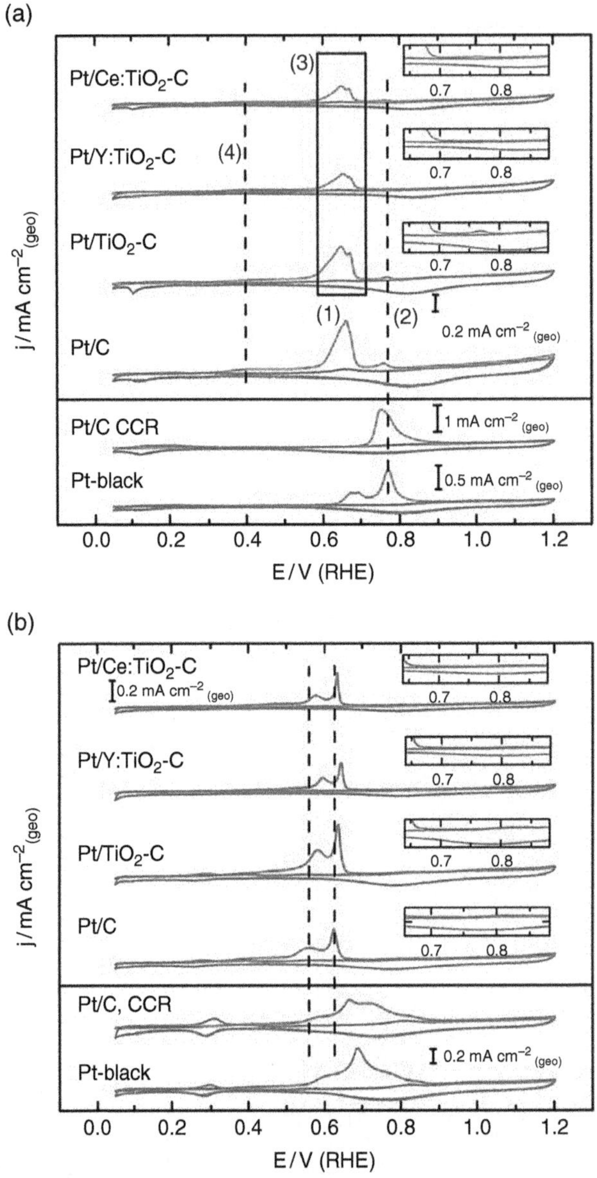

Figure 1.26 CO-stripping characteristic in acid (a) and alkaline (b) for photo-deposited Pt onto C, TiO_2-C, and M: TiO_2-C (M = Y or Ce). Pt/C CCR was obtained by CCR and compared to Pt-black, used as reference. The insets zoom the potential interval corresponding to the interaction with the C-sp^3 domain. *Source:* From [42] / with permission of John Wiley & Sons.

respectively, for H_{upd} or CO monolayers), j^{dl} is the current owing to the double layer (μA), and ν is the scan rate (Vs^{-1}).

Table 1.2 summarizes the ECSA value for carbon and oxide domains supported Pt. We can observe that the ECSA values in the C-sp^2 surface are higher than that of the oxide surface. The electronic effect is indeed induced by the metal–support interaction.

Table 1.2 ECSA for Pt-based materials supported on carbon and oxide domains and obtained for CO stripping. Ref. [a][108] and [b][42].

Catalyst	ECSA in $HClO_4$ [m^2/g_{Pt}]	ECSA in KOH [m^2/g_{Pt}]
[a]Pt/C JM*	48.93	n.a.
[a]Pt/C	24.47	n.a.
[a]Pt/CNT	35.33	n.a.
[a]Pt/p-CNT	42.68	n.a.
[a]Pt/e-CNT	23.63	n.a.
[b]Pt/TiO_2-C	24.8 ± 5.1	25
[b]Pt/Y:TiO_2-C	9 ± 1	12
[b]Pt/Ce: TiO_2-C	10.7 ± 0.2	11

* JM = Johnson Matthey.

1.3.3 Chalcogenide Materials

Chalcogenide materials are an alternative to Pt-based materials. These materials are compounds formed with metals and elements of group 16 (i.e. S, Se, Te) [146]. Like the analysis made on Pt materials, non-precious $CoSe_2$ EC can be also investigated on different supports. Figure 1.27 shows the carbon (XC-72), TiO_2/C, functionalized nanotubes (*f*-CNT) and nitrogen doped-carbon nanohorns (NCNH) supported $CoSe_2$ in acid and alkaline media. The analysis of the current potential characteristics made evident that the behavior is rather different according to the nature of the materials, serving as supports, and according to the nature of the electrolyte. An additional observation is a slight increment in the capacitance of $CoSe_2$/NCNH. This latter phenomenon may be linked to the interaction between NCNH and $CoSe_2$ particles. A similar trend is also observed in 0.1 M KOH.

Moreover, using a modified oxide-carbon composite, as support, does not show any significant changes. Figure 1.28 illustrates the same catalytic center-supported on M: TiO_2-C (M = Y or Ce element) under the same experimental conditions as those of Figure 1.27. The fact that metal–support interaction between $CoSe_2$ and oxide-carbon composite takes place is the formation evidence of interfacial CoO_x onto TiO_2.

The capacitance is related to the electrical conductivity of the support, according to Eq. (1.45). TiO_2 phase has lower conductivity than carbon, cf. Figure 1.20c. The carbon phase enhances the electrical conductivity and capacitance of the oxide in the oxide-carbon composite. Figure 1.21 confirms this behavior by comparing 20 wt.% and 40 wt.% of oxide in the composite. Therefore, due to the low electrical conductivity of CoO_x/TiO_2, this material must show a low capacitance. Thus, the electrochemical behavior between $CoSe_2$ and the oxide-carbon composite confirms that the interaction does not affect the electrical conductivity and further indicating that the $CoSe_2$ phase is not in intimate contact with the oxide.

In the analysis of current-potential characteristics of non-precious materials, one must consider different aspects, namely, the crystallite size, overpotential of a reaction of interest, ECSA, and roughness. Figure 1.29 shows the particle size effect on the surface redox processes [147]. The most significant feature is linked to the reciprocal of the charge

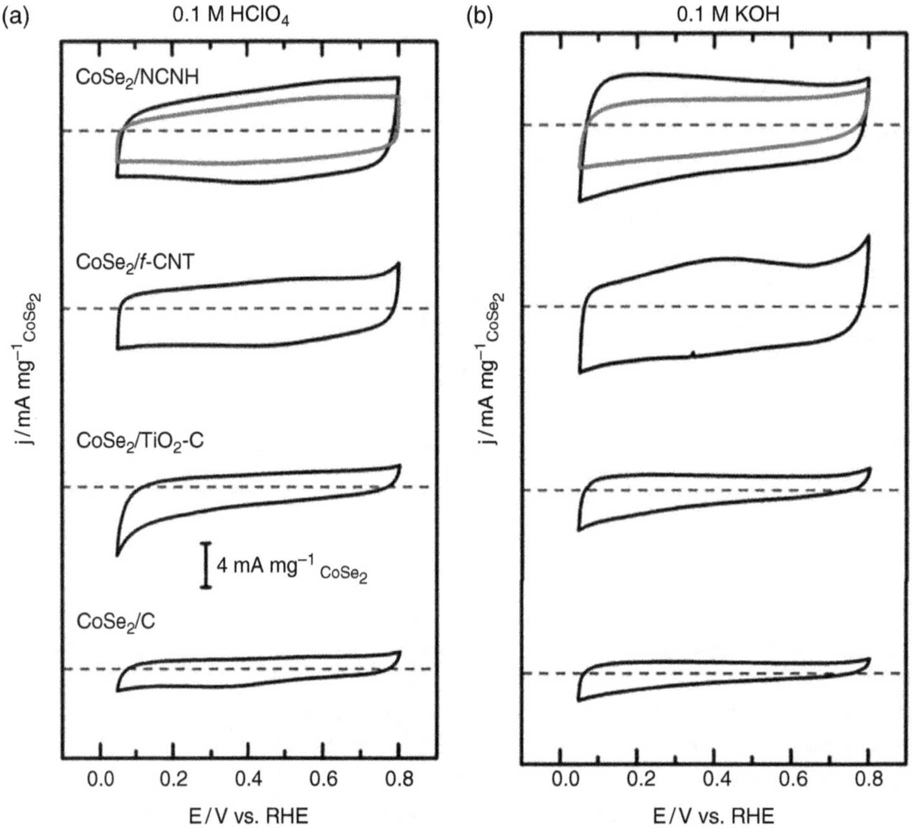

Figure 1.27 (a) Current potential of $CoSe_2$ deposited onto C, TiO_2-C, *f*-CNT, and NCNH in 0.1 M $HClO_4$ and (b) 0.1 M KOH. Scan rate at 50 mVs^{-1}. The curves inserted in the upper panel represent the voltammograms of the NCNH in both 0.1 M $HClO_4$ and 0.1 M KOH, respectively. *Source:* From [54] / with permission of IOP Publishing, Ltd.

transfer resistance, R_{ct}^{-1}, parameter, Figure 1.29b, obtained by potentiodynamic electrochemical impedance spectroscopy (PDEIS). The size effect in the underpotential deposition region ($E < -0.4$ V) affects the charge transfer [148].

The ECSA is the active site where species can be adsorbed, like protons in H_{upd} or CO-stripping processes. With non-precious materials, the ECSA value is likely affected by the surface roughness. This parameter is difficult to determine because of the unclear capacitive behavior. There are two ways to obtain the double layer capacitance on such systems: (i) capacitance current as a function of scan rate and (ii) electrochemical impedance spectroscopy (EIS) experiments.

In the first one, the capacitance current intensity is considered a function of scan rate in a non-faradic region, Eq. (1.45), Figure 1.30a [147].

$$i_c = C_{dl} v \tag{1.45}$$

The capacitance current variation yields a straight line with a slope equal to C_{dl}, Figures 1.30b and c. C_{dl} value of 0.091 mF has been reported for non-precious catalyst [149].

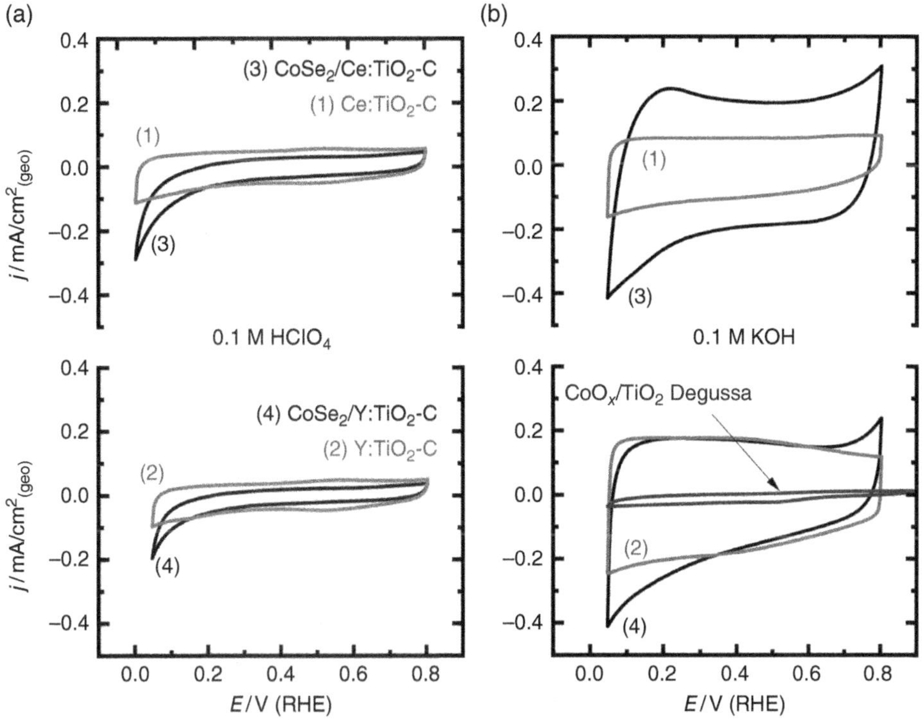

Figure 1.28 (a) Current-potential curves of M: TiO_2/C (M = Y and Ce, 20 wt.%, (1)–(2)) supported $CoSe_2$ in 0.1 M $HClO_4$ and (b) in 0.1 M KOH. The electrochemical behavior of the corresponding oxide-carbon composite is marked as (3) and (4). For comparison, CoO_x/TiO_2 Degussa is added in KOH solution. The scan rate was 50 mVs^{-1}.

The second method considers the execution of (EIS) experiments in the same non-faradic region with a modified Randles circuit, Eq. (1.46), insets on Figure 1.29b [150].

$$z(\omega) = R_s + \frac{R_{ct}}{R_{ct} * Q_0 (j\omega)^{\alpha} + 1} \tag{1.46}$$

R_s and R_{ct} parameters are the resistance of the solution, and the charge-transfer resistance, Q_0, is a constant phase related to the double layer, ω is the frequency. Therefore, Eq. (1.47) shows the dependence of C_{dl} with Q_0 [150].

$$C_{dl} = \left[Q_0 \left(\frac{R_s R_{ct}}{R_s + R_{ct}} \right)^{1-\alpha} \right]^{1/\alpha} \tag{1.47}$$

The ECSA and roughness values of a catalyst are the ratio of C_{dl}/C_s and the ratio of ECSA/A_{geo}. A_{geo} is the geometrical surface area and C_s is the specific capacitance. This latter value is the critical parameter for non-precious catalyst [149].

To summarize, the analysis of current-potential characteristics serves to describe the material and obtain some critical parameters, such as stability [151], MSA, SA [101], and

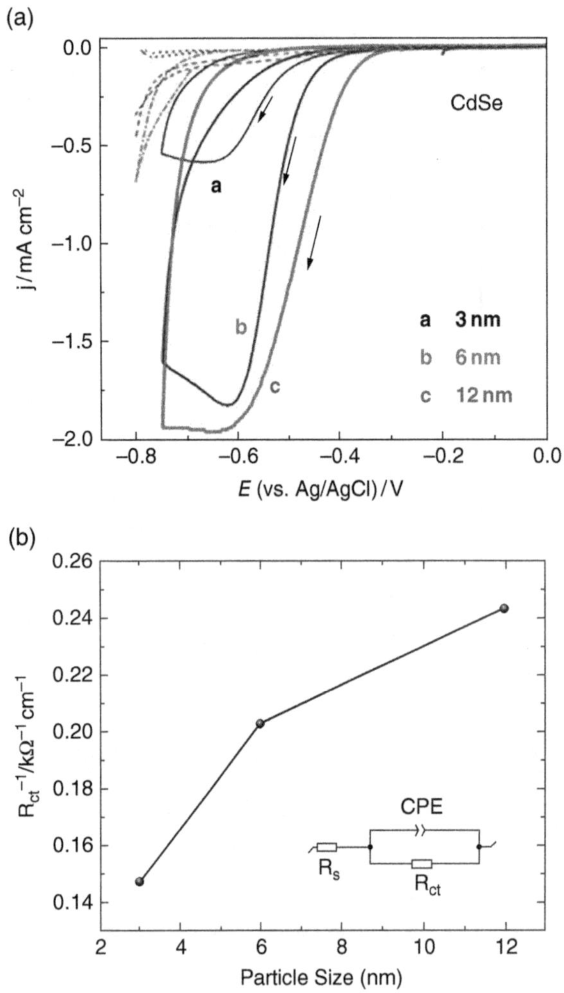

Figure 1.29 (a) Cyclic voltammetry; (b) reciprocal charge transfer resistance R_{ct}^{-1} of Cd upd on CdS and CdSe chemical bath deposition (CBD) films on fluorine-doped tin oxide (FTO) substrates with particles of different size. Electrolyte: 10 mM $CdSO_4$ + 0.1 M Na_2SO_4 (solid), dotted lines correspond to the blank 0.1 M Na_2SO_4. *Source:* Modified figures from reference [148].

the turnover frequency number [23, 152] in Pt-based or non-precious catalysts – a function of microstructural properties; see Section 1.2.2 [23, 42, 147].

1.4 Oxide and Oxide-Carbon PhotoElectrochemistry

Oxides and oxide-carbon composite materials are the subjects of much attention in photochemical and photoelectrochemical systems [153, 154]. Among the advantages of surface electrochemical analysis, photoelectrochemical tool is to be considered [155]. The photoelectrochemical behavior of oxide and oxide-carbon compounds are discussed. As examined in Section 1.2.3, the metal NPs synthesis via photo-assisted deposition induces the metal–support interaction, a phenomenon favorable to boost the electrocatalytic effect in terms of stability and selectivity. As we know, TiO_2 is an excellent photocatalyst [156], and different modifications were made, such as dopants [157] or mixed-oxide [158].

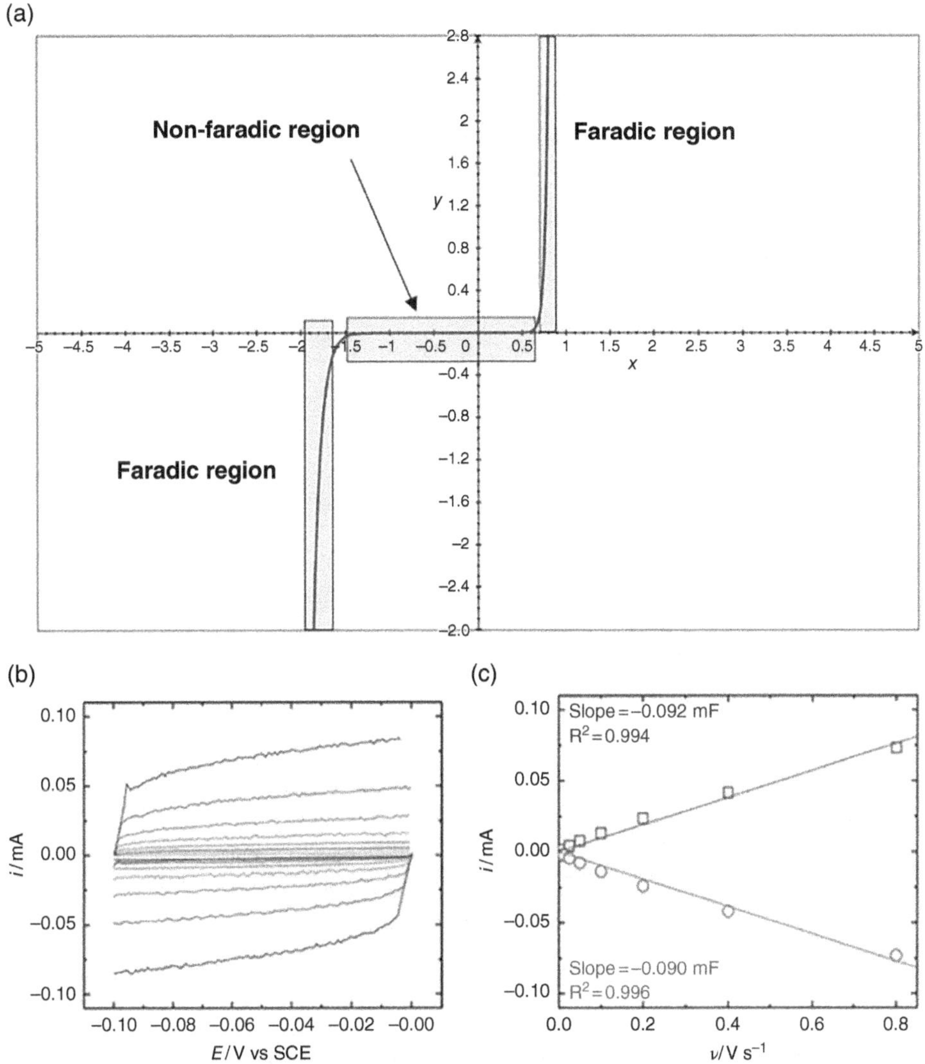

Figure 1.30 (a) Non-faradic and faradic region; (b) cyclic voltammograms were measured in a non-faradaic region of the voltammogram at the following scan rate: 0.005, 0.01, 0.025, 0.05, 0.1, 0.2, 0.4, and 0.8 V/s. (c) The cathodic (open circle) and anodic (open square) charging currents measured at −0.05 V vs. saturated calomel electrode (SCE) plotted as a function of scan rate. *Source:* Figures (B-C) from [149] / with permission of American Chemical Society.

The photoelectrochemical studies reveal hints as to the nature of the photoactive species. Figure 1.31a shows the measurements in darkness and under UV–vis illumination for three oxides, namely, $Ti_{0.95}W_{0.05}O_2$, $Ti_{0.9}W_{0.1}O_2$, and $Ti_{0.8}W_{0.2}O_2$, deposited in SnO_2:F substrate.

The photoelectrochemical behavior shows that the redox peaks of Ti^{4+} and Ti^{3+} are responsible for the current at lower potential (E < 0.6 V(RHE)). However, electron–hole (e^-/h^+) recombination is responsible for increasing or decreasing the photocurrent. In addition, the presence of amorphous and surface photoactive phases in the oxide decreases the

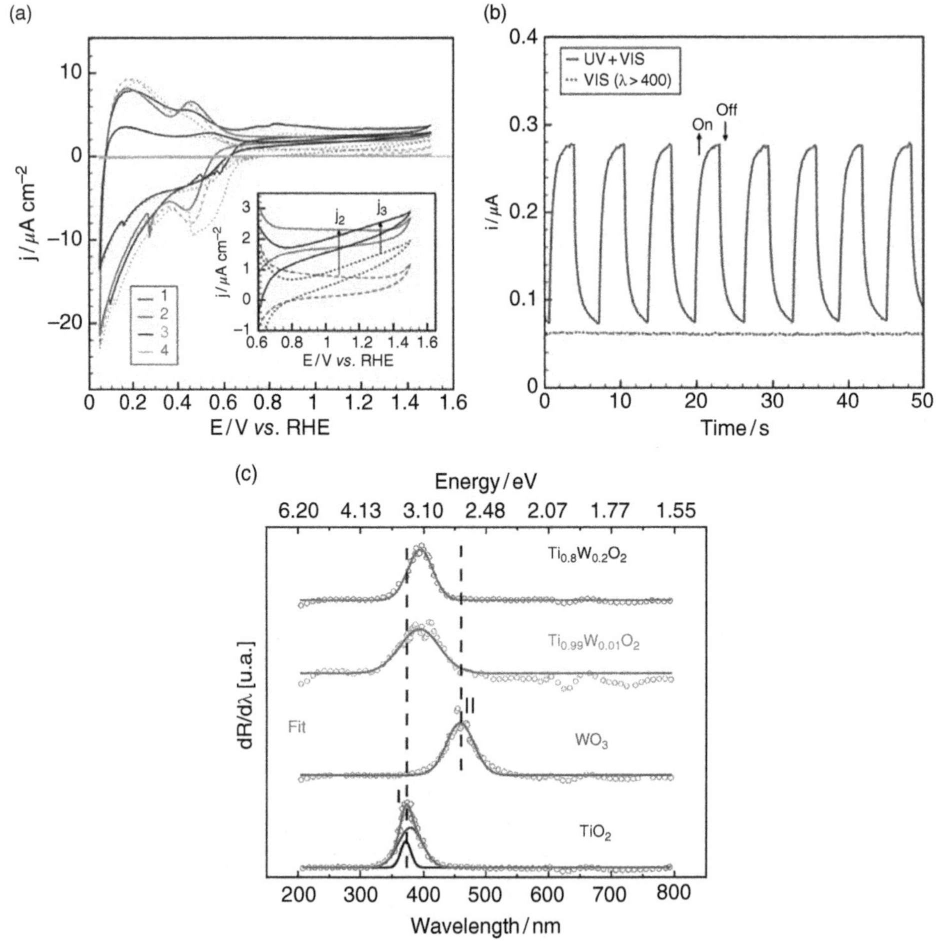

Figure 1.31 (a) Current-potential characteristics of selected samples: 1: $Ti_{0.95}W_{0.05}O_2$, 2: $Ti_{0.9}W_{0.1}O_2$, 3: $Ti_{0.8}W_{0.2}O_2$, and 4: SnO_2:F substrate in darkness (dashed lines) and under UV (full lines). The insert zooms the photo response of j_3: $Ti_{0.8}W_{0.2}O_2$ and j_4: $Ti_{0.9}W_{0.1}O_2$. The chopped on–off photocurrent response of $Ti_{0.8}W_{0.2}O_2$ at E = 1 V vs. RHE, under UV and visible (λ > 400 nm) irradiation is shown in (b). Measurements were carried out in an Argon-saturated 0.5 M H_2SO_4 electrolyte with a scan rate of 50 mVs^{-1}. (c) DPR analysis. *Source:* Figure (A-B) and (C) from [158] / with permission of Elsevier.

photocurrent response. Here, the WO_3 phase is responsible for this effect, and the modification of microstructural properties of the TiO_2 is essential [158]. The modification of the oxide comes out evidently, under UV radiation, on $Ti_{0.8}W_{0.2}O_2$, as shown in Figure 1.31b. In addition, the derivative peak fitting (DPR) of diffuse reflectance UV–visible spectra analysis confirmed that the superficial photoactive phases modified the optical band-gap energies (OBGEs) [157, 159], Figure 1.31c. From this figure, we can observe two OBGE values in the TiO_2 sample, which correspond to anatase (3.34 eV) and rutile (3.27 eV) phases, zone I. The OBGE value of the pure WO_3 oxide (2.71 eV) lies outside the zone I. The OBGE value of the mixed oxide (3.15 eV) is between zone I and II. The electrical conductivity increases from

1.77×10^{-3} to 147.06 mS/cm, between TiO_2 and $Ti_{0.8}W_{0.2}O_2$ materials, where the modification of TiO_2 was confirmed [158]. This analysis indicates that photoactive materials are responsible for the photocurrent in the surface photoelectrochemistry analysis, as confirmed by the chopped illumination shown in Figure 1.31b.

In some cases, the diffuse reflectance method and the photoelectrochemical analysis (action spectra) deliver similar conclusions. Díaz-Real et al. [160] explored that the Kubelka–Munk and photo energy plot analysis can near the OBGE values in the two phases of TiO_2 (anatase and brookite).

Similar values of OBGE from these two methods delivered similar conclusions as to the oxide's surface, Figure 1.32. However, the interaction in the oxide-carbon composite made the complex interaction evident [161]. First, a modification of the TiO_2 with yttrium (Y) and cerium or a heat treatment in a hydrogen atmosphere showed that the grain boundaries' resistivity controls the photochemical process and induced a metal-like behavior [157, 158]. This phenomenon is rather like that found with the oxide-carbon composite. Figure 1.33 exemplifies the chopped behavior for TiO_2/C and Y: TiO_2/C materials in acid conditions. We can observe that the photocurrent decreased on Yttrium modified TiO_2 in the oxide-carbon composite, and the onset potential where the electron–hole separation takes place is shifted to a more positive potential.

These oxide-carbon composites are stable in acid conditions; see carbon composite materials section. However, the dispersion of the oxide in the carbon matrix is still a subject of development [42, 43]. More details are needed to review and optimize the performance of hybrid materials, e.g. the synthesis method. We can nevertheless state that on the oxides' nature [162], the interaction oxide-carbon [161, 163], oxide dispersion [38], grain boundaries' resistivity [157], and the photoactive oxide subsurface species [164] take place. These aspects show an important impact on the photocatalytic effect and stability [157], and pathway mechanism, as revealed by DFT simulation analysis. A discussion regarding the design of carbon/metal oxide hybrid using the two approaches for electrochemical technologies was reviewed [161].

1.5 Case Studies

The HER-HOR and ORR-OER are the most critical reactions in FC technologies. Here, we briefly discuss the microstructural analysis, metal–support interaction, and stability of the aforementioned reactions.

1.5.1 Stability in HER-HOR

As an essential reaction in low-temperature H_2/O_2 FCs, it is known that the HER/HOR reactions proceed via Volmer–Tafel or the Volmer–Heyrovsky mechanisms, Eqns. (1.37)–(1.41) in an acid or alkaline medium. On Pt, the HER/HOR kinetics in alkaline is slower by two orders of magnitude than in acid electrolyte in Pt [140]. Since the OH concentration is a function of pH, it seems evident that a new phenomenon can occur. The so-called hydrogen binding energy (HBE) [102] seems to be involved in this process. Thus, the OH-adsorption capacity of material might represent the oxophilicity character of the EC. For instance, in PtRu/C, the Pt sites preferentially absorb CO, whereas OH from water

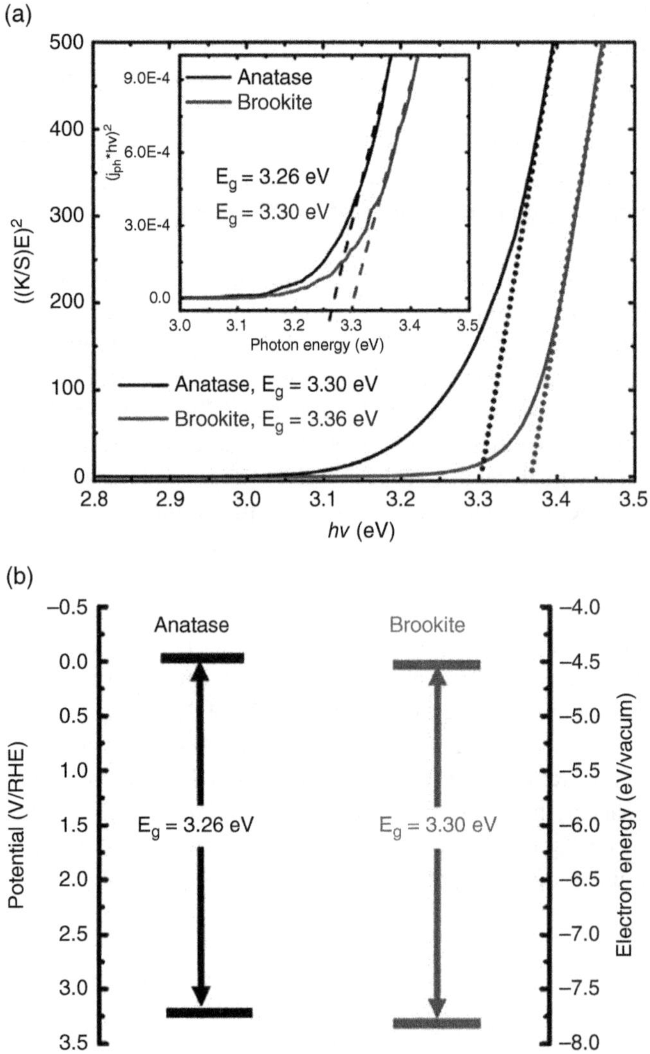

Figure 1.32 (a) Kubelka–Munk vs. photon energy plot for anatase and brookite obtained by diffuse reflectance. Both samples were heat treated at 400 °C. A glassy carbon and RHE served as counter and reference electrodes, respectively. Irradiance at electrode distance = 5.0 mW/cm^2. (b) Relative energy positions for anatase and brookite deduced from electrochemical experiments. *Source:* From [160] / with permission of Elsevier.

dissociation interacts on Ru sites, favoring the bifunctional mechanism, as proved in CO-stripping experiments. In other words, the reaction proceeds via WMSI domains generated by the metal–metal interaction [96] and the carbon support, Figure 1.34.

In the WMCI, non-electronic and electronic effects are present on carbon supports when the catalytic center is either modified or not. For example, Figure 1.35a and b show the Tafel plots and micro-polarization region for the HER/HOR on various nanoalloyed Pt-M materials. Here, it is emphasized that among all Pt-M, PtNi/C showed the highest catalytic

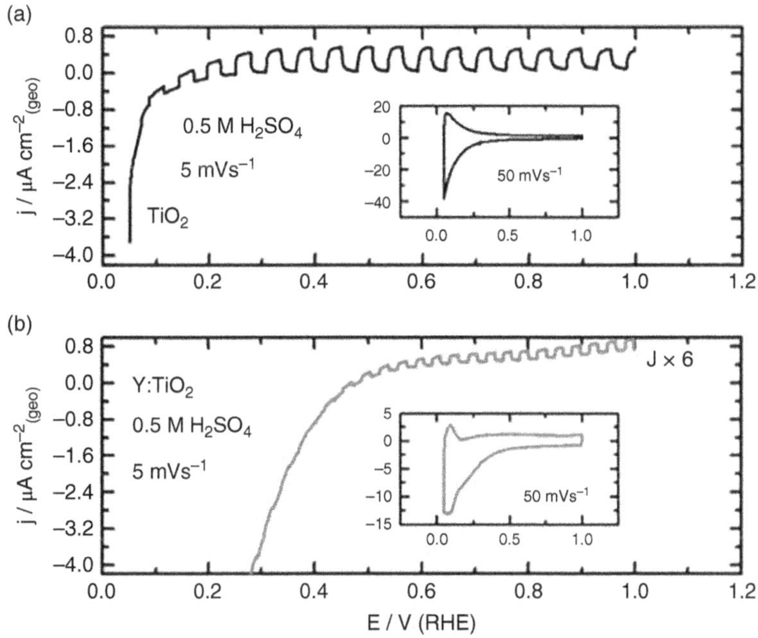

Figure 1.33 (a) The chopped light (on–off) photocurrent response at TiO_2, and (b) Y: TiO_2 oxides in 0.5 M H_2SO_4 at 5 mV/s. Inset: current-potential curves (in dark) in the interval from 0.05 to 1 V (RHE) at 25 °C. *Source:* From [43] / with permission of Elsevier.

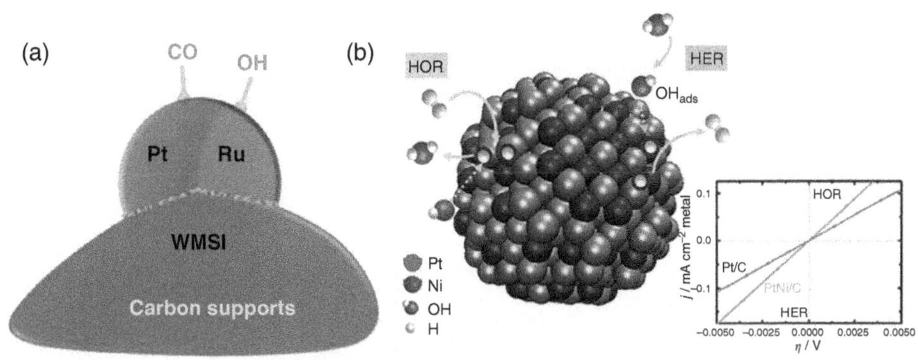

Figure 1.34 (a) The CO-stripping bifunctional mechanism effect on PtRu/C. (b) Pt-M alloys for HER/HOR. Source: Figure (A) from [96] / with permission of Elsevier and Figure (B) from [102] / with permission of Springer Nature.

effect, followed by PtCo/C. Within the experimental error, the results shown in Figure 1.35c assess that the surface-specific exchange current density, $j_{0,s}$, does not depend on the materials' particle size but on the oxophilic capacity of the metal M in PtM, with a similar manner for Pt/pCNT [108]. Although the electronic effect originated from the solid solution, the nanoalloys induce a variation in the lattice constant and Pt-Pt distance that must be considered; see Table 1.3.

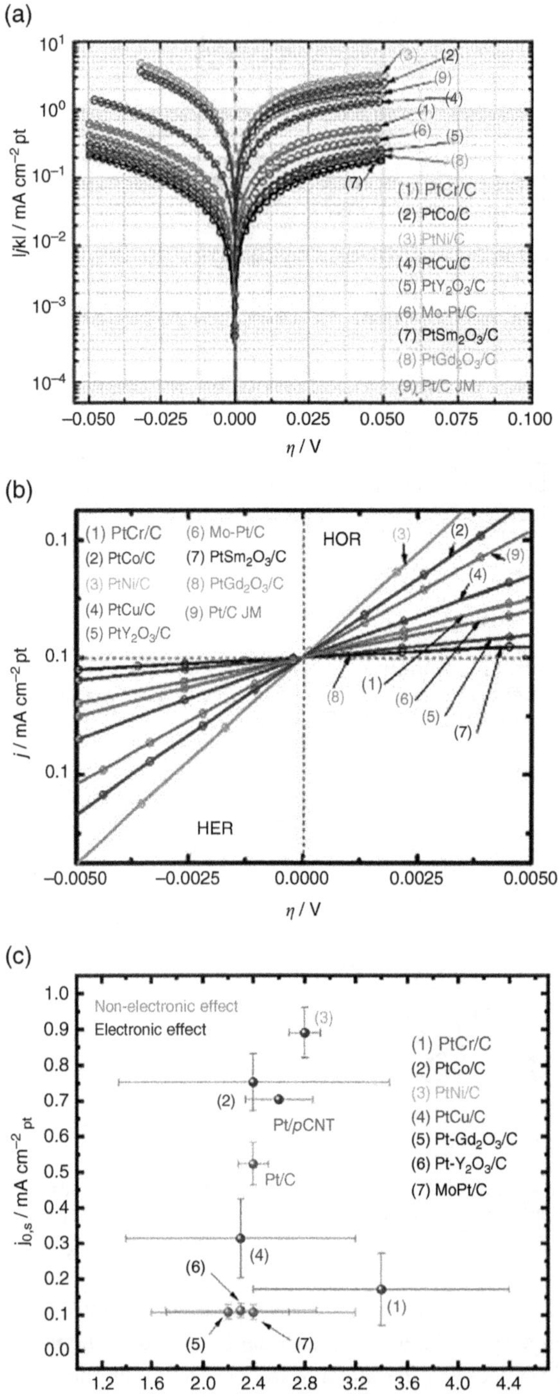

Figure 1.35 (a) Tafel plot on Pt/C, Pt-M/C nanoalloyed structures. (b) Micro-polarization region. Measurements carried out in H_2-saturated in 0.1 M KOH at 25 °C and (c) the HER/HOR surface-specific exchange current density, $j_{0,s}$, as a function of particle size. *Source:* Figure (A-B) from [102] / with permission of Springer Nature.

Table 1.3 Microstructural properties and the HER/HOR surface-specific exchange current density, $j_{0,s}$, and specific mass activity, $j_{0,m}$, for Pt-based materials. Data was taken from [102].

Sample	Particle size nm	Lattice constant Å	$j_{0,s}$/mA/cm$^2_{Pt}$	$j_{0,m}$/A/g$_{Pt}$
Pt/C	2.4 ± 0.12	3.9230	0.523 ± 0.06	106.55
[a] Pt/pCNT	2.6 ± 0.26	n.a.	0.704	n.a.
PtCr/C	3.4 ± 1.00	3.8950	0.171 ± 0.10	34.39
PtCo/C	2.4 ± 1.06	3.8800	0.752 ± 0.08	178.63
PtNi/C	2.8 ± 0.12	3.8250	0.890 ± 0.07	182.75
PtCu/C	2.3 ± 0.90	3.8960	0.314 ± 0.11	62.75
Pt-Y_2O_3/C	2.3 ± 0.59	3.9090	0.112 ± 0.02	24.69
Mo-Pt/C	2.4 ± 0.80	3.9260	0.107 ± 0.02	13.38
Pt-Gd_2O_3/C	2.2 ± 0.48	3.9070	0.108 ± 0.02	24.44

[a] Ref. [108].

On the other hand, the electronic effect induced by the SMSI provides further information. The hydrogen electrode reactions were probed with the SMSI effect encountered on Pt photo-deposited onto oxide sites of the oxide-carbon composites, e.g. Pt/TiO_2-C, Pt/Y: TiO_2-C. The electrocatalytic activity toward HOR/HER is similar on all tested samples, including Pt/C; see Tafel plot in Figure 1.36a. There is no clear trend in the magnitude of the exchange current with particle size. Besides, the turnover frequency parameter, Eq. (1.48), indirectly measures the active sites from a catalyst in a reaction; a similar analysis is done in ORR [23].

$$\mathrm{TOF}\left(e^{-}\mathrm{site}^{-1}s^{-1}\right) = \frac{MW_{Pt}}{F\,L_{Pt}}\,j_{0,s} \tag{1.48}$$

Where MW_{Pt} and L_{Pt} are the atomic mass and mass loading of Pt per geometric surface area (i.e. Pt mass from ICP/cm$^2_{geo}$) in the catalyst, respectively; F is Faraday constant, and $j_{0,s}$ is the surface-specific exchange current density. From Figure 1.36b, the decrease of the turnover frequency (TOF) parameter must be related to the decrease of the catalytic sites with the size of the particles. These experiments suggest that the oxophilic ingredients certainly present on the oxide surface are sites that do not easily release the OH species to favor the hydrogen reaction, which is determined by the Volmer step.

On TiO_2 supported Pt NPs, Pt-O or Pt-Ti bonds are present. Using DFT, Zheng et al. [165] concluded that these two kinds of bonds provide two separate RDSs, Volmer or Tafel. Here, the nature of the oxide is crucial, as a modification with doping changes the charge of Pt NPs, where the ORR reaction probes the stability of Pt [24, 42, 43, 166]. In addition to that, non-precious metals have been studied to replace Pt-based catalyst [11, 119, 167, 168].

1.5.2 Stability in OER-ORR

The ORR is a multi-electron charge-transfer reaction that includes several elementary steps. Figure 1.37 schematizes the different ORR pathway. The k_1 and k_2 rate constants are the direct four-electron [169] and the series two-electron reduction processes [170].

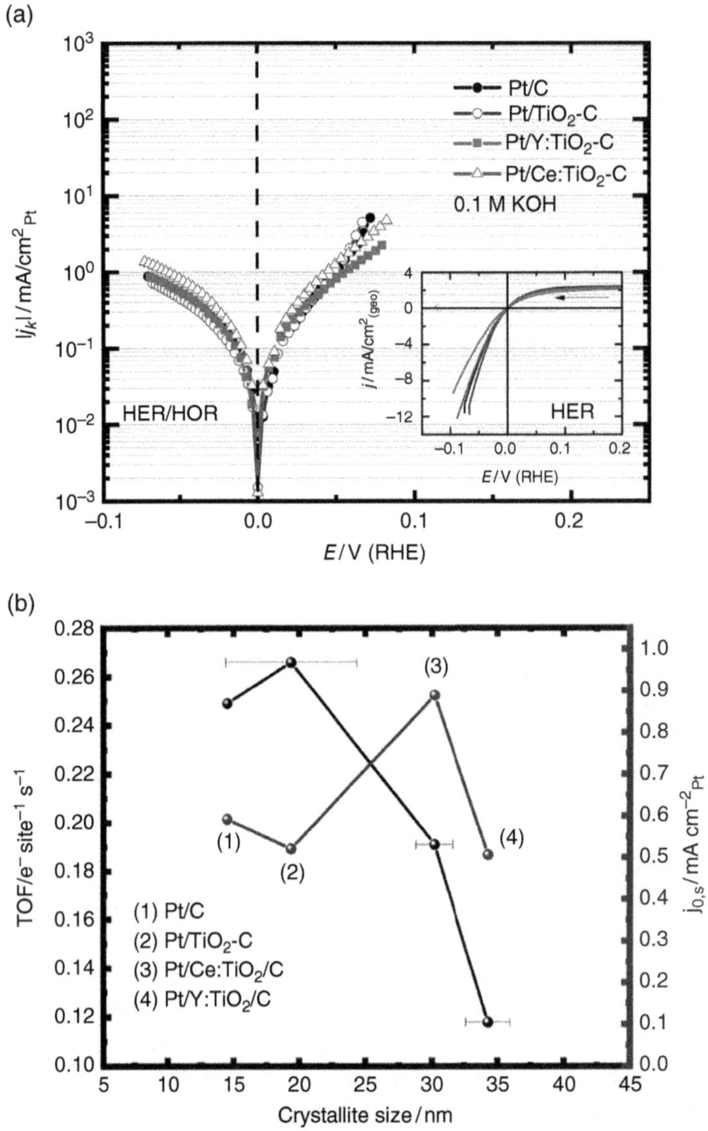

Figure 1.36 (a) Tafel plot for Pt/C, Pt/TiO_2-C, and Pt/M: TiO_2-C (M = Y or Ce); inset micro-polarization and (b) the HER/HOR TOF and surface-specific exchange current density as a function of crystallite size.

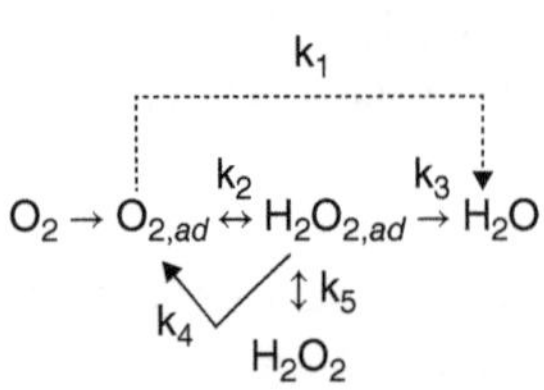

Figure 1.37 The ORR pathways.

The reverse reaction to ORR is the OER [171]. If the catalytic material can perform both ORR and OER processes, it will be named a bifunctional catalyst.

Different metal–support interactions can monitor the intermediates-adsorbed species or deteriorate the catalyst. For the storage of electrochemical energy, the ORR is critical to develop or design novel hybrid materials. Most electrocatalytic materials are based on carbon support. So, the testing protocol comprises two different intervals for testing, namely, (i) 0.6 to 1.0V (RHE) for evaluating the catalytic center and (ii) 0.6 to 1.2V (RHE) for inducing the corrosion of the support. Figure 1.38 compares ECSA and kinetic current density (j_k) obtained after the stability test in the potential interval of 0.6 to 1.0V (RHE) for Pt/CNT, Pt-M/C (M = Cu, Mo), Mo-PtCu/C materials in acid condition. With the Pt/CNT catalyst, both parameters present the same tendency: Pt/*f*-CNT (functionalized) > Pt/CNT > Pt/C. The determined ECSA loss in Pt/*f*-CNT was 19.77%. Pt NPs that sit on C-sp^3 domains are easily deactivated since there is no favorable hybridization bond with this carbon domain with the EC [166].

In Pt-M/C (M = Cu, Mo) and Mo-PtCu/C materials, the Mo-PtCu/C catalyst maintains 85% of its initial kinetic current density and 100% H_{upd} value. The DFT simulation analysis suggested that Mo-doped atoms stabilize the PtCu center avoiding its oxidation. The stabilization of the Pt-alloy center with doping is similar in Mo-Pt_3Ni/C materials [101].

Under the same analysis criterion, over the oxide-carbon composite supported Pt, obtained by the chemical route (CCR), the ECSA value slowly decreases to about 80% in the range of 0.6 to 1.0V (RHE), Figure 1.39a. Moreover, the j_k value follows the trend: Pt/Y:TiO_2-C > Pt/TiO_2-C > Pt/C, within the first potential interval protocol, Figure 1.39c. When corrosion of the oxide-carbon material takes place, the catalytic center is not stable anymore since agglomeration, detachment, or Ostwald ripening effects are promoted [2], Figures 1.39b–d. These processes cause a loss between 50% and 60% of the initial values.

The Pt NPs detachment is inhibited by the SMSI effect (i.e. photo-deposited on oxide domains), when the accelerated stability test, within 0.6 and 1.0V (RHE), cf. Figure 1.40a and c. Moreover, the micro-strain can explain the trend of j_k value: Pt/C > Pt/Y: TiO_2-C > Pt/TiO_2-C, which is similar. Therefore, the microstructural properties of the oxide can modify the stability of the catalytic center. Furthermore, in the range of 0.6 to 1.2V (RHE), the Pt/Y: TiO_2-C loses 40% of its initial value of ECSA and j_k, respectively. The interaction between Pt/Y:TiO_2 prevents the degradation of the catalytic center and favors the stability toward ORR, Figure 1.40b and d. These findings are different from those obtained on Pt NPs when the stability is under WMCI, cf. Figure 1.39b and d.

Bifunctional non-precious catalyst has been tested to replace Pt-based catalysts in ORR-OER reactions, which must be considered as the metal–support interaction [98, 172–175]. Here, considerations that can explain the effect between metal, oxide, and supports are given. These considerations can help to design novel hybrid materials applied to HER-HOR or ORR-OER reactions.

1.6 Conclusion

Several aspects of materials characterization were discussed aimed at model reactions that allow the understanding of the electrocatalytic effect. It is thus clear that physicochemical characterization and its analysis methods are of utmost importance because they provide

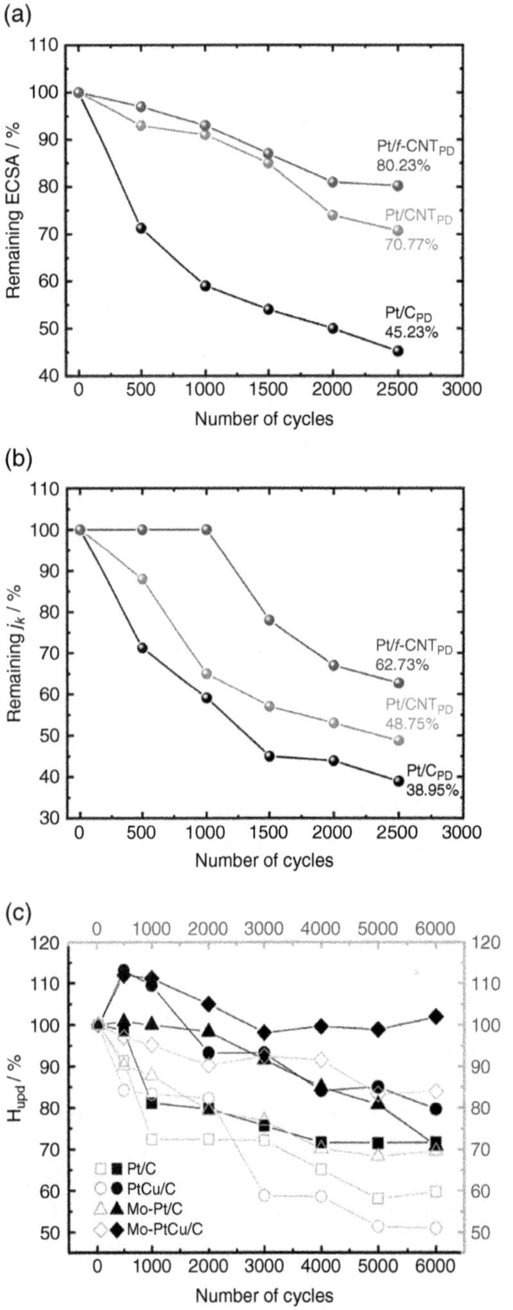

Figure 1.38 Accelerated stability tests (ASTs) performed between 0.6 and 1.0V/RHE for Pt-based material with an electronic effect in (a-b) SMCI and (c) WMCI phenomenon. ECSA was obtained in the H_{upd} region and j_k at 0.9V and 900rpm. *Source:* Figures (A-B) from [166] / with permission of American Chemical Society and Figure (C) From [151] / with permission of John Wiley & Sons.

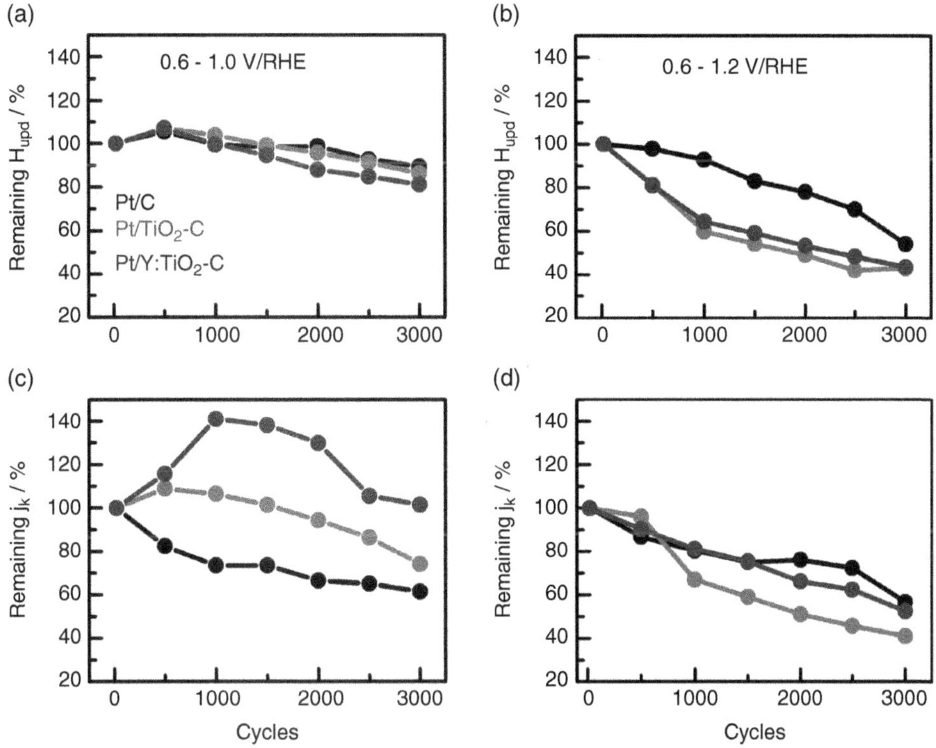

Figure 1.39 Accelerated stability tests (ASTs) for Pt NPs supported onto oxide-carbon composites synthesized by CCR. (a) in the 0.6–1.0 V (RHE), (b) in the 0.6–1.2 V (RHE) potential interval, (c) and (d) show the remaining kinetic current density after (a) and (b) ASTs, respectively.

answers regarding the material's structural properties, e.g. the crystallite size (or particle size), micro-strain, deformations of the phases present, as well as the electronic effect variation of the catalytic center, the type of interaction with the support in the nanoscale domain. The works done on the quantum chemistry calculation (DFT) have shed light to gain insights on the catalytic center–support interaction and the reaction mechanism in terms of adsorption energies.

On the other hand, the surface electrochemical characterization, which is linked to the physicochemical characterization, confirmed the interaction domain between the catalytic center and the support. Indeed, the oxidation–reduction processes manifested at the electrodes' surface serve as a fingerprint of the crystallographic planes of the material in contact with the electrolyte (electrode/electrolyte interface). This can further serve as evidence for the mechanical degradation of support and the catalytic center.

With regard to the photoelectrochemical characterization, the recombination of the photogenerated electron–hole pair (e^-/h^+) at the support is made evident before and after the interaction with the catalytic center. This allows confirmation that the catalytic center lies on anchoring sites, and electrical conductivity is assured and provides potential applications to the photocatalytic processes.

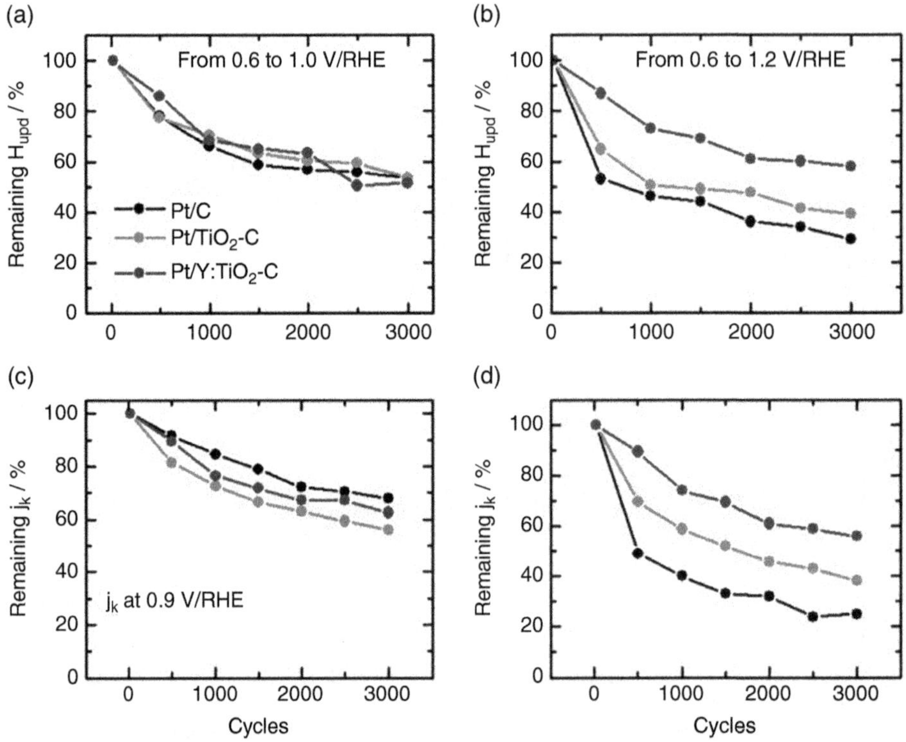

Figure 1.40 Accelerated stability tests (ASTs) for Pt NPs supported oxide-carbon composites presenting the SMSI phenomenon. PDM was used to anchor Pt NPs in oxide domains. (a) in the 0.6–1.0V (RHE), (b) in the 0.6–1.2V (RHE) potential interval, (c) and (d) show the remaining kinetic current density after (a) and (b) ASTs, respectively. *Source:* From [43] / with permission of Elsevier.

References

1 Trogadas, P., Fuller, T.F., and Strasser, P. (2014). Carbon as catalyst and support for electrochemical energy conversion. *Carbon* 75: 5–42.

2 Meier, J.C., Galeano, C., Katsounaros, I. et al. (2014). Design criteria for stable Pt/C fuel cell catalysts. *Beilstein Journal of Nanotechnology* 5: 44–67.

3 Wang, J., Kong, H., Zhang, J. et al. (2021). Carbon-based electrocatalysts for sustainable energy applications. *Progress in Materials Science* 116: 100717.

4 Manzo-Robledo, A., Boucher, A.C., Pastor, E., and Alonso-Vante, N. (2002). Electro-oxidation of carbon monoxide and methanol on carbon-supported Pt—Sn nanoparticles: a DEMS study. *Fuel Cells* 2: 109–116.

5 Schonvogel, D., Hülstede, J., Wagner, P. et al. (2017). Stability of Pt nanoparticles on alternative carbon supports for oxygen reduction reaction. *Journal of The Electrochemical Society* 164: F995–F1004.

6 Zhang, Y., Cao, X., Li, Z., and Zhao, D. (2016). Co-assembly of functional graphene and multiwall carbon nanotubes for supercapacitors by a vertical deposition technique. *Applied Physics A: Materials Science and Processing* 122: 1–7.

7 Cai, Y. and Feng, Y.P. (2016). Review on charge transfer and chemical activity of TiO_2: mechanism and applications. *Progress in Surface Science* 91: 183–202.

8 He, C., Sankarasubramanian, S., Matanovic, I. et al. (2019). Understanding the oxygen reduction reaction activity and oxidative stability of Pt supported on Nb-doped TiO_2. *ChemSusChem* 12: 3468–3480.

9 Ohyama, J., Yamamoto, A., Teramura, K. et al. (2011). Modification of metal nanoparticles with TiO_2 and metal–support interaction in photodeposition. *ACS Catalysis* 1: 187–192.

10 Tofighi, G., Degler, D., Junker, B. et al. (2019). Microfluidically synthesized Au, Pd and AuPd nanoparticles supported on SnO_2 for gas sensing applications. *Sensors and Actuators B: Chemical* 292: 48–56.

11 Wang, Y., Zhao, Y., Ding, X., and Qiao, L. (2021). Recent advances in the electrochemistry of layered post-transition metal chalcogenide nanomaterials for hydrogen evolution reaction. *Journal of Energy Chemistry* 60: 451–479.

12 Paul, B., Sharma, S., Purkayastha, D.D. et al. (2019). Facile synthesis of size-controlled Ag supported on WO_3 nanorods and their application as novel and active catalyst in oxidant-free dehydrogenation of benzyl alcohols. *Catalysis Communications* 132: 105804.

13 Elezovic, N.R., Babic, B.M., Radmilovic, V.R. et al. (2011). Nb–TiO_2 supported platinum nanocatalyst for oxygen reduction reaction in alkaline solutions. *Electrochimica Acta* 56: 9020–9026.

14 Ho, V.T.T., Pan, C.-J., Rick, J. et al. (2011). Nanostructured $Ti_{0.7}Mo_{0.3}O_2$ support enhances electron transfer to Pt: high-performance catalyst for oxygen reduction reaction. *Journal of the American Chemical Society* 133: 11716–11724.

15 Thanh Ho, V.T., Pillai, K.C., Chou, H.-L. et al. (2011). Robust non-carbon $Ti_{0.7}Ru_{0.3}O_2$ support with co-catalytic functionality for Pt: enhances catalytic activity and durability for fuel cells. *Energy and Environmental Science* 4: 4194–4200.

16 Kakinuma, K., Chino, Y., Senoo, Y. et al. (2013). Characterization of Pt catalysts on Nb-doped and Sb-doped SnO_2–δ support materials with aggregated structure by rotating disk electrode and fuel cell measurements. *Electrochimica Acta* 110: 316–324.

17 Ozouf, G., Cognard, G., Maillard, F. et al. (2018). Sb-Doped SnO_2 aerogels based catalysts for proton exchange membrane fuel cells: Pt deposition routes, electrocatalytic activity and durability. *Journal of The Electrochemical Society* 165: F3036–F3044.

18 Shi, Y., Wang, J., and Zhou, R. (2021). Pt-support interaction and nanoparticle size effect in Pt/CeO_2–TiO_2 catalysts for low temperature VOCs removal. *Chemosphere* 265: 129127.

19 Khan, M. and Cao, W. (2013). Preparation of Y-doped TiO_2 by hydrothermal method and investigation of its visible light photocatalytic activity by the degradation of methylene blue. *Journal of Molecular Catalysis A: Chemical* 376: 71–77.

20 Gharaei, S.K., Abbasnejad, M., and Maezono, R. (2018). Bandgap reduction of photocatalytic TiO_2 nanotube by Cu doping. *Scientific Reports* 8: 14192.

21 da Silva, A.G.M., Fernandes, C.G., Hood, Z.D. et al. (2020). PdPt-TiO_2 nanowires: correlating composition, electronic effects and O-vacancies with activities toward water splitting and oxygen reduction. *Applied Catalysis B: Environmental* 277: 119177.

22 Vogel, W., Timperman, L., and Alonso-Vante, N. (2010). Probing metal substrate interaction of Pt nanoparticles: structural XRD analysis and oxygen reduction reaction. *Applied Catalysis A: General* 377: 167–173.

23 Hussain, S., Kongi, N., Erikson, H. et al. (2019). Platinum nanoparticles photo-deposited on SnO_2-C composites: an active and durable electrocatalyst for the oxygen reduction reaction. *Electrochimica Acta* 316: 162–172.

24 Luo, Y., Estudillo-Wong, L.A., and Alonso-Vante, N. (2016). Carbon supported Pt-Y_2O_3 and Pt-Gd_2O_3 nanoparticles prepared via carbonyl chemical route toward oxygen reduction reaction: kinetics and stability. *International Journal of Hydrogen Energy* 41: 19601–19609.

25 Estudillo-Wong, L.A., Arce-Estrada, E.M., and Manzo-Robledo, A. (2014). Interaction of NO during cathodic polarization in alkaline conditions at the interface of Pt-nanostructures supported on C and TiO_2-C. *Electrochimica Acta* 134: 100–106.

26 Timperman, L., Feng, Y.J., Vogel, W., and Alonso-Vante, N. (2010). Substrate effect on oxygen reduction electrocatalysis. *Electrochimica Acta* 55: 7558–7563.

27 Lewera, A., Timperman, L., Roguska, A., and Alonso-Vante, N. (2011). Metal–support interactions between nanosized Pt and metal oxides (WO_3 and TiO_2) studied using X-ray photoelectron spectroscopy. *The Journal of Physical Chemistry C* 115: 20153–20159.

28 Tang, Z., Kong, N., Ouyang, J. et al. (2020). Phosphorus science-oriented design and synthesis of multifunctional nanomaterials for biomedical applications. *Matter* 2: 297–322.

29 Damodharan, J. (2021). Nanomaterials in medicine – An overview. *Materials Today: Proceedings* 37: 383–385.

30 Shrivas, K., Ghosale, A., Bajpai, P.K. et al. (2020). Advances in flexible electronics and electrochemical sensors using conducting nanomaterials: a review. *Microchemical Journal* 156: 104944.

31 Bhanvase, B.A., Pawade, V.B., Dhoble, S.J. et al. (2018). *Nanomaterials for Green Energy*, 1e. Elsevier.

32 Garcia-Segura, S., Ocon, J.D., and Chong, M.N. (2018). Electrochemical oxidation remediation of real wastewater effluents – A review. *Process Safety and Environmental Protection* 113: 48–67.

33 Di, G., Zhu, Z., Zhang, H. et al. (2017). Simultaneous removal of several pharmaceuticals and arsenic on Zn-Fe mixed metal oxides: Combination of photocatalysis and adsorption. *Chemical Engineering Journal* 328: 141–151.

34 Niederberger, M. and Pinna, N. (2009). *Metal Oxide Nanoparticles in Organic Solvents.* Springer London.

35 Chen, D., Huang, F., Cheng, Y.-B., and Caruso, R.A. (2009). Mesoporous anatase TiO_2 beads with high surface areas and controllable pore sizes: a superior candidate for high-performance dye-sensitized solar cells. *Advanced Materials* 21: 2206–2210.

36 Phukhaotong, H., Kiennork, S., Wongwanwattana, P. et al. (2018). Synthesis and characterization of Co_3O_4/SnO_2 composite powders by microwave assisted sol–gel method. *Materials Today: Proceedings* 5: 14099–14104.

37 Xian, F. and Li, X. (2013). Effect of Nd doping level on optical and structural properties of ZnO:Nd thin films synthesized by the sol–gel route. *Optics and Laser Technology* 45: 508–512.

38 Liang, Y., Wang, H., Sanchez Casalongue, H. et al. (2010). TiO_2 nanocrystals grown on graphene as advanced photocatalytic hybrid materials. *Nano Research* 3: 701–705.

39 Huang, F., Guo, Y., Wang, S. et al. (2017). Solgel-hydrothermal synthesis of Tb/Tourmaline/TiO_2 nanotubes and enhanced photocatalytic activity. *Solid State Sciences* 64: 62–68.

40 Chebil, W., Fouzri, A., Fargi, A. et al. (2015). Characterization of ZnO thin films grown on different p-Si substrate elaborated by solgel spin-coating method. *Materials Research Bulletin* 70: 719–727.

41 Ruiz Camacho, B., Morais, C., Valenzuela, M.A., and Alonso-Vante, N. (2013). Enhancing oxygen reduction reaction activity and stability of platinum via oxide-carbon composites. *Catalysis Today* 202: 36–43.

42 Estudillo-Wong, L.A., Ramos-Sanchez, G., Calvillo, L. et al. (2017). Support interaction effect of platinum nanoparticles on Non-, Y-, Ce-doped anatase and its implication on the ORR in acid and alkaline media. *ChemElectroChem* 4: 3264–3275.

43 Estudillo-Wong, L.A., Luo, Y., Díaz-Real, J.A., and Alonso-Vante, N. (2016). Enhanced oxygen reduction reaction stability on platinum nanoparticles photo-deposited onto oxide-carbon composites. *Applied Catalysis B: Environmental* 187: 291–300.

44 Luo, Y., Mora-Hernández, J.M.M., Estudillo-Wong, L.A. et al. (2015). Nanostructured palladium tailored via carbonyl chemical route toward oxygen reduction reaction. *Electrochimica Acta* 173: 771–778.

45 Feng, Y., He, T., and Alonso-Vante, N. (2009). Oxygen reduction reaction on carbon-supported $CoSe_2$ nanoparticles in an acidic medium. *Electrochimica Acta* 54: 5252–5256.

46 Dickinson, A.J., Carrette, L.P.L., Collins, J.A. et al. (2002). Preparation of a Pt-Ru/C catalyst from carbonyl complexes for fuel cell applications. *Electrochimica Acta* 47: 3733–3739.

47 Delacôte, C., Lewera, A., Pisarek, M. et al. (2010). The effect of diluting ruthenium by iron in Ru_xSe_y catalyst for oxygen reduction. *Electrochimica Acta* 55: 7575–7580.

48 Yang, H., Vogel, W., Lamy, C., and Alonso-Vante, N. (2004). Structure and electrocatalytic activity of carbon-supported Pt-Ni alloy nanoparticles toward the oxygen reduction reaction. *The Journal of Physical Chemistry B* 108: 11024–11034.

49 Solorza-Feria, O., Ellmer, K., Giersig, M., and Alonso-Vante, N. (1994). Novel low-temperature synthesis of semiconducting transition metal chalcogenide electrocatalyst for multielectron charge transfer: molecular oxygen reduction. *Electrochimica Acta* 39: 1647–1653.

50 Longoni, G. and Chini, P. (1976). Synthesis and chemical characterization of platinum carbonyl dianions $[Pt_3(CO)_6]_n^{2-}$ (n = .apprx.10,6,5,4,3,2,1). A new series of inorganic oligomers. *Journal of the American Chemical Society* 98: 7225–7231.

51 Alonso-Vante, N. (2006). Carbonyl tailored electrocatalysts. *Fuel Cells* 6: 182–189.

52 Le Rhun, V., Garnier, E., Pronier, S., and Alonso-Vante, N. (2000). Electrocatalysis on nanoscale ruthenium-based material manufactured by carbonyl decomposition. *Electrochemistry Communications* 2: 475–479.

53 Yang, H., Alonso-Vante, N., Leger, J.-M., and Lamy, C. (2004). Tailoring, structure, and activity of carbon-supported nanosized Pt-Cr alloy electrocatalysts for oxygen reduction in pure and methanol-containing electrolytes. *The Journal of Physical Chemistry B* 108: 1938–1947.

54 Unni, S.M., Campos-Roldan, C.A., Mora-Hernandez, J.M. et al. (2015). The effect of carbon-based substrate onto non-precious and precious electrcatalytic center. *ECS Transactions* 1: 35–42.

55 Clark, W.C. and Vondjidis, A.G. (1965). An infrared study of the photocatalytic reaction between titanium dioxide and silver nitrate. *Journal of Catalysis* 4: 691–696.

56 Kraeutler, B. and Bard, A.J. (1978). Heterogeneous photocatalytic preparation of supported catalysts. Photodeposition of platinum on titanium dioxide powder and other substrates. *Journal of the American Chemical Society* 100: 4317–4318.

57 Mori, K., Araki, T., Shironita, S. et al. (2009). Supported Pd and PdAu nanoparticles on Ti-MCM-41 prepared by a photo-assisted deposition method as efficient catalysts for direct synthesis of H_2O_2 from H_2 and O_2. *Catalysis Letters* 131: 337–343.

58 Cunningham, D.A.H., Vogel, W., Sanchez, R.M.T. et al. (1999). Structural analysis of Au/ TiO_2 catalysts by debye function analysis. *Journal of Catalysis* 183: 24–31.

59 Taing, J., Cheng, M.H., and Hemminger, J.C. (2011). Photodeposition of Ag or Pt onto TiO_2 nanoparticles decorated on step edges of HOPG. *ACS Nano* 5: 6325–6333.

60 Oh, S.-D., So, B.-K., Choi, S.-H. et al. (2005). Dispersing of Ag, Pd, and Pt—Ru alloy nanoparticles on single-walled carbon nanotubes by γ-irradiation. *Materials Letters* 59: 1121–1124.

61 McDonald, K.J. and Choi, K.-S. (2011). Photodeposition of co-based oxygen evolution catalysts on α-Fe_2O_3 photoanodes. *Chemistry of Materials* 23: 1686–1693.

62 Hamdy, M.S., Saputera, W.H., Groenen, E.J., and Mul, G. (2014). A novel TiO_2 composite for photocatalytic wastewater treatment. *Journal of Catalysis* 310: 75–83.

63 Luo, J., Zhang, S., Sun, M. et al. (2019). A Critical review on energy conversion and environmental remediation of photocatalysts with remodeling crystal lattice, surface, and interface. *ACS Nano* 13: 9811–9840.

64 Ma, J., Habrioux, A., Pisarek, M. et al. (2013). Induced electronic modification of Pt nanoparticles deposited onto graphitic domains of carbon materials by UV irradiation. *Electrochemistry Communications* 29: 12–16.

65 Rajeshwar, K., Janaky, C., Lin, W.-Y. et al. (2013). Photocatalytically prepared metal nanocluster–oxide semiconductor–carbon nanocomposite electrodes for driving multielectron transfer. *The Journal of Physical Chemistry Letters* 4: 3468–3478.

66 Einaga, H. and Harada, M. (2005). Photochemical preparation of Poly(N-vinyl-2-pyrrolidone)-stabilized platinum colloids and their deposition on titanium dioxide. *Langmuir* 21: 2578–2584.

67 Harada, M. and Einaga, H. (2006). Formation mechanism of Pt particles by photoreduction of Pt ions in polymer solutions. *Langmuir* 22: 2371–2377.

68 Ma, J., Valenzuela, E., Gago, A.S. et al. (2014). Photohole trapping induced platinum cluster nucleation on the surface of TiO_2 nanoparticles. *The Journal of Physical Chemistry C* 118: 1111–1117.

69 Mourdikoudis, S., Pallares, R.M., and Thanh, N.T.K. (2018). Characterization techniques for nanoparticles: comparison and complementarity upon studying nanoparticle properties. *Nanoscale* 10: 12871–12934.

70 Brandon, D. and Kaplan, W.D. (2008). *Microstructural Characterization of Materials*, 1e. Wiley.

71 Singh, J. (2019). *Optical Properties of Materials and Their Applications*. Wiley.

72 Gomez-Villalba, L.S., Sierra-Fernandez, A., Rabanal, M.E., and Fort, R. (2016). TEM-HRTEM study on the dehydration process of nanostructured Mg—Ca hydroxide into Mg—Ca oxide. *Ceramics International* 42: 9455–9466.

73 Hsiung, L., Fluss, M., Tumey, S. et al. (2011). HRTEM study of oxide nanoparticles in K3-ODS ferritic steel developed for radiation tolerance. *Journal of Nuclear Materials* 409: 72–79.

74 Prathibha, V., Karthika, S., Cyriac, J. et al. (2011). Synthesis of pure anatase TiO_2 nanocrystals in SiO_2 host and the determination of crystal planes by ImageJ. *Materials Letters* 65: 664–666.

75 Kirkland, E.J. (2020). *Advanced Computing in Electron Microscopy*. Cham: Springer.

76 Liu, H., Foley, M., Lin, Q., and Liu, J. (2016). EDP2XRD: a computer program for converting electron diffraction patterns into X-ray diffraction patterns. *Journal of Applied Crystallography* 49: 636–641.

77 Botero, M.L., Chen, D., González-Calera, S. et al. (2016). HRTEM evaluation of soot particles produced by the non-premixed combustion of liquid fuels. *Carbon* 96: 459–473.

78 Holder, C.F. and Schaak, R.E. (2019). Tutorial on powder X-ray diffraction for characterizing nanoscale materials. *ACS Nano* 13: 7359–7365.

79 Grazulis, S., Chateigner, D., Downs, R.T. et al. (2009). Crystallography open database – An open-access collection of crystal structures. *Journal of Applied Crystallography* 42: 726–729.

80 Dinnebier, R.E. and Billinge, S.J.L. (2008). *Powder Diffraction*, 1e. RSC Publishing.

81 Kumar, C.S.S.R. (2016). *X-ray and Neutron Techniques for Nanomaterials Characterization*, 1e. Berlin Heidelberg: Springer.

82 Langford, J.I. and Wilson, A.J.C. (1978). Scherrer after sixty years: a survey and some new results in the determination of crystallite size. *Journal of Applied Crystallography* 11: 102–113.

83 Lutterotti, L. and Scardi, P. (1990). Simultaneous structure and size-strain refinement by the Rietveld method. *Journal of Applied Crystallography* 23: 246–252.

84 Wojdyr, M. (2010). Fityk: a general-purpose peak fitting program. *Journal of Applied Crystallography* 43: 1126–1128.

85 Lutterotti, L. (2010). Total pattern fitting for the combined size–strain–stress–texture determination in thin film diffraction. *Nuclear Instruments and Methods in Physics Research Section B: Beam Interactions with Materials and Atoms* 268: 334–340.

86 Doebelin, N. and Kleeberg, R. (2015). Profex: a graphical user interface for the Rietveld refinement program BGMN. *Journal of Applied Crystallography* 48: 1573–1580.

87 Rodríguez-Carvajal, J. (1993). Recent advances in magnetic structure determination by neutron powder diffraction. *Physica B: Condensed Matter* 192: 55–69.

88 Debye, P. (1915). Zerstreuung von Röntgenstrahlen. *Annalen der Physik* 351: 809–823.

89 Bertolotti, F., Moscheni, D., Migliori, A. et al. (2016). A total scattering Debye function analysis study of faulted Pt nanocrystals embedded in a porous matrix. *Acta Crystallographica Section A Foundations and Advances* 72: 632–644.

90 Gnutzmann, V. and Vogel, W. (1989). Surface oxidation and reduction of small platinum particles observed by in-situ X-ray diffraction. *Zeitschrift für Physik D Atoms, Molecules and Clusters* 12: 597–600.

91 Hall, B.D., Flüeli, M., Monot, R., and Borel, J.P. (1989). Electron diffraction on unsupported ultrafine silver particles. *Zeitschrift für Physik D Atoms, Molecules and Clusters* 12: 97–101.

92 Vogel, W. (1998). X-Ray diffraction from clusters. *Crystal Research and Technology* 33: 1141–1154.

93 Cernuto, G., Masciocchi, N., Cervellino, A. et al. (2011). Size and shape dependence of the photocatalytic activity of TiO_2 nanocrystals: a total scattering Debye function study. *Journal of the American Chemical Society* 133: 3114–3119.

94 Cervellino, A., Frison, R., Bertolotti, F., and Guagliardi, A. (2015). DEBUSSY 2.0: the new release of a Debye user system for nanocrystalline and/or disordered materials. *Journal of Applied Crystallography* 48: 2026–2032.

95 Yang, J., Li, W., Wang, D., and Li, Y. (2020). Electronic metal-support interaction of single-atom catalysts and applications in electrocatalysis. *Advanced Materials* 32: e2003300.

96 Pan, C.-J., Tsai, M.-C., Su, W.-N. et al. (2017). Tuning/exploiting Strong Metal-Support Interaction (SMSI) in heterogeneous catalysis. *Journal of the Taiwan Institute of Chemical Engineers* 74: 154–186.

97 Zhang, B. and Su, D.S. (2015). Probing the metal–support interaction in carbon-supported catalysts by using electron microscopy. *ChemCatChem* 7: 3639–3645.

98 Luo, Y. and Alonso-Vante, N. (2015). The effect of support on advanced Pt-based cathodes toward the oxygen reduction reaction. *State of the Art, Electrochimica Acta* 179: 108–118.

99 Shao, M., Chang, Q., Dodelet, J.-P., and Chenitz, R. (2016). Recent advances in electrocatalysts for oxygen reduction reaction. *Chemical Reviews* 116: 3594–3657.

100 Wang, Y.-J., Zhao, N., Fang, B. et al. (2015). Carbon-supported Pt-based alloy electrocatalysts for the oxygen reduction reaction in polymer electrolyte membrane fuel cells: particle size, shape, and composition manipulation and their impact to activity. *Chemical Reviews* 115: 3433–3467.

101 Huang, X., Zhao, Z., Cao, L. et al. (2015). High-performance transition metal–doped Pt_3Ni octahedra for oxygen reduction reaction. *Science* 348: 1230–1234.

102 Campos-Roldán, C.A. and Alonso-Vante, N. (2021). Understanding the oxophilic effect on the hydrogen electrode reaction through PtM nanostructures. *Journal of Solid State Electrochemistry* 25: 187–194.

103 Litkohi, H.R., Bahari, A., and Gatabi, M.P. (2020). Improved oxygen reduction reaction in PEMFCs by functionalized CNTs supported Pt–M (M = Fe, Ni, Fe—Ni) bi- and tri-metallic nanoparticles as efficient electrocatalyst. *International Journal of Hydrogen Energy* 45: 23543–23556.

104 Yu, P., Coll, M., Amade, R. et al. (2020). Homogeneous Fe_2O_3 coatings on carbon nanotube structures for supercapacitors. *Dalton Transactions* 49: 4136–4145.

105 Mahmoodinia, M., Åstrand, P.-O., and Chen, D. (2017). Tuning the electronic properties of single-atom Pt catalysts by functionalization of the carbon support material. *The Journal of Physical Chemistry C* 121: 20802–20812.

106 Wang, T., Yang, J., Chen, J. et al. (2020). Nitrogen-doped carbon nanotube-encapsulated nickel nanoparticles assembled on graphene for efficient CO_2 electroreduction. *Chinese Chemical Letters* 31: 1438–1442.

107 Zhang, B., Shao, L., Zhang, W. et al. (2014). Interaction between palladium nanoparticles and surface-modified carbon nanotubes: role of surface functionalities. *ChemCatChem* 6: 2607–2612.

108 Campos-Roldán, C.A., González-Huerta, R.G., and Alonso-Vante, N. (2018). The oxophilic and electronic effects on anchored platinum nanoparticles on sp^2 carbon sites: the hydrogen evolution and oxidation reactions in alkaline medium. *Electrochimica Acta* 283: 1829–1834.

109 Tauster, S.J., Fung, S.C., and Garten, R.L. (1978). Strong metal-support interactions. Group 8 noble metals supported on titanium dioxide. *Journal of the American Chemical Society* 100: 170–175.

110 Zhang, J., Su, N., Hu, X. et al. (2017). Facile synthesis of Pt nanoparticles supported on anatase TiO_2 nanotubes with good photo-electrocatalysis performance for methanol. *RSC Advances* 7: 56194–56203.

111 Jiang, Z., Jing, M., Feng, X. et al. (2020). Stabilizing platinum atoms on CeO_2 oxygen vacancies by metal-support interaction induced interface distortion: mechanism and application. *Applied Catalysis B: Environmental* 278: 119304.

112 Lu, C., Chen, Z., and Singh, V. (2010). Highly hydrogen-sensitive SnO_2 nanoscale-particle films with platinum electrodes. *Sensors and Actuators B: Chemical* 146: 145–153.

113 Antolini, E. (2010). Composite materials: an emerging class of fuel cell catalyst supports. *Applied Catalysis B: Environmental* 100: 413–426.

114 Ma, J., Habrioux, A., and Alonso-Vante, N. (2013). Enhanced HER and ORR behavior on photodeposited Pt nanoparticles onto oxide–carbon composite. *Journal of Solid State Electrochemistry* 17: 1913–1921.

115 Čolić, V. and Bandarenka, A.S. (2016). Pt alloy electrocatalysts for the oxygen reduction reaction: from model surfaces to nanostructured systems. *ACS Catalysis* 6: 5378–5385.

116 Reier, T., Oezaslan, M., and Strasser, P. (2012). Electrocatalytic oxygen evolution reaction (OER) on Ru, Ir, and Pt catalysts: a comparative study of nanoparticles and bulk materials. *ACS Catalysis* 2: 1765–1772.

117 Greeley, J., Stephens, I.E., Bondarenko, A.S. et al. (2009). Alloys of platinum and early transition metals as oxygen reduction electrocatalysts. *Nature Chemistry* 1: 552–556.

118 Greeley, J., Jaramillo, T.F., Bonde, J. et al. (2006). Computational high-throughput screening of electrocatalytic materials for hydrogen evolution. *Nature Materials* 5: 909–913.

119 Campos-Roldán, C.A. and Alonso-Vante, N. The hydrogen oxidation reaction in alkaline medium: an overview. *Electrochemical Energy Reviews* 2: 312–331.

120 Chun, H.-J., Apaja, V., Clayborne, A. et al. (2017). Atomistic insights into nitrogen-cycle electrochemistry: a combined DFT and kinetic Monte Carlo analysis of NO electrochemical reduction on Pt(100). *ACS Catalysis* 7: 3869–3882.

121 Kortlever, R., Shen, J., Schouten, K.J. et al. (2015). Catalysts and reaction pathways for the electrochemical reduction of carbon dioxide. *The Journal of Physical Chemistry Letters* 6: 4073–4082.

122 Medford, A.J., Vojvodic, A., Hummelshøj, J.S. et al. (2015). From the Sabatier principle to a predictive theory of transition-metal heterogeneous catalysis. *Journal of Catalysis* 328: 36–42.

123 Norskov, J.K., Bligaard, T., Rossmeisl, J., and Christensen, C.H. (2009). Toward the computational design of solid catalysts. *Nature Chemistry* 1: 37–46.

124 Giustino, F. (2014). *Materials Modeling Using Density Functional Theory: Properties and Predictions*, 1e. USA: Oxford University Press.

125 Calle-Vallejo, F. and Koper, M.T.M. (2013). Theoretical considerations on the electroreduction of CO to C_2 species on Cu(100) electrodes. *Angewandte Chemie International Edition* 52: 7282–7285.

126 Mino, L., Spoto, G., and Ferrari, A.M. (2014). CO_2 capture by TiO_2 anatase surfaces: a combined DFT and FTIR study. *The Journal of Physical Chemistry C* 118: 25016–25026.

127 Fang, H., Kamakoti, P., Zang, J. et al. (2012). Prediction of CO_2 adsorption properties in zeolites using force fields derived from periodic dispersion-corrected DFT calculations. *The Journal of Physical Chemistry C* 116: 10692–10701.

128 Skafte, T.L., Guan, Z., Machala, M.L. et al. (2019). Selective high-temperature CO_2 electrolysis enabled by oxidized carbon intermediates, Nature. *Energy* 4: 846–855.

129 Kattel, S., Ramírez, P.J., Chen, J.G. et al. (2017). Active sites for CO_2 hydrogenation to methanol on Cu/ZnO catalysts. *Science* 355: 1296.

130 Calle-Vallejo, F. and Koper, M.T.M. (2012). First-principles computational electrochemistry: achievements and challenges. *Electrochimica Acta* 84: 3–11.

131 Xie, J. and Varadan, V.K. (2005). Synthesis and characterization of high surface area tin oxide/functionalized carbon nanotubes composite as anode materials. *Materials Chemistry and Physics* 91: 274–280.

132 Maiyalagan, T. and Khan, F.N. (2009). Electrochemical oxidation of methanol on Pt/ V_2O_5–C composite catalysts. *Catalysis Communications* 10: 433–436.

133 Chiku, M., Takeda, H., Matsumura, S. et al. (2015). Amorphous vanadium oxide/carbon composite positive electrode for rechargeable aluminum battery. *ACS Applied Materials and Interfaces* 7: 24385–24389.

134 Lota, K., Sierczynska, A., and Lota, G. (2011). Supercapacitors based on nickel oxide/ carbon materials composites. *International Journal of Electrochemistry* 2011: 321473.

135 Chang, J.-K., Lin, C.-T., and Tsai, W.-T. (2004). Manganese oxide/carbon composite electrodes for electrochemical capacitors. *Electrochemistry Communications* 6: 666–671.

136 Zheng, J.P. (1999). Ruthenium oxide-carbon composite electrodes for electrochemical capacitors. *Electrochemical and Solid-State Letters* 2: 359.

137 Kim, J., Kang, Y., Song, S.-W., and Suk, J. (2019). Freestanding sulfur-graphene oxide/ carbon composite paper as a stable cathode for high performance lithium-sulfur batteries. *Electrochimica Acta* 299: 27–33.

138 Liu, Y., Higgins, D.C., Wu, J. et al. (2013). Cubic spinel cobalt oxide/multi-walled carbon nanotube composites as an efficient bifunctionalelectrocatalyst for oxygen reaction. *Electrochemistry Communications* 34: 125–129.

139 Rheinländer, P.J., Herranz, J., Durst, J., and Gasteiger, H.A. (2014). Kinetics of the hydrogen oxidation/evolution reaction on polycrystalline platinum in alkaline electrolyte reaction order with respect to hydrogen pressure. *Journal of The Electrochemical Society* 161: F1448–F1457.

140 Sheng, W., Gasteiger, H.A., and Shao-Horn, Y. (2010). Hydrogen oxidation and evolution reaction kinetics on platinum: acid vs alkaline electrolytes. *Journal of The Electrochemical Society* 157: B1529.

141 García, G. and Koper, M.T.M. (2009). Mechanism of electro-oxidation of carbon monoxide on stepped platinum electrodes in alkaline media: a chronoamperometric and kinetic modeling study. *Physical Chemistry Chemical Physics* 11: 11437–11446.

142 Lai, S.C.S., Lebedeva, N.P., Housmans, T.H.M., and Koper, M.T.M. (2007). Mechanisms of carbon monoxide and methanol oxidation at single-crystal electrodes. *Topics in Catalysis* 46: 320–333.

143 van der Niet, M.J.T.C., Garcia-Araez, N., Hernández, J. et al. (2013). Water dissociation on well-defined platinum surfaces: the electrochemical perspective. *Catalysis Today* 202: 105–113.

144 Vidal-Iglesias, F.J., Arán-Ais, R.M., Solla-Gullón, J. et al. (2012). Electrochemical characterization of shape-controlled Pt nanoparticles in different supporting electrolytes. *ACS Catalysis* 2: 901–910.

145 Suchorski, Y., Spiel, C., Vogel, D. et al. (2010). Local reaction kinetics by imaging: CO oxidation on polycrystalline platinum. *ChemPhysChem* 11: 3231–3235.

146 Alonso-Vante, N. (2018). *Chalcogenide Materials for Energy Conversion*, 1e. Springer International Publishing AG.

147 Kumar Samantara, A. and Ratha, S. (2019). *Metal Oxides/Chalcogenides and Composites*, 1e. Switzerland: Springer Nature.

148 Ragoisha, G.A., Aniskevich, Y.M., Bakavets, A.S., and Streltsov, E.A. (2020). Electrochemistry of metal adlayers on metal chalcogenides. *Journal of Solid State Electrochemistry* 24: 2585–2594.

149 McCrory, C.C.L., Jung, S., Peters, J.C., and Jaramillo, T.F. (2013). Benchmarking heterogeneous electrocatalysts for the oxygen evolution reaction. *Journal of the American Chemical Society* 135: 16977–16987.

150 Orazem, M.E. and Tribollet, B. (2017). *Constant-Phase Elements, Electrochemical Impedance Spectroscopy*, 395–419. Wiley.

151 Luo, Y., Kirchhoff, B., Fantauzzi, D. et al. (2018). Molybdenum-doping augments platinum-copper oxygen reduction electrocatalyst. *ChemSusChem* 11: 193–201.

152 Peuckert, M., Yoneda, T., Betta, R.A.D., and Boudart, M. (1986). Oxygen reduction on small supported platinum particles. *Journal of The Electrochemical Society* 133: 944–947.

153 Janáky, C. and Rajeshwar, K. (2015). The role of (photo)electrochemistry in the rational design of hybrid conducting polymer/semiconductor assemblies: From fundamental concepts to practical applications. *Progress in Polymer Science* 43: 96–135.

154 Hussain, M.Z., Pawar, G.S., Huang, Z. et al. (2019). Porous ZnO/Carbon nanocomposites derived from metal organic frameworks for highly efficient photocatalytic applications - correlational study. *Carbon* 146: 348–363.

155 Chen, Z., Dinh, H.N., and Miller, E. (2013). *Photoelectrochemical Water Splitting*. New York, NY: Springer.

156 Bai, Y., Mora-Seró, I., De Angelis, F. et al. (2014). Titanium dioxide nanomaterials for photovoltaic applications. *Chemical Reviews* 114: 10095–10130.

157 Estudillo-Wong, L.A. and Alonso-Vante, N. (2019). Boosting the mineralization of reactive black 5 dye with Y- or H_2-doped anatase phase: Equivalent induced photocatalytic effect. *Journal of Electroanalytical Chemistry* 852: 113521.

158 Mokrane-Soualah, S., Gago, A.S., Habrioux, A., and Alonso-Vante, N. (2014). Mixed-oxide $Ti_{1-x}W_xO_2$ as support for (photo)-electrochemical processes. *Applied Catalysis B: Environmental* 147: 756–763.

159 Pennington, A.M., Okonmah, A.I., Munoz, D.T. et al. (2018). Changes in polymorph composition in P25-TiO_2 during pretreatment analyzed by differential diffuse reflectance spectral analysis. *The Journal of Physical Chemistry C* 122: 5093–5104.

160 Díaz-Real, J.A., Ma, J., and Alonso-Vante, N. (2016). Highly photoactive Brookite and Anatase with enhanced photocatalytic activity for the degradation of indigo carmine application. *Applied Catalysis B: Environmental* 198: 471–479.

161 Fleischmann, S., Tolosa, A., and Presser, V. (2018). Design of carbon/metal oxide hybrids for electrochemical energy storage. *Chemistry – A European Journal* 24: 12143–12153.

162 Gawande, M.B., Pandey, R.K., and Jayaram, R.V. (2012). Role of mixed metal oxides in catalysis science – versatile applications in organic synthesis. *Catalysis Science and Technology* 2: 1113–1125.

163 Ryu, J., Yoon, S., Park, J. et al. (2020). Fabrication of nickel oxide composites with carbon nanotubes for enhanced charge transport in planar perovskite solar cells. *Applied Surface Science* 516: 146116.

164 Cho, Y., Kim, S., Park, B. et al. (2018). Multiple heterojunction in single titanium dioxide nanoparticles for novel metal-free photocatalysis. *Nano Letters* 18: 4257–4262.

165 Zheng, X., Li, L., Deng, M. et al. (2020). Understanding the effect of interfacial interaction on metal/metal oxide electrocatalysts for hydrogen evolution and hydrogen oxidation reactions on the basis of first-principles calculations. *Catalysis Science and Technology* 10: 4743–4751.

166 Campos-Roldan, C.A., Ramos-Sanchez, G., Gonzalez-Huerta, R.G. et al. (2016). Influence of sp^3-sp^2 carbon nanodomains on metal/support interaction, catalyst durability, and catalytic activity for the oxygen reduction reaction. *ACS Applied Materials and Interfaces* 8: 23260–23269.

167 Wang, M., Yang, H., Shi, J. et al. (2021). Alloying nickel with molybdenum significantly accelerates alkaline hydrogen electrocatalysis. *Angewandte Chemie* 133: 5835–5841.

168 Ni, W., Krammer, A., Hsu, C.-S. et al. (2019). Ni_3N as an active hydrogen oxidation reaction catalyst in alkaline medium. *Angewandte Chemie International Edition* 58: 7445–7449.

169 Marković, N.M., Schmidt, T.J., Stamenković, V., and Ross, P.N. (2001). Oxygen reduction reaction on Pt and Pt bimetallic surfaces: a selective review. *Fuel Cells* 1: 105–116.

170 Timperman, L., Luo, Y., and Alonso-Vante, N. (2016). On the availability of active sites for the hydrogen peroxide and oxygen reduction reactions on highly dispersed platinum nanoparticles. *ChemElectroChem* 3: 1705–1712.

171 Schuler, T., Kimura, T., Schmidt, T.J., and Büchi, F.N. (2020). Toward a generic understanding of oxygen evolution reaction kinetics in polymer electrolyte water electrolysis. *Energy and Environmental Science* 13: 2153–2166.

172 Tahir, M., Pan, L., Idrees, F. et al. (2017). Electrocatalytic oxygen evolution reaction for energy conversion and storage: a comprehensive review. *Nano Energy* 37: 136–157.

173 Song, X.-W., Zhang, S., Zhong, H. et al. (2021). FeCo nanoalloys embedded in nitrogen-doped carbon nanosheets/bamboo-like carbon nanotubes for the oxygen reduction reaction. *Inorganic Chemistry Frontiers* 8: 109–121.

174 Zhong, H., Alberto Estudillo-Wong, L., Gao, Y. et al. (2021). Oxygen vacancies engineering by coordinating oxygen-buffering CeO_2 with CoO nanorods as efficient bifunctional oxygen electrode electrocatalyst. *Journal of Energy Chemistry* 59: 615–625.

175 Zhong, H., Estudillo-Wong, L.A., Gao, Y. et al. (2020). Cobalt-based multicomponent oxygen reduction reaction electrocatalysts generated by melamine thermal pyrolysis with high performance in an alkaline hydrogen/oxygen microfuel cell. *ACS Appl Mater Interfaces* 12: 21605–21615.

2

Nanocatalysis for Renewable Aromatics

Saikat Dutta, Navya Subray Bhat, and Harshitha N. Anchan

Department of Chemistry, National Institute of Technology Karnataka (NITK), Mangalore, Karnataka, India

2.1 Introduction

Aromatic compounds have a ubiquitous presence in synthetic organic chemistry and the organic chemical industry. They have many applications and established markets as high-octane transportation fuels, solvents, surfactants, plasticizers, agrochemicals, personal care products, and pharmaceuticals [1]. At present, aromatic hydrocarbons are primarily derived from fossilized carbon sources like petroleum. However, the excessive use of petroleum for energy, chemicals, and materials has serious consequences, including sociopolitical complications, market volatility, and environmental degradation. The past two decades have seen the astonishing development of emission-free, renewable technologies to produce energy in the form of electricity [2]. However, a carbon-based renewable source is desired to replace or supplant the petrofuels and petrochemicals. In some futuristic scenarios, biomass integrates well as a renewable, carbon-neutral feedstock for fuels and chemical production [3]. Commercial production of biofuels from the first generation of biomass, such as ethanol production from sugars and starchy feedstock and biodiesel production from edible vegetable oils, has already been realized. However, the focus has now shifted to second-generation (lignocellulose) and third-generation (algae) non-food biomass for apparent economic and environmental incentives [4]. The concept of biorefinery has received renewed interest since the 1990s for the effective utilization of diverse biomass feedstock into electricity, fuels, and value-added chemicals in an industrial setting [5]. An ideal biorefinery would be self-reliant in terms of the energy and materials requirement. It must be able to produce a wide spectrum of conventional products as well as their novel and greener alternatives. One of the major technological hurdles in a biorefinery is the molecular-level differences between petroleum and biomass. Petroleum is composed of molecular hydrocarbons with a range of molecular weights. On the contrary, biomass contains highly functionalized, complex biopolymers (e.g. cellulose, lignin) along with inorganic impurities, moisture, and extractives. The production of transportation fuels and chemicals from biomass requires selective depolymerization, defunctionalization, and energy densification [6]. A two-step strategy for biomass value

Heterogeneous Nanocatalysis for Energy and Environmental Sustainability: Energy Applications,
First Edition. Edited by Putla Sudarsanam, Yusuke Yamauchi, and Pankaj Bharali.

addition is often proposed. Initially, the biomass components are converted into a handful of small molecules, acting as renewable chemical platforms, which are then synthetically transformed into molecules with designed structural features and properties. In recent years, there is a coordinated effort between the scientists, technologists, and policymakers to make a smooth transition from petrorefinery to biorefinery, achieving long-anticipated sustainability of the chemical industry [7]. Chemocatalytic value addition of biomass has received immense attention since it is fast, selective, biomass independent, energy efficient, and reagent economic [8]. Nanocatalysts (NCs) have been employed as robust, inexpensive, efficient, and eco-friendly catalysts for various synthetic organic transformations [9]. The efficient preparation of aromatic compounds from biomass must be accomplished to unleash the full potential of biorefinery. This chapter elaborates on the production strategies of various aromatic compounds from the biomass components under nanocatalysis.

2.1.1 Aromatics from Petroleum

Aromatic hydrocarbons are primarily produced by the catalytic reforming processes in the petrorefineries. Steam cracking of naphtha and coke-oven light oil during coke formation from coal are other important routes. The catalytic processes have been optimized over the past several decades. Benzene, toluene, and xylenes (a mixture of isomers) (BTX) are among the chemical building blocks that can be synthetically modified into virtually all the petrochemicals and materials encountered in domestic and industrial settings. The total annual BTX production volume was estimated as 95 million metric tons in 2012, with around 3–4% growth per year [10]. Benzene is principally used for making styrene (via ethylbenzene), phenol (via cumene), and cyclohexanone (via cyclohexane) for various high-volume polymers [11]. Xylenes, especially *p*-xylene (PX), are converted into purified terephthalic acid (PTA) for the terephthalate polymers. Toluene has applications as a solvent, a monomer for polyurethanes, and as a chemical feed to make more of benzene and xylenes. The aromatic hydrocarbons are subjected to catalytic processes for selective oxygenation and functionalization into chemical intermediates. These intermediates are then synthetically transformed into structurally complex molecules for specific applications. Aromatic compounds have found applications as fuels to solvents, to polymers, and to pharmaceuticals. Mononuclear aromatics, such as toluene, are blended with gasoline to improve octane number and other combustion characteristics of the latter. A significant proportion of aromatics are also present in the diesel fraction. Strict environmental regulations are now in place on the upper limit of aromatics in transportation fuels. The incomplete combustion of aromatic fuels forms polyaromatic hydrocarbons (PAHs) associated with severe health and environmental risks. Therefore, more of the aromatic fractions are diverted for the synthesis of chemicals. However, the aromatics coming from petroleum are nonrenewable, and a sustainable source for aromatic compounds must be sought.

2.1.2 Aromatics from Biomass

Fossilized sources such as petroleum have been the primary source of transportation fuels, organic chemicals, and materials (e.g. polymers) for the past several decades. However, the excessive use of petroleum has stressed the already depleted reserves, increased the volatility of the petro-economies, and caused severe damage to the environment. Since the early 1990s, there has been a coordinated effort worldwide to identify a renewable and commercially viable source of carbon that can replace, at least partially, petroleum. Non-food lignocellulosic biomass has been identified as a renewable source of carbon that can be transformed into transportation fuels and chemicals, supplanting those coming from petroleum. The primary environmental benefit of using biofuels is that the release of CO_2 during their combustion gets fixed by the vegetation within a reasonable time. The geographical diversity of biomass, abundance, inexpensiveness, and rapid replenishing are additional benefits of using biomass as feedstock for fuels and chemicals. However, there are technological and economic challenges, too. Lignocellulosic biomass is primarily made of highly oxygenated and structurally diverse biopolymers, such as cellulose, hemicellulose, and lignin. Besides, they contain variable quantities of moisture, extractives, and inorganic compounds. Cellulose and hemicellulose fractions are composed of hexoses and pentoses connected by the ether linkages. The conversion of these fractions into aromatics requires depolymerization, dehydration, and selective deoxygenation steps (Figure 2.1). The lignin fraction is made of phenolic monomers connected primarily by the ether linkages. The synthesis of biochemicals (including aromatics) offers new opportunities. The hydrocarbon-based aromatics (e.g. BTX), obtained from petroleum, are functionalized selectively by specific reagents and/or catalytic processes to make value-added aromatic compounds. The synthetic steps employed for selective functionalization (e.g. oxidation) of the hydrocarbon structures generate waste streams. The biomass-derived aromatic compounds contain oxygen-containing functionalities, which can be directly exploited for downstream value addition pathways.

2.1.3 Nanocatalysis

In the beginning years, the fields of catalysis and synthetic organic chemistry advanced independently. Interestingly, many of the early organic transformations used catalysts knowingly or unknowingly. At present, catalysis is an integral part of synthetic chemistry and materials science and is increasingly being employed both in the laboratory and industrial settings. Catalysis satisfies many of the principles of green chemistry to make a chemical process more efficient and safer. Catalysts allow the formation of unprecedented products by accessing novel mechanistic pathways. They increase the selectivity of a product by decreasing the activation energy of one reaction pathway over another. In addition, they afford product formation under mild reaction conditions. They are broadly classified into homogeneous and heterogeneous catalysts. The homogeneous catalysts generally work under relatively mild conditions with high selectivity since they stay in the same phase as the reactants and products. However, effective separation and recycling of the catalysts continue to challenge the reactions employing homogeneous catalysis. Besides, contamination of the product with a trace amount of the catalyst remains an issue. The heterogeneous catalysts are increasingly popular, especially in the chemical industries, due to their relatively straightforward separation from the reaction mixture and recycling. However, the reactions using heterogeneous catalysts generally require more demanding conditions than their homogeneous counterparts. Besides, the loss of catalytic sites by various deactivation methods (e.g. poisoning, fouling) and leaching of the active catalyst from

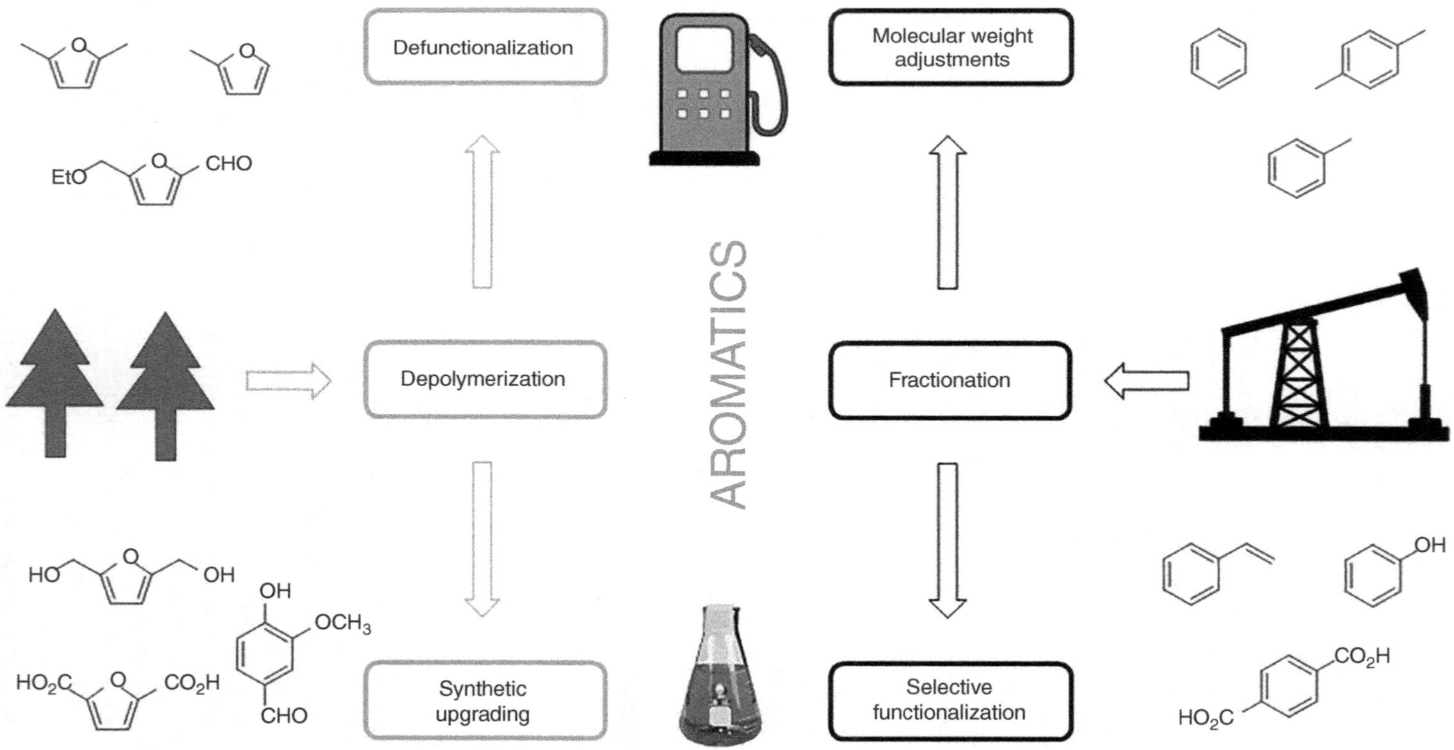

Figure 2.1 Schematic of the processes for making aromatic fuels and chemicals from petroleum and biomass.

the supporting materials continue to challenge the development of efficient heterogeneous catalysts. In this regard, the field of catalysis is reaping the benefits of the tremendous progress of nanoscience and nanotechnology made in the recent years.

Detailed understanding of the change in physicochemical characteristics of catalysts in the nanoscale has helped to develop tailor-made catalysts with superior activity and selectivity. High conversion of the substrate and high selectivity of the product is mainly due to the effective contact between the substrate and active sites of the catalysts. For example, metals that show little or no catalytic activity with the particle size in the micrometer range start showing good catalytic activity as the particle size reaches the nanoscale (<100 nm). The change of properties of materials by particle size reduction has been attributed to the higher surface-to-bulk atom ratio and the higher (unbalanced) energy of the surface atoms compared to those in bulk. The known catalysts have also shown remarkable properties as their particle size enters the nano range. NCs have been entrusted to bridge the gap between homogeneous and heterogeneous catalysts adopting the advantageous properties from both types of catalysts. They are available in various forms, including nanoporous materials, nanocomposites, nanoparticle (NP)-supported catalysts, and nanotubes.

2.2 Selective Preparation of Furfurals from Carbohydrates

The acid-catalyzed concomitant hydrolysis and dehydration of carbohydrates lead to the formation of furfural (FUR) and 5-(hydroxymethyl)furfural (HMF) (Scheme 2.1). These processes are known for more than 100 years and are well-documented in the literature. The process forms furan-based aromatic compounds from carbohydrates. It removes excess oxygen-containing functionalities in the form of water under energy-efficient conditions and without C—C bond-breaking reactions. An acid catalyst of some sort is routinely used for the dehydration process. A significant focus in this chemistry is to develop new acid catalysts that are robust, efficient, inexpensive, and eco-friendly. Various NCs containing acidic sites have been studied for the high-yielding preparation of FUR and HMF from carbohydrates. Both FUR and HMF retain the key functionalities of the parent sugar

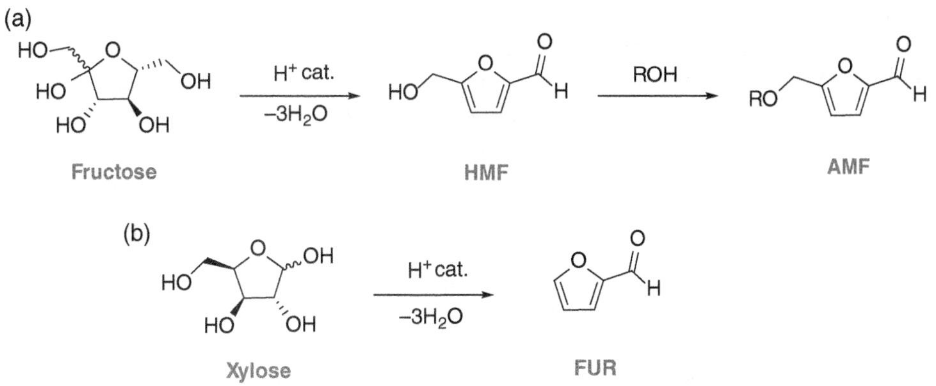

Scheme 2.1 Schematic of the acid-catalyzed dehydration of (a) hexoses to HMF (and AMF) and (b) pentoses to FUR.

molecule, which can be exploited during synthetic upgrading. FUR and HMF have received renewed attention since the late 1990s as chemical building units of renewable origin. Though FUR or HMF cannot be used directly as fuels, they can be synthetically upgraded conveniently to prepare transportation fuels, chemicals, and materials. The acid-catalyzed etherification of HMF using monohydric alkyl alcohols or the one-pot dehydration-etherification of carbohydrates leads to 5-(alkoxymethyl)furfurals (AMFs). For example, using ethanol as the reaction medium forms 5-(ethoxymethyl)furfural (EMF), a promising fuel oxygenate and chemical intermediate.

2.2.1 Synthesis of Furfurals

HMF and FUR are furan-based aromatic compounds produced by the acid-catalyzed dehydration of sugars and polymeric carbohydrates by a series of depolymerization, hydrolysis, and dehydration steps. The recent focus is to develop efficient, heterogeneous, and recyclable acid catalysts to produce FUR and HMF [12]. NCs have shown remarkable activity in this area due to their larger surface area, controlled acidity, and tunable morphology compared to the traditional heterogeneous catalysts. As a result, the past several years have seen an increasingly large number of publications on the production of FUR and HMF from various biomass feedstock using suitable NCs [13]. Recently, they have gained extensive attention due to superior reactivity, better selectivity, and improved recyclability. In this section, the recent literature, published during 2019–2021, using NCs, has been tabulated (Table 2.1) and discussed briefly.

The use of heteropoly acid immobilized ionic liquids-modified organosilica hollow spheres catalyst (Table 2.1, entry 1) and sulfonated NCs (entries 2 & 3) provided excellent yields of HMF due to their high acidity and unique structural morphology. The use of heterogenous tungsten and TiO_2 dual-acid composite gave high glucose conversion with a moderate yield of HMF (entry 4). Also, metal-free carbon-based NCs with different functionality and doping (entries 2, 7, 9, 10) were successfully employed to produce HMF in satisfactory yields from carbohydrates. The use of sulfonated magnetically separable catalysts in the presence of deep eutectic solvent (DES) afforded a good yield of HMF (entry 12). Dimethyl sulfoxide (DMSO) was found to promote the dehydration reaction of sugars to HMF while minimizing the side reactions (entries 1, 3, 5-7). An aqueous–organic biphasic reaction mixture was also frequently used. Non-chlorinated solvents like methyl isobutyl ketone (MIBK), tetrahydrofuran (THF), and γ-valerolactone (GVL) were used as organic extractant. The idea was to sequester HMF into the organic layer and restrict the decomposition reactions triggered by aqueous acid (entries 4, 8–11). The sulfonated palygorskite and zinc-doped copper oxide gave a good FUR yield (Table 2.1, entries 13 & 14) in the aqueous reaction medium. An attempt was made to synthesize FUR directly from bagasse sugarcane using iron oxide NPs (entry 15) in the aqueous medium.

EMF, a promising fuel oxygenate, was also synthesized by direct alcoholysis of carbohydrates using acidic NCs. There are dedicated reviews on the preparation and value addition of EMF (and AMFs in general) [34]. Heterogeneous NCs have gained significant attention in recent years to prepare EMF due to their superior qualities, such as high activity, better selectivity, and convenient separation. A two-step approach to EMF via the HMF intermediate has also been reported using nanocatalysis [35]. Various sulfonated and sulfonic

Table 2.1 The synthesis of furfurals from sugars using NCs.

S/N	Feedstock	Catalyst	Reaction conditions	Product	Yield (%)	References
1	Fructose	PW_{12}-ILs-C8-HNS	100 °C, 2 h, DMSO	HMF	93.7	[14]
2	Glucose	S-GCN	200 °C, 5 h, water	HMF	94	[15]
3	Fructose	N-MON-AS	140 °C, 20 h, DMSO	HMF	98	[16]
4	Glucose	WO_3-TiO_2 (MW)	170 °C, 2 h, H_2O/MIBK	HMF	58	[17]
5	Fructose	Fe_3O_4@SiO_2@ TiO_2-HPW	100 °C, 3 h, DMSO	HMF	83	[18]
6	Fructose	0.75 M-ZrO_2/SO_4^{2-}	120 °C, 1 h, DMSO	HMF	85	[19]
7	Fructose	Heparin-PEI@ PDA@MWCNT	160 °C, 3 h, DMSO	HMF	99.2	[20]
8	Fructose	SA-PESO-SBA-15	120 °C, 7 h, THF/water	HMF	68.7	[21]
9	Fructose	PTSa-CN24	140 °C, 24 h, MIBK/ H_2O	HMF	52	[22]
10	Glucose	NCC-2	120 °C, 6 h, IPA/DMSO	HMF	62.8	[23]
11	Fructose	Ag@$ZnFe_2O_4$-NW	2 h, Xe lamp (600 mW cm^{-2}), water/ DMSO	HMF	64.9[a]	[24]
12	Fructose	SO_3H-Dendrimer-SiO_2@Fe_2O_3	100 °C, 1 h, ChCl:Gly (DES)	HMF	93	[25]
13	Xylose	$Cu_{0.89}Zn_{0.11}O$	150 °C, 12 h, water	FUR	86	[26]
14	Xylose	PAL-SO_3H	180 °C, 1 h, water/GVL	FUR	87	[27]
15	Sugarcane bagasse	Fe_3O_4/SGO	190 °C, 1.5 h, water	FUR	14.0	[28]
16	Fructose	D0.5-SPC	140 °C, 18 h, ethanol/ THF	EMF	68	[29]
17	Fructose	S-PANI-$FeVO_4$	90 °C, 24 h, ethanol	EMF	72.5	[30]
18	Fructose	PSDVB-SO_3H	120 °C, 2.5 h, ethanol	EMF	42.1	[31]
19	Fructose	$H_{1.15}Nb_{1.15}Mo_{0.85}O_6$	100 °C, 12 h, ethanol	EMF	67	[32]
20	Fructose	S-poly	100 °C, 8 h, ethanol/ THF	EMF	48.1	[33]

Abbreviations: DMSO: dimethyl sulfoxide; PW_{12}-ILs-C4-HNS: phosphotungstic acid ($PW_{12} = H_3PW_{12}O_{40}$) immobilized ionic liquids modified organosilica hollow nanosphere; S-GCN: sulfonated graphitic carbon nitride; N-MON-AS: aliphatic sulfonic acid incorporated nano-sized microporous organic network; MW: microwave; MIBK: methyl isobutyl ketone; Fe_3O_4@SiO_2@TiO_2-HPW: phosphotungstic acid immobilized on SiO_2 and TiO_2-coated magnetically separable nanoparticle; Heparin-PEI@PDA@MWCNT: heparin grafted onto polydopamine (PDA) and polyethyleneimine (PEI)-coated multiwalled carbon nanotube (MWCNT); SA-PESO-SBA-15: poly(ethylene sulfoxide) added sulfonic acid on SBA-15 nanocomposite; THF: tetrahydrofuran; PTSa-CN24: *p*-toluenesulfonic acid-functionalized nano graphitic material; IPA: isopropyl alcohol; NCC: nitrogen-doped carbonaceous catalyst; Ag@$ZnFe_2O_4$ NW: $ZnFe_2O_4$-coated Ag-nanowire; PAL-SO_3H: sulfonated palygorskite; Fe_3O_4/SGO: Fe_2O_4 nanoparticle on sulfonated graphene oxide (GO); D0.5-SPC: double hydrogen-bonded sulfonated polymer catalyst; S-PANI-$FeVO_4$: $FeVO_4$ decorated -SO_3H functionalized polyaniline; S-poly: sulfonated HIPE templated polymeric composite; ChCl: choline chloride.
[a] conversion.

acid-functionalized NCs on different heterogeneous supporting materials have been used for the ethanolysis of carbohydrates (Table 2.1, entries 16, 17, 18, 20). Mixed metal oxide (MO)-based NCs have also been employed (entry 19). The challenges of preparing AMF directly from carbohydrates include difficulty in purifying the NCs from the reaction broth and their deactivation by poisoning and leaching of active catalysts from the supporting material.

In general, attempts should be made to develop robust and inexpensive NCs, providing FUR and HMF in satisfactory yields starting from untreated or minimally treated cellulosic biomass under energy-efficient conditions. The process should be scalable, and the catalyst must be recyclable without losing its activity. More research is warranted for a detailed understanding of the mechanistic details of HMF formation on the surface of NCs. The effect of the type and concentration of acidic sites on the NC surface, pore size, pore structure, and morphology of the catalysts must be discerned to design and synthesize better NCs for this chemistry.

2.2.2 Transforming Sugars into Furanics

Among multifarious platform chemicals derived from biomass-derived feedstocks, downstream value addition to furanics has paramount importance. In this context, oxidation and reduction of sugars and platform chemicals have great importance since the transformed products have a myriad of importance in industry and academia [36]. The nanostructured metal or MO is highly selective for transforming sugars into furanics. Hence, several precious and non-precious monometallic and bimetallic NPs on different organic, inorganic, and polymer-supported materials were studied as heterogeneous catalysts.

2.2.2.1 Oxidation Products of Furfurals

The selective oxidation of biomass-derived carbohydrates into furan-based aromatic compounds of commercial value is an attractive proposition. FUR and HMF can be oxidized into a number of oxidized furanic compounds with huge commercial potential. Inexpensive and innocuous oxidants like molecular oxygen and aqueous H_2O_2 are preferred for significant environmental benefits, such as better atom economy and less waste formation. Alternatively, the carbohydrates can directly be converted into oxidized furanics in a "one-pot" process using a combination of acidic and oxidation catalysts. Bifunctional catalysts having both acidic sites (for dehydration reaction) and metal sites (for oxidation reaction) have also been studied. There are certain advantages and disadvantages in both the strategies discussed earlier. While a one-pot strategy is recommended to avoid extra synthetic steps, the by-products (e.g. humin) formed in the reaction often render the catalyst inactive and challenging to recover. On the other hand, the two-step strategy provides better yield in the individual steps. The recovery of metal-based catalysts in the oxidation reaction (second step) is relatively more straightforward. However, the process requires the isolation and purification of FUR or HMF in the dehydration reaction (first step).

The oxidation of HMF or dehydration followed by oxidation of hexose sugars like glucose leads to 2,5-furandicarboxylic acid (FDCA), 2,5-diformylfuran (DFF), and 5-(hydroxymethyl)-2-furancarboxylic acid (HMFCA) (Scheme 2.2) [37]. The compounds are potential monomers for various biorenewable homopolymers or copolymers. FDCA, for

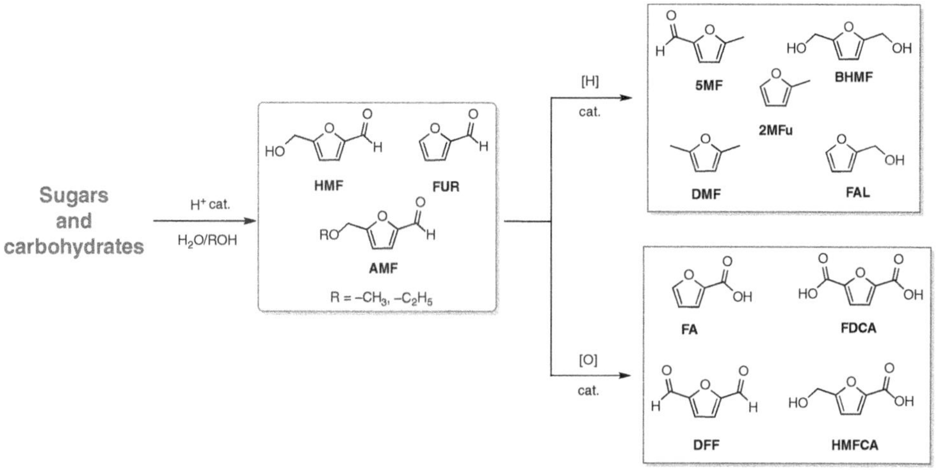

Scheme 2.2 The chemocatalytic conversion of carbohydrates into furanic chemicals.

example, is a promising replacement for PTA in the terephthalate markets. The oxidation of pentosan-derived FUR forms 2-furoic acid (FA) [38]. FA is a promising renewable chemical feedstock for fragrances and various other compounds in the paint and polymer industry.

Nanocatalysis has proved to be particularly useful in the selective oxidation of HMF to FDCA. One-pot, two-step preparation of FDCA from carbohydrates using a combination of an acid catalyst for the dehydration reaction and a metal-based catalyst for the subsequent oxidation reaction are also reported. The polymerization between FDCA and ethylene glycol (EG) leads to polyethylene furoate (PEF) polymer. If ethylene for EG production is sourced from bioethanol, PEF would be 100% biorenewable polymer. Importantly, PEF does not compromise on the performance and shows comparable physicochemical properties to those of poly(ethylene terephthalate) (PET). There are several literatures available on the oxidation of purified HMF to FDCA using NCs [39]. However, for obvious reasons, a one-pot preparation of FDCA from carbohydrates is more desirable. Therefore, recent literature on the direct synthesis of FDCA from carbohydrates using NCs is tabulated in Table 2.2. The triphasic reactor was used for one-pot synthesis of FDCA using noble Au-Pd-based catalysts (Table 2.2, entry 2). The magnetically separable iron oxide-based NCs were synthesized for direct conversion of fructose (entry 1). Pd NPs (5 and 10 nm) supported on carbonaceous material with a high surface area (68.3 m^2/g) were prepared and used as a catalyst for the direct conversion of fructose to FDCA (entry 3). The non-noble zirconium and iron mixed oxide with a particle size around 6.9 nm was successfully used to synthesize FDCA in 46.4% overall yield using ionic liquid as the reaction medium (entry 4).

DFF is a potential monomer for polymeric applications and a chemical intermediate for pharmaceuticals, antifungal agents, and macrocyclic ligand [55]. Table 2.2 lists the direct synthesis of DFF from carbohydrates. The non-noble metal NPs and MO nanocomposites were used for DFF preparation (Table 2.2, entries 5, 7–9). The monometallic and bimetallic catalysts using noble metals were employed for the one-pot preparation of DFF from carbohydrates (entries 10, 12, 14). The heteropoly acid-based catalyst supported on silica

Table 2.2 The targeted synthesis of oxidized furanics from hexose sugars.

S/N	Catalyst	Reaction conditions	Product	Yield (%)	References
1	Nano-Fe_2O_3-CoO_x	1. Fructose, Fe_2O_3@SiO_2-SO_3H, DMSO, 100 °C, 1 h 2. Nano-Fe_2O_3-CoO_x, tBuOOH, 80 °C, 12 h	FDCA	59.8	[40]
2	Au-Pd/HT	1. Fructose, Amberlyst-15, TEAB, water, MIBK, 95 °C, 0.5 h 2. Au-Pd/HT, Na_2CO_3, water, MIBK, O_2 bubbling, 95 °C, 20 h	FDCA	78	[41]
3	Pd/CC	1. Fructose, water, Amberlyst-15, Pd/CC, 100 °C, 9 h 2. K_3CO_3, 20 mL/min O_2, 140 °C, 30 h	FDCA	64	[42]
4	$Fe_{0.6}Zr_{0.4}O_2$	Fructose, [Bmim]Cl, Amberlyst-15, 160 °C, 24 h, 2 MPa O_2	FDCA	46.4	[43]
5	Fe-S/C	1. Fructose, Fe-S/C, ethanol, 0.1 MPa N_2, 100 °C, 2 h 2. 0.1 MPa O_2, 100 °C, 3 h	DFF	> 99	[44]
6	GO	Fructose, DMSO, 20 mL/min O_2, 140 °C, 24 h	DFF	53	[45]
7	WO_3HO-VO(salten)-SiO_2@Fe_3O_4	1. Fructose, WO_3HO-VO(salten)-SiO_2@Fe_3O_4, IPA, 80 °C, 1 h 2. H_2O_2, 60 °C, 15 h	DFF	71.1	[46]
8	α-CuV_2O_6	Fructose, P(EVPI-Br), DMSO, 0.1 MPa O_2 balloon, 135 °C, 3.5 h	DFF	63.1	[47]
9	f-$Ce_9Mo_1O\delta$	Fructose, DMSO, 10 mL/min O_2, 120 °C, 12 h	DFF	45	[48]
10	Ru(3%)/H-Beta	1. Sucrose, Ru(3%)/H-Beta, DMSO, 10 mL/min N_2, 120 °C, 4 h 2. 20 mL/min O_2, 140 °C, 24 h	DFF	79.7	[49]
11	$H_5PMo_{10}V_2O_{40}/SiO_2$	Starch, water, 30% H_2O_2, 70 °C, 10 h	DFF	37.9	[50]
12	Au-Ru/rGO	1. Fructose, $Au_{0.5}Ru_{0.25}$/rGO, DMSO, 2 h 2. 0.5 MPa O_2, 5 h	DFF	86	[51]
13	MoO_x/CS	Fructose, DMSO, 20 mL/min O_2, 130 °C, 8 h	DFF	78	[52]
14	Ru-MMT	Fructose, DMSO, 0.3 MPa O_2, 120 °C, 6 h	DFF	81	[53]
15	GN-NS	Fructose, DMSO, 20 mL/min O_2, 150 °C, 25 h	DFF	70.3	[54]

Abbreviations: HT: hydrotalcite; CC: carbonaceous catalyst; Fe-S/C: Fe_3O_4 nanoparticle with a metal–organic framework (MIL-88B) and sulfur powder as a dopant; GO: graphene oxide nanosheet; P(EVPI-Br): mesoporous poly(ionic liquid) derived solid acid; MoO_x/CS: carbon sphere supported on molybdenum oxide; MMT: montmorillonite; GN-NS: bifunctionalized (nitro, carboxyl, and sulfonic group) carbon nanoplatelet; TEAB: triethylammonium bromide; rGO: reduced graphene oxide.

(entry 11) and carbon-based heterogeneous NCs (entries 6, 13, 15) were also reported. In some instances, the bifunctional NCs helped in the dehydration reaction of carbohydrates to HMF as well as its selective oxidation to FDCA. Fructose was chosen as the substrate for most studies since its fructofuranose form dehydrates at greater ease and forms HMF in excellent yields under relatively mild reaction conditions. DMSO has been used as the medium as it participates in the dehydration mechanism and increases the selectivity toward HMF. Water and ionic liquids have been employed as green reaction media.

Apart from the direct synthesis of oxidized furfurals from carbohydrates, many publications reported for the synthesis of FDCA, DFF, HMFCA, and FA from HMF and FUR using NCs [56]. Partial oxidation of HMF produces compounds like HMFCA. HMFCA has been recognized as potential monomers for renewable polymers and reactive intermediates for downstream value addition [57]. Selective oxidation of the aldehyde group alone in HMF into a carboxylic acid forms HMFCA. The selective synthesis of HMFCA was reported using Ag, Au, and Ru NPs on different supports with particle sizes between 2 and 15 nm. On the other hand, oxidation of the hydroxymethyl group to an aldehyde group forms DFF. Most of the literature reports the liquid phase oxidation of HMF under moderate temperature using molecular oxygen or air as an oxidant and an organic solvent as the reaction medium. The catalytic oxidation of FUR gives rise to FA. It can also be synthesized from FUR using the Cannizzaro reaction. Attempts have been made to synthesize FA from FUR using NCs in the presence and absence of a base additive. FA can be synthesized in satisfactory yields using noble metal and non-noble metal-based monometallic and bimetallic heterogeneous NCs and O_2 as the oxidant. Au-based NCs have been particularly useful for this purpose and have been studied extensively over the past years.

The NPs of Au, Ag, Ru, Pt, Pd, and Rh on different organic and inorganic supports were extensively used to transform HMF into various oxidized furanics. However, the earth-abundant, inexpensive non-noble metal-based catalysts have received particular attention from both the economic and environmental perspectives. O_2 was employed as an inexpensive and green oxidant in most cases. Although some earlier processes used an inorganic base as an additive, the recent processes attempt to synthesize FDCA without using any base additive to avoid forming waste streams in the form of salt.

Figure 2.2 displays a correlation between particle size of noble metal NPs used to oxidize HMF into FDCA [58]. Interestingly, most NPs were relatively smaller in size (1–10 nm).

2.2.2.2 Reduction Products of Furfurals

The reduction of furfurals allows selective removal of the oxygen-containing functionalities and energy densification of the furanic derivative for their applications as biofuels. For example, the catalytic reduction of HMF leads to the formation of 2,5-dimethylfuran (DMF). The process passes through a series of hydrogenation and hydrogenolysis steps forming various interesting intermediates in between. DMF has comparable physicochemical properties to those of gasoline, high energy density (30 MJ/kg), and a research octane number of 119. It is a cleaner-burning fuel with a better air-fuel ratio due to the presence of an oxygen atom in its structure. The major objective of the hydrogenation process is to develop inexpensive, robust, and preferably heterogeneous catalysts that can provide DMF with excellent selectivity and yield under eco-friendly conditions. Over the past two

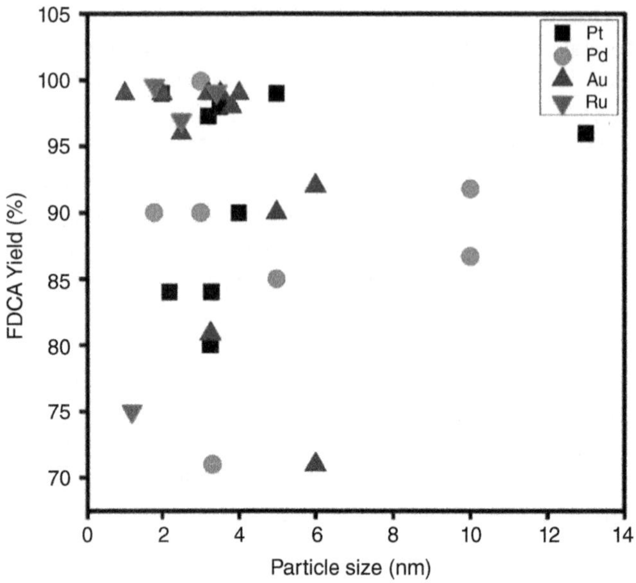

Figure 2.2 Synthesis of FDCA from HMF using noble metal nanoparticles. *Source:* Adapted from [58].

decades, 311 publications appeared in the literature reporting the synthesis of DMF by the catalytic hydrogenation of HMF [59]. DMF can also be converted to PX by reacting with ethylene.

Sugars and carbohydrates can be converted directly into DMF via a series of dehydration and hydrogenation steps in the liquid or gas phase. DMF was synthesized from fructose in a continuous-flow system using a combination of Amberlyst-15 catalyst for the dehydration reaction and Ni NP on tungsten carbide (Ni@WC) for the hydrogenation step. The combination of an acid catalyst and metal-based NC gave 38.5% yield of DMF within a residence time of only 20 s at 150 °C [60]. Li et al. synthesized carbon-based multifunctional *p*-toluenesulfonic acid functionalized, encapsulated non-noble bimetallic CuCo NCs (CuCo@C-TsOH-IM) for the direct synthesis of DMF. The particle size of CuCo was found to be 34.7 nm. DMF was obtained in a 71.1% yield at the near quantitative conversion of fructose when the reaction was performed at 220 °C for 10 hours in the THF medium under 2 MPa of O_2 pressure [61]. In the same year, Ji et al. had developed sulfonate functionalized polydivinylbenzene support on Pd NP (Pd/PDVB-S-X, where X = contact angle of the catalyst and water droplet) for selective DMF synthesis. Under optimized reaction conditions (120 °C and 6 hours), the Pd/PDVB-S-143 catalyst gave 94.2% of DMF using polymethylhydrosiloxane as a hydrogen donor in 1-butanol [62].

A number of publications reported the synthesis of reduced furan-based aromatic compounds like DMF, 2-methylfuran (2MFu), 5-methylfurfural (5MF), 2,5-bis(hydroxymethyl) furan (BHMF), etc., from furfurals (i.e. FUR and HMF) using NCs. Selective reduction of the ring substituents of isolated HMF forms DMF [63]. Several reports are available on the synthesis of DMF from HMF using NCs. In the case of noble metal-based catalyst systems, Ru- and Pd-based catalysts showed excellent activity. DMF has been synthesized from

HMF using non-noble metal-based NCs as well as bimetallic catalysts. Selective hydrogenolysis of the -CH_2OH group in HMF produces 5MF with potential applications as a food additive and a chemical intermediate. On the other hand, the reduction of FUR using NCs (monometallic and bimetallic) gives furfuryl alcohol (FAL) and 2MFu having promising applications for making resin, lubricants, and fibers. One-pot preparation of the partially reduced furanic compounds can also be envisioned. However, the processes often suffer from one or more issues, such as low selectivity toward the targeted product, challenging separation of the catalyst due to the formation of humins as the side product.

The reduced furanics can be converted into the BTX fraction selectively and in satisfactory yields under catalytic reaction conditions. For example, the Diels–Alder (DA) reaction between DMF and ethylene forms the oxanorbornene adduct, which on dehydrative aromatization under acid catalysis forms PX. Recent reviews have summarized the achievements in this field [64, 65]. PX can be oxidized to PTA. PTA can be polymerized with EG to form PET, a drop-in equivalent of the petroleum-derived PET. Alternatively, HMF can be oxidized to FDCA. Polymerization of FDCA with EG produces PEF, a greener alternative to PET. Scheme 2.3 shows the synthesis of PET and PEF polymers from biomass-derived renewable aromatics.

The heterogeneous NCs have played a determining role in moderating the reaction conditions and fine-tuning the PX selectivity. Table 2.3 lists the literature reports using NCs for the preparation of PX starting from DMF.

Heterogeneous NCs with high surface area, porosity, uniform particle size distribution, and controlled acidity showed the best catalytic activity. The use of tin phosphate and bimetallic Au—Pd catalyst showed excellent catalytic activity compared to other metal-based catalysts (Table 2.3, entries 1,3). The W and Nb-based NCs were used for transformation (entries 2, 4, 7, 9). Zeolite-based NCs were also used (entries 6, 8). The synthesis of toluene and benzene has been attempted starting from 2MFu and furan. The 5-substituted furfurals have been converted into functionalized benzene derivatives by reacting with suitable dienophiles under DA reaction conditions followed by aromatization of the corresponding adducts. Well-ordered nanocrystalline WO_x on ZrO_2 was used as a catalyst to prepare toluene and benzene from 2MFu and furan, respectively. Up to 18 and 34% yield

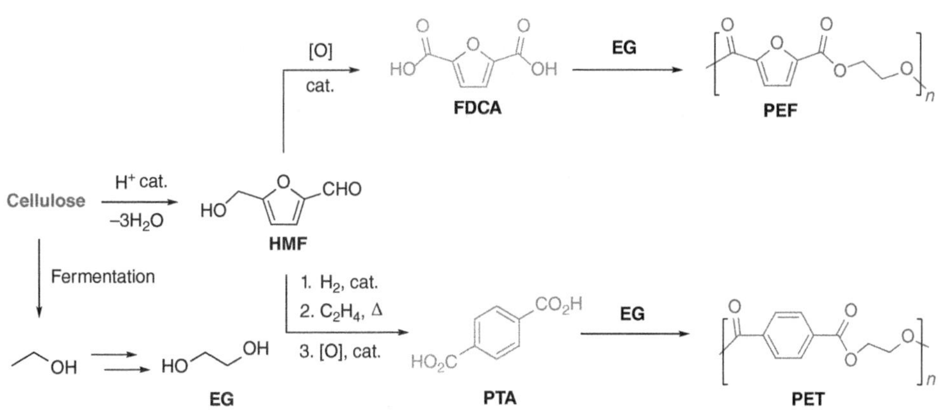

Scheme 2.3 The preparation of aromatic polyesters (drop-in and renewable equivalent of PET) from biomass components.

Table 2.3 Synthesis of PX from DMF starting using nanocatalyst.

S/N	Catalyst	Reaction conditions	Yield (%)	References
1	SnPO	250 °C, 18 h, 2 MPa ethylene	93	[66]
2	NbO_x	250 °C, 6 h, 5.4 MPa ethylene	80.8	[67]
3	$Au^\wedge Pd/ZrO_2$	275 °C, 18 h, n-heptane, 4 MPa ethylene	88	[68]
4	WO_x-ZrO_2	250 °C, 24 h, n-heptane, 5 MPa ethylene	82.6	[69]
5	SO_3H-SiO_2	250 °C, 6 h, 4.5 MPa ethylene	68	[70]
6	MFI zeolite nanosheet	250 °C, 24 h, n-heptane, 5 MPa ethylene	75.8	[71]
7	WO_3/SBA-15	250 °C, 24 h, n-heptane, 5.4 MPa ethylene	56.6	[72]
8	NSP-BEA	250 °C, 24 h, n-heptane, 5 MPa ethylene	79.4	[73]
9	WO_x-ZrO_2	250 °C, 1.5–6 h, anhydrous hexadecane, 2 MPa ethylene	80	[74]

Abbreviations: SnPO: tin phosphate; $Au^\wedge Pd/ZrO_2$: bimetallic Pd-decorated Au clusters anchored on tetragonal-phase zirconia; NSP-BEA: nano sponge-mesoporous zeolite beta.

of benzene and toluene were achieved when the reactions were performed at 250 °C under 2 MPa of ethylene pressure [74].

2.2.3 Biomass to Aromatics by Catalytic Pyrolysis

Thermochemical value addition of biomass is fast (seconds to minutes), biomass independent, technologically familiar, and could be incorporated into the petrorefinery infrastructure. The biopolymers in lignocellulosic biomass can be transformed into a hydrocarbon-rich mixture including aromatics by treating them at high temperatures in the absence of O_2. The catalyst used during the pyrolysis reaction helps to moderate the reaction parameters while improving the yield of pyrolysis oil. However, pyrolysis oil is a complex mixture of compounds and often unstable to be used directly in downstream processing. Therefore, they are usually stabilized first by processes such as catalytic hydrogenation and hydrodeoxygenation (HDO). Aromatic fractions can be separated from the mixture using the fractional distillation method adopted in the petrorefineries. The naphtha fractions can also be separated and converted into aromatics following the catalytic reformation pathway. A bimetallic Ni_3Fe NC was found to be effective for the HDO of pyrolysis oil derived from date seeds, resulting in the formation of important aromatic compounds like DMF and FAL [75].

Zeolites are the most widely used catalyst for the pyrolysis process because of their high heat resistance, good porosity, easy recyclability, controlled acidity, etc. By fine-tuning the surface acidity and pore size of zeolites, high selectivity toward desired products can be achieved [76, 77]. Shahsavari et al. showed that by altering metal (i.e. Sn & Rh) loading on the ZSM-5 zeolite, the process could be made selective toward aromatics. Under optimized metal loadings (i.e. 1.58% Rh and 3% Sn), the selectivity toward aromatic compounds reached 51.27% [78].

Other NCs are also being explored for the catalytic pyrolysis process. The fast pyrolysis of bagasse produced CH_3COOH as the main product under non-catalyzed conditions. When MoS_2 over GO was used as an NC, phenols were obtained in a major fraction [79]. Because of their low cost, char materials have been widely studied as support for the NC due to their high surface area and the ability to stabilize the metal NPs [80]. Oxygenated aromatic compounds (i.e. phenols) and light tar compounds (i.e. substituted benzene and naphthalene) were mainly formed after cracking of tar using Cu NC supported on char material [81]. Though nano-sized MOs have attracted greater attention in various catalytic processes, the research on the application of nano MOs in biomass pyrolysis is still in the nascent stage [82]. Pyrolysis of cellulose over nano SnO_2 revealed that the liquid fraction of the product mixture mostly consists of furan derivatives [83]. The pyrolysis of hazelnut shells by SnO_2 NC produced a liquid mixture containing phenol derivatives, formed by depolymerization of lignin in the feedstock [84].

In an attempt to improve both the quality and quantity of aromatic fraction in bio-oil, co-pyrolysis of biomass with hydrocarbon-rich starting materials has shown promise [85]. Co-pyrolysis of biomass with other wastes (plastic, tires, and lubricant oil) allows the hydrogen transfer to occur in one reactor and offers more effective waste treatment. The red-mud-based NC employed in the co-pyrolysis of pinewood waste and low-density polyethylene waste resulted in aromatic-rich pyrolysis oil. Further, it exhibited a considerable aromatization potential compared to the thermal co-pyrolysis, as the selectivity toward monocyclic aromatic hydrocarbons increased from 12.7% to 19.6%. It was concluded that Fe_2O_3 and TiO_2 present in the red-mud catalyst were responsible for the aromatization of alkenes [86].

The challenges in forming aromatic fuels and chemicals through biomass pyrolysis using NCs include catalyst deactivation, energy-intensive reaction conditions, and poor selectivity toward aromatic products. The mechanism of forming aromatic compounds and the interaction of the feedstock and the reaction intermediates with the NCs under pyrolytic conditions must be studied in detail. The economic feasibility of the scalable production of aromatics by biomass pyrolysis using NCs should be thoroughly investigated.

2.2.4 Aromatics from Biomass Gasification

Gasification of biomass is another crucial strategy for the value addition of biomass. The lignocellulosic biomass is treated at high temperatures (higher than pyrolytic processes) in a limited O_2 supply. A gaseous mixture primarily consisting of CO and H_2 is produced along with some other gases as impurities. The gaseous stream is then cleaned and adjusted for a suitable ratio of CO and H_2 to form syngas. Finally, the gaseous stream is subjected to the Fischer–Tropsch process using a suitable heterogeneous catalyst (Scheme 2.4). The product composition is similar to crude oil that can be separated into individual fractions, such as naphtha by fractional distillation. The naphtha fraction can then be transformed into aromatic compounds such as BTX by using a suitable zeolite catalyst. During the Fischer–Tropsch synthesis (FTS), syngas is converted into a pool of compounds like paraffins, olefins, aromatics, and oxygenated compounds. Tuning catalysts structure in nanoscale to improve the selectivity toward the aromatics is highly attractive. The integration of

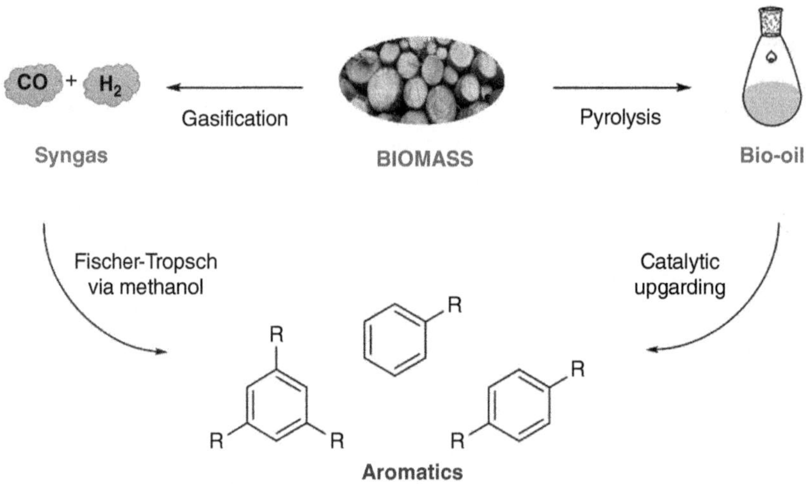

Scheme 2.4 Synthesis of aromatic compounds from biomass via pyrolysis and gasification routes.

$ZnO-ZrO_2$ NPs with HZSM-5 allowed for the efficient conversion of syngas into aromatics via CH_3OH. It was proposed that CH_3OH formed initially over the $ZnO-ZrO_2$ NPs, which was then converted to aromatics over the HZSM-5 catalyst [87].

Catalytic hydrogenation of CO_2 to chemicals, especially aromatics, can be achieved by combining reverse water gas shift reaction (RWGS) and FTS. The idea is to sequester CO_2 and convert it into fuels and chemicals without exhausting other resources. However, the approach is more challenging due to the chemical inertness of CO_2 and the high activation energy required for cleaving C—O bonds. CO is an important intermediate in the hydrogenation of CO_2. Thus, bifunctional catalysts that reduce CO_2 to CO and then catalyze FTS of the latter are of immense interest. Oxide-zeolite nanoscale composites have shown promising catalytic activities in such transformations [88]. Also, decent selectivity toward aromatics was obtained by direct hydrogenation of CO_2 using nanocomposites comprising Cu-promoted Fe_2O_3 and HZSM-5 [89].

Another strategy involves a two-stage process involving CO_2 to CH_3OH followed by the methanol to aromatics (MTA) process, wherein the oxide-zeolite NCs have shown excellent selectivity toward aromatics [90, 91]. Using a nano-sized ZSM-5 zeolite, 98% selectivity toward aromatic compounds was obtained in the MTA process [92]. It was recently found that CO_2 chemisorbed on the surface of $ZnZrO_x$ was initially hydrogenated to CH_3OH and then was converted to aromatics at 295 °C with 80.1% selectivity over the HZSM-5 catalyst. Hydrogenation of CO_2 to aromatics over the $ZnZrO_x$/HZSM-5 catalyst is favored by the lower reaction temperature compared to the traditional MTA [93].

A major technological hurdle in transforming CO_2 to fuels and chemicals (including aromatics) is the high activation energy of the process. For a truly green process, the energy required for the process must be sourced from a renewable energy source. More research is required to transform CO_2 into aromatics using a scalable process that works under operationally simple conditions and affords satisfactory yields of targeted products.

2.2.5 Phenolics from Lignin

Around 15–30% (by mass) of the 170 billion tons of lignocellulosic biomass produced annually in the world is lignin. Lignin is typically burnt to generate the process heat required for the value addition of other fractions of biomass (e.g. paper industry). The heating value of lignin is significantly higher than cellulose and hemicellulosic fractions due to the presence of aromatic core structures and a higher C/O ratio. The depolymerization of lignin into simple aromatics and other small molecules has received tremendous interest from academia and industry alike. The polyphenolic lignin structure makes it an attractive feedstock for functionalized aromatic compounds like phenols, aromatic carboxylic acids, esters, and anisoles [94].

However, the complex polymeric structure of lignin limits its direct utilization and relies on strategies that allow its efficient deconstruction into small molecules by cleaving the C—O (e.g. β-O-4, α-O-4, and 4-O-5) and C—C (e.g. β-5 and 5–5′) bonds while keeping the aromatic core structure intact (Scheme 2.5). Selective C—C bond cleavage reaction at the 5–5′ position is more challenging than the C—O bond cleavage due to the former's higher bond dissociation energy [95]. Model compounds that mimic the structural features and functional groups present in lignin are routinely used to evaluate the efficiency of a reagent, catalyst, or process on native or industrial lignin. Lignin is divided into various classes based on its extraction strategy from lignocellulosic biomass (e.g. wood). The sulfur-bearing processes include Kraft lignin (Kraft process using NaOH and Na_2S) and Lignosulfonate lignin (sulfite pulping). The sulfur-free techniques include Organosolv lignin (solvent pulping) and Soda lignin (soda pulping) [96].

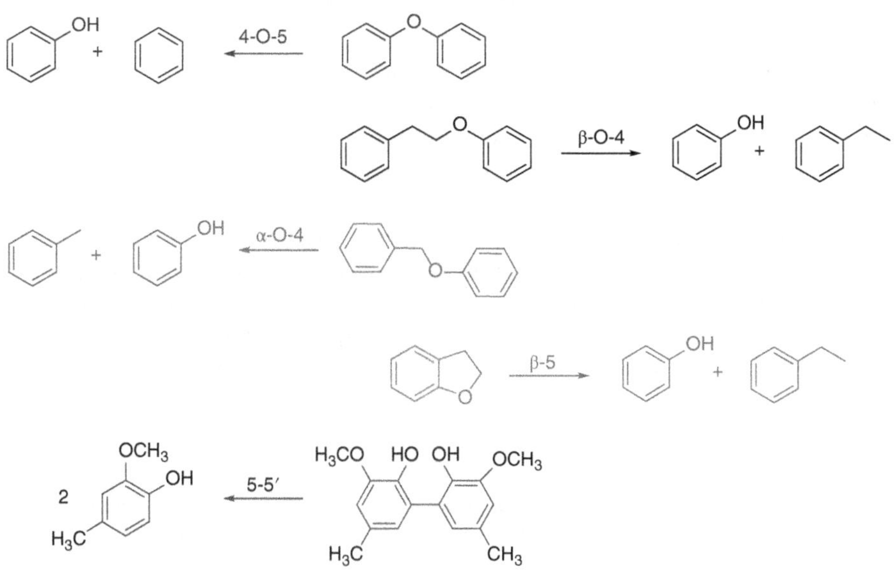

Scheme 2.5 Catalytic preparation of aromatic hydrocarbons and phenol from lignin model compounds under pyrolysis conditions.

Several strategies have been adopted for the depolymerization of lignin, such as oxidative depolymerization, catalytic hydroprocessing (or hydrotreating), thermal deconstruction, acid- or base-catalyzed depolymerization [97]. Selective hydrogenolysis of the ether linkages that interconnect the phenolics in lignin is a promising strategy to derive aromatic compounds. The reaction is typically conducted using an overpressure of H_2 and a suitable metal-based, heterogeneous catalyst. The hydrogenation process also helps to stabilize the reactive intermediates (e.g. free radicals, carbocations, etc.) and minimize repolymerization and char formation. Various lignin model compounds such as benzyl phenyl ether (α-O-4 linkage), 2-phenylethyl phenyl ether (β-O-4 linkage), diphenyl ether (4-O-5 linkage), etc., are used to mimic various C—O bonds in lignin [98]. The hydrogenolysis of lignin model compounds using various types of Ni-based NCs like Ni/SiO_2, Ni NPs supported on $ZnO\text{-}Al_2O_3$, bimetallic nano Ru—Ni, and Rh—Ni resulted in a mixture of various aromatic compounds [99–101]. The NC particle size also affects the C—O bond cleavage reactions. The smaller Ru (1.2 nm, 1.8 nm) showed higher selectivity toward aromatic hydrocarbons due to preferential cleavage of both $C(sp^3)$—O and $C(sp^2)$—O bonds, while larger particle sizes (2.5 nm, 5.3 nm) afforded higher selectivity for phenolics, during the hydrogenolysis of β-O-4 lignin model compounds [102]. A noble metal catalyst supported on Nb_2O_5 was used for the hydrogenolysis of lignin and various lignin model compounds. The conversion of 2-phenylethyl phenyl ether reached 99.3% with 98.9% selectivity toward aromatic products when the hydrogenolysis reaction was conducted for 4 hours at 260 °C under 0.1 MPa of H_2 pressure [103]. Although noble metal catalysts show promising activity for lignin depolymerization, they significantly increase capital expenditure [104]. Besides, many noble metal catalysts catalyze the aromatic ring hydrogenation lowering the yield of targeted aromatic compounds. A non-noble metal like Ni influences the reactivity of noble metals in a bimetallic catalyst and displays improved product selectivity when compared to the constituent metal catalysts. For example, a bimetallic $Ni_{85}Ru_{15}$ catalyst showed higher selectivity toward aromatic compounds during organosolv lignin hydroprocessing. The aromatic ring hydrogenation reaction was substantially suppressed in the case of $Ni_{85}Ru_{15}$ catalyst, unlike the pure Ru catalyst [99]. Similar observations were reported for the NiAu catalyst that is used to deconstruct lignin and lignin model compounds by hydrogenolysis reaction [105].

Zhang et al. investigated the hydrogenolysis of lignin model compounds that mimicked the β-O-4 linkages and obtained excellent selectivity toward aromatics (including phenolics) under ambient H_2 pressure over a $PdNi/ZrO_2$ catalyst. The hydrogenation of produced phenol is controlled by fine-tuning the amount of $NaBH_4$ used. Alloying noble and non-noble metals at nanoscale led to enhanced catalytic activity due to synergistic effect [106, 107]. It was concluded that the superior catalytic activity of a bimetallic catalyst is due to the change of the surface electronic properties in the former triggered by the interaction between constituent metal atoms [108].

Song et al. reported the catalytic hydrogenolysis of lignosulfonate into phenols (e.g. 4-propyl guaiacol, 4-ethyl guaiacol) over heterogeneous Ni-based catalysts. The catalysts gave satisfactory yields of phenolics by inhibiting the ring hydrogenation of arenes [109]. Ni NPs supported on niobic acid-activated carbon composite showed remarkable activity for hydrogenolysis of lignin model aryl ethers. The TiN-Ni nanocomposite displayed superior activity in the formation of phenol and toluene as the sole reaction products at 125 °C, whereas the individual components gave no conversion [110, 111]. In general, lignin depolymerization by metal-catalyzed hydrogenolysis requires overpressure of H_2. The use of lower H_2 pressure or eliminating the requirement of external H_2 altogether would be a greener and more sustainable approach. In 2019, Hossain et al. showed that the aliphatic alcohols produced during the Ru-catalyzed hydrogenolysis of the lignin model compounds mimicking the β-O-4 bond in lignin could act as hydrogen donor [112]. The hydrogenolysis of lignin forms various monomers, dimers, and oligomers depending on the type of catalyst employed. The monomers include guaiacyl and syringyl derivatives (Scheme 2.6), whereas dimers include resinol, phenylcoumaran, diarylpropane, and biphenyl [113].

Alcoholysis under catalytic conditions is another novel approach for the chemocatalytic value addition of lignin. Bai et al. reported one-pot depolymerization of lignin-rich corncob residue (LRCR) in supercritical C_2H_5OH over various Ni- and Mo-based monometallic and bimetallic catalysts supported on Al_2O_3. The reaction was performed for 7.5 hours at 320 °C under a 2.76 MPa of H_2 pressure. A mass yield of 255.4 mg of aromatics per gram of LRCR was obtained (57.9 wt.% alkylphenols) using the NiMo/Al catalyst [114]. Propylguaiacol and propylsyringol were obtained as the main products by depolymerizing organosolv lignin in CH_3OH medium with 2 MPa H_2 pressure over a bimetallic Ni-Fe/AC catalyst at 200 °C [115]. Very recently, Sang et al. showed that the depolymerization of enzymatic hydrolysis lignin in C_2H_5OH over unsupported Ni catalyst produced aromatic esters and substituted phenols as main liquid products [116]. Various Mo-based catalysts (e.g. α-MoC/AC, MoO_3, and Mo/Al_2O_3) were employed for the ethanolysis of Kraft lignin [117, 118]. The Mo_2C and Mo_2C/Al_2O_3 catalysts provided 425 mg and 419 mg of the total mass yield of aromatic compounds, respectively, per gram of lignin. The yield of benzyl alcohols reached 162 mg per gram of lignin when the reaction was performed at 280 °C using the Mo/Al_2O_3 NC [119]. The major products of the ethanolysis of lignin using Mo-based NCs include C8–C10 arenes, phenols, benzyl alcohols, aliphatic esters, and alcohols [120–122]. The alcoholysis of lignin has certain advantages over the pyrolysis strategy, such as the suppression of repolymerization and reduced char formation. Supercritical C_2H_5OH, a biorenewable alcohol, is particularly advantageous for the depolymerization of lignin. However, low molecular weight aliphatic compounds also formed as side products by the self-condensation of ethanol.

A two-step oxidation–hydrogenolysis strategy was adopted to depolymerize lignin and convert it to phenolic compounds with satisfactory selectivity [123, 124]. The pretreatment of lignin under oxidative conditions inhibits the generation of β-O-4 carbocations, originating from the Cα–OH of β-O-4 linkages, which are responsible for polymerization of the products and char formation [125, 126]. A Co-based NC, derived from the pyrolysis of vitamin B_{12} supported on N-doped carbon, was employed for the oxidative cleavage of 2-phenoxy-1-phenylethanol to study the cleavage of the β-O-4 linkage that resulted in the high yields of phenol and aromatic esters. Various other model compounds, including the β-O-4 ketone model compounds, also provided good yields of phenol. More importantly, the catalyst was effective in depolymerizing organosolv lignin [127]. Photodegradation of lignin or its model compounds in the presence of a suitable catalyst has also received

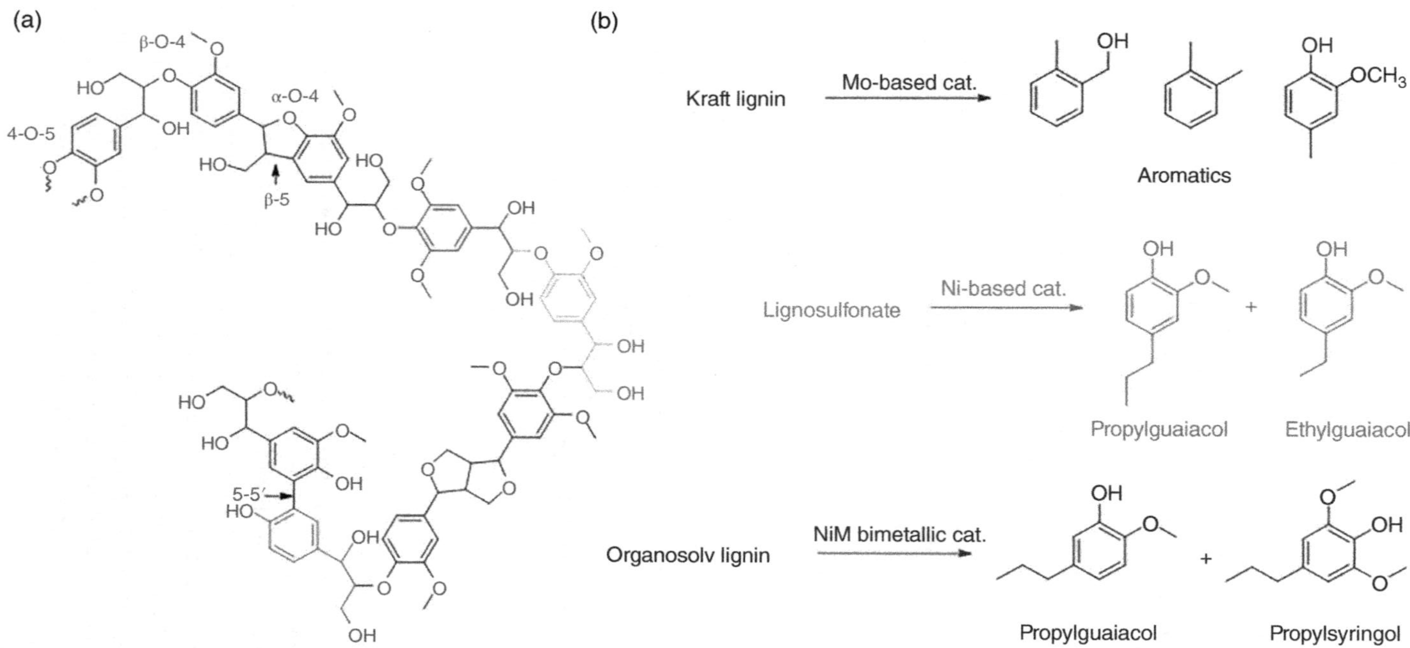

Scheme 2.6 Deconstruction of lignin into aromatic compounds under nanocatalysis. (a) The polymeric structure of lignin highlighting coumaryl alcohol, coniferyl alcohol, and sinapyl alcohol fragments, and (b) deconstruction of lignin into aromatic compounds using nanocatalysis.

immense attention. A mild photocatalytic oxidative strategy was developed to cleave C—C bonds in lignin model compounds using a mesoporous graphitic carbon nitride catalyst [128]. The multifunctional Ru/$NbOPO_4$ catalyst cleaved both C—C and C—O bonds in lignin resulting in monocyclic hydrocarbons, including aromatics [129].

The base-catalyzed depolymerization of lignin requires relatively milder reaction conditions. For example, depolymerization of lignin in a water–ethanol system at 250 °C using a Na-exchanged X zeolite (NaX) catalyst gave 51% yield of low molecular weight products like guaiacol, vanillin, homovanillic acid, 2,6-dimethoxy phenol, 3,4-dimethoxyphenol, and 1,2,4-trimethoxybenzene [130]. Cs-exchanged Keggin-type phosphotungstic acid NCs efficiently deconstructed 2,3-dihydrobenzofuran, a lignin model compound, to phenol, ethylbenzene, and 2-ethylphenol [131]. $Cs_{2.5}H_{0.5}PW_{12}O_{40}$ with high surface acidity showed the best catalytic activity. The catalyst provided noticeably better yields of the desired aromatic products than HZSM-5.

In all the strategies adopted, special care was taken to minimize unwarranted polycondensation of the small aromatic molecules produced. The phenolic compounds can then be subjected to further functionalization or defunctionalization processes. Though there are multiple novel strategies reported regarding the depolymerization of lignin, the production of aromatic compounds in the desired selectivity starting from native lignin under economically acceptable reaction parameters remains elusive. A thorough understanding of mechanistic details for lignin deconstruction on the heterogeneous catalyst surface is the key to develop better processes and catalysts.

2.2.6 Aromatics from Biomethanol

MTA is a corollary of the methanol to hydrocarbon (MTH) technology. CH_3OH is generally produced from coal and natural gas via syngas. The process can be renewable if the syngas or CH_3OH is sourced from biomass. The MTA process is conducted at elevated temperatures in the presence of a heterogeneous catalyst containing acidic sites. Interestingly, preparation of CH_3OH itself can be promoted under nanocatalysis [132, 133]. A nano-sized H[Zn, Al]ZSM-5 zeolite was applied for the aromatization of CH_3OH to BTX [134]. When the reaction was performed at 437 °C, the yield of BTX reached 48% with the weight hourly space velocity (WHSV) of 0.8 h^{-1}. The zeolite catalyst was composed of aggregates of 250 nm × 50 nm × 25 nm crystals with around 2.1 nm inter-crystal voids. CH_3OH has also been employed as an alkylation agent for benzene and toluene to ethylbenzene and xylenes under nanocatalysis. For example, a Pt-modified nano-sized HZSM-5 catalyst was prepared for the shape-selective methylation of toluene to PX with >98% selectivity in a continuous-flow fixed-bed reactor [135]. Similarly, mobil-type five (MFI) zeolite nanosponge (2.5 nm thickness) with a high mesoporous structure was employed for the alkylation of toluene to xylene. The catalyst showed superior activity, selectivity, and stability compared to bulk MFI zeolite [136]. A porous Ti-ZSM-5 catalyst was prepared by substituting part of Al by Ti and used for the alkylation of benzene to xylene. The diminished acidity of the catalyst helped to minimize the formation of olefins and ethylbenzene as the side product [137]. CH_3OH has also been used for the side-chain alkylation of toluene to styrene [138].

2.3 Conclusion and Future Prospects

There are multiple competing technologies available for the sustainable preparation of aromatic compounds from renewable biomass. Highly active and selective NCs are increasingly being used for the chemocatalytic value addition of the biomass components into renewable aromatics. The aromatic compounds derived from biomass can be drop-in equivalent as well as functional replacement of those derived from fossilized carbon sources. Some of the deconstruction strategies of lignin and the reduction of furanics require H_2 to be used. NCs have been proven effective for the efficient preparation of H_2 from biomass itself. Some of the biomass-to-aromatics technologies discussed earlier (e.g. MTA) are established in the petrorefinery setting and could directly be extended to biomass feedstock. However, some technologies require more research (e.g. lignin to phenolics) before their commercial implementation. Feedstock flexibility is crucial in a biorefinery setting, and a detailed analysis of the process must be conducted to comprehend the process's economic and environmental benefits. Although remarkable advancements have been made over the past years in the molecular-level understanding of the catalyst behavior and the effect of reaction parameters on the product selectivity, more research is needed in developing more superior catalysts. Elucidation of the reaction mechanism with theoretical calculations and experimental data is essential to improve the product selectivity and fine-tuning of catalyst activity. Emphasis should be given to the physical and chemical stability under the reaction conditions, cost, selectivity, recyclability, and environmental compatibility to design new NCs. Thorough studies on the toxicity and life cycle analysis of the NCs developed for the industrial processes must also be undertaken.

References

1 Franck, H.-G. and Stadelhofer, J.W. (2012). *Industrial Aromatic Chemistry: Raw Materials·Processes·Products*. Springer Science & Business Media.

2 Owusu, P.A. and Asumadu-Sarkodie, S. (2016). A review of renewable energy sources, sustainability issues and climate change mitigation. *Cogent Eng.* 3: 1167990.

3 Ahorsu, R., Medina, F., and Constantí, M. (2018). Significance and challenges of biomass as a suitable feedstock for bioenergy and biochemical production: a review. *Energies* 11: 3366.

4 Lee, R.A. and Lavoie, J.M. (2013). From first-to third-generation biofuels: challenges of producing a commodity from a biomass of increasing complexity. *Anim. Front* 3: 6–11.

5 Berntsson, T., Sandén, B., Olsson, L. and Åsblad, A. (2021). What is a biorefinery? Core. https://core.ac.uk/download/pdf/70603511.pdf (accessed 15 March 2021).

6 Takkellapati, S., Li, T., and Gonzalez, M.A. (2018). An overview of biorefinery-derived platform chemicals from a cellulose and hemicellulose biorefinery. *Clean Technol. Environ. Policy* 20: 1615–1630.

7 Clark, J.H., Budarin, V., Deswarte, F.E. et al. (2006). Green chemistry and the biorefinery: a partnership for a sustainable future. *Green Chem.* 8: 853–860.

8 Jing, Y., Guo, Y., Xia, Q. et al. (2019). Catalytic production of value-added chemicals and liquid fuels from lignocellulosic biomass. *Chem* 5: 2520–2546.

9 Astruc, D. (2020). Introduction: nanoparticles in catalysis. *Chem. Rev.* 120: 461–463.

10 Bender, M. (2013). Global aromatics supply – today and tomorrow. https://www.osti.gov/etdeweb/servlets/purl/22176034 (accessed 15 March 2021).

11 https://www.energy.gov/eere/amo/downloads/itp-chemicals-energy-and-environmental-profile-us-chemical-industry-may-2000 (accessed 15 March 2021).

12 Xu, C., Paone, E., Rodríguez-Padrón, D. et al. (2020). Recent catalytic routes for the preparation and the upgrading of biomass derived furfural and 5-hydroxymethylfurfural. *Chem. Soc. Rev.* 49: 4273–4306.

13 Sudarsanam, P., Peeters, E., Makshina, E.V. et al. (2019). Advances in porous and nanoscale catalysts for viable biomass conversion. *Chem. Soc. Rev.* 48: 2366–2421.

14 An, S., Wang, Z., Zhang, H. et al. (2019). Self-organization of ionic liquid-modified organosilica hollow nanospheres and heteropolyacids: efficient preparation of 5-HMF under mild conditions. *ChemCatChem* 11: 2526–2536.

15 Chhabra, T., Bahuguna, A., Dhankhar, S.S. et al. (2019). Sulfonated graphitic carbon nitride as a highly selective and efficient heterogeneous catalyst for the conversion of biomass-derived saccharides to 5-hydroxymethylfurfural in green solvents. *Green Chem.* 21: 6012–6026.

16 Cho, K., Lee, S.M., Kim, H.J. et al. (2019). Nanoparticulate and microporous solid acid catalysts bearing aliphatic sulfonic acids for biomass conversion. *Chem. Commun.* 55: 3697–3700.

17 Ganji, P. and Roy, S. (2019). Trade-off between acidic sites and crystallinity of the WO_3–TiO_2 catalyst toward dehydration of glucose to 5-hydroxymethylfurfural. *Energy Fuels* 33: 5293–5303.

18 Zhang, L., Shah, A., and Michel, F.C. Jr. (2019). Synthesis of 5-hydroxymethylfurfural from fructose and inulin catalyzed by magnetically-recoverable Fe_3O_4@SiO_2@TiO_2–HPW nanoparticles. *J. Chem. Technol. Biotechnol.* 94: 3393–3402.

19 Zhou, Y., Zhang, L., and Tao, S. (2019). Mesoporous ZrO_2 nanopowder catalysts for the synthesis of 5-hydroxymethylfurfural. *ACS Appl. Nano Mater.* 2: 5125–5131.

20 Wang, C., Gong, W., Lu, X. et al. (2019). Heparin immobilized on multiwalled carbon nanotubes for catalytic conversion of fructose in water with high yield and selectivity. *ACS Omega* 4: 16808–16815.

21 Whitaker, M.R., Parulkar, A., and Brunelli, N.A. (2020). Selective production of 5-hydroxymethylfurfural from fructose in the presence of an acid-functionalized SBA-15 catalyst modified with a sulfoxide polymer. *Mol. Syst. Des. Eng.* 5: 257–268.

22 Martin, G.D., Bounoukta, C.E., Ammari, F. et al. (2020). Fructose dehydration reaction over functionalized nanographitic catalysts in MIBK/H_2O biphasic system. *Catal. Today* 366: 68–76.

23 Zhang, Y., Li, B., Guan, W. et al. (2020). One-pot synthesis of HMF from carbohydrates over acid-base bi-functional carbonaceous catalyst supported on halloysite nanotubes. *Cellulose* 27: 3037–3054.

24 Fu, X., Li, S., Wen, J. et al. (2021). Visible light-induced photo-fenton dehydration of fructose into 5-hydroxymethylfurfural over $ZnFe_2O_4$-coated Ag nanowires. *Colloids Surf., A* 609: 125685.

25 Karimi, S., Shekaari, H., Halimehjani, A.Z., and Niakan, M. (2021). Solvent-free production of 5-hydroxymethylfurfural from deep eutectic substrate reaction mixtures over a magnetically recoverable solid acid catalyst. *ACS Sustainable Chem. Eng.* 9: 326–336.

26 Mishra, R.K., Kumar, V.B., Victor, A. et al. (2019). Selective production of furfural from the dehydration of xylose using Zn doped CuO catalyst. *Ultrason. Sonochem.* 56: 55–62.

27 Wang, R., Liang, X., Shen, F. et al. (2019). Mechanochemical synthesis of sulfonated palygorskite solid acid catalysts for selective catalytic conversion of xylose to furfural. *ACS Sustainable Chem. Eng.* 8: 1163–1170.

28 Trung, T.Q., Thinh, D.B., Anh, T.N.M. et al. (2020). Synthesis of furfural from sugarcane bagasse by hydrolysis method using magnetic sulfonated graphene oxide catalyst. *Vietnam J. Chem.* 58: 245–250.

29 Dai, J., Liu, Z., Hu, Y. et al. (2019). Adjusting the acidity of sulfonated organocatalyst for the one-pot production of 5-ethoxymethylfurfural from fructose. *Catal. Sci. Technol.* 9: 483–492.

30 Kumar, A. and Srivastava, R. (2019). $FeVO_4$ decorated–SO_3H functionalized polyaniline for direct conversion of sucrose to 2,5-diformylfuran & 5-ethoxymethylfurfural and selective oxidation reaction. *Mol. Catal.* 465: 68–79.

31 Zhang, L., Zhu, Y., Tian, L. et al. (2019). One-pot alcoholysis of carbohydrates to biofuel 5-ethoxymethylfufural and 5-methoxymethylfufural via a sulfonic porous polymer. *Fuel Process. Technol.* 193: 39–47.

32 Yang, F., Tang, J., Ou, R. et al. (2019). Fully catalytic upgrading synthesis of 5-ethoxymethylfurfural from biomass-derived 5-hydroxymethylfurfural over recyclable layered-niobium-molybdate solid acid. *Appl. Catal. B* 256: 117786.

33 Chen, Y., Guan, W., Zhang, Y. et al. (2020). One-pot synthesis of the biofuel 5-ethoxymethylfurfural from carbohydrates using a bifunctional catalyst prepared through a pickering hipe template and pore-filled strategy. *Energy Fuels* 34: 14264–14274.

34 Anchan, H.N. and Dutta, S. (2021). Recent advances in the production and value addition of selected hydrophobic analogs of biomass-derived 5-(hydroxymethyl)furfural. *Biomass Convers. Biorefin.* https://doi.org/10.1007/s13399-021-01315-1.

35 Wang, S., Zhang, Z., Liu, B., and Li, J. (2013). Silica coated magnetic Fe_3O_4 nanoparticles supported phosphotungstic acid: a novel environmentally friendly catalyst for the synthesis of 5-ethoxymethylfurfural from 5-hydroxymethylfurfural and fructose. *Catal. Sci. Technol.* 3: 2104–2112.

36 Gupta, K., Rai, R.K., and Singh, S.K. (2018). Metal catalysts for the efficient transformation of biomass-derived HMF and furfural to value added chemicals. *ChemCatChem* 10: 2326–2349.

37 Zhao, D., Su, T., Wang, Y. et al. (2020). Recent advances in catalytic oxidation of 5-hydroxymethylfurfural. *Mol. Catal.* 495: 111133.

38 Corma, A., Iborra, S., and Velty, A. (2007). Chemical routes for the transformation of biomass into chemicals. *Chem. Rev.* 107: 2411–2502.

39 Dutta, S., Bhat, N.S., and Vinod, N. (2020). Oxidation and reduction of biomass-derived 5-(hydroxymethyl)furfural and levulinic acid by nanocatalysis. In: *Advanced Heterogeneous Catalysts Volume 1: Applications at the Nano-Scale*, 239–259. ACS Publications.

40 Wang, S., Zhang, Z., and Liu, B. (2015). Catalytic conversion of fructose and 5-hydroxymethylfurfural into 2,5-furandicarboxylic acid over a recyclable Fe_3O_4–CoO_x magnetite nanocatalyst. *ACS Sustainable Chem. Eng.* 3: 406–412.

41 Yi, G., Teong, S.P., and Zhang, Y. (2015). The direct conversion of sugars into 2,5-furandicarboxylic acid in a triphasic system. *ChemSusChem* 8: 1151–1155.

42 Rathod, P.V. and Jadhav, V.H. (2018). Efficient method for synthesis of 2,5-furandicarboxylic acid from 5-hydroxymethylfurfural and fructose using Pd/CC catalyst under aqueous conditions. *ACS Sustainable Chem. Eng.* 6: 5766–5771.

43 Yan, D., Wang, G., Gao, K. et al. (2018). One-pot synthesis of 2,5-furandicarboxylic acid from fructose in ionic liquids. *Ind. Eng. Chem. Res.* 57: 1851–1858.

44 Fang, R., Luque, R., and Li, Y. (2017). Efficient one-pot fructose to DFF conversion using sulfonated magnetically separable MOF-derived $Fe_3O_4(111)$ catalysts. *Green Chem.* 19: 647–655.

45 Lv, G., Wang, H., Yang, Y. et al. (2016). Direct synthesis of 2,5-diformylfuran from fructose with graphene oxide as a bifunctional and metal-free catalyst. *Green Chem.* 18: 2302–2307.

46 Mittal, N., Nisola, G.M., Malihan, L.B. et al. (2016). One-pot synthesis of 2,5-diformylfuran from fructose using a magnetic bi-functional catalyst. *RSC Adv.* 6: 25678–25688.

47. Hou, W., Wang, Q., Guo, Z. et al. (2017). Nanobelt α-CuV_2O_6 with hydrophilic mesoporous poly(ionic liquid): a binary catalyst for synthesis of 2,5-diformylfuran from fructose. *Catal. Sci. Technol.* 7: 1006–1016.

48 Yang, Z., Qi, W., Su, R., and He, Z. (2017). 3D flower-like micro/nano Ce–Mo composite oxides as effective bifunctional catalysts for one-pot conversion of fructose to 2,5-diformylfuran. *ACS Sustainable Chem. Eng.* 5: 4179–4187.

49 Sarmah, B., Satpati, B., and Srivastava, R. (2018). One-pot tandem conversion of monosaccharides and disaccharides to 2,5-diformylfuran using a Ru nanoparticle-supported H-beta catalyst. *Catal. Sci. Technol.* 8: 2870–2882.

50 Yan, S., Li, Y., Li, P. et al. (2018). Fabrication of mesoporous POMs/SiO_2 nanofibers through electrospinning for oxidative conversion of biomass by H_2O_2 and oxygen. *RSC Adv.* 8: 3499–3511.

51 Ma, B., Wang, Y., Guo, X. et al. (2018). Photocatalytic synthesis of 2,5-diformylfuran from 5-hydroxymethyfurfural or fructose over bimetallic Au-Ru nanoparticles supported on reduced graphene oxides. *Appl. Catal., A* 552: 70–76.

52 Zhou, C., Zhao, J., Sun, H. et al. (2018). One-step approach to 2,5-diformylfuran from fructose over molybdenum oxides supported on carbon spheres. *ACS Sustainable Chem. Eng.* 7: 315–323.

53 Raut, A.B., Shende, V.S., and Bhanage, B.M. (2020). The one-step transformation of fructose to 2,5-diformylfuran over Ru metal supported on montmorillonite. *New J. Chem.* 44: 13659–13668.

54 Zhao, J., Yan, Y., Hu, Z.-T. et al. (2020). Bifunctional carbon nanoplatelets as metal-free catalysts for direct conversion of fructose to 2,5-diformylfuran. *Catal. Sci. Technol.* 10: 4179–4183.

55 Dai, J. (2021). Synthesis of 2,5-diformylfuran from renewable carbohydrates and its applications: a review. *Green Energy Environ.* 6:22–32. https://doi.org/10.1016/j.gee.2020.06.013

56 Hu, L., Zhao, G., Hao, W. et al. (2012). Catalytic conversion of biomass-derived carbohydrates into fuels and chemicals via furanic aldehydes. *RSC Adv.* 2: 11184–11206.

57 Zhao, D., Rodriguez-Padron, D., Luque, R., and Len, C. (2020). Insights into the selective oxidation of 5-hydroxymethylfurfural to 5-hydroxymethyl-2-furancarboxylic acid using silver oxide. *ACS Sustainable Chem. Eng.* 8: 8486–8495.

58 Sajid, M., Zhao, X., and Liu, D. (2018). Production of 2,5-furandicarboxylic acid (FDCA) from 5-hydroxymethylfurfural (HMF): recent progress focusing on the chemical-catalytic routes. *Green Chem.* 20: 5427–5453. and the references cited therein.

59 Scifinder reaction search results of 5-(hydroxymethyl)furfural (HMF) to 2,5-dimethylfuran (DMF) (accessed 15 March 2021).

60 Braun, M. and Antonietti, M. (2017). A continuous flow process for the production of 2,5-dimethylfuran from fructose using (non-noble metal based) heterogeneous catalysis. *Green Chem.* 19: 3813–3819.

61 Li, J., Song, Z., Hou, Y. et al. (2019). Direct production of 2,5-dimethylfuran with high yield from fructose over a carbon-based solid acid-coated CuCo bimetallic catalyst. *ACS Appl. Mater. Interfaces* 11: 12481–12491.

62 Ji, K., Shen, C., Yin, J. et al. (2019). Highly selective production of 2,5-dimethylfuran from fructose through tailoring of catalyst wettability. *Ind. Eng. Chem. Res.* 58: 10844–10854.

63 Wang, X., Liang, X., Li, J., and Li, Q. (2019). Catalytic hydrogenolysis of biomass-derived 5-hydroxymethylfurfural to biofuel 2,5-dimethylfuran. *Appl. Catal., A* 576: 85–95.

64 Mika, L.T., Csefalvay, E., and Nemeth, A. (2018). Catalytic conversion of carbohydrates to initial platform chemicals: chemistry and sustainability. *Chem. Rev.* 118: 505–613.

65 Maneffa, A., Priecel, P., and Lopez-Sanchez, J.A. (2016). Biomass-derived renewable aromatics: selective routes and outlook for *p*-xylene commercialisation. *ChemSusChem* 9: 2736–2748.

66 Feng, X., Cui, Z., Ji, K. et al. (2019). Ultra-selective p-xylene production through cycloaddition and dehydration of 2,5-dimethylfuran and ethylene over tin phosphate. *Appl. Catal., B* 259 (118): 108.

67 Yin, J., Shen, C., Feng, X. et al. (2018). Highly selective production of *p*-xylene from 2,5-dimethylfuran over hierarchical NbO_x-based catalyst. *ACS Sustainable Chem. Eng.* 6: 1891–1899.

68 Tao, L., Yan, T.-H., Li, W. et al. (2018). Toward an integrated conversion of 5-hydroxymethylfurfural and ethylene for the production of renewable *p*-xylene. *Chem* 4: 2212–2227.

69 Lee, J.-S., Kim, S.-Y., Kwon, S.-J. et al. (2018). Selective production of *p*-xylene from dimethylfuran and ethylene over tungstated zirconia catalysts. *J. Nanosci. Nanotechnol.* 18: 1419–1422.

70 Feng, X., Shen, C., Tian, C., and Tan, T. (2017). Highly selective production of biobased *p*-xylene from 2,5-dimethylfuran over SiO_2–SO_3H catalysts. *Ind. Eng. Chem. Res.* 56: 5852–5859.

71 Kim, J.-C., Kim, T.-W., Kim, Y. et al. (2017). Mesoporous MFI zeolites as high performance catalysts for Diels-Alder cycloaddition of bio-derived dimethylfuran and ethylene to renewable *p*-xylene. *Appl. Catal., B* 206: 490–500.

72 Feng, X., Shen, C., Ji, K. et al. (2017). Production of *p*-xylene from bio-based 2,5-dimethylfuran over high performance catalyst WO_3/SBA-15. *Catal. Sci. Technol.* 7: 5540–5549.

73 Kim, T.-W., Kim, S.-Y., Kim, J.-C. et al. (2016). Selective *p*-xylene production from biomass-derived dimethylfuran and ethylene over zeolite beta nanosponge catalysts. *Appl. Catal., B* 185: 100–109.

74 Wang, D., Osmundsen, C.M., Taarning, E., and Dumesic, J.A. (2013). Selective production of aromatics from alkylfurans over solid acid catalysts. *ChemCatChem* 5: 2044–2050.

75 Bharath, G., Rambabu, K., Hai, A. et al. (2020). Catalytic hydrodeoxygenation of biomass-derived pyrolysis oil over alloyed bimetallic Ni_3Fe nanocatalyst for high-grade biofuel production. *Energy Convers. Manage.* 213: 112859.

76 Jae, J., Tompsett, G.A., Foster, A.J. et al. (2011). Investigation into the shape selectivity of zeolite catalysts for biomass conversion. *J. Catal.* 279: 257–268.

77 Lazaridis, P.A., Fotopoulos, A.P., Karakoulia, S.A., and Triantafyllidis, K.S. (2018). Catalytic fast pyrolysis of Kraft lignin with conventional, mesoporous and nanosized ZSM-5 zeolite for the production of alkyl-phenols and aromatics. *Front. Chem.* 6: 295.

78 Shahsavari, S. and Sadrameli, S.M. (2020). Production of renewable aromatics and heterocycles by catalytic pyrolysis of biomass resources using rhenium and tin promoted ZSM-5 zeolite catalysts. *Process Saf. Environ. Prot.* 141: 305–320.

79 Khodafarin, R., Tavasoli, A., and Rashidi, A. (2020). Single-step conversion of sugarcane bagasse to biofuel over Mo-supported graphene oxide nanocatalyst. *Biomass Convers. Biorefin.* https://doi.org/10.1007/s13399-020-01037-w.

80 Shen, Y. and Yoshikawa, K. (2014). Tar conversion and vapor upgrading via in situ catalysis using silica-based nickel nanoparticles embedded in rice husk char for biomass pyrolysis/gasification. *Ind. Eng. Chem. Res.* 53: 10929–10942.

81 Guo, F., Peng, K., Liang, S. et al. (2019). One-step synthesis of biomass activated char supported copper nanoparticles for catalytic cracking of biomass primary tar. *Energy* 180: 584–593.

82 Lu, Q., Zhang, Z.-F., Dong, C.-Q., and Zhu, X.-F. (2010). Catalytic upgrading of biomass fast pyrolysis vapors with nano metal oxides: an analytical Py-GC/MS study. *Energies* 3: 1805–1820.

83 Donar, Y.O. and Sınağ, A. (2016). Catalytic effect of tin oxide nanoparticles on cellulose pyrolysis. *J. Anal. Appl. Pyrolysis* 119: 69–74.

84 Gökdai, Z., Sınağ, A., and Yumak, T. (2010). Comparison of the catalytic efficiency of synthesized nano tin oxide particles and various catalysts for the pyrolysis of hazelnut shell. *Biomass Bioenergy* 34: 402–410.

85 Xu, D., Yang, S., Su, Y. et al. (2020). Simultaneous production of aromatics-rich bio-oil and carbon nanomaterials from catalytic co-pyrolysis of biomass/plastic wastes and in-line catalytic upgrading of pyrolysis gas. *Waste Manage.* 121: 95–104.

86 Ahmed, M.H., Batalha, N., Qiu, T. et al. (2020). Red-mud based porous nanocatalysts for valorisation of municipal solid waste. *J. Hazard. Mater.* 396: 122711.

87 Zhou, W., Zhou, C., Yin, H. et al. (2020). Direct conversion of syngas into aromatics over a bifunctional catalyst: inhibiting net CO_2 release. *Chem. Commun.* 56: 5239–5242.

88 Ramirez, A., Dutta Chowdhury, A., Dokania, A. et al. (2019). Effect of zeolite topology and reactor configuration on the direct conversion of CO_2 to light olefins and aromatics. *ACS Catal.* 9: 6320–6334.

89 Song, G., Li, M., Yan, P. et al. (2020). High conversion to aromatics via CO_2-FT over a co-reduced Cu-Fe_2O_3 catalyst integrated with HZSM-5. *ACS Catal.* 10: 11268–11279.

90 Li, Z., Qu, Y., Wang, J. et al. (2019). Highly selective conversion of carbon dioxide to aromatics over tandem catalysts. *Joule* 3: 570–583.

91 Ni, Y., Chen, Z., Fu, Y. et al. (2018). Selective conversion of CO_2 and H_2 into aromatics. *Nat. Commun.* 9: 1–7.

92 Ma, Y., Cai, D., Li, Y. et al. (2016). The influence of straight pore blockage on the selectivity of methanol to aromatics in nanosized Zn/ZSM-5: an atomic Cs-corrected STEM analysis study. *RSC Adv.* 6: 74797–74801.

93 Wang, T., Yang, C., Gao, P. et al. (2021). $ZnZrO_x$ integrated with chain-like nanocrystal HZSM-5 as efficient catalysts for aromatics synthesis from CO_2 hydrogenation. *Appl. Catal., B* 286: 119929.

94 Sudarsanam, P., Ruijten, D., Liao, Y. et al. (2020). Towards lignin-derived chemicals using atom-efficient catalytic routes. *Trends Chem.* 2: 898–913.

95 Zhu, J., Wang, J., and Dong, G. (2019). Catalytic activation of unstrained C(aryl)–C(aryl) bonds in 2,2′-biphenols. *Nat. Chem.* 11: 45–51.

96 Mandlekar, N., Cayla, A., Rault, F. et al. (2018). An overview on the use of lignin and its derivatives in fire retardant polymer systems. *Lignin-Trends Appl.* 207: 231.

97 Sudarsanam, P., Duolikun, T., Babu, P.S. et al. (2019). Recent developments in selective catalytic conversion of lignin into aromatics and their derivatives. *Biomass Convers. Bioref.* 10:873–883.

98 Jiang, L., Guo, H., Li, C. et al. (2019). Selective cleavage of lignin and lignin model compounds without external hydrogen, catalyzed by heterogeneous nickel catalysts. *Chem. Sci.* 10: 4458–4468.

99 Zhang, J., Teo, J., Chen, X. et al. (2016). A series of NiM (M=Ru, Rh, and Pd) bimetallic catalysts for effective lignin hydrogenolysis in water. *ACS Catal.* 4:1574–1583.

100 He, J., Zhao, C., and Lercher, J.A. (2012). Ni-catalyzed cleavage of aryl ethers in the aqueous phase. *J. Am. Chem. Soc.* 134: 20768–20775.

101 Bulut, S., Saikevich, S., van Muyden, A.P. et al. (2018). Efficient cleavage of aryl ether C–O linkages by Rh–Ni and Ru–Ni nanoscale catalysts operating in water. *Chem. Sci.* 9: 5530–5535.

102 Dong, L., Yin, L.-L., Xia, Q. et al. (2018). Size-dependent catalytic performance of ruthenium nanoparticles in the hydrogenolysis of a β-O-4 lignin model compound. *Catal. Sci. Technol.* 8: 735–745.

103 Guan, W., Chen, X., Hu, H. et al. (2020). Catalytic hydrogenolysis of lignin β-O-4 aryl ether compound and lignin to aromatics over Rh/Nb_2O_5 under low H_2 pressure. *Fuel Process. Technol.* 203: 106392.

104 Toledano, A., Serrano, L., Pineda, A. et al. (2014). Microwave-assisted depolymerisation of organosolv lignin via mild hydrogen-free hydrogenolysis: catalyst screening. *Appl. Catal., B* 145: 43–55.

105 Zhang, J., Asakura, H., van Rijn, J. et al. (2014). Highly efficient, NiAu-catalyzed hydrogenolysis of lignin into phenolic chemical. *Green Chem.* 16: 2432–2437.

106 Kim, J. K., Lee, J. K., Kang, K. H. et al. (2015). Selective cleavage of CO bond in benzyl phenyl ether to aromatics over Pd–Fe bimetallic catalyst supported on ordered mesoporous carbon. *Appl. Catal., A 498*: 142–149.

107 Zhang, J., Cai, Y., Lu, G., and Cai, C. (2016). Facile and selective hydrogenolysis of β-O-4 linkages in lignin catalyzed by Pd–Ni bimetallic nanoparticles supported on ZrO_2. *Green Chem.* 18: 6229–6235.

108 Kim, J.-Y., Park, S.Y., Choi, I.-G., and Choi, J.W. (2018). Evaluation of Ru_xNi_{1-x}/SBA-15 catalysts for depolymerization features of lignin macromolecule into monomeric phenols. *Chem. Eng. J* 336: 640–648.

109 Song, Q., Wang, F., and Xu, J. (2012). Hydrogenolysis of lignosulfonate into phenols over heterogeneous nickel catalysts. *Chem. Commun.* 48: 7019–7021.

110 Molinari, V., Giordano, C., Antonietti, M., and Esposito, D. (2014). Titanium nitride-nickel nanocomposite as heterogeneous catalyst for the hydrogenolysis of aryl ethers. *J. Am. Chem. Soc.* 136: 1758–1761.

111 Jin, S., Xiao, Z., Chen, X. et al. (2015). Cleavage of lignin-derived 4-O-5 aryl ethers over nickel nanoparticles supported on niobic acid-activated carbon composites. *Ind. Eng. Chem. Res.* 54: 2302–2310.

112 Hossain, M.A., Phung, T.K., Rahaman, M.S. et al. (2019). Catalytic cleavage of the β-O-4 aryl ether bonds of lignin model compounds by Ru/C catalyst. *Appl. Catal., A* 582: 117100.

113 Verziu, M., Tirsoaga, A., Cojocaru, B. et al. (2018). Hydrogenolysis of lignin over Ru-based catalysts: the role of the ruthenium in a lignin fragmentation process. *Mol. Catal.* 450: 65–76.

114 Bai, Y., Cui, K., Sang, Y. et al. (2019). Catalytic depolymerization of a lignin-rich corncob residue into aromatics in supercritical ethanol over an alumina-supported NiMo alloy catalyst. *Energy Fuels* 33: 8657–8665.

115 Zhai, Y., Li, C., Xu, G. et al. (2017). Depolymerization of lignin via a non-precious Ni–Fe alloy catalyst supported on activated carbon. *Green Chem.* 19: 1895–1903.

116 Sang, Y., Chen, M., Yan, F. et al. (2020). Catalytic depolymerization of enzymatic hydrolysis lignin into monomers over an unsupported nickel catalyst in supercritical ethanol. *Ind. Eng. Chem. Res.* 59: 7466–7474.

117 Ma, X., Ma, R., Hao, W. et al. (2015). Common pathways in ethanolysis of Kraft lignin to platform chemicals over molybdenum-based catalysts. *ACS Catal.* 5: 4803–4813.

118 Chen, M., Hao, W., Ma, R. et al. (2017). Catalytic ethanolysis of Kraft lignin to small-molecular liquid products over an alumina supported molybdenum nitride catalyst. *Catal. Today* 298: 9–15.

119 Cattelan, L., Yuen, A., Lui, M. et al. (2017). Renewable aromatics from Kraft lignin with molybdenum-based catalysts. *CatChemCat* 9: 2717–2726.

120 Ma, R., Hao, W., Ma, X. et al. (2014). Catalytic ethanolysis of Kraft lignin into high-value small-molecular chemicals over a nanostructured α-molybdenum carbide catalyst. *Angew. Chem.* 126: 7438–7443.

121 Chen, M., Ma, X., Ma, R. et al. (2017). Ethanolysis of kraft lignin over a reduction-modified MoO_3 catalyst. *Ind. Eng. Chem. Res.* 56: 14025–14033.

122 Ma, X., Cui, K., Hao, W. et al. (2015). Alumina supported molybdenum catalyst for lignin valorization: effect of reduction temperature. *Bioresour. Technol.* 192: 17–22.

123 Qi, S.-C., Hayashi, J.-I, Kudo, S., and Zhang, L. (2017). Catalytic hydrogenolysis of Kraft lignin to monomers at high yield in alkaline water. *Green Chem.* 19: 2636–2645.

124 Wang, Z., Ying, A., Fan, Z. et al. (2017). Tertiary amino group in cationic gold catalyst: tethered frustrated Lewis pairs that enable ligand-controlled regiodivergent and stereoselective isomerizations of propargylic esters. *ACS Catal.* 7: 3676–3680.

125 Rashidi, M., Konarova, M., Aslam, W., and Beltramini, J.N. (2018). TiN-Cu heterogeneous nanocatalysts for effective depolymerisation of oxidised lignin. *ChemistrySelect* 3: 3379–3385.

126 Wang, M., Zhang, X., Li, H. et al. (2018). Carbon modification of nickel catalyst for depolymerization of oxidized lignin to aromatics. *ACS Catal.* 8: 1614–1620.

127 Luo, H., Wang, L., Li, G. et al. (2018). Nitrogen-doped carbon-modified cobalt-nanoparticle-catalyzed oxidative cleavage of lignin β-O-4 model compounds under mild conditions. *ACS Sustainable Chem. Eng.* 6: 14188–14196.

128 Liu, H., Li, H., Lu, J. et al. (2018). Photocatalytic cleavage of C–C bond in lignin models under visible light on mesoporous graphitic carbon nitride through π–π stacking interaction. *ACS Catal.* 8: 4761–4771.

129 Dong, L., Lin, L., Han, X. et al. (2019). Breaking the limit of lignin monomer production via cleavage of interunit carbon–carbon linkages. *Chem* 5: 1521–1536.

130 Chaudhary, R. and Dhepe, P.L. (2017). Solid base catalyzed depolymerization of lignin into low molecular weight products. *Green Chem.* 19: 778–788.

131 Kim, J.K., Park, H.W., Hong, U.G. et al. (2014). $Cs_xH_{3.0-x}PW_{12}O_{40}$ (x=2.0–3.0) heteropolyacid nano-catalysts for catalytic decomposition of 2,3-dihydrobenzofuran to aromatics. *J. Nanosci. Nanotechnol.* 14: 8884–8890.

132 Guil-López, R., Mota, N., Llorente, J. et al. (2019). Methanol synthesis from CO_2: a review of the latest developments in heterogeneous catalysis. *Materials* 12: 3902.

133 Behrens, M. and Armbrüster, M. (2011). *Methanol Steam Reforming, Catalysis for Alternative Energy Generation*. New York: Springer.

134 Ni, Y., Sun, A., Wu, X. et al. (2011). The preparation of nano-sized H [Zn, Al] ZSM-5 zeolite and its application in the aromatization of methanol. *Microporous Mesoporous Mater.* 143: 435–442.

135 Zhao, Y., Tan, W., Wu, H. et al. (2011). Effect of Pt on stability of nano-scale ZSM-5 catalyst for toluene alkylation with methanol into *p*-xylene. *Catal. Today* 160: 179–183.

136 Lee, C., Lee, S., Kim, W., and Ryoo, R. (2018). High utilization of methanol in toluene methylation using MFI zeolite nanosponge catalyst. *Catal. Today* 303: 143–149.

137 Cen, J., Zhang, N., Hu, H. et al. (2020). Synthesis of hierarchical porous Ti-ZSM-5: a high active catalyst for benzene alkylation with methanol. *Catalysts* 10: 693.

138 Hong, Z., Xiong, C., Zhao, G., and Zhu, Z. (2019). Side-chain alkylation of toluene with methanol to produce styrene: an overview. *Catal. Sci. Technol.* 9: 6828–6840.

3

Synthesis, Characterization and Applications of Solid-Based Heterogeneous Nanocatalysts for Biodiesel Production

Kiran Mustafa[1], *Nadeem Iqbal*[2], *Xianfeng Fan*[3], *Sarah Farrukh*[4], *and Sara Musaddiq*[1]

[1] *Department of Chemistry, The Women University Multan, Multan, Pakistan*
[2] *Centre for Undergraduate Studies, University of the Punjab, Lahore, Pakistan*
[3] *School of Engineering, The University of Edinburgh, Edinburgh, United Kingdom*
[4] *School of Chemical and Materials Engineering, National University of Science and Technology, Islamabad, Pakistan*

3.1 Introduction

The ever-increasing population is escalating the energy demand continuously. The current consumption and supply of fossil fuels indicated that the oil reserves would be diminished by the year 2050 [1, 2]. This resulted in a huge increase in the fuel price along with the continuous and fast exploration of alternative fuels, such as biodiesel and sustainable hydrogen. A wide variety of biomass feedstocks can be utilized for competing with the current and future demands of renewable and sustainable fuel. Biofuels include biodiesel, biohydrogen, bioethanol, and biomethane. Bioethanol is produced based on a three-step process: pretreatment, saccharification, and fermentation. All these fuels are obtained from different substrates and by different chemical reactions. For instance, bioethanol is generated by wheat, sugarcane, potatoes, and corn. This originates the moral concern of using food crops for fuel production. That is why the lignocellulose biomass is used as the alternative feedstock for bioethanol production. Biodiesel is obtained through different chemical reactions such as transesterification; in which long-chain triglycerides are converted into fatty acid methyl esters. Biodiesel is an important biofuel as it is biodegradable, nontoxic, and renewable. The combustion emission profile of biodiesel is also suitable owing to the low emissions of carbon, sulfur, and nitrogen oxides along with very low particulate matter [3]. Biodiesel has several benefits over conventional fuels, such as larger energy security and minimal social and economic problems. The reduction in the greenhouse effect is also expected from the utilization of biodiesel [4, 5]. Additionally, biodiesel has exhibited a high combustion efficiency, cetane number, flash point, and, and improved lubricant efficiency as compared to the petroleum-diesel. Biodiesel can be made from a variety of renewable sources, including edible oil, nonedible oil, and waste edible oil. Vegetable oil is considered edible oil, and it is not well suited for biodiesel production due to its high cost and use as food [6].

Heterogeneous Nanocatalysis for Energy and Environmental Sustainability: Energy Applications,
First Edition. Edited by Putla Sudarsanam, Yusuke Yamauchi, and Pankaj Bharali.

The biodiesel formation from oil or other raw materials occurred through a variety of chemical processes such as transesterification or esterification. All the processes use various catalysts to perform well. The catalysts range from being homogenous to heterogeneous catalysts. For instance, sodium methoxide or sodium hydroxide behave as homogenous catalysts, whereas sodium silicate serves as the heterogeneous catalyst in transesterification [7–9]. The solid catalysts are extensively employed for the production of biodiesel mainly because of their recyclable nature [10]. Owing to the far-reaching applications of heterogeneous catalysts in biodiesel production, they have been extensively studied. The study of heterogeneous catalysis involves the exploration of the various materials which can be used for the purpose, their fabrication, characterization, and utilization. The heterogeneous catalyst can be of alkaline, acidic, or enzymatic nature. Different materials serve as heterogeneous solid-based catalysts, such as oxides of different metals, alkaline zeolites, clays, and sulfonated carbons [11]. A detailed account of the catalysts that are involved in biodiesel production is presented in Table 3.1.

Recently, nanomaterials have proven their worth in the production of biodiesel. Metal-based nanomaterials consisting of metals, metal oxides, and their composites are all involved in the production of biodiesel. The heterogeneous catalysts are not limited to metal-based materials. Other important catalysts for biodiesel production are zeolites, hydrotalcites, and carbon-based nanomaterials. This chapter will comprehensively explain a variety of heterogeneous nanocatalysts that are involved in the production of biodiesel.

Table 3.1 Different solid-based heterogeneous catalysts involved in the production of biodiesel.

Type of catalyst	Examples	Advantages	Disadvantages
Acid catalyst – homogeneous	Sulfuric acid, hydrochloric acid, and sulfonic acid	Better catalysis	Corrosion of equipment and slower rate of reaction
Acid catalysts – heterogeneous	Bronsted-type acid catalyst, Lewis-type acid catalyst, and Keggin-type heteropolyacids	The catalyst can be reusable and easy to separate	Reaction is severe and expensive to prepare catalyst
Base catalyst – homogeneous	Sodium hydroxide, potassium hydroxide, and sodium methylate	High rate of catalysis, noncorrosive, cheap, and wide availability	Soap formation at high FFA and cannot be reused
Base catalyst – heterogeneous	Alkaline metal carbonates, alkaline earth metal carbonates, and alkaline earth metal oxides	Catalyst can be reused, minimal generation of waste, and easier purification of product	Expensive and leaching of active sites during recovery
Enzyme catalysts	Lipases	Highly purified product, easy to separate, and reaction is mild	Methanol may deactivate the enzyme activity and high cost of operation

Source: From [12] / with permission of Elsevier.

3.2 Zeolites-Based Nanoparticles

Among several other catalysts, zeolites are vital candidates for biodiesel production. Their excellent catalytic activity can be associated with the large surface area, strong acidic sites, promising molecular sieving capability, and selectivity. These make them important catalytic reagents not only for biodiesel production but also for several other chemical reactions [13]. Owing to the useful applications of the zeolites as catalysts, researchers have tried to improve their catalytic capabilities by designing the nanoscale zeolites. Several successful attempts have been made for the above-mentioned purpose. These materials are hydrophobic, with strong stability in both polar and nonpolar solvents. They are also capable of supporting enzymes, such as lipase. The immobilized enzymes are highly useful catalysts for chemical reactions like esterification. Zeolites enhance the catalytic activity of the enzymes to several folds, as they offer strong resistance to the heat reaching the enzyme. The most important aspect of the catalysts in any chemical reaction is their ability to be reused. The restriction of the enzymes on the surface of zeolites could provide a similar activity after several consecutive cycles during transesterification of triglycerides or other raw materials in the production of the biodiesel [13, 14].

In this regard, researchers have studied the prospects of nanozeolites with varying crystallographic materials modified with 3-aminopropyl trimethoxysilane (APTMS), followed by the crosslinking with the glutaraldehyde (GA). The modified catalyst serves as the solid support for the restriction of lipase originated from *Thermomyces lanuginosus*. The role of the catalyst in the ethanolysis transesterification was also disclosed in the study. The results of the research exhibited that the nanozeolites modified with the APTMS and GA crosslinking can restrict larger amounts of lipase. The nanozeolite-immobilized enzymes have higher enzymatic capabilities as compared to the free enzyme. The nanozeolite and the enzyme complex catalyst have shown the biodiesel yield greater than 93%. The catalytic complex has remarkable reusability up to five consecutive cycles of ethanolysis. The promising reusability indicates the extraordinary stability of the catalyst. Additionally, it has been found that the enzyme restriction through the covalent associations is more effective as compared to the physical adsorption on the surface of the catalyst. The performance of the enzyme-zeolite complex has shown that the chemical modification of the zeolite surface is an efficient process for the fabrication of the effective biocatalyst for transesterification [14].

Another group has studied the mesoporous intracrystalline basic Faujasite-type catalysts for biodiesel production. The catalyst was prepared via post-synthetic alterations and ion exchange of basic cation-enriched hierarchical zeolites. The study has evaluated the performance efficiency of the prepared catalyst in the production of biodiesel from the conversion of triglycerides of vegetable oils through transesterification [15]. Similarly, ion-exchanged zeolite P nanocatalyst has proven worth in the production of biodiesel [16].

3.3 Nanosized Metal Oxide-Based Catalysts

Metal nanoparticles serve as the vital solid-based heterogeneous catalysts for various chemical reactions. Among different catalysts that are involved in the production, the metallic nanoparticles secure a very promising spot as they are involved in the production of biodiesel from a wide variety of feedstocks. Different metal oxides are involved in the production of biodiesel. For instance, nanosized SiO_2/ZrO_2 has been employed in the conversion of soybean oil to biodiesel. These metal oxide nanocatalysts with a large surface area have been prepared via the sol–gel technique. The catalyst depicted remarkable catalytic activity with the maximum conversion of ~96.2% after three hours of reaction. Additionally, the catalyst exhibited good reusability, as they were assessed for six cycles of transesterification, with 84.1% yield of the final cycle [17].

$C_4H_4O_6HK$ (potassium bitartrate)-restrained ZrO_2 nanomaterials with the size varying from 10 to 40 nm, have been prepared and evaluated for the production of biodiesel from soybean oil and methanol through transesterification. At the methanol/oil ratio of 16 : 1 with 6 wt% of the catalyst and at 60 °C temperature, the maximum yield of 98.03% was obtained after two hours of the reaction [18].

TiO_2 and Zn mixed oxide-based nanoparticles, i.e. TiO_2–ZnO is also used as a nanocatalyst for the transesterification reaction. They have been employed for the production of biodiesel via the transesterification of palm oil. The defects formation, owing to the Zn substitution with the Ti on the surface of the Zn lattice, leads to the enhanced catalytic ability and stability. Biodiesel yield of 92% and oil conversion of 98% were obtained with 200 mg of catalyst, 6 : 1 methanol/oil molar ratio, 60 °C after five hours of reaction. The sulfate-enhanced TiO_2 solid acid nanocatalyst has also been employed in biodiesel production. These nanomaterials are actively involved in the various processes of petroleum refineries. The SO_4^{2-}/TiO_2 nanoparticles have exhibited better performance as compared to the other sulfated metal oxide catalysts. The better performance of the SO_4^{2-}/TiO_2 complex can be associated with the improved acidic strength of the TiO_2 nanoparticles with the surface loading of the SO_4^{2-} groups. The SO_4^{2-}/TiO_2 nanoparticles were employed in the biodiesel formation from the cooking oil [19, 20]. Other than the above-mentioned metallic nanoparticles, several catalysts are also used in biodiesel formation. Some of them are listed in Table 3.2.

The bimetallic catalysts, such as Cu–Ni restrained over the surface of varying metal oxides (like TiO_2, ZrO_2, and CeO_2), have also found their applications in the production of biodiesel [21, 22]. The nanocatalysts formed by the doping of Mn over the surface of ZnO nanomaterial and calcined at 600 °C have been employed for the transesterification of mahua oil. At the optimized temperature of 50 °C, with the methanol/oil ratio of 7 : 1, 8 wt% of catalyst amount, the maximum yield of 97% was observed after the 50 minutes of the reaction. The kinetic analysis of the transesterification of mahua oil with methanol has shown that the activation energy of the reaction is 181.91 kJ/mol with Mn/ZnO catalyst [23]. The Fe and Fe oxide-based magnetic nanoparticles functionalized with varying acids, such as sulfamic and sulfonic encased with the SiO_2 coverings, have been employed for the transesterification of glyceryl trioleate [24]. The mixed Fe/Cd and Fe/Sn oxide nanoparticles have also found their applications in the production of biodiesel from soybean oil as a feasible approach [25].

Alkaline earth metal oxide nanocatalysts, such as CaO-based nanocatalysts, have served as promising catalysts of the transesterification process for biodiesel production from different sorts of feedstock [26]. The CaO nanocatalysts are of four major types such as pristine CaO, loaded Cao, doped CaO, and waste CaO, where all of them are normally employed as nanocatalysts for transesterification. Moreover, it has been found that the loaded and doped CaO-based nanocatalysts are preferred over the pristine CaO as they can form

Table 3.2 The optimized parameters and metallic nanocatalyst for the production of biodiesel from different feedstocks.

				Operating parameters		
Heterogenous nanocatalyst	Oil source	Synthesis method	Size (nm)	Catalyst concentration	Reaction time	Reaction temperature (°C)
Zinc oxide nanocatalyst						
TiO_2-ZnO	Palm oil	Combustion	34.2	200 mg	5 h	60
Manganese-ZnO	Mahua oil	Co-precipitation	24.18	8%	50 min	50
Heteropolyacid-coated ZnO	Madhuca oil	Co-precipitation and calcination	5–29	0.6%	5 h	55
Copper-doped ZnO nanorods	Waste cooking oil	Co-precipitation	80	12%	50 min	55
Copper-doped ZnO nanorods	Neem oil	Co-precipitation	40.62	10%	60 min	55
Ag-ZnO	Simarouba oil	Solution combustion	42	1.50%	2 h	64
Iron (II)-doped ZnO	Castor oil	Co-precipitation	52.8	14%	50 min	55
Copper ion coupled zinc oxide nanoferrites	Soybean oil	Combustion	28–32	4%	1 h	180
Nickel-doped ZnO	Castor oil	Co-precipitation	35.1	11%	60 min	55

Source: From [12] / with permission of Elsevier.

hydrogen bonds with glycerol and methanol. Different researchers have employed pristine CaO for transesterification reactions. For instance, Reddy et al. have employed the pristine CaO nanocatalysts for the production of biodiesel with the poultry fat as the feedstock. The CaO-based nanocatalyst showed 100% conversion at room temperature with alcohol/oil ratio of 10 : 3 [27]. Similarly, the same catalyst has been employed with palm oil as the substrate, where 95.7% conversion yield was achieved when the alcohol/oil molar ratio was kept at 15 : 1 and the temperature was maintained at 60 °C [28].

Another CaO-based nanomaterial KF/CaO has been fabricated via the impregnation process with a size between 30 and 100 nm. Furthermore, the prepared nanomaterials were evaluated for their catalytic activity in the production of biodiesel from the Chinese tallow seed oil. The results have exhibited a yield of 96.8%, which is very suitable for industrial usage [29]. The same microporous solid-based catalyst has also been employed by another group of researchers for biodiesel production from rapeseed oil. The reaction conditions were optimized, and the alcohol/oil ratio was kept between 6 : 1 and 12 : 1 at the temperature of 70–90 °C. At these conditions, a maximum yield of biodiesel (93%) was obtained [30].

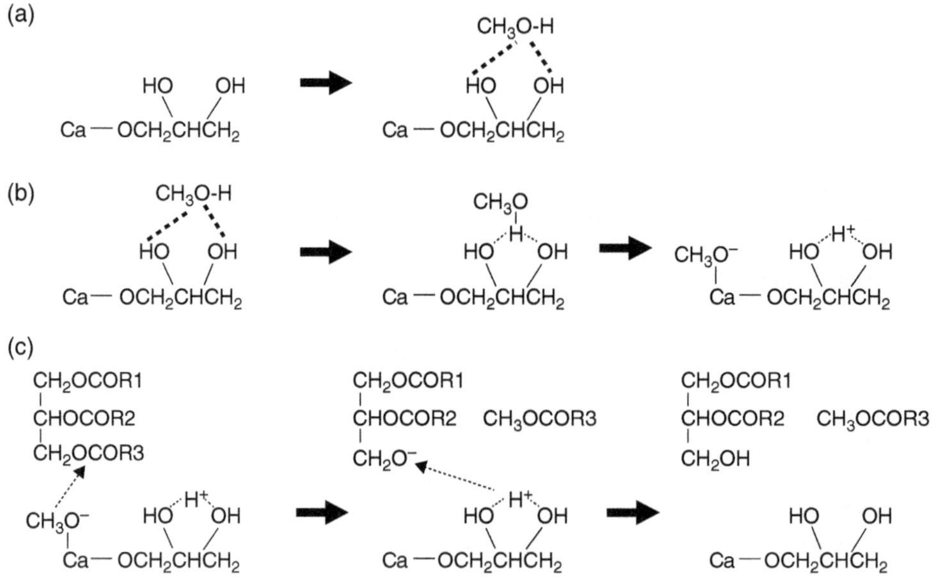

Figure 3.1 CaO-based nanocatalyst mechanism for transesterification. (a) Binding of the catalyst with the alcohol, (b) abstraction of the proton by a hydroxyl group, (c) substation with methoxide anion followed by the subsequent stabilization of the anion. *Source:* From [31] / with permission of Elsevier.

The mechanism of the CaO-based nanocatalysts for the production of biodiesel can be explained in few steps. In the first step, the catalyst binds with the alcohol, followed by the proton abstraction by basic sites. Finally, the nucleophilic reaction occurs with the methoxide anion and the anion is stabilized by the proton. The general mechanism of the process is shown in Figure 3.1.

Graphene oxide is also involved in the modification of the CaO nanocatalysts. This nanocomplex was used in the production of biodiesel from Jatropha and soybean oil. The reaction condition was optimized, and the alcohol/oil molar ratio was maintained at 70 °C for Jatropha oil, and 99% yield was obtained. In the case of soyabean oil, similar percentage of the yield was obtained at the same molar ratio of alcohol/oil at 60 °C [32]. MgO and CaO nanoparticles fabricated via sol–gel self-combustion methods and sol–gel process, respectively, are used for biodiesel production from cooking oils. In practice, the CaO nanoparticles have shown a substantial increase in the biodiesel yield as compared to MgO nanoparticles [33].

3.4 Nano-Hydrotalcites

Hydrotalcites are important catalytic materials that are actively involved in the production of biodiesel. They secure a significant spot in the areas of catalysis and materials science research owing to their widespread applications. In order to improve the catalytic ability of the hydrotalcites, scientists have thought of converting them into the domain, which adds to their abilities due to inherent characteristics, such as high surface area. The widely used nano-hydrotalcites consist of Al–Mg layered double hydroxides or anionic clays. The nano-hydrotalcite belongs to the category of 2D nanostructured materials, in which the anionic clays of the hydrotalcites contain different metallic ions which are held in the brucite-like layers with the hydroxyl groups [34, 35].

Mg/Al hydrotalcite nanoparticles fabricated with co-precipitation method using urea have been employed as the catalyst for biodiesel production via transesterification with Jatropha oil. These nanomaterials have proven to be very effective catalytically with a biodiesel yield of 95% [36]. Cerium-modified Mg/Al hydrotalcites have been employed in the methanolysis reaction of soybean oil. These hydrotalcites have shown 90% yield [37]. A study has reported the production of biodiesel from Pongamia oil through the transesterification process with the Mg/Al-based nano-hydrotalcite. The study has reported the maximum yield of 90.8% with six parts of methanol and one part of the catalyst at 65 °C [38]. Another study suggested that the macroporous hydrotalcites when transformed into nanoscale, significantly improved the formation of biodiesel. These porous Mg/Al nano-hydrotalcites were prepared through the alkali-free co-precipitation process. They showed the remarkable catalytic performance in the transesterification of C4–C18 triglycerides to produce biodiesel. Additionally, it was proposed that the enhanced catalytic performance of the nano-hydrotalcite catalyst is due to the porous structure of the material, allowing the efficient diffusion of the massive triglycerides, leading to the excellent access and binding of the substrates with the catalyst [88]. Similarly, Manivannan et al. have designed the Mg/Al nano-hydrotalcites and employed them as the solid-based heterogeneous catalysts during the methanolysis of neem oil to biodiesel. Additionally, it was determined that the efficiency of nano-hydrotalcite is affected by certain reaction conditions, such as the amount of the catalyst, reaction time, temperature, and the molar ratio of methanol/oil. The results revealed that the temperature of the reaction is very important for biodiesel production. At the lower temperatures (room temperatures), the yield of the biodiesel is less, and it increases with the increase in the temperature. The maximum yield of 84% was observed at the optimum temperature of 65 °C. The further increase in the temperature results in the decline of biodiesel production [39].

Modified hydrotalcites have also been prepared and used as the catalyst for the production of biodiesel. In a study, KF/Ca–Mg–Al hydrotalcite with varying cation ratios have been employed for the purpose. Furthermore, they evaluated the effects of different cation ratio of the Ca–Mg–Al hydrotalcite and the methanol/oil molar ratios on biodiesel yield from palm oil through transesterification. The study exhibited that all the modified hydrotalcite have the significantly high catalytic ability. About 90% biodiesel yield was obtained with a 12 : 1 molar ratio of the methanol and oil within 10 minutes of the reaction, whereas the 5% (wt/wt) catalyst was used at 65 °C. Additionally, it was found that the improved yield of 99.6% was obtained with the KF/Ca-Mg-Al hydrotalcite after a similar time. The nano-hydrotalcite allows the production of the biodiesel at ambient temperatures, hence keeping the reaction environmentally friendly. The nanosized Zn/Mg/Al hydrotalcite is used as a solid-based heterogeneous catalyst for biodiesel production from neem oil via transesterification. The incorporation of the Zn in the nano-hydrotalcite has improved the catalytic ability of the nanomaterial. A 90.5% biodiesel yield was obtained at 65 °C after four hours of the reaction [40].

Another nano-hydrotalcite, i.e. Mg/Al/Zn-based hydrotalcite/SBA-15 composite material has also been studied for biodiesel production via transesterification with soybean oil as the substrate. The integration of the Zn in Mg–Al hydrotalcite has critically influenced the basic site densities, the textural features, and the catalytic capability of the nanosized hydrotalcite composite. The maximum yield of 90% (biodiesel) was achieved by applying the modified material in the transesterification process. In the Mg/Al/Zn-based hydrotalcite/SBA-15 catalyst, the varying ratios of Mg/Al ranging from 1.5 to 5 have shown remarkable catalytic efficacy [41]. In another study, it was claimed that the calcined Li/Al hydrotalcites are more effective for glycerolysis of fatty acids as compared to the Mg/Al hydrotalcites owing to their high Lewis basicity [42]. The Co/Fe-mixed oxides nano-hydrotalcites of different types, varying each other on the basis of the chemical modifications, are all useful in the production of biodiesel from various sorts of oils [43]. Some of the most commonly employed hydrotalcites with optimized reaction conditions are summarized in Table 3.3 [31].

Table 3.3 The reaction conditions and percentage yield of biodiesel with different hydrotalcite as catalysts.

	Calcination		Reaction conditions			
Heterogeneous catalyst	**temperature (°C)**	**Time (h)**	**Molar ratio (methanol to oil)**	**Reaction time (h), temperature (°C)**	**Catalyst amount (wt%)**	**Conversion (C), yield (ϒ) (%)**
Mg–Al hydrotalcite $Mg_6Al_2(OH_{16})CO_3{\cdot}4H_2O$	500	8	15 : 1	9; reflux of methanol	7.5	C = 67
Na/hydrotalcite with soybean oil	500	8	15 : 1	8; 60	7.0	ϒ = 88
Na/hydrotalcite with used frying oil			9 : 1	8; 60	7.0	ϒ = 67
Zn/Al complex oxide derived from hydrotalcite	450	8	24 : 1	1.5; 200; Pressure= 2.5 MPa	1.4	C = 83.6
Mg–Al–CO_3 hydrotalcite	350	6	6 : 1	3; 180–200	1	C = 99
Mg-Al hydrotalcite	500	12	6 : 1	4; 65	1.5	C = 90.5
Mg–Al hydrotalcite	450	35	30 : 1	6; 100	7	C = 96.9,
	500 (after loading with potassium acetate)	2				ϒ = 86.6
KF/hydrotalcite	450	3	12 : 1	5; 65	3.0	ϒ = 92
KF/Ca–Al hydrotalcite	550	5	12 : 1	3 : 65	5.0	ϒ = 99.74
$(Zn_5(OH)_8(NO_3)_2{\cdot}2H_2O)$	–	–	6 : 1	(Time not)	6.0	C = 95.7

Source: From [31] / with permission of Elsevier.

3.5 Magnetic Nanoparticles-Based Catalysts

Many nanoparticles have secured a prime spot in the field of solid-based heterogeneous catalysis, owing to their magnetic capabilities. The magnetic properties of these materials allow the easy recovery of the catalysts and ensure their reusability in several cycles of transesterification reactions, hence enabling them economically effective. Efforts have been made in the area of magnetic nanoparticle fabrication, and the magnetic nanoparticles have been extensively employed in the transesterification process of different feedstocks for biodiesel production. In a study, novel magnetic-nanoparticles of $KF/CaO\text{-}Fe_3O_4$ have been prepared. They have a porous structure and employed for the production of biodiesel with stillingia oil. These nanoparticles have a diameter of about 50 nm with the Brunauer-Emmett-Teller (BET) surface area of 20.08 m^2/g. This nanocatalyst has shown the maximum yield of 95% (biodiesel) at the optimal temperature of 65 °C, the oil/methanol ratio of 1 : 12, the catalyst concentration of 4 wt% for three hours reaction. These magnetic nanoparticles have been found to depict similar activity after the 14 cycles of transesterification reaction [44].

In a relatively simple, fast, and economically favorable method, biodiesel has been prepared from soybean oil. For this purpose, Fe/Cd- and Fe/Sn-oxide nanoparticles with magnetic characteristics were employed as the heterogeneous catalysts for transesterification and hydrolysis of fatty acids from soybean oil. The catalysts were fabricated via a co-precipitation process. The maximum yield of 84% (biodiesel) was achieved when Fe/Sn-oxide nanoparticles were employed as the catalyst for transesterification at the optimal temperature of 200 °C after one hour of reaction. The catalysts were recovered after the reaction and tested for the evaluation of reusability; similar activity was observed for four cycles in the case of Fe/Sn-oxide, whereas the other set of the catalysts depicted a reduction in the catalytic ability [25]. Feyzi et al. have prepared $Ca/Fe_3O_4@SiO_2$ nanoparticles as the potential catalytic candidate for biodiesel production. These nanomaterials were fabricated by sol–gel and incipient wetness impregnation techniques collectively. These nanoparticles have strong magnetic capabilities that ensure the material's reusability several times, without the loss of catalytic ability. These materials have exhibited remarkable catalytic activity with a maximum yield of 97% (biodiesel) at the optimum conditions [45].

The $MgO/MgFe_2O_4$ nanoparticles have magnetic characteristics; they have been fabricated and evaluated as the heterogeneous solid-based catalysts for the production of biodiesel from sunflower oil. The maximum conversion of ~91.2% (biodiesel) was achieved using this catalyst at 110 °C, with a methanol/oil molar ratio of 1 : 2, and the catalyst concentration of 4 wt% after four hours of reaction. This magnesium-based nanocatalyst has also shown good reusability, as they were evaluated in five consecutive cycles of transesterification. In the last cycle, the lowest yield of biodiesel (82.4%) was achieved [46].

$ZnO/BiFeO_3$ nanoparticles are also equipped with magnetic properties. $ZnO/BiFeO_3$ nanoparticles have been fabricated via a co-precipitation technique and were analyzed for their ability to produce biodiesel from canola oil via transesterification reaction. The use of these nanoparticles for biodiesel production is economically favorable as the complete recovery of the material is possible (by the use of an external magnetic field). The materials have shown yields of 95.43 and 95.02% (biodiesel) in two consecutive cycles, respectively; at the optimum conditions, such as 15 : 1 molar ratio of methanol/oil, 65 °C with 4 wt% of catalyst. The catalyst was reused five times with the final yield of 92.08% [47]. The magnetic core-mesoporous of KOH/Fe_3O_4@γ-Al_2O_3 nanocatalyst was prepared using the Fe_3O_4@γ-Al_2O_3 core-shell structure as support and KOH as an active component. The catalytic efficiency of this magnetic catalyst was indicated by the high biodiesel yield (97.4%), which was achieved under specific optimum reaction conditions through transesterification of canola oil [48].

Acid-functionalized magnetic nanosubstances are also actively involved in the production of biodiesel from a wide variety of feed materials. In an investigation, highly effective, recyclable nanoscale magnetic solid acid catalysts have been fabricated. The core of these nanocatalysts comprises iron oxide magnetic nanoparticles enclosed by the shell of poly(glycidyl methacrylate), whereas phosphotungstic acid was employed as the surface modifier acidic moiety. The prepared acidic magnetic nanoparticles have been employed as the solid-based heterogeneous catalysts for the conversion of grease to biodiesel. In this investigation, it was observed that the magnetic nanocatalyst generates biodiesel from grease via both transesterification and esterification of triglycerides (with methanol) and free fatty acids, respectively. At optimum reaction conditions, a maximum yield of 98% was obtained. The catalysts were recovered easily and were evaluated in the 10 consecutive cycles of the reaction, and their reusability was promising as the final cycle exhibited a 95% yield of biodiesel [49].

In another study, the acid-functionalized magnetic particles have been prepared with sulfamic acid and sulfonic acid, for their potential use in the production of biodiesel from the waste cooking oil. In practice crystalline, Fe/Fe_3O_4 core/shell magnetic nanoparticles were fabricated, and their catalytic performance was evaluated in the transesterification reaction of glyceryl trioleate. The results of catalytic activity revealed that both the acid moieties considerably enhance the catalytic ability of the nanomaterials; however, the nanoparticles modified by sulfamic acid showed high catalytic capability as compared to the sulfonic acid-modified nanocatalysts [24].

In a comparatively novel approach, modified iron oxide nanoparticles have been fabricated via the Stöber method. The nanoparticles were functionalized with chlorosulfonic acid, followed by the subsequent encasing with silica layers. The modified nanoparticles have good magnetic properties and were found to be effective catalysts for biodiesel production. Additionally, the catalytic ability of both coated and noncoated solid-based heterogeneous nanocatalysts was assessed in the esterification of palmitic acid and methanol. The study has explained that the coating of the silica layers over the surface of magnetic nanoparticles reduces the magnetic properties of the nanoparticles but assisted the mass transportation. This can be associated with the fact that covering the magnetic nanoparticles with silica provides a stable acidic reaction medium to the relatively porous structure and magnetic core [50].

3.6 Carbon-Based Nanomaterials

Carbon-based nanomaterials are actively involved in the production of biodiesel production. These materials serve as the solid-based heterogeneous catalysts for catalyzing various processes, such as transesterification, ethanolysis. For this purpose, single-walled carbon nanotubes (SWCNTs), multi-walled carbon nanotubes (MWCNTs), functionalized

CNTs, conjugates of the carbon-based nanomaterials with different metals, and biological enzymes are prepared using different methods. CNTs have gained an increased interest in biodiesel production primarily due to their constructive properties, such as a greater specific surface area, greater porosity, particular electronic properties, superior affordability, and hollow/layered structures [51].

The major disadvantage of CNTs is their low dispersibility in aqueous medium or organic solvent and problematic recovery after the completion of the chemical reaction. Researchers have also found that the immobilization of certain enzymes like lipase is also facilitated with the acid treatment of the CNTs; these enzymes impart other useful properties to the carbon catalyst. Similarly, maximum recovery can be achieved by introducing magnetic properties to the catalysts [52–54].

3.7 Enzymes-Based Heterogeneous Catalysts for Biodiesel Production

Enzyme-immobilized various nanomaterials have been used as catalysts for biodiesel production. Enzymes like lipases can catalyze the reactions such as esterification, transesterification, as well as hydrolysis reactions of triglycerides to fatty acids and glycerol. In short, they are valuable reagents serving as catalysts in biodiesel production. The major drawback of using lipase as the catalyst is its high cost and low reusability. The immobilization of the lipases on the surface of the nanomaterials offers the solution to these problems as this technique improves the stability, activity, and reusability of the enzymes. Significant cost reduction is also achieved through the immobilized enzyme-based immobilized catalysts. Several studies have investigated the immobilization of varying origin lipases on a variety of nanomaterials (which serve as the carrier of the enzyme activity) and observed the differences in their catalytic efficiency in the production of biodiesel from different substrates. Enzymes have been immobilized on various nanomaterials, such as inorganic nanoparticles, metal composites-based nanomaterials, CNTs.

Nanoparticles of various materials, such as polystyrene, chitosan, zirconia, silica, and polylactic acid, are used for enzyme immobilization for heterogeneous catalysts formation. These catalysts are actively involved in biodiesel production. One of the setbacks of these enzyme-immobilized catalysts is their recovery, production of which needs vigorous centrifugation. However, the magnetic nanoparticle-based catalysts are comparatively easily recovered [55–59]. The restriction of the enzyme on hydrophilic surfaces is weak owing to the isolation of the catalytic site by the polypeptide chain of the enzyme [60]. In a study, researchers have grafted various carboxylic acids such as stearic acid and valeric acid on zirconia nanoparticles, followed by the immobilization of the *Pseudomonas cepacia* lipase. Best activity and enantioselectivity were found in the case of stearic acid grafting. The primary activity of the Lipase/stearic/ZrO_2 was 16.6 greater than simple ZrO_2/lipase and 10.5 times greater than the pristine lipase powder. This can be associated with the fact that the long hydrophobic chain of the acid reduces the inactivation of the catalytic site caused by interfacial interaction of the enzyme and enzyme polypeptide chain [55]. The restriction of the lipase enzyme is also studied on the silica nanoparticles. The study was conducted by the use of *Mucor janaicus* lipase. The enzyme restriction was done via the covalent attachment of ethylene

Table 3.4 Lipase from different origins immobilized on varying nanomaterials have varying percentage efficiency and reusability with divergent substrates.

Nanomaterials	Strain	Carrier	Type of binding	Special feature
Nanoparticles	*Pseudomonas cepacia*	Zirconia	Covalent	Increased activity and enantioselectivity
	Mucor japonicus	Silica	Covalent	Enhanced enzyme loading and enzyme stability
	Candida antarctica	Polystyrene	Adsorption	High hydrolytic activity
	Candida rugosa	Chitosan	Covalent	High enzyme loading and activity retention
	C. rugosa	Polylactic acid	Adsorption	Enhanced activity and stability
	C. rugosa	γ-Fe_2O_3	Covalent	Enhanced stability
	Porcine pancreas	Magnetic	Adsorption	Good reusability

Source: Adapted from [14].

diamine (EDA) over the surface of activated silica nanoparticles, where the activation was achieved by the application of the 1,4-phenylene diisothicyanate (NCS) or GA. NCS or GA serves as a coupling agent, which prevents the site binding of the carriers and the enzymes. Thus, the EDA-GA- and EDA-NCS-activated nanoparticles could access surface areas for enzyme immobilization. The modified enzyme/nanoparticle complex improves the loading and the activity of the enzyme, along with high thermal stability and vast pH tolerance. The GA and NCS-modified enzyme nanoparticle complex exhibited 115 and 107% greater catalytic activity, respectively, as compared to the pristine mobilized enzyme [57]. A brief description of some inorganic nanoparticles for lipase immobilization is given in Table 3.4.

CNTs are excellent candidates for enzyme restriction. CNTs based on graphitic sheets have a unique structure that rolls into a cylindrical shape. In the past years, extensive research has been done on the enzyme restriction on various selections of CNTs. Both SWNTs and MWNTs are involved in the utilization of enzymes for the production of biodiesel. A brief account of such CNTs is depicted in Table 3.5.

In a study, lipase was immobilized onto SWCNTs via two solvent systems, where one is ionic and the other is buffer. Buffer is incapable of dissolving the CNTs due to the van der Waals interactions of SWCNTs. Ionic liquid facilitates better dispersion of the CNTs along with the enhanced restriction activity of the enzymes [67]. MWCNTs have the greater ability to hold the enzymes [68]. The CNTs have also been employed as additives for the enzymes for prevention against enzyme inactivation during the sol–gel process. 2.7% (wt/wt) MWCNTs offer the 10-fold increase in the efficiency of *Candida rugosa*, *Candida antarctia*, and *T. lanuginosus* lipase during esterification. The reusability of the enzymes was maintained to 96% after five cycles, in contrast to the loss of activity with pristine enzymes at similar conditions [53, 62, 64]. Lipases from *C. rugosa* was restrained on MWCNT via physical adsorption, resulting in 97% activity. The immobilized enzyme depicted 14- and 2.2-fold increase in transesterification reaction initial rate in water-immiscible ionic liquids and

Table 3.5 Enzyme based CNTs heterogenous catalysts for biodiesel production.

Enzyme strain	CNTs	Enzyme activity	Substrates	Procedure	Time (h)	Temperature (°C)	% conversion	References
Candida antarctica Lipase B	SWCNTS		Sunflower oil	Ethanolysis	4	35	83.4>90% was achieved after 10 cycles	[61]
Candida rugosa, C. antarctica B, Thermomyces lanuginosus	MWCNTs	Improved stability		Esterification			96% same activity after 5 cycles	([62])
Rhizomucor miehei lipase	(3-Aminopropyl) triethoxysilane (APTES)-modified magnetic MWCNTs		Waste vegetable oil	Transesterification	10	50	94% Same activity after 10 cycle reuses	[63]
C. antarctica	MWCNTs	Improved storage and operational ability				60	55% activity retained after 6 mo at 4 °C	[64]
Burkholderia cepacia Lipase	Superparamagnetic MWCNTS/Fe_3O_4 NPs			Esterification	2.5	30	92.8 Same activity after 20 cycles	[65]
Rhizopus arrhizus	MWCNTs	Improved Resolution				35–60		[53]
C. rugosa	MWCNTs	Greater activity of enzymes						[66]

hexane, respectively [66]. The simple adsorption method is used for the immobilization of the lipases on the surface of acid-modified MWCNTs. The acidic moiety (COO−) is integrated on the surface of MWCNTs by stirring them with the acidic mixture consisting of H_2SO_4 and HNO_3. The acidic moiety is capable of bonding with the basic polar substituents like NH_2 and OH of the enzyme. The restriction of the enzyme on the surface of the nanotubes leads to enhanced mechanical strength, thermal stability, activity, and structural integrity [69].

3.8 Characterization and Recovery of Nanocatalysts

The nature and the particular characteristics of the nanocatalysts can affect the yield and quality of the biodiesel in a reaction. Consequently, any nanocatalyst should be meticulously characterized before its application in the actual reaction. Different techniques have been employed for the characterization of the nanomaterials. One of the commonly employed techniques is X-ray diffraction (XRD). It is used for the analysis of crystalline substances. It can provide information about the crystallinity and the composition of the material under observation. Scanning electron microscope [70] is another promising characterization technique. It provides information about the surface morphologies of fabricated nanomaterials as well as their precursors. For the determination of the particle diameter and structure, transmission electron microscopy (TEM) is used. Other than these, there are several methods that are employed for the characterization of the nanocatalysts, such as Fourier transform infrared (FTIR) spectroscopy, which provide the information about the nature of the functionalization in the sample material. The surface area of the nanoparticles can be evaluated by BET and Barrett Joyner Hlenda (BJH) techniques. The thermal stability of the samples is determined by thermogravimetric analysis (TGA). These techniques have been employed by several scientists for the examination of various nanocatalysts [18, 71–75]. For instance, Kambiz et al., studied the morphological characterization of the MgO nanomaterial via scanning electron microscope (SEM). Similarly, the structure of the a heteropolyacid-coated ZnO nanocatalyst and heteropolyacid was determined simultaneously by Baskar et al. [33, 76].

The biodiesel's economic value depends on the cost of the catalysts and the raw materials employed for the reaction. Therefore, the reusability of the catalyst is very crucial factor for the inexpensive and sustainable production of biodiesel. The catalyst is lost if not recovered properly or may be inactivated due to the contamination of the impurities. This problem can be resolved by altering the reaction mixture. The sintering of the catalyst reduces the number of active sites, decreasing its catalytic ability [77, 78].

For the evaluation of the reusability of catalyst, it is first separated from the glycerol and fatty acid and then employed in several cycles. In an investigation, transesterification of the algal oil was studied. The used catalyst was washed, but the activity was gradually decreased with the successive cycles. This can be related to the blockage of the active sites of the catalyst by phospholipids of microalgae (S. H. [79]). From this case, it can be conferred that the solvent washing is not enough for the recovery of the catalysts. The magnetic nanoparticles can be separated easily by the application of an external magnet and can be reused. Shengyang et al. recovered and reused the magnetic nanoparticles of KF/CaO-Fe_3O_4

14 times for transesterification reaction of stillingia [44]. Additionally, Baskar et al., employed heteropolyacid-coated ZnO as a nanocatalyst for biodiesel production from *Madhuca indica*. For the determination of the changes in the characteristics of the catalyst, the nanocatalyst was recovered and compared with the newly fabricated catalyst utilizing several characterization techniques [76]. It is evident from the above discussion that nanocatalysts can be recovered and reutilized. In order to efficiently use the nanocatalysts for economical and sustainable biodiesel production, the recovery and reutilization of the nanocatalysts should be explored thoroughly.

3.9 Conclusion

In the modern era, human beings are heavily dependent on energy. As fossil fuels are depleting, alternate sources of energy are very high in demand. Biodiesel is one of the most promising alternative fuels, as it is renewable and green, unlike conventional fuels that are extremely harmful to the environment. Biodiesel is generated from a variety of renewable feedstocks such as animal fats, plant seeds, and vegetable oils via different chemical reactions. These chemical reactions employed different catalysts for overcoming the activation energy barriers. Solid-based heterogeneous nanocatalysts promisingly serve the purpose. This chapter elaborates on the fabrication and utilization of several such catalysts in an organized manner. The optimized working conditions for biodiesel production have also been discussed.

References

1 Harun, R., Danquah, M.K., and Forde, G.M. (2010). Microalgal biomass as a fermentation feedstock for bioethanol production. *Journal of Chemical Technology & Biotechnology* 85 (2): 199–203.

2 Ho, S.-H., Chen, C.-Y., Lee, D.-J., and Chang, J.-S. (2011). Perspectives on microalgal CO_2-emission mitigation systems – a review. *Biotechnology Advances* 29 (2): 189–198.

3 Al-Zuhair, S. (2007). Production of biodiesel: possibilities and challenges. *Biofuels, Bioproducts and Biorefining: Innovation for a Sustainable Economy* 1 (1): 57–66.

4 Abbaszaadeh, A., Ghobadian, B., Omidkhah, M.R., and Najafi, G. (2012). Current biodiesel production technologies: a comparative review. *Energy Conversion and Management* 63: 138–148.

5 Calero, J., Luna, D., Sancho, E.D. et al. (2015). An overview on glycerol-free processes for the production of renewable liquid biofuels, applicable in diesel engines. *Renewable and Sustainable Energy Reviews* 42: 1437–1452.

6 Gharat, N. and Rathod, V.K. (2013). Ultrasound assisted enzyme catalyzed transesterification of waste cooking oil with dimethyl carbonate. *Ultrasonics Sonochemistry* 20 (3): 900–905.

7 Guo, F., Peng, Z.-G., Dai, J.-Y., and Xiu, Z.-L. (2010). Calcined sodium silicate as solid base catalyst for biodiesel production. *Fuel Processing Technology* 91 (3): 322–328. https://doi.org/10.1016/j.fuproc.2009.11.003.

8 Luque, R., Lovett, J.C., Datta, B. et al. (2010). Biodiesel as feasible petrol fuel replacement: a multidisciplinary overview. *Energy & Environmental Science* 3 (11): 1706–1721. https://doi.org/10.1039/C0EE00085J.

9 Zhang, F., Fang, Z., and Wang, Y.-T. (2015). Biodiesel production directly from oils with high acid value by magnetic $Na_2SiO_3@Fe_3O_4/C$ catalyst and ultrasound. *Fuel* 150: 370–377. https://doi.org/10.1016/j.fuel.2015.02.032.

10 Chouhan, A.P.S. and Sarma, A.K. (2011). Modern heterogeneous catalysts for biodiesel production: a comprehensive review. *Renewable and Sustainable Energy Reviews* 15 (9): 4378–4399. https://doi.org/10.1016/j.rser.2011.07.112.

11 Ingle, A.P., Chandel, A.K., Philippini, R. et al. (2020). Advances in nanocatalysts mediated biodiesel production: a critical appraisal. *Symmetry* 12 (2): 256. https://doi.org/10.3390/sym12020256.

12 Narasimhan, M., Chandrasekaran, M., Govindasamy, S., and Aravamudhan, A. (2021). Heterogeneous nanocatalysts for sustainable biodiesel production: a review. *Journal of Environmental Chemical Engineering* 9 (1): 104876. https://doi.org/10.1016/j.jece.2020.104876.

13 de Vasconcellos, A., Miller, A.H., Aranda, D.A.G., and Nery, J.G. (2018). Biocatalysts based on nanozeolite-enzyme complexes: effects of alkoxysilane surface functionalization and biofuel production using microalgae lipids feedstock. *Colloids and Surfaces B: Biointerfaces* 165: 150–157. https://doi.org/10.1016/j.colsurfb.2018.02.029.

14 Kim, K.H., Lee, O.K., and Lee, E.Y. (2018). Nano-immobilized biocatalysts for biodiesel production from renewable and sustainable resources. *Catalysts* 8 (2): 68.

15 Al-Ani, A., Darton, R.J., Sneddon, S., and Zholobenko, V. (2018). Nanostructured zeolites: the introduction of intracrystalline mesoporosity in basic Faujasite-type catalysts. *ACS Applied Nano Materials* 1 (1): 310–318. https://doi.org/10.1021/acsanm.7b00169.

16 Al-Ani, A., Mordvinova, N.E., Lebedev, O.I. et al. (2019). Ion-exchanged zeolite P as a nanostructured catalyst for biodiesel production. *Energy Reports* 5: 357–363. https://doi.org/10.1016/j.egyr.2019.03.003.

17 Faria, E.A., Marques, J.S., Dias, I.M. et al. (2009). Nanosized and reusable SiO_2/ZrO_2 catalyst for highly efficient biodiesel production by soybean transesterification. *Journal of the Brazilian Chemical Society* 20 (9): 1732–1737.

18 Qiu, F., Li, Y., Yang, D. et al. (2011). Heterogeneous solid base nanocatalyst: preparation, characterization and application in biodiesel production. *Bioresource Technology* 102 (5): 4150–4156. https://doi.org/10.1016/j.biortech.2010.12.071.

19 Gardy, J., Hassanpour, A., Lai, X., and Ahmed, M.H. (2016). Synthesis of $Ti(SO_4)O$ solid acid nano-catalyst and its application for biodiesel production from used cooking oil. *Applied Catalysis A: General* 527: 81–95. https://doi.org/10.1016/j.apcata.2016.08.031.

20 Gardy, J., Hassanpour, A., Lai, X. et al. (2017). Biodiesel production from used cooking oil using a novel surface functionalised TiO_2 nano-catalyst. *Applied Catalysis B: Environmental* 207: 297–310. https://doi.org/10.1016/j.apcatb.2017.01.080.

21 Ambursa, M.M., Ali, T.H., Lee, H.V. et al. (2016). Hydrodeoxygenation of dibenzofuran to bicyclic hydrocarbons using bimetallic Cu–Ni catalysts supported on metal oxides. *Fuel* 180: 767–776. https://doi.org/10.1016/j.fuel.2016.04.045.

22 Ambursa, M.M., Sudarsanam, P., Voon, L.H. et al. (2017). Bimetallic Cu-Ni catalysts supported on MCM-41 and Ti-MCM-41 porous materials for hydrodeoxygenation of lignin model compound into transportation fuels. *Fuel Processing Technology* 162: 87–97. https://doi.org/10.1016/j.fuproc.2017.03.008.

23 Baskar, G., Gurugulladevi, A., Nishanthini, T. et al. (2017). Optimization and kinetics of biodiesel production from Mahua oil using manganese doped zinc oxide nanocatalyst. *Renewable Energy* 103: 641–646. https://doi.org/10.1016/j.renene.2016.10.077.

24 Wang, H., Covarrubias, J., Prock, H. et al. (2015). Acid-functionalized magnetic nanoparticle as heterogeneous catalysts for biodiesel synthesis. *The Journal of Physical Chemistry C* 119 (46): 26020–26028. https://doi.org/10.1021/acs.jpcc.5b08743.

25 Alves, M.B., Medeiros, F., Sousa, M.H. et al. (2014). Cadmium and tin magnetic nanocatalysts useful for biodiesel production. *Journal of the Brazilian Chemical Society* 25 (12): 2304–2313.

26 Kumar, D., Sharma, S., Srivastava, N. et al. (2018). Advancement in the utilization of nanocatalyst for transesterification of triglycerides. *Journal of Nanoscience and Technology* 4 (3): 374–379.

27 Venkat Reddy, C.R., Oshel, R., and Verkade, J.G. (2006). Room-temperature conversion of soybean oil and poultry fat to biodiesel catalyzed by nanocrystalline calcium oxides. *Energy & Fuels* 20 (3): 1310–1314. https://doi.org/10.1021/ef050435d.

28 Yoosuk, B., Udomsap, P., Puttasawat, B., and Krasae, P. (2010). Modification of calcite by hydration–dehydration method for heterogeneous biodiesel production process: the effects of water on properties and activity. *Chemical Engineering Journal* 162 (1): 135–141. https://doi.org/10.1016/j.cej.2010.05.013.

29 Wen, L., Wang, Y., Lu, D. et al. (2010). Preparation of KF/CaO nanocatalyst and its application in biodiesel production from Chinese tallow seed oil. *Fuel* 89 (9): 2267–2271. https://doi.org/10.1016/j.fuel.2010.01.028.

30 Hu, S., Wen, L., Wang, Y. et al. (2012). Gas–liquid countercurrent integration process for continuous biodiesel production using a microporous solid base KF/CaO as catalyst. *Bioresource Technology* 123: 413–418. https://doi.org/10.1016/j.biortech.2012.05.143.

31 Sharma, Y.C., Singh, B., and Korstad, J. (2011). Latest developments on application of heterogeneous basic catalysts for an efficient and eco friendly synthesis of biodiesel: a review. *Fuel* 90 (4): 1309–1324. https://doi.org/10.1016/j.fuel.2010.10.015.

32 Zu, Y., Tang, J., Zhu, W. et al. (2011). Graphite oxide-supported CaO catalysts for transesterification of soybean oil with methanol. *Bioresource Technology* 102 (19): 8939–8944. https://doi.org/10.1016/j.biortech.2011.07.032.

33 Tahvildari, K., Anaraki, Y.N., Fazaeli, R. et al. (2015a). The study of CaO and MgO heterogenic nano-catalyst coupling on transesterification reaction efficacy in the production of biodiesel from recycled cooking oil. *Journal of Environmental Health Science and Engineering* 13 (1): 1–9. https://doi.org/10.1186/s40201-015-0226-7.

34 Chelladurai, K. and Rajamanickam, M. (2014). Environmentally benign neem biodiesel synthesis using nano-Zn-Mg-Al hydrotalcite as solid base catalysts. *Journal of Catalysts* 2014: 326575. https://doi.org/10.1155/2014/326575.

35 Shekoohi, K., Hosseini, F.S., Haghighi, A.H., and Sahrayian, A. (2017). Synthesis of some Mg/Co-Al type nano hydrotalcites and characterization. *MethodsX* 4: 86–94. https://doi.org/10.1016/j.mex.2017.01.003.

36 Deng, X., Fang, Z., Liu, Y.-h., and Yu, C.-L. (2011). Production of biodiesel from Jatropha oil catalyzed by nanosized solid basic catalyst. *Energy* 36 (2): 777–784. https://doi.org/10.1016/j.energy.2010.12.043.

37 Soares Dias, A.P., Bernardo, J., Felizardo, P., and Neiva Correia, M.J. (2012). Biodiesel production over thermal activated cerium modified Mg-Al hydrotalcites. *Energy* 41 (1): 344–353. https://doi.org/10.1016/j.energy.2012.03.005.

38 Obadiah, A., Kannan, R., Ravichandran, P. et al. (2012). Nano hydrotalcite as a novel catalyst for biodiesel conversion. *Digest Journal of Nanomaterials & Biostructures (DJNB)* 7 (1): 321–327.

39 Manivannan, R. and Karthikeyan, C. (2013). Synthesis of biodiesel from neem oil using Mg-Al nano hydrotalcite. *Advanced Materials Research* 678: 268–272.

40 Prabu, M., Manikandan, M., Kandasamy, P. et al. (2019). Synthesis of biodiesel using the Mg/Al/Zn hydrotalcite/SBA-15 nanocomposite catalyst. *ACS Omega* 4 (2): 3500–3507. https://doi.org/10.1021/acsomega.8b02547.

41 Thangaraj, B., Solomon, P.R., Muniyandi, B. et al. (2019). Catalysis in biodiesel production – a review. *Clean Energy* 3 (1): 2–23.

42 Di Serio, M., Cozzolino, M., Giordano, M. et al. (2007). From homogeneous to heterogeneous catalysts in biodiesel production. *Industrial & Engineering Chemistry Research* 46 (20): 6379–6384. https://doi.org/10.1021/ie070663q.

43 Gutiérrez-Ortega, N., Ramos-Ramírez, E., Serafín-Muñoz, A. et al. (2019). Use of Co/Fe-mixed oxides as heterogeneous catalysts in obtaining biodiesel. *Catalysts* 9 (5): 403. https://doi.org/10.3390/catal9050403.

44 Hu, S., Guan, Y., Wang, Y., and Han, H. (2011). Nano-magnetic catalyst KF/CaO–Fe_3O_4 for biodiesel production. *Applied Energy* 88 (8): 2685–2690. https://doi.org/10.1016/j.apenergy.2011.02.012.

45 Feyzi, M. and Norouzi, L. (2016). Preparation and kinetic study of magnetic Ca/Fe_3O_4@SiO_2 nanocatalysts for biodiesel production. *Renewable Energy* 94: 579–586. https://doi.org/10.1016/j.renene.2016.03.086.

46 Alaei, S., Haghighi, M., Toghiani, J., and Rahmani Vahid, B. (2018). Magnetic and reusable MgO/$MgFe_2O_4$ nanocatalyst for biodiesel production from sunflower oil: Influence of fuel ratio in combustion synthesis on catalytic properties and performance. *Industrial Crops and Products* 117: 322–332. https://doi.org/10.1016/j.indcrop.2018.03.015.

47 Salimi, Z. and Hosseini, S.A. (2019). Study and optimization of conditions of biodiesel production from edible oils using ZnO/$BiFeO_3$ nano magnetic catalyst. *Fuel* 239: 1204–1212. https://doi.org/10.1016/j.fuel.2018.11.125.

48 Ghalandari, A., Taghizadeh, M., and Rahmani, M. (2019). Statistical optimization of the biodiesel production process using a magnetic core-mesoporous shell KOH/Fe_3O_4@γ-Al_2O_3 nanocatalyst. *Chemical Engineering & Technology* 42 (1): 89–99. https://doi.org/10.1002/ceat.201700658.

49 Zillillah, Z.L., Ngu, T.A., and Li, Z. (2014). Phosphotungstic acid-functionalized magnetic nanoparticles as an efficient and recyclable catalyst for the one-pot production of biodiesel from grease via esterification and transesterification. *Green Chemistry* 16 (3): 1202–1210. https://doi.org/10.1039/C3GC41379A.

50 Erdem, S., Erdem, B., and Öksüzoğlu, R.M. (2018). Magnetic nano-sized solid acid catalyst bearing sulfonic acid groups for biodiesel synthesis. *Open Chemistry* 16 (1): 923–929.

51 Lu, W., Zu, M., Byun, J.H. et al. (2012). State of the art of carbon nanotube fibers: opportunities and challenges. *Advanced Materials* 24 (14): 1805–1833.

52 Chen, C.L., Wang, X.K., and Nagatsu, M. (2009). Europium adsorption on multiwall carbon nanotube/iron oxide magnetic composite in the presence of polyacrylic acid. *Environmental Science & Technology* 43 (7): 2362–2367.

53 Ji, P., Tan, H., Xu, X., and Feng, W. (2010). Lipase covalently attached to multiwalled carbon nanotubes as an efficient catalyst in organic solvent. *AIChE Journal* 56 (11): 3005–3011.

54 Ke, C., Li, X., Huang, S. et al. (2014). Enhancing enzyme activity and enantioselectivity of *Burkholderia cepacia* lipase via immobilization on modified multi-walled carbon nanotubes. *RSC Advances* 4 (101): 57810–57818.

55 Chen, Y.Z., Ching, C.B., and Xu, R. (2009). Lipase immobilization on modified zirconia nanoparticles: studies on the effects of modifiers. *Process Biochemistry* 44 (11): 1245–1251. https://doi.org/10.1016/j.procbio.2009.06.023.

56 Chronopoulou, L., Kamel, G., Sparago, C. et al. (2011). Structure–activity relationships of Candida rugosalipase immobilized on polylactic acid nanoparticles. *Soft Matter* 7 (6): 2653–2662. https://doi.org/10.1039/C0SM00712A.

57 Kim, M.I., Ham, H.O., Oh, S.-D. et al. (2006). Immobilization of *Mucor javanicus* lipase on effectively functionalized silica nanoparticles. *Journal of Molecular Catalysis B: Enzymatic* 39 (1): 62–68. https://doi.org/10.1016/j.molcatb.2006.01.028.

58 Miletić, N., Abetz, V., Ebert, K., and Loos, K. (2010). Immobilization of Candida antarctica lipase B on polystyrene nanoparticles. *Macromolecular Rapid Communications* 31 (1): 71–74. https://doi.org/10.1002/marc.200900497.

59 Wu, Y., Wang, Y., Luo, G., and Dai, Y. (2010). Effect of solvents and precipitant on the properties of chitosan nanoparticles in a water-in-oil microemulsion and its lipase immobilization performance. *Bioresource Technology* 101 (3): 841–844. https://doi.org/10.1016/j.biortech.2009.08.103.

60 Chen, Y.Z., Yang, C.T., Ching, C.B., and Xu, R. (2008). Immobilization of lipases on hydrophobilized zirconia nanoparticles: highly enantioselective and reusable biocatalysts. *Langmuir* 24 (16): 8877–8884. https://doi.org/10.1021/la801384c.

61 Bencze, L.C., Bartha-Vári, J.H., Katona, G. et al. (2016). Nanobioconjugates of Candida antarctica lipase B and single-walled carbon nanotubes in biodiesel production. *Bioresource Technology* 200: 853–860. https://doi.org/10.1016/j.biortech.2015.10.072.

62 Lee, S.H., Doan, T.T.N., Won, K. et al. (2010). Immobilization of lipase within carbon nanotube–silica composites for non-aqueous reaction systems. *Journal of Molecular Catalysis B: Enzymatic* 62 (2): 169–172.

63 Fan, Y., Wu, G., Su, F. et al. (2016). Lipase oriented-immobilized on dendrimer-coated magnetic multi-walled carbon nanotubes toward catalyzing biodiesel production from waste vegetable oil. *Fuel* 178: 172–178. https://doi.org/10.1016/j.fuel.2016.03.071.

64 Pavlidis, I.V., Tsoufis, T., Enotiadis, A. et al. (2010a). Functionalized multi-wall carbon nanotubes for lipase immobilization. *Advanced Engineering Materials* 12 (5): B179–B183. https://doi.org/10.1002/adem.200980021.

65 Fan, Y., Su, F., Li, K. et al. (2017). Carbon nanotube filled with magnetic iron oxide and modified with polyamidoamine dendrimers for immobilizing lipase toward application in biodiesel production. *Scientific Reports* 7: 45643–45643. https://doi.org/10.1038/srep45643.

66 Shah, S., Solanki, K., and Gupta, M.N. (2007). Enhancement of lipase activity in non-aqueous media upon immobilization on multi-walled carbon nanotubes. *Chemistry Central Journal* 1 (1): 30. https://doi.org/10.1186/1752-153X-1-30.

67 Lee, H.-K., Lee, J.-K., Kim, M.-J., and Lee, C.-J. (2010). Immobilization of lipase on single walled carbon nanotubes in ionic liquid. *Bulletin of the Korean Chemical Society* 31 (3): 650–652.

68 Yang, N., Chen, X., Ren, T. et al. (2015). Carbon nanotube based biosensors. *Sensors and Actuators B: Chemical* 207: 690–715. https://doi.org/10.1016/j.snb.2014.10.040.

69 Mohamad, N., Buang, N.A., Mahat, N.A. et al. (2015). Simple adsorption of Candida rugosa lipase onto multi-walled carbon nanotubes for sustainable production of the flavor ester geranyl propionate. *Journal of Industrial and Engineering Chemistry* 32: 99–108. https://doi.org/10.1016/j.jiec.2015.08.001.

70 Li, R., Liu, J., Shi, A. et al. (2019). A facile method to modify polypropylene membrane by polydopamine coating via inkjet printing technique for superior performance. *Journal of Colloid and Interface Science* 552: 719–727. https://doi.org/10.1016/j.jcis.2019.05.108.

71 Gurunathan, B. and Ravi, A. (2015). Process optimization and kinetics of biodiesel production from neem oil using copper doped zinc oxide heterogeneous nanocatalyst. *Bioresource Technology* 190: 424–428.

72 Kaur, M. and Ali, A. (2011). Lithium ion impregnated calcium oxide as nano catalyst for the biodiesel production from Karanja and Jatropha oils. *Renewable Energy* 36 (11): 2866–2871.

73 Li, Y., Qiu, F., Yang, D. et al. (2011). Preparation, characterization and application of heterogeneous solid base catalyst for biodiesel production from soybean oil. *Biomass and Bioenergy* 35 (7): 2787–2795.

74 Rafiee, H.R., Feyzi, M., Jafari, F., and Safari, B. (2013). Preparation and characterization of promoted Fe-V/SiO_2 nanocatalysts for oxidation of alcohols. *Journal of Chemistry* 2013: 412308.

75 Teo, S.H., Islam, A., and Taufiq-Yap, Y.H. (2016). Algae derived biodiesel using nanocatalytic transesterification process. *Chemical Engineering Research and Design* 111: 362–370.

76 Thangaraj, B. and Piraman, S. (2016). Heteropoly acid coated ZnO nanocatalyst for *Madhuca indica* biodiesel synthesis. *Biofuels* 7 (1): 13–20.

77 Baskar, G. and Aiswarya, R. (2016). Trends in catalytic production of biodiesel from various feedstocks. *Renewable and Sustainable Energy Reviews* 57: 496–504.

78 Prescott, W.V. and Schwartz, A.I. (2008). *Nanorods, Nanotubes, and Nanomaterials Research Progress*. Nova Publishers.

79 Teo, S.H., Islam, A., Ng, F.L., and Taufiq-Yap, Y.H. (2015). Biodiesel synthesis from photoautotrophic cultivated oleoginous microalgae using a sand dollar catalyst. *RSC Advances* 5 (58): 47140–47152.

4

Hybrid Electrocatalysts with Oxide/Oxide and Oxide/Hydroxide Interfaces for Oxygen Electrode Reactions

*Suranjana Patowary, Bhugendra Chutia, Kumar Kashyap Hazarika, and Pankaj Bharali**

Department of Chemical Sciences, Tezpur University, Nappam, Assam, India

4.1 General Introduction

4.1.1 A Brief Overview of Hydrogen Cars, Fuel Cells, and Oxygen Electrochemistry

We all know that with the increase in the global human population, the energy demand increases manifold. This demand shall keep on rising with time as industrialization has taken over the earth. The mother sources of this energy production are, currently, fossil fuels. But the burning of fossil fuels emits a lot of hazardous gases like carbon monoxide (CO) and carbon dioxide (CO_2). These gases are responsible for all kinds of environmental issues like global warming, drastic climate changes, and melting of icebergs and glaciers, to name a few. Also, there is a major concern that there might not be enough fossil fuel left in the earth-crust for the future. A major portion of these greenhouse gases comes from automobile emissions. In this regard, the concept of "hydrogen car" is a greener technology choice [1, 2] that takes up hydrogen (H_2) from a prefilled H_2 cylinder and oxygen (O_2) directly from the air. Both the gases combine within a fuel cell (FC) setup to produce H_2O as the only by-product with the generation of a large amount of electrical energy sufficient to run the vehicle. The production cost of H_2 gas is cheaper than gasoline per unit energy, and such vehicles can exhibit higher mileage at a more affordable rate. And with H_2 generated from biomass, the cost would be even cheaper; this, in turn, benefits the farmers and agro-based sectors too. With unitized regenerative FCs, there shall be no requirement for frequent battery replacement as in conventional cars, thereby demanding low maintenance. Hence, with a view to switching toward sustainable energy domain, switching from conventional petrol/diesel vehicles to greener H_2 cars can bring upon a revolutionary change. However, such automobiles are currently manufactured on an experimental basis only and have not yet achieved proper commercialization. Despite that, in 1960, NASA's

*Stands for the corresponding author

Heterogeneous Nanocatalysis for Energy and Environmental Sustainability: Energy Applications, First Edition. Edited by Putla Sudarsanam, Yusuke Yamauchi, and Pankaj Bharali.

Apollo journey to the moon deployed an H_2-based FC powerhouse to its onboard electrical circuit successfully. The major bottleneck for practical application of H_2 automobiles is that the cathode reaction "oxygen reduction reaction" (ORR) in the FC setup occurs at a very sluggish rate than the anode reaction "hydrogen oxidation reaction" (HOR) [2]. The basic anatomy of an FC is shown in Figure 4.1.

In 1839, the first successful operation of traditional FC was carried out by Sir William Grove [4] using Pt electrode as the EC. ORR is generally associated with its reversible reaction, oxygen evolution reaction, abbreviated as OER, which is a half-reaction of water splitting and finds application in rechargeable metal-air batteries [5]. This reaction is of major importance because green plants perform water splitting during photosynthesis, giving out O_2 via OER [6]. These two reactions, viz. ORR and OER, are the main O_2 reactions involved in most energy-efficient devices like supercapacitors, FCs, metal-air batteries, etc.

4.1.2 Reaction Kinetics and Mechanism of ORR and OER

The fundamental reaction mechanism for ORR and OER is somewhat ambiguous. It is difficult to find a clear explanation of the elementary steps involved in these reactions. Unlike the anodic HOR that involves only one intermediate (H*), the O_2 electrode reactions involve multiple intermediates, viz. O_2*, O*, OH*, and OOH*. Although ORR and OER are oppositely directed reactions, they do not follow proper microscopic reversibility and require different experimental platforms to occur. In general, ORR occurs at pure metal surfaces while OER needs metal oxides (MOs). Figure 4.2 shows a volcano plot of the electrochemical activities of different metals toward ORR at various applied potentials.

From Figure 4.2, it is seen that noble metals like Pt, Pd, and Ir race toward the peak of the volcano with Pt sitting almost at the summit. For OER, on the other hand, noble MOs such

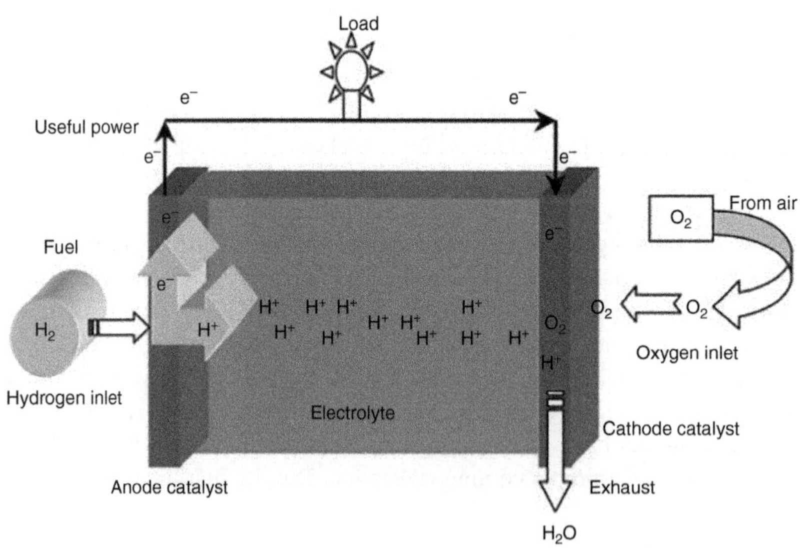

Figure 4.1 A typical hydrogen-based fuel cell (FC). *Source:* From [3] / with permission of Elsevier.

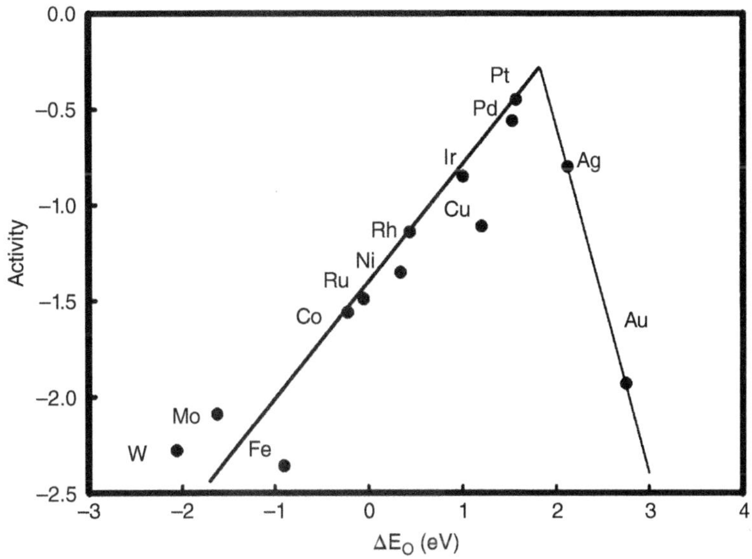

Figure 4.2 Volcano plot for ORR. *Source:* From [7] / with permission of American Chemical Society.

as IrO_2 [8], RuO_2 [8], $IrO_x/SrIrO_3$ [9], PtO_2 [10], and various lanthanum-based perovskites of the type $LaMO_3$ [11, 12] as well as a few non-noble transition MOs like CoO_x, $NiCoO_x$, $CoFeO_x$ [13], FeMoW [14], and NiO_x [15] act as catalyst materials.

The kinetics of ORR is regulated by binding energies of the attaching/detaching mode of the adsorbed O_2. A facile reaction is possible only when the energies of adsorption/desorption on the catalyst surface are optimum. A surface that binds O_2 too tightly or too loosely is likely to exhibit slow kinetics. The overpotential (generally not less than 0.3 V) [16] that limits facile ORR presumably arises from the fact that the binding energies for all the intermediates involved (O_2*, O*, OH*, and OOH*) fall in a narrow range and are difficult to deconvolute. This, in turn, makes probing into the reaction mechanism difficult.

The O_2 molecule can hold on to the catalyst surface by either bidentate mode or monodentate mode. Obviously, bidentate O_2 binding makes the reduction process faster as more units of O_2 are reduced, with a lesser number of bonds to be broken. Again, since it is a complex proton-coupled electron-transfer reaction, the number of electrons transferred plays a significant role in influencing its rate. The overall ORR in acidic and alkaline media can be represented as shown in Table 4.1.

Table 4.1 ORR mechanism in both acidic and alkaline media.

Pathway	Acidic medium	Alkaline medium
$4e^-$	$O_2 + 4H^+ + 4e^- \rightarrow 2H_2O$	$O_2 + 2H_2O + 4e^- \rightarrow 4OH^-$
$2e^-$	$O_2 + 2H^+ + 2e^- \rightarrow H_2O_2$ $H_2O_2 + 2H^+ + 2e^- \rightarrow 2H_2O$	$O_2 + H_2O + 2e^- \rightarrow HO_2^- + OH^-$ $H_2O + HO_2^- + 2e^- \rightarrow 3OH^-$

ORR can occur in acidic as well as in basic medium. However, it is practically more feasible to carry it out under alkaline conditions because of the critical stability of non-noble metal-based ECs [17]. Note that the $2e^-$ mechanism is an indirect process that involves the formation of peroxide intermediate, HO_2^- and consecutively leading to reduced oxygen products. On the contrary, the $4e^-$ pathway does not form any such unwanted intermediate; it directly forms H_2O or OH^- ions, which is why, it is often called "direct $4e^-$ pathway" and the $2e^-$ pathway is called "indirect $2e^-$ pathway." For a material to be called a good O_2 EC, it should undoubtedly favor the direct $4e^-$ pathway. Benchmark catalysts like Pt/C favor the direct $4e^-$ pathway [18].

In the case of OER, elementary steps include adsorption of OH^-, formation of active oxygen intermediates like O* and OOH*, and subsequent desorption of O_2 molecule [19]. The simplest depiction of this reaction in alkaline medium is:

$$2H_2O \rightarrow O_2 + 4H^+ + 4e$$

It is a half-reaction of water-splitting reaction, which is considered to be one of the most important ones since the world survives on O_2. Plants perform water splitting by photosynthesis and deliver O_2 to the biosphere. The other half-reaction associated with OER in water splitting is the hydrogen evolution reaction (HER) which is also equally important since H_2 gas is a cleaner and potential substitute for conventional fossil fuel. However, OER, due to its highly sluggish kinetics (generally not less than 1.23 V overpotential) [16], has gained more importance than HER in terms of studying design principles and developing ECs.

4.1.3 Benchmark Catalysts and their Evolution Over Time

Since the invention of Pt EC by Sir William Grove, the scientific community has been tirelessly involved in designing cost-effective ECs by alloying Pt with non-noble metals and MOs as well as Pt-free ECs. Figure 4.3 gives a summary of developments and modifications of Pt-based ECs over time.

By far, Pt nanoparticles (NPs) supported on multidimensional carbon networks like nanosheets, nanotubes, and core–shell structures, broadly represented as Pt/C, have served as the benchmark catalysts required for satisfactory ORR kinetics in FCs [21–23]. However, its low natural abundance and expensive nature limit it from being the most accessible ECs.

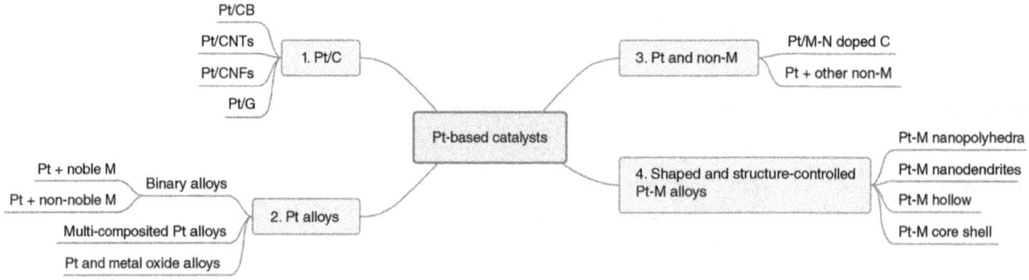

Figure 4.3 A schematic of recent developments to Pt-based ECs. *Source:* From [20] / with permission of Royal Society of Chemistry.

Moreover, there are certain intrinsic demerits of Pt-based ECs; the carbonaceous support undergoes corrosion due to electro-oxidation, leaving the Pt-NPs agglomerated [17]. Hence, all Pt-based catalysts are undoubtedly the best ECs for O_2 electrode reactions, but only under the best operating conditions. In order to bring down the dependency on such precious metals, designing other budget-friendly and competent alternatives became a hotspot of research for electrochemists.

For the past few decades, rational engineering of interfaces is regarded as an efficient strategy to fine-tune the electronic and catalytic properties of the materials. It also helps prevent quick agglomeration, which is common to all pure-metallic ECs. Today we have numerous literature works reported on the successful synthesis of noble as well as non-noble, intermetallic ECs hybridized with oxide and/or hydroxide hetero-phases, mostly supported on a highly conductive carbonaceous bed. Metal oxyhydroxides, in the form of vacancy-modified layered double hydroxides (LDHs) are an established class of O_2 ECs [24–26]. The carbonaceous bed, such as graphitic carbon-nitride (g-C_3N_4) and reduced graphene oxide (rGO) provide a smooth platform to ensure even distribution of the catalyst [27]. This helps in improving proton-electron conductivity and reduces intrinsic Ohmic resistance. Further modifications, such as surface-etching, de-lamination, sulfurization, introducing oxygen vacancies into MO structures, add to the decrease in Ohmic resistance and increase proton-electron conductivity of the electrolyte media (acidic or basic). The details in this regard shall be discussed in later sections of the chapter. Owing to promising ORR and OER functionalization, non-noble MOs are considered potential candidates for cost-effective electrocatalysis. Specially, transition MOs of all types of crystallinity (rock-salt MO, rutile MO_2 or M_2O_3, perovskite ABO_3, and spinel AB_2O_4) are efficient O_2 ECs. Grimaud and co-workers experimented upon various MOs for OER by in-situ ^{18}O isotope labeling mass spectrometry and found that the O_2 generated comes from lattice-oxygen oxidation [28]. There are reports of electronic perturbation (hot electron dynamics) at inverse metal/oxide interface sites that promote catalytic activity [29]. Also, metal/oxide/oxyhydroxide heterogeneous and inverse interface sites have reports of being outstanding catalytic performance in running important reactions like water–gas shift reactions [30], CO oxidation [31, 32], and water splitting [33]. A prompt and effective way of enhancing the electrocatalytic activity is by hybridizing its structural network by adding dopants or creating interfaces by forming oxygen vacancies and interstitial defects. Modification or hybridization at the molecular level brings about electronic changes to the architecture of the catalyst material. For example, rock-salt oxides like MnO, CoO, and NiO have fairly good intrinsic OER tendency. However, when these are intermixed to form MnO-CoO having oxide/oxide heterophases, the intrinsic activity is further enhanced [34].

4.2 Role of Interfaces in Nano-Electrocatalysis

At this point of discussion, the role of the interface (or the concept of surface hybridization) demands sincere attention. The presence of interface and surface boundaries seem to directly influence the material properties of an EC. To begin with the simplest instances, single perovskites like $LaFeO_3$, $LaCoO_3$, and $CaMnO_3$ that are an established class of O_2 ECs are predicted to be good [35–37] because of intermetallic boundaries of A/B, A/O, and

B/O (A, B = metal, O = oxygen). Also, the crystal structure with distorted [AO_{12}] and [BO_6] conjoint units have a direct impact upon its enhanced catalytic properties [38]. Again, spinel oxides such as $NiCo_2O_4$ also show improved activity toward ORR and OER compared to their monologs Co_3O_4 and NiO [39–41]. This added electro-activity can be attributed to the development of interfaces in the nanostructure of the $NiCo_2O_4$ hybrid EC. Likewise, numerous reports in the scientific research database are found to acknowledge the importance of hybrid ECs and defects or interface engineering that has shown extraordinary soar than their intrinsic electro-activity toward O_2 redox reactions.

After understanding the immense influence of proper interfacial hybridization into a catalyst framework, the next task, i.e. to incorporate dopants like nitrogen, sulfur, phosphorus (N/S/P) or generate anion vacancies in a controllable manner, remains somewhat complicated to accomplish. Not only that, but there also is no unanimous agreement among the research fraternity around the globe as to what mechanism is followed that leads to an increased extent of the catalyst performance [42]. Hence, it is an established fact that all kinds of MOs have gained supreme importance as potential O_2 ECs due to their inherent anionic oxygen defects, which can be further elevated by doping and interface tuning. Defect-generation methods like thermal annealing under a reducing environment can be a very effective strategy as it creates oxygen vacancies maintaining the original crystallinity of the material. Kim et al. reported that the oxygen-deficient double perovskite $Ca_2Mn_2O_5$ generated by thermal annealing showed a higher OER rate compared to $CaMnO_3$. They attribute this enhanced OER to the vacancy-originated porosity for the conducting ions like OH^- [43]. Thermal annealing can sometimes effectively bring about the transformation of phases into more active secondary phases. Chen et al. reported remarkably enhanced O_2 electrocatalysis by thermally annealed pristine $Pr_{0.5}Ba_{0.5}MnO_{3-}\delta$, which is attributed to the newly transformed mix-phase catalyst comprising pristine and layered $PrBaMn_2O_{5+}\delta$ (H-PBM) [44]. Besides this, cationic doping with strontium [45], anionic doping [46, 47] with N/P/S, strain-induced oxygen vacancy can also be an effective strategy in enhancing the ORR/OER activity. In fact, the coupling of strain and anion defects has a great role to play. Techniques such as vacuum annealing impose strain into the crystal structure and shuffle the electronic configuration of the material [48]. The bandgap, which is a key factor in influencing electrocatalytic properties, gets squeezed, thereby filling up the e_g levels of the metal. Often, the electroneutrality after oxygen vacancy generation is maintained by concurrent reduction of multivalent cations. The induced defects in the crystal structure can be confirmed from various material characterization techniques such as X-ray diffraction (XRD), transmission electron microscopy (TEM), scanning transmission electron microscopy (STEM), X-ray absorption fine structure spectroscopy (EXAFS), etc., with XRD being the most chosen technique. After understanding the positive influences of oxygen vacancy concentration toward ORR/OER, a similar observation is expected from another closely related oxygen-vacant oxide/hydroxide interphasic ECs. Cai and group reported that DFT + U study suggests performance enhancement of ultrathin Co_3O_4 EC toward OER due to increased oxygen vacancy concentration and concurrent charge balance by surface reduction of Co sites from +3 to +1 state [49].

Moving one step forward, introducing H-atoms into some of the oxygen voids can be thought of as another modified version of a hybrid EC, i.e. exploring metal oxy-hydroxides as O_2 ECs. Oxide/hydroxide is also a type of oxygen vacancy interface under a reduced atmosphere. There are reports of successful vacancy engineering at the oxide/hydroxide interface developing due to the local electron potential at the interface of oxide/hydroxide [50]. Pt-alloyed bi-oxyhydroxide can enhance the anti-poisoning nature of pure Pt EC. With the development of the oxide/hydroxide interface into its molecular framework, the unwanted adsorption of CO, which previously led to CO-poisoning of the Pt catalyst, decreased. On the contrary, the OH adsorption increased, which enhanced the overall efficiency of the catalyst. In this regard, LDH has attracted tremendous attention in different electrochemical reactions. Their composition and ion exchangeability can be tuned to get better results in different applications.

4.2.1 Oxide/Oxide Interfaces

Interface modification of catalysts has become a crucial methodology to enfold the challenges of modern catalysis phenomenon for optimized chemical reactions as it plays a significant role in materials sciences, electrochemistry, chemical industries, and photochemical reactions [16, 51, 52]. In recent years, interfacial catalysis methodology has been widely emerging in the field of electrochemistry to achieve the adequate kinetics of electrode reactions. To design and develop a high-performance electrode catalyst, the interfacial mechanism has attained considerably superior characteristics to replace noble metal-based ECs. MO moiety on MO (i.e. mixed metal oxide; MMO) heterostructures is a typical class of catalytic systems, which are found to be highly active for diverse applications, particularly in energy conversion, petroleum refinement, and hydrogenation processes [53–55]. The synergy among its various elemental compositions brings about fundamental electronic changes due to which MMOs can act as low-cost and environment-friendly roles in EC-based energy conversion devices [56–58]. It is noteworthy that a potential electrode catalyst should exhibit characteristic properties such as low cost, high stability, efficiency, and longer durability along with high electrochemical performances. To accumulate all these properties in a single catalyst, many researchers have been working on oxide/oxide interface-modulated MO hybrid EC.

Based on the relative locations of species in a multicomponent hybrid catalytic system, materials can be classified into three major categories, viz. (i) core@shell structure, (ii) heterostructure, and (iii) supported structures [51]. Consequently, there exist different interfacial properties with distinct features corresponding to each model nanostructures. Additionally, the interface structure can also be modulated for an optimized reaction process by employing a characteristic synthesis approach.

As the regulation of catalyst functionalities through oxide/oxide interface modulating seems optimistic, many researchers have been working on tailoring oxide/oxide interface to design advanced electrode catalysts for ORR/OER. For instance, Goswami et al. reported hydrothermally synthesized highly rich and well-dispersed oxide/oxide interfacial CuO_x-CeO_2/C hetero-nanostructured EC for ORR/OER [59]. The HRTEM image as shown in Figure 4.4 (e,f) clearly displays the interfacial structures between different facets of CuO_x-CeO_2, CeO_2-CeO_2 and oxide–carbon. The white dotted line indicates the oxide/oxide interface between monoclinic CuO-002 planes (d spacing; 0.25 nm) and fcc CeO_2-111 planes (d spacing; 0.31 nm), while the yellow dotted line indicates the oxide/oxide interface between CeO_2 phases.

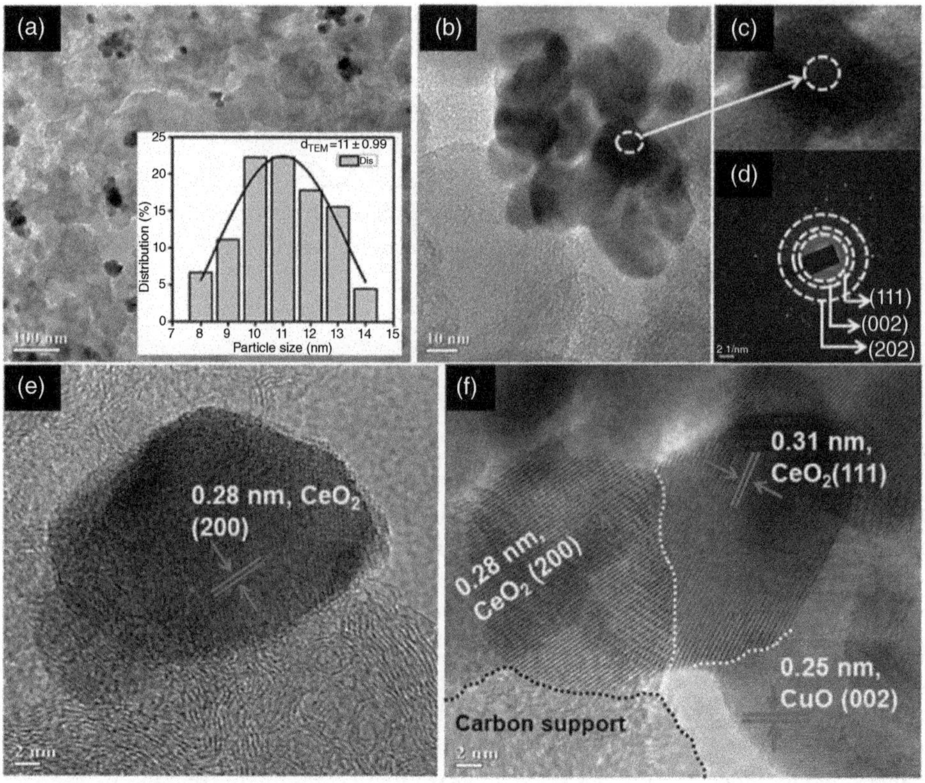

Figure 4.4 (a–c) TEM and (e, f) HRTEM images with distinguished lattice fringes, particle size distribution curve (inset image [a]), and (d) SAED pattern for the CuO_x-CeO_2/C EC. *Source:* From [59] / with permission of American Chemical Society.

The EC exhibits a direct $4e^-$ ORR pathway demonstrating higher current density ($-5.03\,mA\,cm^{-2}$) and a smaller Tafel slope (65 mV/dec) with onset potential (−0.10 V vs. Ag/AgCl), which is comparable to benchmark Pt/C. In addition, it shows better stability with a loss of 23% of its initial current after 6 hours compared to 28% as observed for Pt/C. The EC also exhibited significant OER activity in an alkaline medium with a less positive onset potential (0.20 V) and a lower Tafel slope (177 mV/dec). The strong oxide/oxide interfacial contact between CuO_x and CeO_2 and the favorable electronic perturbations at their interface is primarily reported for the observed enhanced activity. Chen et al. have synthesized a hybrid of MnO_x–CeO_2/KB (KB: Ketjen Black) by a continuous two-step process. KB was dispersed into the $Ce(NO_3)_3$ solution followed by uniform addition of CeO_2 NPs after a facile hydrothermal process, then MnO_x NPs were deposited on CeO_2/KB via sacrificial oxidation of carbon in the presence of $KMnO_4$. The hybrid was obtained by simple heating in an inert atmosphere. The resulting MnO_x–CeO_2/KB catalyst was evaluated for ORR activity. The observed electrocatalytic ORR performance of the catalyst is better than MnO_x/KB and CeO_2/KB. Thus, non-noble metal-based hybrid EC displayed comparable ORR activity and significantly higher durability to the benchmark Pt/C [60].

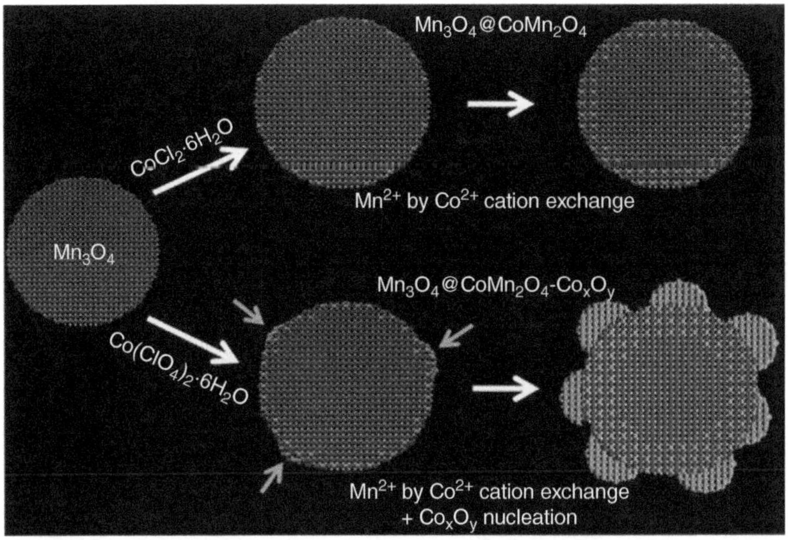

Figure 4.5 Scheme of the formation of the Mn_3O_4@$CoMn_2O_4$ and Mn_3O_4@$CoMn_2O_4-Co_xO_y$ NPs using different cobalt precursor solutions. Mn_3O_4@$CoMn_2O_4$ NPs were obtained by partial cation exchange between Mn^{2+} and Co^{2+} using a cobalt chloride precursor. Mn_3O_4@$CoMn_2O_4-Co_xO_y$ NPs were obtained by identical reaction and simultaneous nucleation of Co_xO_y using cobalt perchlorate. Green arrows in the bottom cartoon point at Co_xO_y nucleation sites. *Source:* From [61] / with permission of American Chemical Society.

Cabot and co-workers have designed an interface-modulated Mn_3O_4@$CoMn_2O_4-Co_xO_y$ (with [Co]/[Mn] =1) NPs entailing the cation exchange between Mn and Co in preformed Mn_3O_4 NPs via a one-pot two-step synthesis method as shown in Figure 4.5. The Mn_3O_4@$CoMn_2O_4-Co_xO_y$ was investigated for ORR/OER in an alkaline medium. It exhibited a lower Tafel slope of 52 mV/dec and low overpotentials of 0.31 V at −3 mA/cm² for ORR. A remarkable OER performance has also been observed with a Tafel slope of 81 mV/dec and overpotentials of 0.31 V at 10 mA/cm². The overall activity of the catalyst was observed to be superior to commercial Pt-, IrO_2-based, and previously reported TMO-based catalysts. This cation exchange-based interfacial modification strategy offers an insightful strategy for designing novel heterostructured NPs as efficient bifunctional O_2 ECs [61].

Zhang et al. have synthesized porous $NiO/NiCo_2O_4$ nanofibers containing rich oxide/oxide interfacial active sites, which exhibited superior electrocatalytic activity. They adopted a facile electro-spinning strategy to synthesize the heterostructured $NiO/NiCo_2O_4$ with precursor transition metal salts in a nonstoichiometric ratio to get a rich oxide/oxide interfacial area. The feasibility of chemical bonds in the oxide/oxide heterostructure, which may accelerate the charge transfer process, has been established via DFT calculation. XPS analysis of the interface organized reveals the existence of enormous Ni^{3+} and Co^{3+} species coupling with the oxide/oxide interface. Consequently, $NiO/NiCo_2O_4$ heterostructure exhibit outstanding and durable performances toward OER and ORR in KOH solution. Figure 4.6 displays the electrochemical activities exhibited by the catalyst. The catalyst possesses an ORR half-wave potential of 0.73 V and a lower OER overpotential of 357 mV at 10 mA/cm². Besides, testing for zinc-air battery, the $NiO/NiCo_2O_4$ shows remarkable specific capacities of 814.4 mA/h/g, and reliable cycling stability of 175 hours. This new

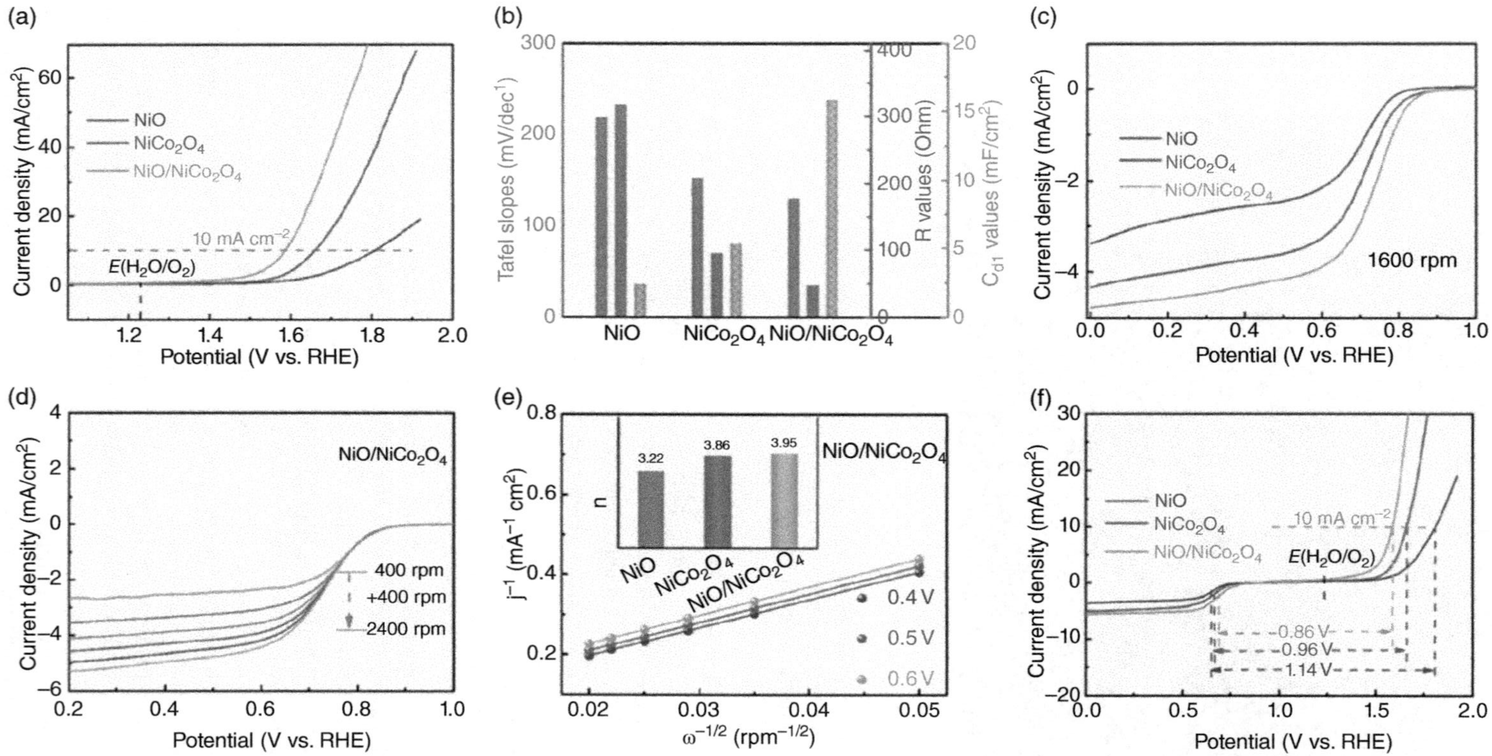

Figure 4.6 OER: (a) LSV, and (b) Tafel slopes (purple), resistance (dark cyan), and double-layer capacitances values (orange). ORR: (c) comparison of LSVs of various ECs, (d) LSV of NiO/$NiCo_2O_4$ nanofibers at different rpm, (e) corresponding K – L plots of NiO/$NiCo_2O_4$ nanofibers, inset: "n" values for ORR, and (f) bifunctional ORR/OER performance of various ECs. *Source:* From [62] / with permission of American Chemical Society.

strategy toward modifying oxide/oxide interface over MMO heterostructure could optimize them as potential EC for energy-converting devices [62].

Liu et al. incorporated Co and CeO_2 into KB through a facile two-step hydrothermal method and established $Co_3O_4-CeO_2$/KB as an advanced ORR EC for Al–air batteries. Thc investigated electrochemical performances with the catalysts are shown in Figure 4.7.

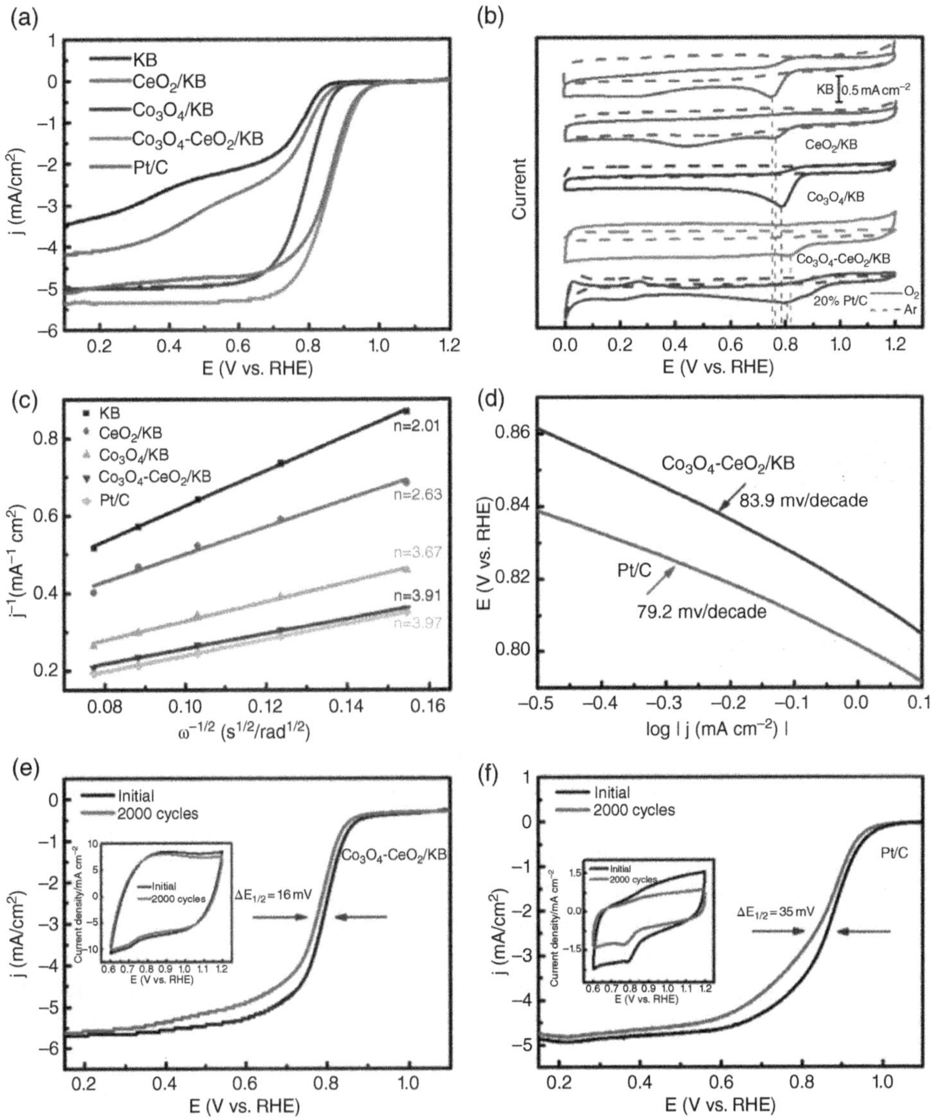

Figure 4.7 (a) ORR polarization curves for different catalysts in O_2-saturated 0.1 M KOH solution at 1600 rpm; (b) CVs of KB, CeO_2/KB, Co_3O_4/KB, $Co_3O_4-CeO_2$/KB composite, and Pt/C in Ar and O_2-saturated 0.1 M KOH solution; (c) K–L plots of different samples at 0.5 V; (d) Tafel plots of kinetic current for $Co_3O_4-CeO_2$/KB and Pt/C; the ORR activity of (e) $Co_3O_4-CeO_2$/KB and (f) Pt/C, measured before and after 2000 cycles. Insets: CVs for ORR at the $Co_3O_4-CeO_2$/KB (e) and Pt–C (f) before cycling and after 2000 cycles. *Source:* From [63] / with permission of American Chemical Society.

The observed ORR activity was significantly higher as compared to that of a single Co_3O_4/KB. Furthermore, along with the favorable direct $4e^-$ pathway, $Co_3O_4-CeO_2$/KB exhibited remarkable stability than benchmark Pt/C in alkaline electrolytes. The test for the catalyst's cathodic discharge voltage is higher than that of Co_3O_4/KB and CeO_2/KB [63].

Song and co-workers have successfully constructed nano-hybrid (minimum particle size ~7.1 nm) with unique oxide/oxide interfacial structures by using lanthanide precursors. They designed $La_2O_2CO_3$ encapsulated La_2O_3 NPs on Vulcan carbon (Vulcan XC-72), by using a chemical precipitation process in an aqueous medium with the aid of cetyltrimethylammonium bromide (CTAB), followed by calcination at 750 °C. The electrochemical activity analysis for the interface-engineered $La_2O_2CO_3$@La_2O_3/C hybrid displays a remarkably enhanced performance compared to individual $La_2O_2CO_3$, La_2O_3, and carbon in an alkaline medium. In addition to that, the stability test for the EC surpasses commercial Pt/C catalysts. The enhanced ORR kinetics is ascribed to the sandwich-like $La_2O_2CO_3$ layer at the interface of La_2O_3 and carbon that facilitates suitable surface-adsorbed −OH substitution and endorses active O_2 adsorption at the interfaces. The characteristic −C−O−C(=O)−O−La−O− bonds, constructed at the interfaces might act as active sites on the hybrid EC, thereby opening the path toward fabrication of La-based MO hybrid as a next-generation EC for ORR [64].

4.2.2 Oxide/Hydroxide Interfaces

Hybrid materials composed of MO/hydroxides have been recently suggested as highly impressive ECs toward ORR and OER [65–67]. The physicochemical properties of these MO/hydroxides can be tuned by engineering their anion chemistry. Controlling the composition and morphology of these engineered nanostructures is an effective way to improve their catalytic performance than commercial catalysts [68–70]. The interactions between these constituents can lead to excellent electronic properties. Therefore, by designing multimetallic structures, the performance of the ECs can be enhanced. The performances of these materials have been further modified by using different supports. Carbonaceous materials, viz. graphene sheets, carbon nanotubes, etc., are used to increase electrocatalytic activity. Moreover, carbon support provides a high BET surface area to obtain remarkable ORR/OER performances [71–74]. This unique interface of MO/hydroxide/support provides ample effective electron pathways through the framework.

Liu and Hu used Fe oxides and oxyhydroxides to synthesize four Fe oxide/oxyhydroxide/GO composites [72]. They have systemically investigated these materials for ORR in an alkaline medium. The synthetic procedures of these oxides are presented in Figure 4.8. These composite oxides display impressive ORR activity compared to benchmark catalysts. The α-Fe_2O_3/GO shows superior ORR activity compared to other composites and benchmark catalysts. It displays better onset potential, current density, and electron transfer number in alkaline electrolytes. The composite also unveils its excellent stability and CH_3OH tolerance ability.

Li et al. have synthesized NiFe LDH over rGO Ni foam (NiFe-LDH/RGO/NF) for OER in an alkaline medium. A two-step electrodeposition process was employed for the synthesis. The rGO modifies the NF surface. The modification results in increased interface conductivity and electrochemically active surface area. The material shows higher catalytic activity with a current density of 10 mA/cm^2 and a lower overpotential of 150 mV. The higher

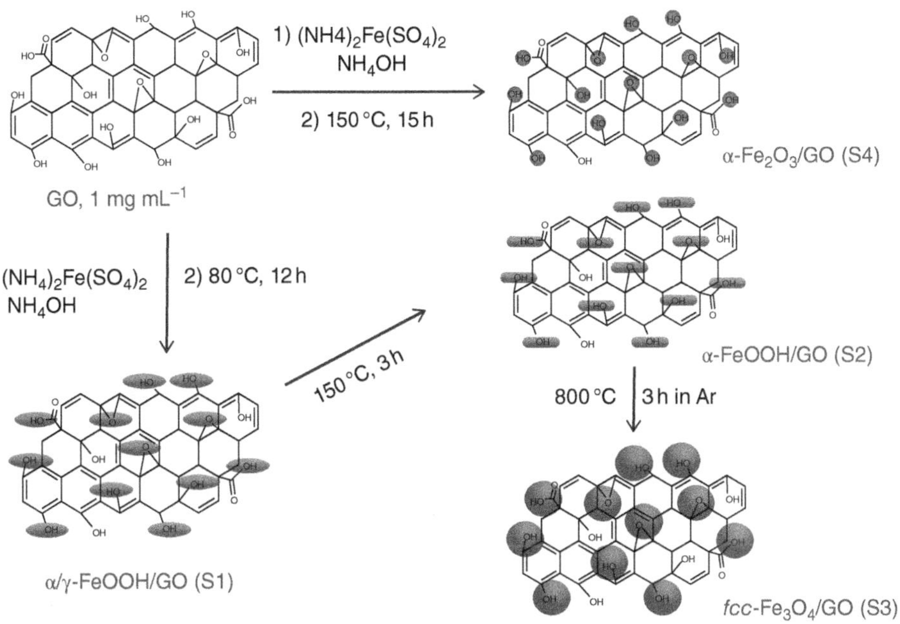

Figure 4.8 Schematic illustration of the synthesis of various Fe oxide or oxyhydroxide/GO composites. *Source:* From [72] / with permission of Royal Society of Chemistry.

activity is due to the surface-active groups of rGO and also the improved interface conductivity [75]. Cao et al. designed a flower-like three-dimensional NiFe-LDH over GO (NiFe-LDH/GO) by a one-step hydrothermal method and applied for ORR in alkaline electrolyte. The designed material has a large BET surface area, good conductivity, and more active sites [76]. It has a smaller Tafel slope of 82 mV/dec with a high onset potential. The interaction between the LDH and GO was responsible for the excellent ORR performance. Figure 4.9 shows the XPS of the composite. Prominent peaks are observed at 712.8, and 726.0 eV (Figure 4.9c), which confirms the existence of Fe(III), while sharp peaks at 856.9 eV and 874.5 eV (Figure 4.9d) imply the presence of Ni(II). The content of bivalent Ni is found to be ~3 times that of trivalent Fe, unveiling another important aspect for the formation of sheet-like structures.

Garcia et al. synthesized an Ni oxyhydroxide (NiOOH) EC by the precipitation of $Ni(OH)_2$ on polycrystalline Au by an electrochemical method for OER. The effect of various alkali cations, e.g. Na^+, K^+, Li^+, and Cs^+ with and without impurities of Fe were investigated for boosting the electrocatalytic activity of NiOOH for OER. The catalytic activity for OER was effected when impurities of Fe were present. The better OER activity is due to the interactions of the cation with the produced charged intermediates. However, computational studies were needed to confirm their proposal [77]. Jia et al. have designed a ternary CoNiMn- LDH/polypyrrole (PPy)/rGO composite by a one-step method. The electrocatalytic performance of this composite was explored for both OER and ORR in an alkaline medium. They used the rGO and PPy to increase the electron transfer rate toward the reactions. It displays excellent bifunctional electrocatalytic activity due to the synergistic effect

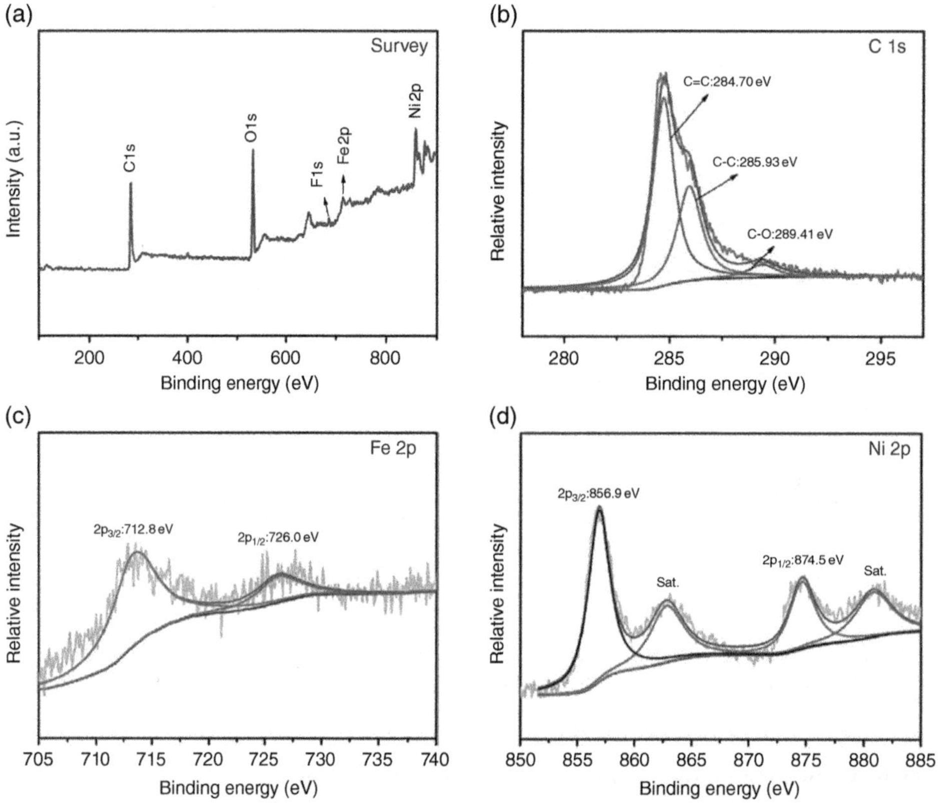

Figure 4.9 (a) XPS survey spectra and high-resolution scans of (b) C 1s, (c) Fe 2p, and (d) Ni 2p of NiFe-LDH/GO. *Source:* From [76] / with permission of Elsevier.

between different arrangements. The extent of bifunctionality was presented in terms of ΔE (difference in onset potentials for ORR and OER). The prepared composite also shows remarkable durability toward both reactions [73].

The trimetallic oxyhydroxides are a very important class of materials for ORR and OER. However, their structures and surfaces still need to be tuned to get much higher activity for these reactions. A coralloid-like trimetallic oxyhydroxide was synthesized over NF or carbon nanotubes by Huang and co-workers [78]. A series of trimetallic WCoFe-oxyhydroxides was synthesized by a wet-chemical method. The trimetallic structures were optimized based on the Fe content. The optimized catalyst ($W_{0.5}Co_{0.4}Fe_{0.1}$/NF) exhibits excellent OER activity. The OER activity of the catalyst is directly dependent on the Fe content with a lower Tafel slope of 32 mV dec^{-1}. XPS spectra reveal the presence of both Co^{2+} and Co^{3+}. The ratio of Co^{2+}/Co^{3+} was calculated from XPS spectra and it increased with Fe content added. The OER activity has been correlated to the ratio of $Co^{2+}/Co^{3+}$and the amount of Fe added. Thus, better OER activity is directly related to the Co^{2+}/Co^{3+} ratio and the eventually $W_{0.5}Co_{0.4}Fe_{0.1}$ catalyst shows the highest activity.

Not only bifunctionality but also trifunctionality is being observed with certain two-dimensional layered materials that have been studied lately. Lim et al. have studied the trifunctionality of layered transition metal for HER, OER, and ORR [79]. The electrocatalytic properties of several first-row transition-metal oxyhydroxides with the general formula MO(OH), where M represents the transition metal, e.g. CoO(OH), MnO(OH), FeO(OH), NiO(OH), and CrO(OH) for the above-mentioned reactions. Their electrochemical studies show that the Co and Ni oxyhydroxides display superior performance compared to other oxyhydroxides for OER and HER. An NiFe LDH supported over rGO was reported by Lee and co-workers by a one-pot solvothermal method and studied for electrochemical OER and photoelectrochemical H_2O splitting reactions [74]. It displays superior OER activity and stability compared to pristine-NiFe and benchmark IrO_2 catalyst. The material shows 97% Faradaic efficiency for OER. The better performance of the material could be due to the synergistic effect. In another study, Zhan et al. have reported the synthesis of an NiFe LDH supported on rGO, which tested successfully as a bifunctional catalyst for ORR and OER in an alkaline medium. It follows a $4e^-$ pathway for ORR in 0.1 M KOH (inset of Figure 4.10). The results are compared with pristine NiFe-LDH, NiFe-LDH/GO, and NiFe-LDH/rGO. The detailed ORR polarization curves are presented in Figure 4.10. From Figure 4.10a, it is clearly seen that the O_2 reduction peak appears at ~0.78 V for all three catalysts. However, more positive onset potential and higher current densities (Figure 4.10b) are exhibited by NiFe-LDH/rGO, rendering it the best among the three catalysts. Additionally, it is observed that pristine NiFe-LDH showed the least favorable output for all parameters involved. This puts into light the importance of strong and conductive carbon support and also proper vacancy modulation induced by the presence of rGO. The excellent bifunctionality was due to the synergistic effect and the porous structure of the nanohybrid, which was beneficial for fast electron transfer [80].

In 2017, the same group had reported N-doped NiFe LDH supported over rGO as an excellent bifunctional EC for ORR and OER. The solvothermal method was used to synthesize the mesoporous NiFe-LDH/NrGO nanosphere. It displays desirable performance toward both ORR and OER in an alkaline medium compared to individual NiFe-LDH and N-free NiFe-LDH/rGO. The better performance of the nanosphere could be due to the hybridization of NiFe-LDH with rGO and N-doping. The hybridization can result in improved conductivity and stability [81]. Bimetallic Fe/Co double hydroxide/oxide NPs over N-doped carbon nanotubes (FeCo-DHO/NCNTs) were reported by Sun and co-workers [82]. They optimized the synthesis procedure so that FeCo-DHO was uniformly anchored over the CNT surface. The prepared FeCo-DHO/NCNTs display significant electrochemical activity toward both ORR and OER. The ORR and OER kinetics were studied in 1.0 M KOH solution. The performance of the catalyst was much better in 1.0 M KOH electrolyte than 0.1 M KOH electrolyte. In addition to high catalytic activity, the prepared FeCo-DHO/NCNTs display very high stability. The LSV curves before and after the accelerated durability test (ADT) in 0.1 M KOH O_2-saturated solution for ORR and OER displayed high activity and stability of FeCo-DHO/NCNTs. The oxide/hydroxide interfaces play an important role in boosting the electrocatalytic performance. The composition, structure, and surfaces of these interfaces can be tuned to obtain excellent activity. The interactions between these components led to tunable electronic properties and multimetallic structure, which helps in the enhancement of the electrocatalytic performance.

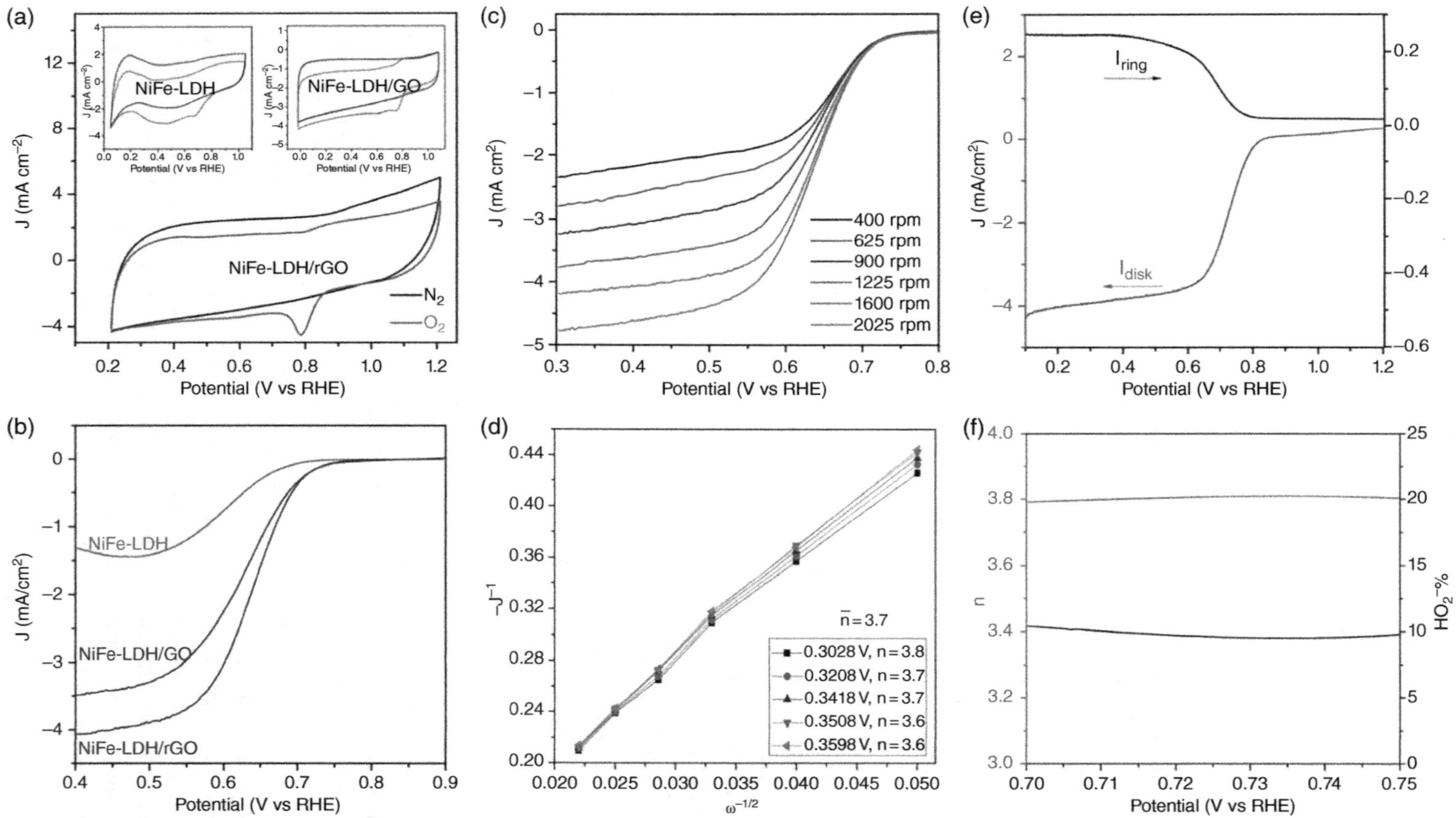

Figure 4.10 (a) CVs in N_2- and O_2-saturated and (b) LSV in O_2-saturated 0.1 M KOH for NiFe-LDH, NiFe-LDH/GO, and NiFe-LDH/rGO composites at 1600 rpm, (c) RDE voltammograms of rGO-LDH in O_2-saturated 0.1 M KOH at different rotating speed, and (d) the corresponding K-L plot of J^{-1} vs. $\omega^{-1/2}$. The inset shows the electron transfer number (n), (e) RRDE voltammograms of NiFe-LDH/rGO in O_2-saturated 0.1 M KOH. The disk potential is scanned at 10 mV/s and the ring potential is constant at 1.3 V, and (f) The "n" and percentage of HO_2^- at certain potentials based on the corresponding RRDE results. *Source:* From [80] / with permission of Elsevier.

4.3 Future Aspects and Concluding Remarks

Realizing the harmful impacts on the environment from the burning of fossil fuels and the need for an alternative sustainable energy domain, the practical utilization of FC technologies is to be given supreme priority. As of now, there is enough natural stock of fossils to sustain the earth for several more years and taking this time as a period of grace, electrochemical ORR and OER shall contribute massive support toward speeding up FC kinetics. It is still a challenging task to synthesize oxide/oxide or oxide/hydroxide biphasic hybrid ECs. It demands controlled redox conditions, especially for oxides and hydroxides to form in the near-equimolar ratio in a single framework. Hence, it is scarce to find numerous reports on such work. However, continuous efforts are made by a modification to LDHs, with controlled calcination at appropriate temperatures to generate the oxyhydroxides. As such, oxide/hydroxide can be considered a subset of LDHs. The study of LDHs serves as a pioneering guide to synthetic measures for the generation of oxyhydroxide catalysts. These, being hybrids of fully (oxidic) and partially oxidized (hydroxide), feature the wide potential for O_2 electrocatalysis, sometimes overpowering their corresponding oxide/oxide analogues. Such redox hybrids display tremendous scope for interfacial electrocatalysis. Hence, understanding their molecular structure and behavioral phenomena is of pivotal importance. Development and design synthesis of practically viable ECs with minimal external input, less hazardous starting materials, and optimum tuning of elemental compositions is expected to give an outstanding acceleration to oxygen kinetics; this is the scope of the study. In yet another perspective, oxide/hydroxide can be considered precursors to oxide/oxide catalyst, and as such, a study of their comparison analysis can be remarkable guidance in the design-synthesis of interfacial ECs at the industrial level. While there is no solid explanation to how the interfacial sites render more sensitivity and activity toward enhanced electrokinetics, there is a unanimous agreement among scientists worldwide that better synergy between nonidentical borders bring about differences in *d*-band gaps and, hence, in-situ electronic modification happens due to which a lot of behavioral properties are observed compared to their pristine non-hybrid analogues. Also, the feasibility of adsorption and desorption of O_2 molecules at anion vacancies due to the uneven redox nature of the material adds to better conductivity of electrons between various phases inside the EC. In a nutshell, hybrid and modified multi-interfacial oxide/oxide, or oxide/hydroxide, holds immense potential as revolutionary material is shifting toward sustainable energy domain, and consequently, their study is given careful note at present times.

Acknowledgments

The authors thank Tezpur University (TU) and the Department of Science and Technology (DST No: INT/RUS/RFBR/P-322), New Delhi, for generous financial support. SP acknowledges TU for institutional fellowship. BC thanks CSIR for senior research fellowship. Research and Innovation grants awarded to SP and BC by TU are gratefully acknowledged.

References

1 Edwards, P.P., Kuznetsov, V.L., David, W.I.F., and Brandon, N.P. (2008). Hydrogen and fuel cells: towards a sustainable energy future. *Energy Policy* 36 (12): 4356–4362.

2 Debe, M.K. (2012). Electrocatalyst approaches and challenges for automotive fuel cells. *Nature* 486 (7401): 43–51.

3 Boudghene Stambouli, A. and Traversa, E. (2002). Fuel cells, an alternative to standard sources of energy. *Renew. Sustain. Energy Rev.* 6 (3): 295–304.

4 Appleby, A.J. (1990). From Sir William Grove to today: fuel cells and the future. *J. Power Sources* 29 (1–2): 3–11.

5 Cao, R., Lee, J.S., Liu, M., and Cho, J. (2012). Recent progress in nonprecious catalysts for metal-air batteries. *Adv. Energy Mater.* 2 (7): 816–829.

6 Renger, G. (2011). Light induced oxidative water splitting in photosynthesis: energetics, kinetics and mechanism. *J. Photochem. Photobiol. B Biol.* 104 (1–2): 35–43.

7 Nørskov, J.K., Rossmeisl, J., Logadottir, A. et al. (2004). Origin of the overpotential for oxygen reduction at a fuel-cell cathode. *J. Phys. Chem. B* 108 (46): 17886–17892.

8 Stoerzinger, K.A., Qiao, L., Biegalski, M.D., and Shao-Horn, Y. (2014). Origin of the overpotential for oxygen reduction at a fuel-cell cathode. *J. Phys. Chem. Lett.* 5 (10): 1636–1641.

9 Seitz, L.C., Dickens, C.F., Nishio, K. et al. (2016). A highly active and stable $IrO_x/SrIrO_3$ catalyst for the oxygen evolution reaction. *Science* 353 (6303): 1011–1014.

10 Iwakura, C. and Fukuda, K. (1976). The anodic evolution of oxygen on platinum oxide electrode in alkaline solutions. *Electrochim. Acta* 21 (7): 501–508.

11 Suntivich, J., May, K.J., Gasteiger, H.A. et al. (2011). A perovskite oxide optimized for oxygen evolution catalysis from molecular orbital principles. *Science* 334 (6061): 1383–1385.

12 Lee, Y.L., Gadre, M.J., Shao-Horn, Y., and Morgan, D. (2015). Ab initio GGA+ U study of oxygen evolution and oxygen reduction electrocatalysis on the (001) surfaces of lanthanum transition metal perovskites $LaBO_3$ (B= Cr, Mn, Fe, Co and Ni). *Phys. Chem. Chem. Phys.* 17 (33): 21643–21663.

13 Bajdich, M., García-Mota, M., Vojvodic, A. et al. (2013). Theoretical investigation of the activity of cobalt oxides for the electrochemical oxidation of water. *J. Am. Chem. Soc.* 135 (36): 13521–13530.

14 Zhang, B., Zheng, X., Voznyy, O. et al. (2016). Homogeneously dispersed multimetal oxygen-evolving catalysts. *Science* 352 (6283): 333–337.

15 McCrory, C.C.L., Jung, S., Peters, J.C., and Jaramillo, T.F. (2013). Benchmarking heterogeneous electrocatalysts for the oxygen evolution reaction. *J. Am. Chem. Soc.* 135 (45): 16977–16987.

16 Yang, Y., Luo, M., Zhang, W. et al. (2018). Metal surface and interface energy electrocatalysis: fundamentals, performance engineering, and opportunities. *Chem* 4 (9): 2054–2083.

17 Wang, H., Wei, L., Yang, C. et al. (2020). A pyridine-Fe gel with an ultralow-loading Pt derivative as ORR catalyst in microbial fuel cells with long-term stability and high output voltage. *Bioelectrochemistry* 131 (107): 370.

18 Huang, Z.F., Wang, J., Peng, Y. et al. (2017). Design of efficient bifunctional oxygen reduction/evolution electrocatalyst: recent advances and perspectives. *Adv. Energy Mater.* 7 (23): 1700544.

19 Ping, Y., Nielsen, R.J., and Goddard, W.A. (2017). The reaction mechanism with free energy barriers at constant potentials for the oxygen evolution reaction at the IrO_2 (110) surface. *J. Am. Chem. Soc.* 139 (1): 149–155.

20 Ren, X., Lv, Q., Liu, L. et al. (2019). Current progress of Pt and Pt-based electrocatalysts used for fuel cells. *Sustain. Energy Fuels* 4 (1): 15–30.

21 Li, M., Zhao, Z., Cheng, T. et al. (2016). Ultrafine jagged platinum nanowires enable ultrahigh mass activity for the oxygen reduction reaction. *Science* 354 (6318): 1414–1419.

22 Gasteiger, H.A., Kocha, S.S., Sompalli, B., and Wagner, F.T. (2005). Activity benchmarks and requirements for Pt, Pt-alloy, and non-Pt oxygen reduction catalysts for PEMFCs. *Appl. Catal. B Environ.* 56 (1-2): 9–35.

23 Kou, R., Shao, Y., Wang, D. et al. (2009). Enhanced activity and stability of Pt catalysts on functionalized graphene sheets for electrocatalytic oxygen reduction. *Electrochem. Commun.* 11 (5): 954–957.

24 Wang, Y., Yan, D., El Hankari, S. et al. (2018). Recent progress on layered double hydroxides and their derivatives for electrocatalytic water splitting. *Adv. Sci.* 5 (8): 1800064.

25 Xiang, K., Guo, J., Xu, J. et al. (2018). Surface sulfurization of NiCo-layered double hydroxide nanosheets enable superior and durable oxygen evolution electrocatalysis. *Energy Mater.* 1 (8): 4040–4049.

26 Wang, Y., Zhang, Y., Liu, Z. et al. (2017). Layered double hydroxide nanosheets with multiple vacancies obtained by dry exfoliation as highly efficient oxygen evolution electrocatalysts. *Angew. Chemie - Int. Ed.* 56 (21): 5867–5871.

27 Duan, X., Xu, J., Wei, Z. et al. (2017). Metal-free carbon materials for CO_2 electrochemical reduction. *Adv. Mater.* 29 (41): 1701784.

28 Grimaud, A., Diaz-morales, O., Han, B. et al. (2017). Activating lattice oxygen redox reactions in metal oxides to catalyse oxygen evolution. *Nat. Chem.* 9 (5): 457–465.

29 Lee, H., Yoon, S., Jo, J. et al. (2019). Enhanced hot electron generation by inverse metal–oxide interfaces on catalytic nanodiode. *Faraday Discuss.* 214: 353–364.

30 Rodriguez, J.A., Ma, S., Liu, P. et al. (2007). Activity of CeO_x and TiO_x nanoparticles grown on Au (111) in the water-gas shift reaction. *Science* 318 (5857): 1757–1760.

31 Fihri, A., Sougrat, R., Rakhi, R.B. et al. (2012). Nanoroses of nickel oxides: synthesis, electron tomography study, and application in CO oxidation and energy storage. *ChemSusChem* 5 (7): 1241–1248.

32 Feng, Y., Li, L., Niu, S. et al. (2012). Controlled synthesis of highly active mesoporous Co_3O_4 polycrystals for low temperature CO oxidation. *Appl. Catal. B Environ.* 111: 461–466.

33 Long, X., Lin, H., Zhou, D. et al. (2018). Enhancing full water-splitting performance of transition metal bifunctional electrocatalysts in alkaline solutions by tailoring CeO_2–transition metal oxides–Ni nanointerfaces. *ACS Energy Lett.* 3 (2): 290–296.

34 Zhu, Y., Liu, X., Jin, S. et al. (2019). Anionic defect engineering of transition metal oxides for oxygen reduction and evolution reactions. *J. Mater. Chem. A* 7 (11): 5875–5897.

35 Sennu, P., Aravindan, V., Nahm, K.S., and Lee, Y.S. (2017). Exceptional catalytic activity of hollow structured $La_{0.6}Sr_{0.4}CoO_{3-\delta}$ perovskite spheres in aqueous media and aprotic Li–O_2 batteries. *J. Mater. Chem. A* 5 (34): 18029–18037.

36 Wang, Z., Zhang, F., Jin, C. et al. (2017). La_2O_3-NCNTs hybrids in-situ derived from $LaNi_{0.9}Fe_{0.1}O_3$-C composites as novel robust bifunctional oxygen electrocatalysts. *Carbon N. Y.* 115: 261–270.

37 Grimaud, A., Carlton, C.E., Risch, M. et al. (2013). Oxygen evolution activity and stability of $Ba_6Mn_5O_{16}$, $Sr_4Mn_2CoO_9$, and $Sr_6Co_5O_{15}$: the influence of transition metal coordination. *J. Phys. Chem. C* 117 (49): 25926–25932.

38 Yin, W.J., Weng, B., Ge, J. et al. (2019). Oxide perovskites, double perovskites and derivatives for electrocatalysis, photocatalysis, and photovoltaics. *Energy Environ. Sci.* 12 (2): 442–462.

39 Li, Y., Li, F.M., Meng, X.Y. et al. (2018). Ultrathin Co_3O_4 nanomeshes for the oxygen evolution reaction. *ACS Catal.* 8 (3): 1913–1920.

40 Han, L., Dong, S., and Wang, E. (2016). Transition-metal (Co, Ni, and Fe)-based electrocatalysts for the water oxidation reaction. *Adv. Mater.* 28 (42): 9266–9291.

41 Wang, J., Fu, Y., Xu, Y. et al. (2016). Hierarchical $NiCo_2O_4$ hollow nanospheres as high efficient bi-functional catalysts for oxygen reduction and evolution reactions. *Int. J. Hydrogen Energy* 41 (21): 8847–8854.

42 Lv, Y., Yang, L., and Cao, D. (2017). Nitrogen and fluorine-codoped porous carbons as efficient metal-free electrocatalysts for oxygen reduction reaction in fuel cells. *ACS Appl. Mater. Interfaces* 9 (38): 32859–32867.

43 Kim, J., Yin, X., Tsao, K.C. et al. (2014). $Ca_2Mn_2O_5$ as oxygen-deficient perovskite electrocatalyst for oxygen evolution reaction. *J. Am. Chem. Soc.* 136 (42): 14646–14649.

44 Chen, D., Wang, J., Zhang, Z. et al. (2016). Boosting oxygen reduction/evolution reaction activities with layered perovskite catalysts. *Chem. Commun.* 52 (71): 10739–10742.

45 Mefford, J.T., Rong, X., Abakumov, A.M. et al. (2016). Water electrolysis on $La_{1-x}Sr_xCoO_{3-\delta}$ perovskite electrocatalysts. *Nat. Commun.* 7 (1): 1–11.

46 Meng, H., Chen, X., Gong, T. et al. (2019). N, P, S/Fe-codoped carbon derived from feculae bombycis as an efficient electrocatalyst for oxygen reduction reaction. *ChemCatChem* 11 (24): 6015–6021.

47 Kuang, X., Kang, B., Wang, Z. et al. (2018). Sulfur-doped CoO nanoflakes with loosely packed structure realizing enhanced oxygen evolution reaction. *Chem. Eur. J.* 24 (65): 17288–17292.

48 Liu, X., Zhang, L., Zheng, Y. et al. (2019). Uncovering the effect of lattice strain and oxygen deficiency on electrocatalytic activity of perovskite cobaltite thin films. *Adv.Sci.* 6 (6): 1801898.

49 Cai, Z., Bi, Y., Hu, E. et al. (2018). Single-crystalline ultrathin Co_3O_4 nanosheets with massive vacancy defects for enhanced electrocatalysis. *Adv. Energy Mater.* 8 (3): 1701694.

50 Wang, X., Yang, Y., Diao, L. et al. (2018). CeO_x-decorated NiFe-layered double hydroxide for efficient alkaline hydrogen evolution by oxygen vacancy engineering. *ACS Appl. Mater. Interfaces* 10 (41): 35145–35153.

51 Xiang, R., Peng, L., and Wei, Z. (2019). Tuning interfacial structures for better catalysis of water electrolysis. *Chem. - A Eur. J.* 25 (42): 9799–9815.

52 Gong, J. and Bao, X. (2017). Fundamental insights into interfacial catalysis. *Chem. Soc. Rev.* 46 (7): 1770–1771.

53 Reich, C.M., Kaiser, A., and Irvine, J.T.S. (2001). Niobia based rutile materials as SOFC anodes. *Fuel Cells* 1 (3-4): 249–255.

54 Gao, W., Hood, Z.D., and Chi, M. (2017). Interfaces in heterogeneous catalysts: advancing mechanistic understanding through atomic-scale measurements. *Acc. Chem. Res.* 50 (4): 787–795.

55 Konsolakis, M. and Lykaki, M. (2020). Recent advances on the rational design of nonprecious metal oxide catalysts exemplified by CuO_x/CeO_2 binary system: implications of size, shape and electronic effects on intrinsic reactivity and metal-support interactions. *Catalysts* 10 (2): 160.

56 Zhang, Z., Liu, J., Gu, J. et al. (2014). An overview of metal oxide materials as electrocatalysts and supports for polymer electrolyte fuel cells. *Energy Environ. Sci.* 7 (8): 2535–2558.

57 Wang, Q., Xue, Y., Sun, S. et al. (2017). $La_{0.8}Sr_{0.2}Co_{1-x}Mn_xO_3$ perovskites as efficient bi-functional cathode catalysts for rechargeable zinc-air batteries. *Electrochim. Acta* 254: 14–24.

58 Liang, Y., Wang, H., Zhou, J. et al. (2012). Covalent hybrid of spinel manganese-cobalt oxide and graphene as advanced oxygen reduction electrocatalysts. *J. Am. Chem. Soc.* 134 (7): 3517–3523.

59 Goswami, C., Yamada, Y., Matus, E.V. et al. (2020). Elucidating the role of oxide-oxide/carbon interfaces of CuO_x-CeO_2/C in boosting electrocatalytic performance. *Langmuir* 36 (49): 15141–15152.

60 Chen, J., Zhou, N., Wang, H. et al. (2015). Synergistically enhanced oxygen reduction activity of MnO_x–CeO_2/Ketjenblack composites. *Chem. Commun.* 51 (50): 10123–10126.

61 Luo, Z., Irtem, E., Ibánez, M. et al. (2016). Mn_3O_4@$CoMn_2O_4$–Co_xO_y nanoparticles: partial cation exchange synthesis and electrocatalytic properties toward the oxygen reduction and evolution reactions. *ACS Appl. Mater. Interfaces* 8 (27): 17435–17444.

62 Zhang, Z., Liang, X., Li, J. et al. (2020). Interfacial engineering of $NiO/NiCo_2O_4$ porous nanofibers as efficient bifunctional catalysts for rechargeable zinc-air batteries. *ACS Appl. Mater. Interfaces* 12 (19): 21661–21669.

63 Liu, K., Huang, X., Wang, H. et al. (2016). Co_3O_4-CeO_2/C as a highly active electrocatalyst for oxygen reduction reaction in Al-air batteries. *ACS Appl. Mater. Interfaces* 8 (50): 34422–34430.

64 Gu, W., Liu, J., Hu, M. et al. (2015). $La_2O_2CO_3$ encapsulated La_2O_3 nanoparticles supported on carbon as superior electrocatalysts for oxygen reduction reaction. *ACS Appl. Mater. Interfaces* 7 (48): 26914–26922.

65 Sonoki, H., Mori, D., Taminato, S. et al. (2019). A hydrated strontium cobalt oxyhydroxide Ruddlesden-Popper phase as an oxygen electrocatalyst for aqueous lithium-oxygen rechargeable batteries. *Chem. Commun.* 55 (52): 7454–7457.

66 Zhu, Y.P., Guo, C., Zheng, Y., and Qiao, S.Z. (2017). Surface and interface engineering of noble-metal-free electrocatalysts for efficient energy conversion processes. *Acc. Chem. Res.* 50 (4): 915–923.

67 Burke, M.S., Kast, M.G., Trotochaud, L. et al. (2015). Cobalt-iron (oxy)hydroxide oxygen evolution electrocatalysts: the role of structure and composition on activity, stability, and mechanism. *J. Am. Chem. Soc.* 137 (10): 3638–3648.

68 Wang, Y., Zhang, D., and Liu, H. (2010). A study of the catalysis of cobalt hydroxide towards the oxygen reduction in alkaline media. *J. Power Sources* 195 (10): 3135–3139.

69 Tang, T., Jiang, W.J., Niu, S. et al. (2017). Electronic and morphological dual modulation of cobalt carbonate hydroxides by mn doping toward highly efficient and stable bifunctional electrocatalysts for overall water splitting. *J. Am. Chem. Soc.* 139 (24): 8320–8328.

70 Jia, Q., Ghoshal, S., Li, J. et al. (2017). Metal and metal oxide interactions and their catalytic consequences for oxygen reduction reaction. *J. Am. Chem. Soc.* 139 (23): 7893–7903.

71 Puigdollers, A.R., Schlexer, P., Tosoni, S., and Pacchioni, G. (2017). The role of metal/oxide interfaces in the formation of oxygen vacancies. *ACS Catal.* 7 (10): 6493–6513.

72 Liu, X. and Hu, W. (2016). Iron oxide/oxyhydroxide decorated graphene oxides for oxygen reduction reaction catalysis: a comparison study. *RSC Adv.* 6 (35): 29848–29854.

73 Jia, X., Gao, S., Liu, T. et al. (2017). Fabrication and bifunctional electrocatalytic performance of ternary CoNiMn layered double hydroxides/polypyrrole/reduced graphene oxide composite for oxygen reduction and evolution reactions. *Electrochim. Acta* 245: 59–68.

74 Youn, D.H., Park, Y.B., Kim, J.Y. et al. (2015). One-pot synthesis of NiFe layered double hydroxide/reduced graphene oxide composite as an efficient electrocatalyst for electrochemical and photoelectrochemical water oxidation. *J. Power Sources* 294: 437–443.

75 Li, C., Chen, J., Wu, Y. et al. (2019). Enhanced oxygen evolution reaction activity of NiFe layered double hydroxide on nickel foam-reduced graphene oxide interfaces. *Int. J. Hydrogen Energy* 44 (5): 2656–2663.

76 Cao, L., Ma, Y., Song, A. et al. (2019). Stable composite of flower-like NiFe-layered double hydroxide nucleated on graphene oxide as an effective catalyst for oxygen reduction reaction. *Int. J. Hydrogen Energy* 44 (12): 5912–5920.

77 Garcia, A.C., Touzalin, T., Nieuwland, C. et al. (2019). Enhancement of oxygen evolution activity of nickel oxyhydroxide by electrolyte alkali cations. *Angew. Chemie* 131 (37): 13133–13137.

78 Pi, Y., Shao, Q., Wang, P. et al. (2017). Trimetallic oxyhydroxide coralloids for efficient oxygen evolution electrocatalysis. *Angew. Chemie - Int. Ed.* 56 (16): 4502–4506.

79 Lim, C.S., Chua, C.K., Sofer, Z. et al. (2015). Layered transition metal oxyhydroxides as tri-functional electrocatalysts. *J. Mater. Chem. A* 3 (22): 11920–11929.

80 Zhan, T., Zhang, Y., Liu, X. et al. (2016). NiFe layered double hydroxide/reduced graphene oxide nanohybrid as an efficient bifunctional electrocatalyst for oxygen evolution and reduction reactions. *J. Power Sources* 333: 53–60.

81 Zhan, T., Liu, X., Lu, S.S., and Hou, W. (2017). Nitrogen doped NiFe layered double hydroxide/reduced graphene oxide mesoporous nanosphere as an effective bifunctional electrocatalyst for oxygen reduction and evolution reactions. *Appl. Catal. B Environ.* 205: 551–558.

82 Wu, M., Wei, Q., Zhang, G. et al. (2018). Fe/Co double hydroxide/oxide nanoparticles on N-Doped CNTs as highly efficient electrocatalyst for rechargeable liquid and quasi-solid-state zinc–air batteries. *Adv. Energy Mater.* 8 (30): 1801836.

5

Porous Graphitic Carbon Nitride Nanostructures and Their Application in Photocatalytic Hydrogen Evolution Reaction

Bindu Antil and Sasanka Deka

Department of Chemistry, University of Delhi, Delhi, India

5.1 Introduction

In modern times, there is a critical need to understand the energy and environmental issues at a wide global level. With the increasing demands of energy and concerning environmental problems, there is an indispensable need to formulate ideas that can help in the production of clean energy fuel to meet sustainable energy developments [1]. Solar energy has been considered a promising renewable resource and its utilization in the production of clean energy fuels has aroused extensive attention over decades [2]. Recently, in the presence of appropriate semiconductor photocatalysts, the generation of clean hydrogen (H_2) fuel from water using renewable solar energy source has been considered as one of the most scalable and sustainable approach [3]. For the first time, Fujishima and Honda in 1972 introduced the concept of semiconductor photocatalysis on TiO_2 electrodes for photoelectrochemical water splitting [4]. Later, in 1976, Carey et al. demonstrated the photocatalytic degradation of organic pollutants in the presence of a TiO_2 photocatalyst. Subsequently, the photocatalytic reduction of CO_2 in aqueous suspensions of semiconductor photocatalyst, such as CdS, TiO_2, GaP, SiC, and ZnO, was demonstrated in 1979 [5]. Thereafter, a lot of work has been done to develop highly efficient, affordable, and stable semiconductor photocatalysts. For photocatalytic water splitting for efficient production of hydrogen, it is highly desirable to develop an active and stable light-harvesting semiconductor that possesses a suitable energy bandgap and positions for promoting charge separation and charge transfer processes, and water dissociation simultaneously [6]. Retrospections into the past reports, several novel functional materials such as inorganic semiconductors such as TiO_2 [7], ZnO [8], Cu_2ZnSnS_4 [9, 10], Fe_2O_3 [11], SnS_2 [12], $BiVO_4$ [13], Cu_2O [14], $CuCo_2S_4$ [15], and CdS [16] and molecular assemblies have been devised [17] for photocatalytic reactions. All photocatalytic reactions are a combination of three-step processes where the first step is the photon absorption followed by electron–hole pair generation and separation, and lastly catalytic surface reactions [18]. Therefore, for improvement in the photocatalytic performance of a photocatalyst, there should be an enhancement in the

Heterogeneous Nanocatalysis for Energy and Environmental Sustainability: Energy Applications,
First Edition. Edited by Putla Sudarsanam, Yusuke Yamauchi, and Pankaj Bharali.

three aforementioned steps. Transition-metal ions with d^0 electronic configuration or post-transition metal ions with d^{10} configuration-based semiconductors along with group VA or VIA ions as counter-anions components are explored so far [19]. For example, $Y_2Ta_2O_5N_2$ and $In_{1-x}Ni_xTaO_4$ are reported as active novel visible-light water-splitting photocatalysts by Zou et al. [20] and Liu et al. [21] Similarly, $Ga_{1-x}Zn_xO_{1-x}N_x$ solid solution can efficiently split water under visible light, as reported by Maeda et al. [22] However, the existing photocatalysts usually show low photocatalytic activity due to faster recombination of charge carriers. Thus, it is highly desirable to develop a suitable semiconductor or semiconductor composite that assures the migration of electrons and holes into the conduction band (CB) and valence band (VB), respectively [23]. Immobilization of the co-catalysts onto the surface of the photocatalyst improves the charge separation and also favors surface catalytic reactions by reducing the activation energy [24].

A metal-free polymeric graphitic carbon nitride (g-C_3N_4) semiconductor was utilized for the first time as a photocatalytic hydrogen evolution (PHE) photocatalyst by Wang and his group [3]. Since then, g-C_3N_4 increasingly stands out as the most typical of polymeric semiconductor. Carbon nitrides can be easily prepared using cheap feedstocks and under mild conditions and possess high stability with resourceful physical and chemical properties, which triggered the researchers to put considerable efforts into the examination of g-C_3N_4, resulting in in-depth fundamental knowledge acquisition [25]. Thereafter, arrays of reports on carbon nitrides are found from history, signifying the synthesis, structure, and its multifunctional applications. Although carbon nitride materials deliver unique electronic and mechanical properties, due to their non-porous nature, they acquire a low surface area ($<10\,m^2/g$). The heterogeneous catalysis and adsorption both are surface-dependent phenomena where the surface area plays an important role in controlling the activity. Therefore, the researchers have put a lot of efforts to flourish porosity in the carbon nitride matrix through various methods while retaining the carbon and nitrogen coordination and the building units of bulk g-C_3N_4 [26]. This chapter gives a brief discussion on porous g-C_3N_4 nanostructures and their applications in PHE reaction.

5.2 Graphitic Carbon Nitrides

Carbon nitrides are a family of polymeric materials mainly consisting of carbon and nitrogen with a general formula near to C_xN_y [27] and can be readily obtained from carbon materials by substituting carbon atoms with nitrogen atoms. Carbon nitride materials exist in five different kinds of structures: one two-dimensional (2D) graphitic C_3N_4 and four three-dimensional (3D) (cubic–C_3N_4, pseudocubic-C_3N_4, α-C_3N_4, and β-C_3N_4) [28]. Among these, 2D g-C_3N_4 is the most stable layered allotrope with a structure similar to the graphitic nanostructure. Since the structure of g-C_3N_4 is similar to graphite, the theoretical specific surface area of an ideal monolayer of g-C_3N_4 could be as high as $2500\,m^2/g$ [29]. The surface of carbon nitride comprises solely a π-conjugated polymeric structure (interconnected tri-s-triazine/heptazine units via secondary nitrogen depending on reaction conditions) with an assembly of C–N bonds with different degrees of condensation [30]. It is believed that the synthesis of carbon nitride under the name "melon" was already reported by Berzelius and Liebig [31] in 1834, but it did not gain much attention for several decades until recently due to its insolubility in acidic, neutral, or basic solvents; its

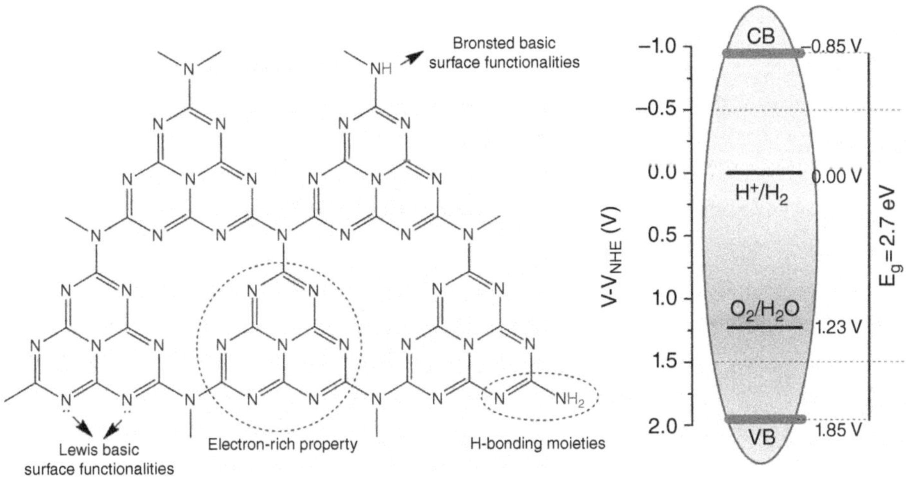

Figure 5.1 Structure of g-C_3N_4 minted with multiple surface properties (left side) and band energy diagram of g-C_3N_4 (right side).

chemical inertness; and its undisclosed structure. "Melon" consists of linear polymers of connected tri-s-triazine units connected through secondary nitrogen atoms. In real, g-C_3N_4 is prepared by thermal polycondensation. During the process, apart from C and N, small amounts of hydrogen ($C_3N_{4+y}H_y$) are also found as primary and/or secondary amine-terminating edges due to incomplete condensation, as seen in Figure 5.1 [32]. Furthermore, due to the polymeric nature of g-C_3N_4, the material's surface chemistry via modification at molecular level and surface engineering can be tuned efficiently. Moreover, the flexibility of the layered structure where it can act as an eminent host for several inorganic nanoparticles is also attributed to polymeric structure. These superlative and unique characteristics of g-C_3N_4 make this material a very promising photocatalyst for various applications.

In literature, there are several methods available for the synthesis of carbon nitride materials, such as direct thermal polymerization of nitrogen-containing precursors and exfoliation methods [33]. The crystallinity and the nitrogen content of the carbon nitride material both depend on the experimental conditions, carbon–nitrogen content in the precursor, and degree of polymerization of the precursor.

5.3 Various Nanostructures of g-C_3N_4

In this section, we discuss various nanostructures of g-C_3N_4. Different morphologies of g-C_3N_4, such as nanotubes, nanosheets, quantum dots, nanowires, and 3D structures, are reported as displayed in Figure 5.2.

5.3.1 Bulk g-C_3N_4

Bulk g-C_3N_4 is structurally similar to graphite and acquires a 3D-layered structure. Mostly, bulk g-C_3N_4 is synthesized by direct thermal polymerization of the carbon and nitrogen-rich precursors [34]. Bulk g-C_3N_4 exhibits an optical bandgap of 2.7 eV and a low surface

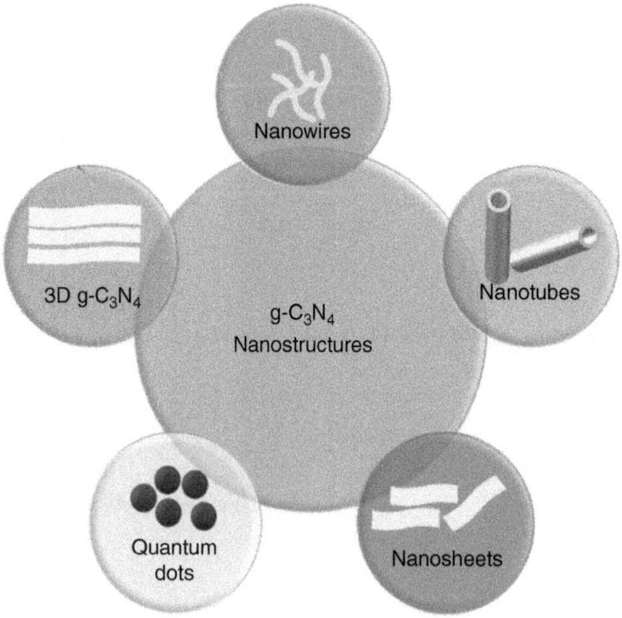

Figure 5.2 Schematic illustration showing various morphologies of g-C_3N_4.

area with no observable porosity due to the stacking of polycondensed layers (Figure 5.2). The apparent quantum yield is also low because of the high recombination rate of excitons and limited or no evident electronic transitions. Due to these drawbacks, it is highly desirable to do modifications to improve its structure/morphology and photocatalytic activity. Several morphologies, such as nanotubes, nanowires, quantum dots, and nanosheets, can be prepared from bulk g-C_3N_4.

5.3.2 Nanowires

Graphitic carbon nitrides nanowires present a one-dimensional (1D) structure resembling rod-like morphology (Figure 5.2) [35]. Recently, the g-C_3N_4 nanowire structure has garnered a lot of attention as they provide high surface area and consequently a large number of active sites, which are vital for catalytic reactions.

5.3.3 Nanotubes

Graphitic carbon nitride nanotubes resemble 1D hollow or tubular-type morphology [36]. g-C_3N_4 nanotubes show exquisite properties compared to the bulk, such as improved light absorption, high surface area, faster charge transport, and low recombination rate of charge carriers.

5.3.4 Quantum Dots

Graphitic carbon nitride quantum dots have a zero-dimensional (0D) structure and show several extravagant properties such as chemical stability, large specific BET surface area, bright fluorescence, valuable biocompatibility, and favorable bandgap with suitable band positions [34].

5.3.5 3D g-C_3N_4

Although g-C_3N_4 has been widely used in photocatalytic processes, it still suffers from several shortcomings such as faster recombination rates of charge carriers, low surface area, and meager light absorption ability. Therefore, to overcome these shortcomings, 3D mesoporous/macroporous g-C_3N_4 can be designed. As visualized, 3D mesoporous g-C_3N_4 [37] prepared using silica template showed 6.5 times higher photocatalytic hydrogen production compared to bulk g-C_3N_4. The 3D open framework structure offered more photocatalytic active sites resulting in better light absorption.

5.3.6 g-C_3N_4 Nanosheets

As it is clearly stated in the above section that bulk g-C_3N_4 bears a resemblance to the layered structure similar to graphite [38]. However, in contrast to graphite, here each layer comprises of tri-s-triazine/s-triazine units, which are composed of carbon and nitrogen atoms held together by covalent bonds, whereas each adjacent layer is held together by weak van der Waals forces of attraction [3]. Thus, each layer can be separated from the bulk g-C_3N_4 layered structure by providing sufficient energy, which breaks the weak van der Waals forces of attraction between the layers generating 2D nanosheets.

5.4 Physical, Chemical, Optical, and Electronic Properties

Graphitic carbon nitride is a highly stable material where the color varies from white to yellow, dark brown. The high thermal stability (up to 600 °C) [39] in the air is due to its aromatic C–N heterocycle and hydrothermal stability is due to the strong van der Waals interactions between the subsequent layers [3], making it a potent material for the application in the field of heterogeneous catalysis. The presence of hydrogen atoms generates a large number of surface defects and promotes the relocalization of electrons on the surface, inducing Lewis-base character. Similarly, N-atoms have one more electron than C-atoms; so both these above factors demonstrate rich-surface properties of g-C_3N_4 such as H-bonding motifs, surface conductivity, electron-rich properties, and basic surface functionalities as displayed in Figure 5.1, which are crucial for catalytic applications. Moreover, g-C_3N_4 has a favorable bandgap of 2.7 eV, showing an intrinsic semiconductor-like absorption in the visible region [3]. The bottom level of the CB (−0.85 V) is more negative than the redox potential of H^+/H_2 (0 V vs NHE), whereas the top level of the VB (+1.85 V) is more positive than the redox potential of O_2/H_2O (1.23 V) [40], and the subsequent band positions are displayed on the right side in Figure 5.1. This energy gap is sufficient enough to overcome the endothermic character (1.23 eV theoretically) of the water-splitting reaction. This property enables the material to be a visible-light-active photocatalyst for a wide range of reactions. g-C_3N_4 exhibits several such unique properties and, therefore, has drawn much attention as a material for photocatalytic water splitting.

5.5 Limitations in the Structure and the Photocatalytic Activity

Unfortunately, pure bulk g-C_3N_4, without optimization of its structure and properties, suffers from several disadvantages such as poor quantum efficiency, the high recombination rate of photogenerated charge carriers, and low surface area, thus significantly limits its application in water splitting [41]. It is generally synthesized by simple thermal polycondensation of melamine or other triazine precursors, and due to agglomeration, it acquires large grain size morphology with a small specific surface area, low charge separation, and thus relatively low photocatalytic activity [42]. Therefore, several efforts have been put in order to develop new functional g-C_3N_4 materials, including exfoliation, structure modification, heteroatom doping, heterojunction, and nano-microstructure designing.

5.6 Modifications in the Nanostructure Design

The advancement in the nanostructure design has witnessed a significant growth in the development of porous g-C_3N_4 materials. Porous carbon nitride materials have been under scrutiny due to their extraordinary surface properties with high surface area, generating an ample number of active sites on the surface and thus, excellent photocatalytic hydrogen production activity in many aspects. Also, highly porous structures gain narrow bandwidth with absorption of visible light in the broad range compared to conventional g-C_3N_4, leading to efficient charge separation. Porous carbon nitride material presents irrefutable properties, which helps in the enhancement of photocatalytic activity. In this chapter, we have carefully focused on the advancement in porous carbon nitride materials, including synthetic methods, structure, properties, and their photocatalytic applications.

5.7 Synthetic Strategies

Porous g-C_3N_4materials can be synthesized using two approaches: bottom-up approach and top-down approach. Bottom-up approach includes the template/nanocasting approach and supramolecular assembly approach, whereas top-down consists of exfoliation approach and thermal oxidation methods. In this section, we discuss each of the synthetic methods in detail.

5.7.1 Bottom-Up Approach: Templating Method

This approach is an effective method for the direct acquisition of porous carbon nitrides from N-containing precursors without damage to the planar structure of g-C_3N_4. The bottom-up templating approach is divided into two categories:

a) Hard-template method
b) Soft-template method

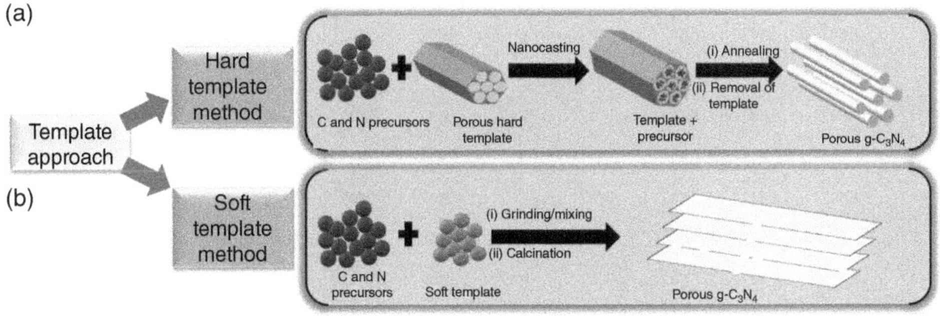

Figure 5.3 Schematic representation of the synthesis of porous g-C_3N_4 using template-assisted approaches. (a) Hard template method and (b) Soft template method.

Both these methods are physical methods to formulate porous carbon nitride materials. The typical schematic diagrams for both methods are depicted in Figure 5.3.

5.7.1.1 Hard-Template Method

This synthetic approach involves the thermal polymerization of N-based precursor in the presence of hard templates (generally silica templates like SBA-15, KIT-6, Ni foams, and porous anodic alumina oxide membranes [AAOM] templates) to generate a uniform porous carbon nitride structure as clearly depicted in Figure 5.3a. Groenewalt et al. [43] in 2006, for the first time, prepared porous carbon nitride using this method. Here, the hard templates are first prepared or purchased; the desired precursors are infused physically over the hard template without disturbing the planar structure and crystallinity as the hard templates are chemically inert in nature; and finally, the hard template is successfully removed in the presence of strong acid or base such as HF, NaOH, and NH_4HF_2. The porous architecture obtained using this method has an ultrahigh accessible specific surface area with a large number of active sites and good light-harvesting capability. Also, the obtained open framework uniformly interconnected porous structure facilitates efficient charge separation. Thus, the use of a hard template to obtain porous nanostructures is quite useful.

Zhao et al. [44] successfully prepared highly ordered mesoporous carbon nitride using SBA-15 as the hard template and hexamethylenetetramine as the precursor where the template was efficiently removed by HF treatment. Figure 5.4 demonstrates the typical preparation process of highly ordered mesoporous carbon nitride with a very high surface area and larger pore volume and pore size. Therefore, this report clearly reveals the advantage of using silica templates to obtain porous materials of specific size and structure.

Moreover, clinging to the hard template is another method of precursor-coating approach where another porous photocatalytic material is used as a self-template to prepare porous carbon nitride. Both the self-template and the as-prepared material can later form a heterojunction resulting in improved photocatalytic activity. For example, Wei et al. [45] successfully designed a heterojunction semiconductor (mesoporous TiO_2/g-C_3N_4) using a nanocasting approach where porous TiO_2 acts as an active scaffold/self-template and g-C_3N_4 (3 wt%) as a visible-light photosensitizer. As compared to other hard template methods, this method is more eco-friendly as no removers are required here, but its usage is still limited because the self-template chosen needs to be porous, of good thermal stability, and

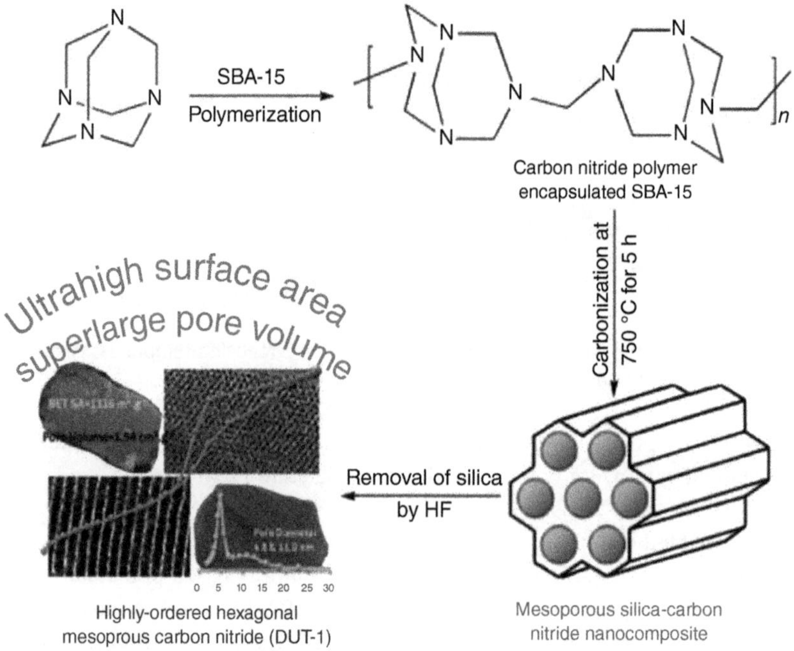

Figure 5.4 Schematic illustration for the synthesis of mesoporous carbon nitride using SBA-15 as the template. *Source:* Reprinted with permission from Zhao et al. [44], Copyright 2014, American Chemical Society.

to withstand its porous structure before the coating of precursor. Therefore, due to these high requirements, there are only a few reports with this method to date [46, 47]. The use of a hard template to synthesize porous carbon nitride materials is beneficial as this method generates controlled architecture of the material. However, the synthetic procedures are quite tedious to perform and require toxic chemicals to remove the hard template. Thus, this limits the use of this mentioned method.

5.7.1.2 Soft Template Method

A soft template approach has widely attracted attention for the synthesis of porous g-C_3N_4 due to its easy use in the synthetic procedures without the worry of removal of the template. The general synthetic procedure is shown in Figure 5.3b. Dai et al. [48] and Ikkala et al. [49] were the first ones to realize soft template strategy, whereas Zhao et al. first reported the synthesis of porous carbon structures and films using this approach [50]. To date, soft-templating strategy for the preparation of mesoporous carbon nitrides is quite limited due to thermodynamics of the synthetic procedures and physicochemical reasons such as soft template decomposition during pyrolysis. However, Antonietti et al. were the first to prepare mesoporous carbon nitride (mpg-C_xN_y) using a variety of block copolymers, non-ionic surfactants, and ionic liquids (ILs) (e.g. Triton X-100, Brij30, Brij58, Brij76, P123 and F127,) as soft templates and dicyandiamide as the precursors [51]. Herein, the self-condensation of the dicyandiamide using soft and direct template strategy results in the formation of nanoporous carbon nitride. Most of the as-prepared carbon nitride materials showed high surface area, but the exploitation of a useful pore system is only observed with Triton X-100 and ILs. The step-by-step annealing process helped to prevent the collapse of the mesoporous structure during the template removal. Intriguingly, the mesoporous carbon nitride materials prepared using soft template strategy show better conducting properties as compared to those prepared using the hard template approach. Recently, Peer et al. [52] prepared mesoporous carbon nitride via a one-pot soft template approach using different non-ionic surfactants (Triton X-100, Pluronic P-123, and Pluronic F-127) and melamine-cyanurate hydrogen-bonded complex and water as a solvent. As a result, several 3D nanostructures with different morphologies (hollow spherical, sheetlike, and tubular or highly porous networks) were obtained by altering the surfactant/precursor ratio as displayed in Figure 5.5(a–d).

Similarly, using 1-butyl-3-methylimidazolium tetrafluoroborate ($BmimBF_4$) IL as soft template and dicyandiamide as an organic precursor, boron and fluorine enriched mesoporous carbon nitride was prepared [53]. Here, no step-wise calcinations were

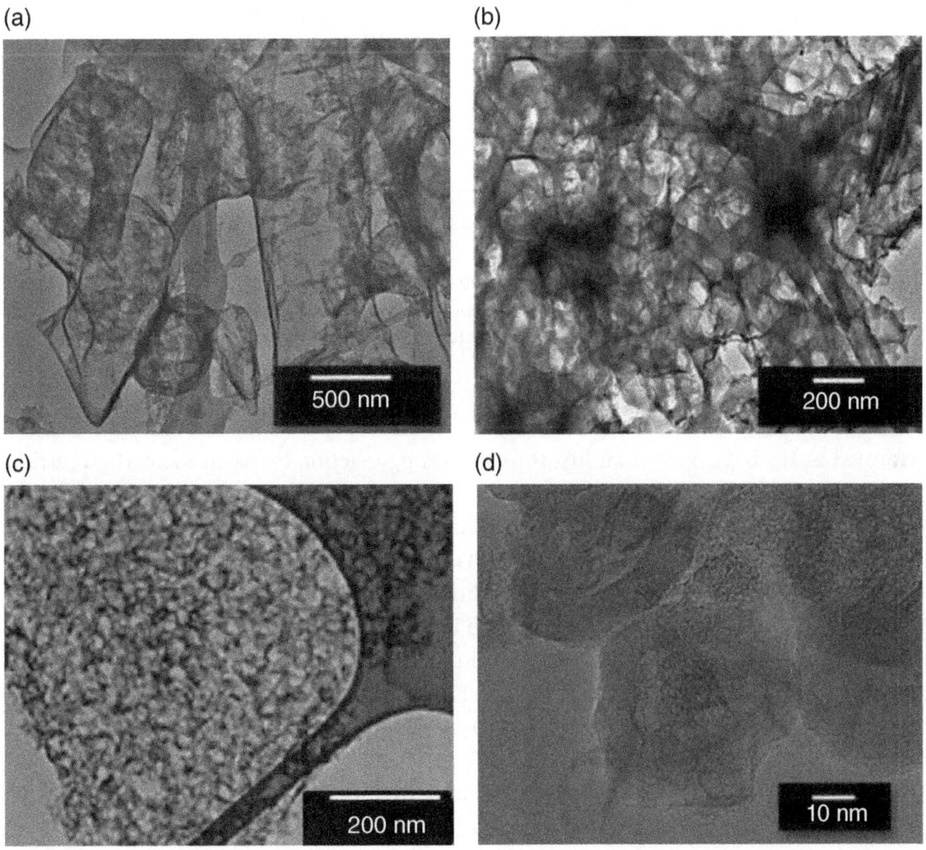

Figure 5.5 (a, b) High-resolution TEM images of samples prepared using Pluronic F-127 surfactant showed hollow spherical and tubular structure and sponge-like porous morphology. (c, d) High-resolution TEM images of samples prepared using Triton X-100 surfactant and Pluronic P-123 surfactant, respectively, showed a highly porous structure. *Source:* Reprinted with permission from Peer et al. [52], Copyright 2017, American Chemical Society.

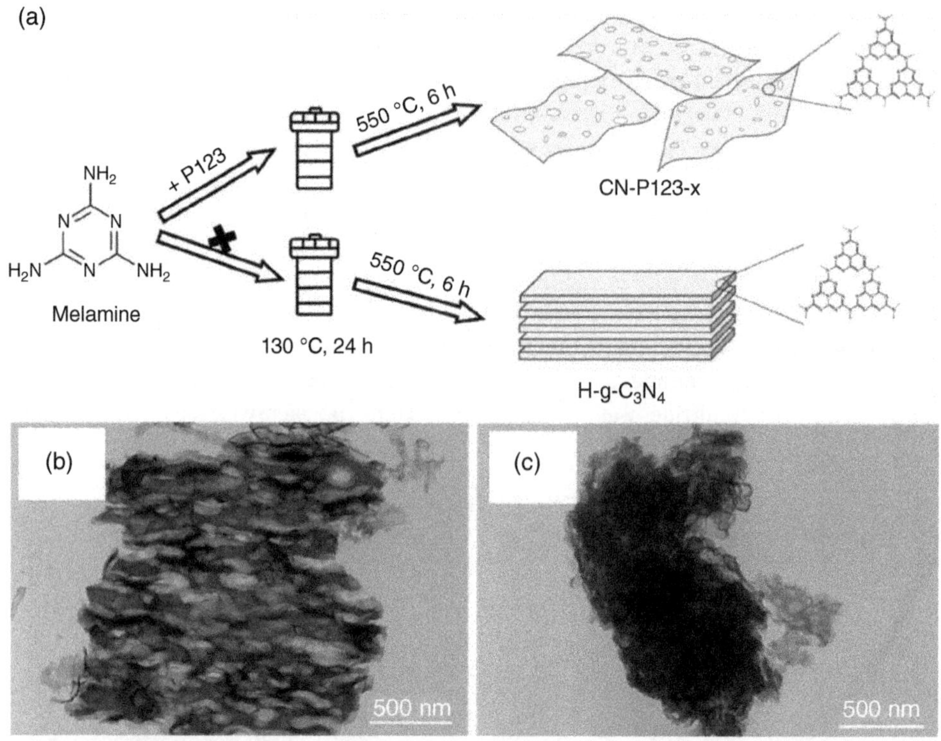

Figure 5.6 (a) Diagrammatic representation of the synthetic procedure of CN P123-x and H-g-C_3N_4. *Source:* Yan et al. [54] / with permission of American Chemical Society. (b) TEM image of CN P123-x showing non-uniform porous structure. (c) TEM image of H-g-C_3N_4 showing non-porous nature of the material.

performed as ILs have poor volatility, thus the strong reaction between the carbon nitride network and the IL-generated nanoporous structure in mesoporous carbon nitrides with high carbon content. Similarly, thin-sliced non-uniformly porous g-C_3N_4 (coded as CN P-123-x) was prepared using Pluronic P-123 as the template as reported by Yan et al. [54] The reaction was carried with and without Pluronic P-123 as displayed in Figure 5.6a. The TEM images present a lamellar structure with irregular pores (Figure 5.6b) for CN P-123-x in the presence of Pluronic P-123, whereas a dense blocky structure is observed for H-g-C_3N_4 prepared in the absence of Pluronic P-123 (Figure 5.6c). Due to the highly porous structure, CN P-123-x acquires a high surface area and thus showed excellent photocatalytic activity.

The above results demonstrate that the soft-templating approach is easy, of low cost, and eco-friendly, but there are only fewer synthetic procedures in comparison to the hard template approach. Also, the mesoporous carbon nitrides prepared using this approach generally offer disordered structure, thus there is no control over the size and morphology of prepared porous samples using this approach, limiting its usage in catalytic applications. Moreover, the soft templates are chemically unstable, due to which several side reactions take place, hampering the photocatalytic activity. Accordingly, the development of greener and easier soft-template-based synthetic approaches is much needed to enhance photocatalytic activity, but it is challenging to accomplish.

5.7.1.3 Supramolecular Assembly Approach

It is another technique providing a great opportunity for the preparation of porous carbon nitrides, where the morphology is carefully controlled through hydrogen bonds without any further templating techniques. Herein, the non-covalent interactions such as hydrogen bonding play a pivotal role in molecular self-assembly due to their specificity, reversibility, and directionality of these interactions [55], and upon annealing, these hydrogen bonds are broken, which increases the degrees of freedom resulting in a highly porous structure [56]. This method has gained much attention as it involves less toxic solvents as compared to template methods. In the supramolecular approach, the appropriate monomers and solvents are chosen. The monomers are organized into different structures via their ability of forming hydrogen bonds in a given solvent resulting in ordered and stable complexes/aggregates, thus forming different structures in various solvents as displayed in Figure 5.7.

Xiao et al. [57] in 2019 successfully synthesized porous few-layer C_3N_4 material using a simple bottom-up method that involves self-assembly of molecules into layered precursors in the presence of glycerol and ethanol. The resultant material acquires a high specific surface area (164.2 m^2/g), high electrical conductivity, and ultrathin porous nanosheet morphology with 2–15 nm width pores. Thus, supramolecular self-assembly approach is considered a potential method for synthesizing various porous microstructures as the approach involves highly saturated and directional intermolecular hydrogen bonds [58]. However, the use of this method is still limited as the molecular self-assembly tends to form stable and large-sized supramolecular aggregates through hydrogen bonds in water, but the self-assembly of the supramolecular aggregates takes place in the presence of organic solvents [59]. Thus, replacing water with various solvents such as dimethylsulfoxide (DMSO) and chloroform makes this whole approach toxic. Hence, it is still challenging to prepare porous carbon nitride using an easy and green synthetic supramolecular assembly method.

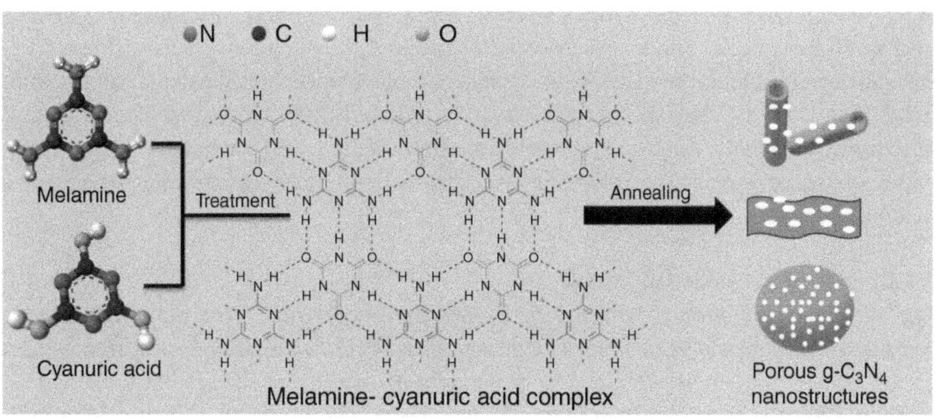

Figure 5.7 Schematic view of the preparation of various kinds of porous g-C_3N_4 nanostructures from melamine-cyanuric acid complex using supramolecular assembly approach.

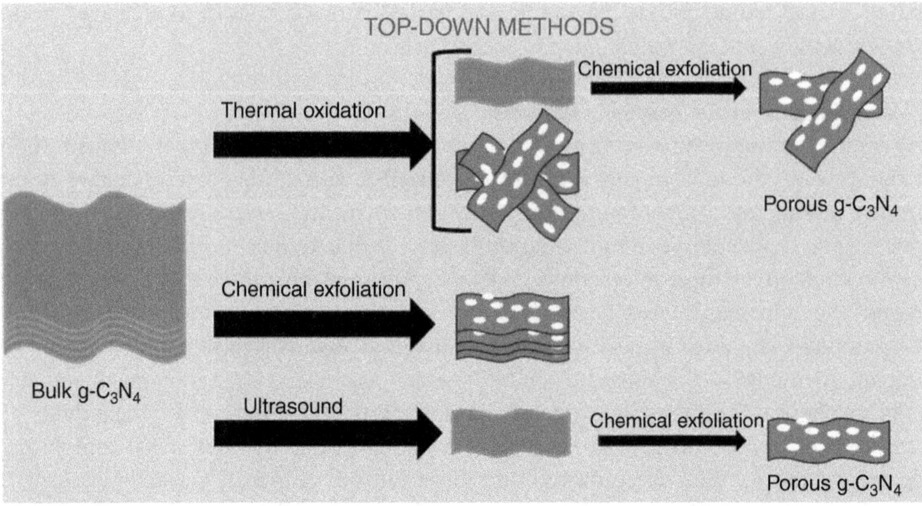

Figure 5.8 Diagrammatic representation of the top-down methods for the preparation of porous g-C_3N_4.

5.7.2 Top-Down Approach: Exfoliation Methods

Exfoliation is a top-down approach where single atomic layers/few atomic layers of g-C_3N_4 are synthesized from its bulk counterparts, as shown in Figure 5.8. The exfoliation process is possible due to the presence of weak van der Waals force of attraction between the layers of bulk g-C_3N_4 and strong covalent bonds between carbon and nitrogen [34]. The idea of the exfoliation came from atomic crystalline graphene, which presents many fascinating properties that are different from bulk graphite. Hence motivated by the exfoliation of graphite into graphene layers using the Hummers method [60], efforts have been put to prepare nanosheets of g-C_3N_4. Exfoliation process induces exotic electronic properties accompanied by high surface area. Hence, encouraged by the Hummers approach, Liu et al. [61] followed the same approach for the exfoliation of bulk g-C_3N_4, but they failed as instead of nanosheets, nanoparticles were formed and the planar structure of the g-C_3N_4 was destroyed. In literature, there are various methods reported for the exfoliation of bulk g-C_3N_4 (Figure 5.8). They are thermal oxidation, thermal exfoliation, chemical exfoliation utilizing strong acids, liquid ultrasonic exfoliation method, and chemical blowing using various blowing agents. The following sections of the chapter discuss these approaches in detail.

5.7.2.1 Chemical Exfoliation Method

The chemical exfoliation approach is performed in the presence of reagents, such as concentrated hydrochloric acid, sulfuric acid, and nitric acid. Xu et al. [62] prepared g-C_3N_4 nanosheets with a single atomic layer (thickness 0.4 nm) using the chemical exfoliation method. Typically, the monolayer of g-C_3N_4 nanosheets was prepared by heating dicyandiamide precursor at 550°C for four hours to obtain bulk g-C_3N_4. The resultant product was mixed with concentrated H_2SO_4 (98 wt%), stirred for eight hours, and finally added water

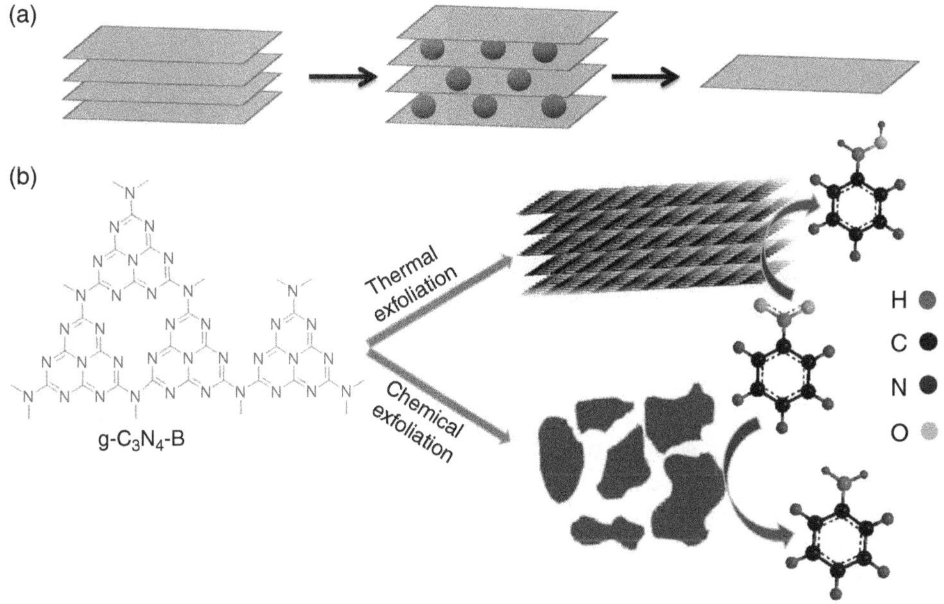

Figure 5.9 (a) Schematic illustration of the chemical exfoliation method to prepare monolayer g-C_3N_4 nanosheets from bulk g-C_3N, where the first step involved intercalation of concentrated sulfuric acid between the layers and later the second step leads to the formation of an exfoliated monolayer of g-C_3N_4 nanosheet. *Source:* Xu et al. [62] / with permission of Royal Society of Chemistry. (b) Schematic representation of the bulk and thermally and chemically exfoliated g-C_3N_4. *Source:* Challagulla et al. [63] / with permission of Royal Society of Chemistry.

followed by sonication to obtain exfoliated g-C_3N_4 nanosheets. Due to the sonication process, the concentrated sulfuric acid results in the intercalation between the adjacent layers of bulk g-C_3N_4 and subsequently on reaction with water, weak van der Waals forces of attraction between the layers are broken, finally generating the monolayer of g-C_3N_4 nanosheets as shown in Figure 5.9a. Although this exfoliation approach for the preparation of ultrathin nanosheets from bulk g-C_3N_4 leads to give high specific surface area nanosheets, poor yields of product with poor controllability are realized and utilization of corrosion acids make this process hazardous for the environment.

5.7.2.2 Thermal Exfoliation Method

Typically, thermal exfoliation involves the exfoliation of bulk g-C_3N_4 in the presence of heat that will generate the weak van der Waals forces of attraction between the layers break, resulting in exfoliated g-C_3N_4 [64]. This process is beneficial as during the heating process, the hydrogen bonded to the tri-s-triazine or s-triazine motifs reacts with oxygen to form a gas that escapes out and generates porosity into the material. Thus, we obtain highly porous and high surface area exfoliated g-C_3N_4 sheets. Cheng et al. devised a simple top-down strategy, i.e. direct thermal etching process in the air in order to synthesize highly anisotropic 2D g-C_3N_4 nanosheets with the thickness of around 2 nm from its bulk counterpart [61]. The obtained nanosheets possess a high surface area of 306 m^2/g and a larger bandgap with improved electron mobility resulting in excellent photocatalytic hydrogen generation. Here, the nanosheets were obtained with a yield of 6% by heating the bulk g-C_3N_4 in an open ceramic container at 500 °C for two hours with a ramp rate of 5°C/min. Recently, Challagulla et al. used both thermal and chemical etching processes to form nanosheets from the bulk sample, where the bulk sample was first thermally etched at 500 °C for two hours in the air and then it was chemically etched with H_2SO_4 (Figure 5.9b) [63]. In conclusion, this method presents several advantages such as cheap, no use of toxic solvents, and a high yield of the product with several structural defects incorporated onto the exfoliated sample. However, the crystallinity of the product is low and a relatively low surface area is acquired, which are essential for the surface-driven photocatalytic reactions.

5.7.2.3 Liquid Exfoliation Method

The liquid exfoliation method embroils the exfoliation of a bulk sample in the presence of a suitable solvent with continuous ultrasonication [64]. Yuan et al. prepared g-C_3N_4 nanosheets using a simple probe sonication-assisted liquid exfoliation method and finally constructed 2D-2D MoS_2/g-C_3N_4 composite photocatalyst for photocatalytic hydrogen production [65]. Here, urea was used as the precursor to prepare bulk g-C_3N_4, which was further dispersed in N-methylpyrolidone and later ultrasonicated to obtain exfoliated g-C_3N_4 (Figure 5.10a). Similarly, g-C_3N_4 nanosheets were synthesized from melamine which was heated first into a muffle furnace, and then the product was dispersed in deionized water under sonication to form highly thin exfoliated g-C_3N_4 nanosheets with an average thickness of 0.6 nm as displayed in Figure 5.10b [66].

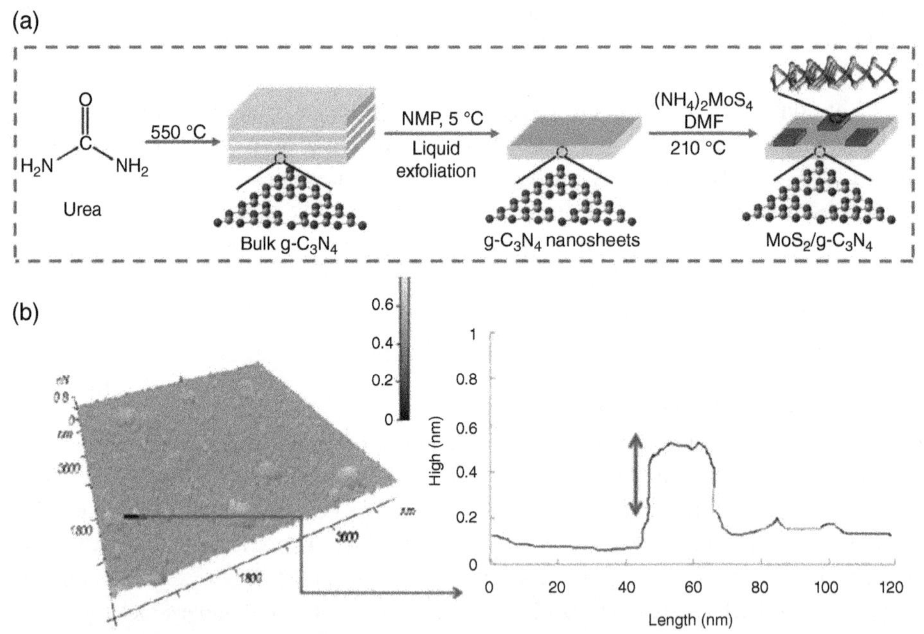

Figure 5.10 (a) Schematic illustration of the exfoliation process of the synthesis of MoS_2/g-C_3N_4 photocatalysts. *Source:* Yuan et al. [65] / with permission of Elsevier. (b) AFM images of the prepared g-C_3N_4 nanosheets and its height profile of the nanosheets. *Source:* Hatamie et al. [66] / With permission of Elsevier.

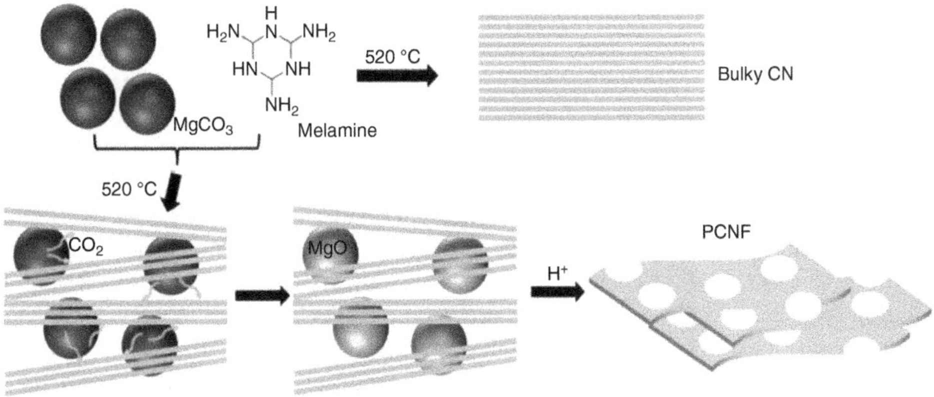

Figure 5.11 Diagrammatic representation of the formation mechanism of porous g-C_3N_4 flakes. *Source:* Yan et al. [69] / with permission of Elsevier.

In conclusion, liquid exfoliation approach has several advantages, such as usage of green solvent like water, producing material with high surface area, and high crystallinity, which are vital for photocatalytic reactions. But the tedious and lengthy synthetic procedure, high cost, and low yield limit the use of this method.

5.7.2.4 Chemical Blowing Method or Direct Thermal Polymerization Approach

In this method, the addition of chemical blowing agents, such as ammonium carbonate [67], ammonium chloride [68], and magnesium carbonate [69], with the precursor's result in the evolution of gas when subjected to heat which later gets trapped in between the layers and leads to the generation of porous exfoliated nanosheet structure. Porous g-C_3N_4 flakes were synthesized via thermally polymerizing melamine in the presence of a chemical blowing agent, i.e. $MgCO_3$ [69]. It was observed that during the reaction process, CO_2 gas was released due to the decomposition of $MgCO_3$ and further inhibits the polymerization of carbon nitride resulting in exfoliated layers of carbon nitride with flake-like morphology. Figure 5.11 clearly demonstrates the synthetic procedure and the mechanism of formation of porous g-C_3N_4 flakes.

Broadly, this approach is very useful in several ways. It is a single-step procedure saving a lot of time and generates porosity into the samples, which is vital for surface-dependent phenomena. But, the selection of a proper blowing agent for different precursors without affecting its optoelectronic properties is highly challenging.

5.8 Applications

Traditional fuels, such as coal, oil, and other energy resources, due to their limited abundance, have failed to meet the current energy demands. As compared to conventional catalysis, photocatalysis based on semiconductors is a more effective technology as photons energy is used here to activate the photocatalyst to catalyze the photocatalytic reaction. Solar energy is considered to be the new source that can be indefinitely used. Solar energy harvesting to generate a clean and high energy density hydrogen fuel has aroused significant interest [70]. There are several inorganic semiconductor photocatalysts, such as titania, metal sulfides, and phosphides, which have been widely used for photocatalytic water splitting [71]. But, still modifications are required in these photocatalysts so that the VB and CB of the catalyst matches that required for splitting water molecules under UV and visible light. Conventional bulk g-C_3N_4, due to its intrinsic structural defects, high exciton binding energy, a smaller number of active sites and small specific surface area, too shows weak photocatalytic hydrogen production activity. But metal-free porous nanostructured g-C_3N_4 semiconductor photocatalyst presents not only a suitable bandgap (2.7 eV) for water splitting but also a highly crystalline structure with several surface functional groups. There are only a few reports on photocatalytic oxygen evolution [3, 72] and overall water splitting [73] using g-C_3N_4. Since oxygen evolution is a four-electron process, thus it is quite complicated to take place as compared to hydrogen evolution. Due to the excellent thermal conductivity of the H_2, the quantitative determination of the produced H_2 is done suitably using a gas-chromatograph instrument which is coupled with a thermal conductivity detector. The porous structure along with high crystallinity is expected to reinforce the photocatalytic hydrogen reactions from water, but the faster recombination rate of electrons and holes limits the activity. Porous carbon nitrides are considered to be the robust photocatalyst as they can efficiently gather photons from the visible-light range. In this section, we will discuss a series of work on the PHE using porous carbon nitride materials possessing different nanostructures and nitrogen contents.

5.8.1 Photocatalytic Water Splitting

In recent years, the usage of g-C_3N_4 as a hydrogen evolution photocatalyst has been broadly studied. As discussed in earlier sections, the photocatalytic activity of pure g-C_3N_4 is not noticeable, thus Wang et al. in 2009 [3], for the first time, used polymeric non-porous g-C_3N_4 as photocatalyst for the production of hydrogen via water-splitting reaction under visible light. This report became a breakthrough in the field of materials chemistry and opened new pavements for carbon nitride materials as metal-free photocatalyst. It is manifested that g-C_3N_4 synthesized from numerous feedstocks [74–76] results in displaying disparate PHE activity as the products are of various structures. A lot of significant achievements have been obtained in the photocatalytic behavior and the reaction mechanism of non-porous carbon nitrides. However, the photocatalytic activity of non-porous carbon nitrides observed is very low due to the faster recombination rates and poor textural parameters.

Actually, there are several ways/factors to enhance the PHE activity of g-C_3N_4. For example, the introduction of nitrogen vacancies [77], surface hydrogen-bonding networks [78], the correct set-up of a photoreactor [79], exposure of photocatalytic systems to different atmospheres [80], decreasing the thickness of g-C_3N_4 nanosheets [75], introduction of porosity which not only increases the specific surface area but also enhances mass–diffusion channels [37, 81], elemental doping and molecular doping to narrow the bandgap for improvement of redox ability [82, 83], formation of semiconductor-semiconductor heterojunctions promoting charge separation [84, 85] are few of them. All these factors help in the momentous improvement in the PHE activity of g-C_3N_4, but still the activity was quite low. This persuades the researchers to explore porous carbon nitrides photocatalysts as they present a high surface area with a large number of active sites, leading to the higher light-harvesting capability and enhanced mass transfer [86].

In 2009 Wang et al., for the first time, prepared mesoporous carbon nitride using a silica-hard-templated self-polymerization route and by carefully optimizing the synthesis conditions [37]. Photocatalytic hydrogen activity was examined under visible-light irradiation. As observed, the rate of photochemical reduction of water of mpg-C_3N_4 increases by ~1 order of magnitude via introducing mesoporosity as compared to non-porous carbon nitride as shown in Figure 5.12a, b. However, the sample possessing a large surface area does not mean it will give higher hydrogen evolution activity because this mpg-C_3N_4 sample having a low surface area showed the highest hydrogen production rate. Thus, these results show the superiority of porous carbon nitrides over its non-porous analogues. Thereafter, a lot of reports discussed the PHE of porous carbon nitride materials prepared from different precursors. Chen et al. prepared mesoporous carbon nitride semiconductor host structure using SBA-15 template for PHE with visible light [81]. This report carefully examined the effect of light and mesoporous carbon nitrides on the H_2 evolution activity. No H_2 evolution was observed in presence of light and absence of the photocatalyst or vice versa. Moreover, the porous nature

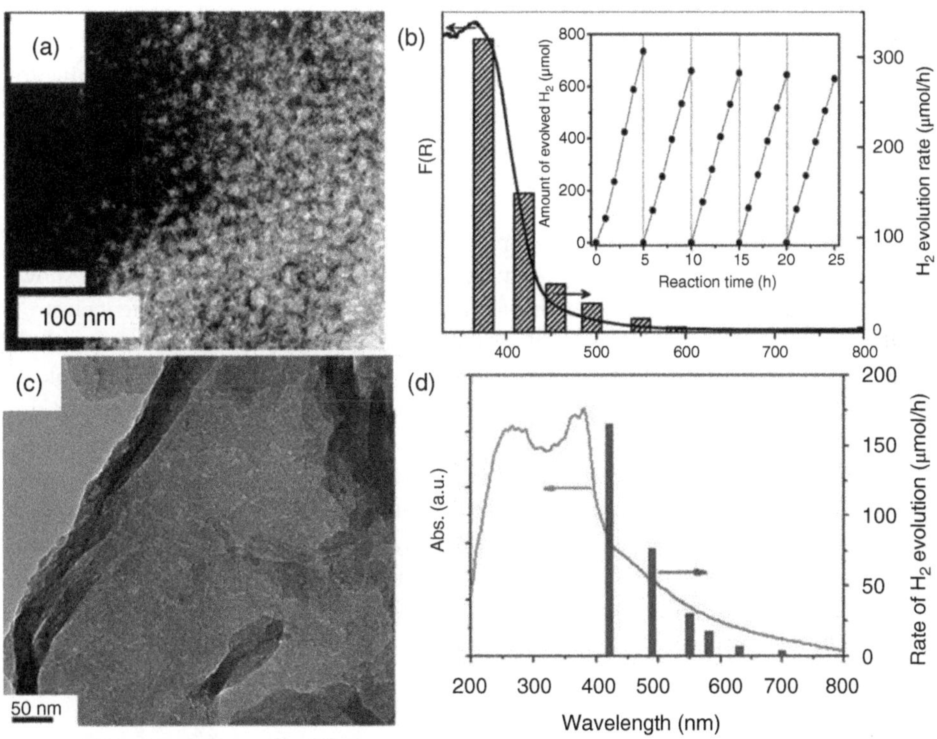

Figure 5.12 Mesoporous carbon nitrides photocatalyst for water splitting. (a) TEM image of mpg-C_3N_4 photocatalyst showing 3D porous structure and (b) wavelength dependence curve while the inset shows the stability test for Pt/mpg-C_3N_4 under visible-light range. *Source:* Wang et al. [37] / with permission of American Chemical Society. (c) High-resolution TEM image of g-C_3N_4 prepared by using Pluronic P-123 surfactant as a soft template. (d) Wavelength dependence of hydrogen evolution rate on g-C_3N_4. *Source:* Yan [87] / with permission of Royal Society of Chemistry.

of photocatalyst delivers a shorter charge separation time than bulk sample (non-porous), resulting in five times enhanced H_2 production activity than bulk carbon nitride.

Later, Cui et al. [88] prepared ordered mesoporous graphitic carbon nitride (ompg-C_3N_4) using ammonium thiocyanate and SBA-15 as the CN_x–precursor and hard template, respectively. Here, the involvement of a sulfur species within the reaction system offers additional chemical control of the size, morphology, pore size, and electronic structures. Also, the PHE activity of the synthesized g-C_3N_4 was improved as compared to g-C_3N_4 obtained from dicyandiamide while keeping high stability at the same time. Yan in 2012 prepared mesoporous g-C_3N_4 with worm-like pores and narrow pore size distribution using a novel soft-templating strategy (Figure 5.12c) [87]. There was a significant improvement in the PHE rate as displayed in Figure 5.12d. The worm-like porous g-C_3N_4 material not only acquires a high specific surface area but also presents an absorption edge extending up to 800 nm (Figure 5.12d). Using soft template Pluronic P-123 to prepare g-C_3N_4 catalyst leads to an enhanced PHE rate even when the irradiation light is $\lambda > 700$ nm. However, the apparent quantum efficiency of the photocatalyst at 420 nm was ca. 1.8% which is still low.

Furthermore, Zhang et al. in 2012 too [76] prepared porous g-C_3N_4 via a template-free approach using direct thermal polymerization of urea. Dicyandiamine and thiourea derived g-C_3N_4 were also prepared by the same method. One-step polymerization of the single precursor induces porosity into the samples. Under simulated sunlight irradiation, the urea-derived porous graphitic carbon nitride (UCN) shows a higher volume of hydrogen production than thiourea (TCN) and dicyandiamide (DCN) derived g-C_3N_4 as shown in Figure 5.13a, b. It was believed that the presence of an oxygen atom plays a key role in adjusting the textural properties and enhancing the photocatalytic activity. Han et al. [89] also prepared atomically thin mesoporous nanomesh of g-C_3N_4 using solvothermal exfoliation of mesoporous g-C_3N_4 fabricated through thermal polymerization of freeze-drying assembled dicyandiamide nanostructure precursor (Figure 5.13c–e). The material shows unique structural advantages of favorable alignment of the energy levels, efficient electron transfer, dynamic light-harvesting, and availability of highly active reaction sites. As a result, the 2D single atom-sheets of mesoporous g-C_3N_4 exhibits an excellent PHE rate of 8510 μmol/h/g under $\lambda > 420$ nm with an apparent efficiency of 5.1% at 420 nm (Figure 5.13f, g).

Similarly, Mane et al. [90] designed nitrogen-rich mesoporous carbon nitride using 3-amino-1,2,4-triazole precursor and KIT-6 silica template, which enabled the formation of well-ordered mesoporous carbon nitride with C_3N_5 stoichiometry (MCN-8). The prepared material possesses a 3D porous structure, low bandgap energy (2.2 eV), and graphitic structure with both tri-s-triazine framework that is similar to that of g-C_3N_4, but its surface adheres to triazole motifs. This report came as a breakthrough as this novel class of material without any dopants showed a superior photocatalytic performance as compared to other previous reports on g-C_3N_4 with a total of 801 μmol of H_2 under visible-light irradiation for three hours. This report opens up the opportunities where more nitrogen-rich samples can be prepared by varying the precursors. Additionally, Talapaneni et al. [91] developed a novel diaminotetrazine-based mesoporous C_3N_6 with highly ordered 3D mesoporous carbon nitride (MCN-9) through a self-polymerization of aminoguanidine hydrochloride infused into the nanochannels of KIT-6 template (Figure 5.14a), whose structure is different from normal g-C_3N_4 (Figure 5.14c, d). The prepared material

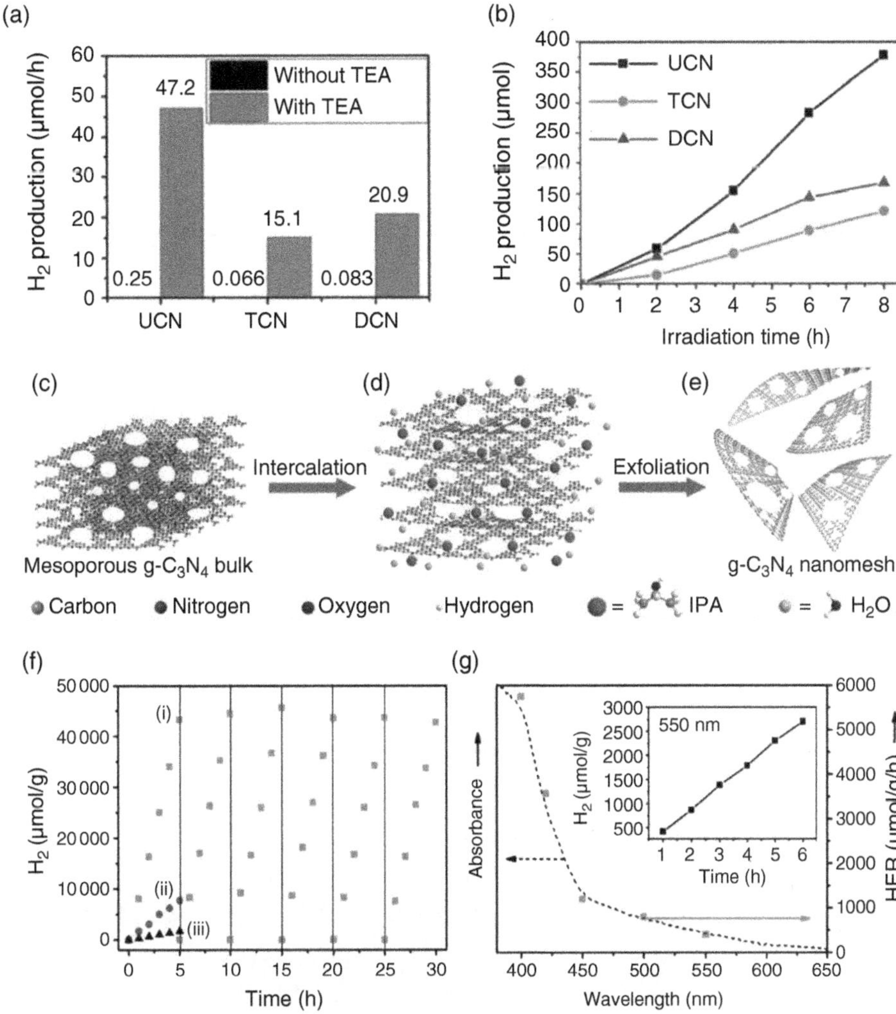

Figure 5.13 Hydrogen production rate from water by UCN, TCN, and DCN (a) without and with triethanolamine (TEA) (~10 vol%) as the sacrificial agent under simulated sunlight by Pt-loaded g-C_3N_4. (b) Stability study of the hydrogen production rate for eight hours with TEA. *Source:* Zhang et al. [76] / with permission of Royal Society of Chemistry. (c–e) Diagrammatic representation of solvothermal exfoliation process from mesoporous g-C_3N_4 bulk to nanomesh. (f) Photocatalytic activity study of (i) the single atomic layer mesoporous g-C_3N_4 nanomesh, (ii) the bulk counterpart (iii) the traditional g-C_3N_4 bulk with cycling stability study of mesoporous g-C_3N_4 nanomesh. (g) Wavelength dependence H_2 evolution rate on the monolayer g-C_3N_4 nanomesh. *Source:* Han et al. [89] / with permission of American Chemical Society.

presents unique structural properties such as high specific surface area, large pore size and volume, and a tunable low bandgap (2.25–2.50 eV). The photocatalytic experiments revealed that MCN-9 with 3 wt% Pt cocatalyst and 10 vol % triethanolamine produces 31.5 μmol/h of H_2 (Figure 5.14b), which was five times higher than non-porous C_3N_6 sample and apparent quantum efficiency of 0.21% at 405 nm. The time course of H_2 evolution

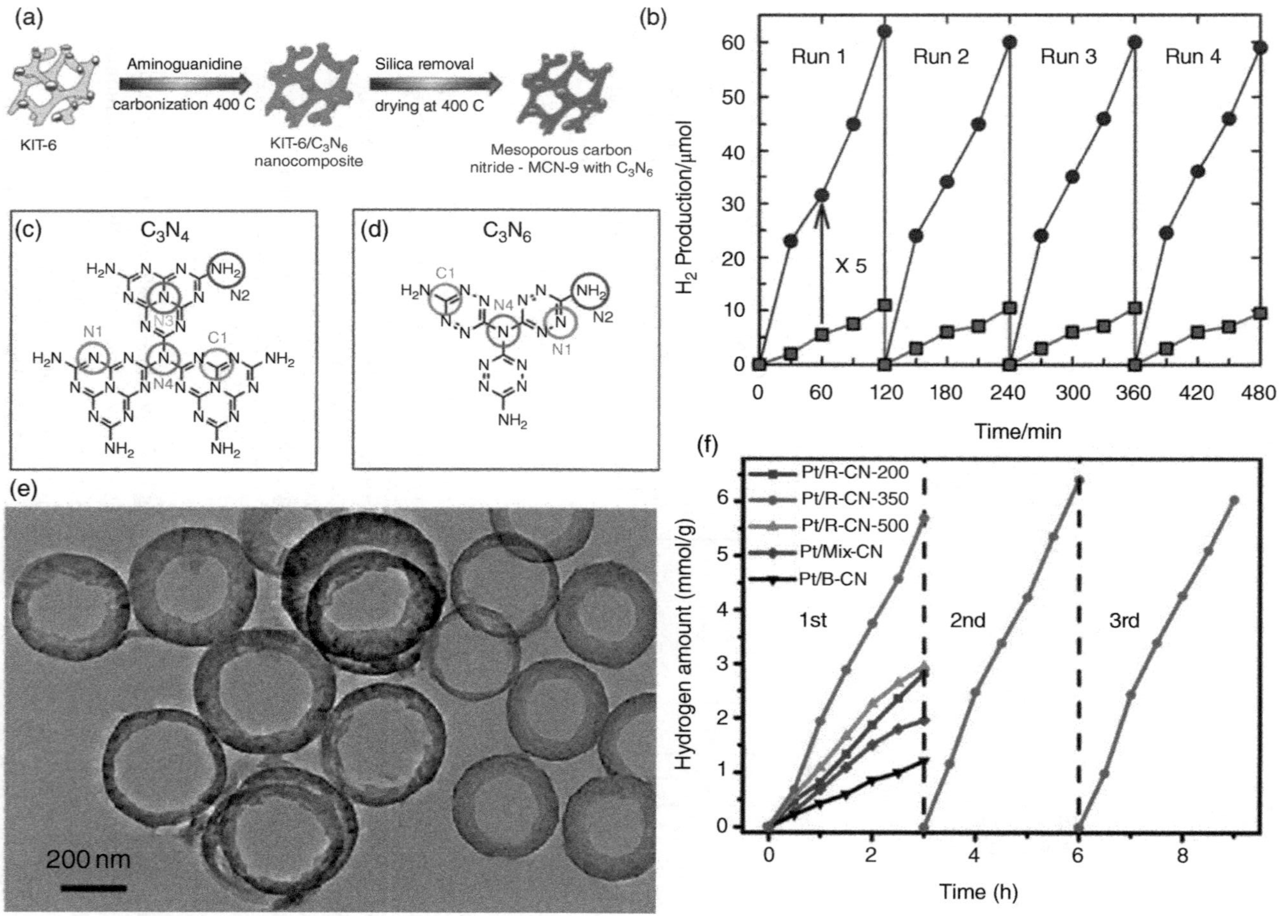

Figure 5.14 (a) Diagrammatic illustration of synthetic procedure of MCN-9 using KIT-6. Schematic molecular structure of (b) g-C_3N_4 and (c) MCN-9 with the stoichiometry of C_3N_6. (d) Hydrogen evolution experiment using MCN-9 photocatalyst (circles) and bulk non-porous C_3N_6 (squares) prepared from aminoguanidine hydrochloride without any template. *Source:* Talapaneni et al. [91] / Royal Society of Chemistry / CC BY-3.0. (e) TEM images of onion-ring-like g-C_3N_4. (f) Time course of H_2 evolution and photocatalytic stability over multiple reaction cycles. *Source:* Cui et al. [92] / with permission of American Chemical Society.

for 12 hours (Figure 5.13f) states the excellent stability of the MCN-9 sample as an organic photocatalyst for solar H_2 production. Recently, Cui et al. [92] in 2018 prepared highly uniform porous onion-ring-like g-C_3N_4 (Figure 5.14e) using an in-air chemical vapor deposition (CVD) method and SiO_2 microspheres (350 nm) as a hard template and melamine as a CVD precursor. The onion-ring-like g-C_3N_4 presents a narrower bandgap (2.58 eV) as compared to bulk g-C_3N_4 (2.70 eV). The as-prepared material in the presence of 3 wt% cocatalyst and 10% triethanolamine as sacrificial agent shows an excellent H_2 evolution rate of 1900 μmol/h/g under visible-light irradiation, which is five times higher than that of bulk (406.5 μmol/h/g) (Figure 5.14f). The excellent activity is corroborated to the efficient charge separation and extended lifetime of photoinduced charge carriers due to the highly porous structure.

Liu et al. [93] prepared 3D mesoporous g-C_3N_4 nanosheets by calcining melamine-cyanaurate supramolecular precursor complex, where the material acquires a thin mesoporous network with enhanced high surface area, increased optical absorption, and higher number of active sites. The material in the presence of 0.2 wt% Pt co-catalysts shows an outstanding PHE performance with HER rate of 136.9 μmol/h and apparent quantum efficiency of 10.6% at 420 nm. Analogously, using acetic acid-treated melamine precursor nitrogen-deficient g-C_3N_4 has been prepared [94]. The as-prepared material possesses a loose porous structure with a specific surface area of 19.24 m^2/g, an ample number of active sites, restrained recombination of charge carriers, and efficient visible-light utilization. The material shows a superior PHE activity (257.5 μmol/h/g) ~2.5 times higher than that of pristine g-C_3N_4. Later in 2019, ultrathin holey non-stoichiometric nitrogen-rich g-$C_3N_{4.8}$ nanosheets were prepared by Deka et al. for the first time using cheap feedstocks via a direct thermal polymerization and chemical etching process [67]. The as-prepared material with unique structural and morphological advantages, such as a highly porous structure (Figure 5.15c), exhibits excellent photocatalytic water-splitting activity of H_2 evolution (2620 μmol/h/g) with an apparent quantum efficiency of 8.5% at 427 nm without any help of any co-catalysts under visible light in the presence of 20% triethanolamine (Figure 5.15a, b). The superlative performance of the sample is due to the high accessible surface area giving a higher number of active sites and the presence of high nitrogen content at the surface generates more basic sites resulting in excellent PHE.

Chen et al. [95] prepared 3D porous g-C_3N_4 ultrathin nanosheets using a facile bottom-up supramolecular self-assembly route (shown in Figure 5.15d), which could directly split pure water in the presence of Pt co-catalysts into H_2 and O_2 with a high evolution rate up to 101.4 and 49.1 μmol/g/h under visible-light irradiation, respectively. These evolutions are approximately 11.8 and 5.1 times higher than bulk g-C_3N_4 and g-C_3N_4 nanosheets and previously reported Pt/g-C_3N_4 with a remarkable stability of more than 100 hours of overall water-splitting reaction. The 3D interconnected open-framework constructed from crystalline ultrathin nanosheets caters faster charge carrier transport leading to frivolous photocatalytic activity. Similarly, porous g-C_3N_4 nanosheets were too developed from bulk g-C_3N_4 using urea-assisted precursor pretreatment by the hydrothermal method [96]. This modified porous g-C_3N_4 (BCN-U) results in more number of active sites for reaction, higher surface area, broader visible-light response, and much lower bandgap (1.96 eV) as compared to bulk (2.71 eV) (BCN) (Figure 5.16). The broadened visible-light response of the porous carbon nitride sample is responsible for the enhanced photocatalytic activity.

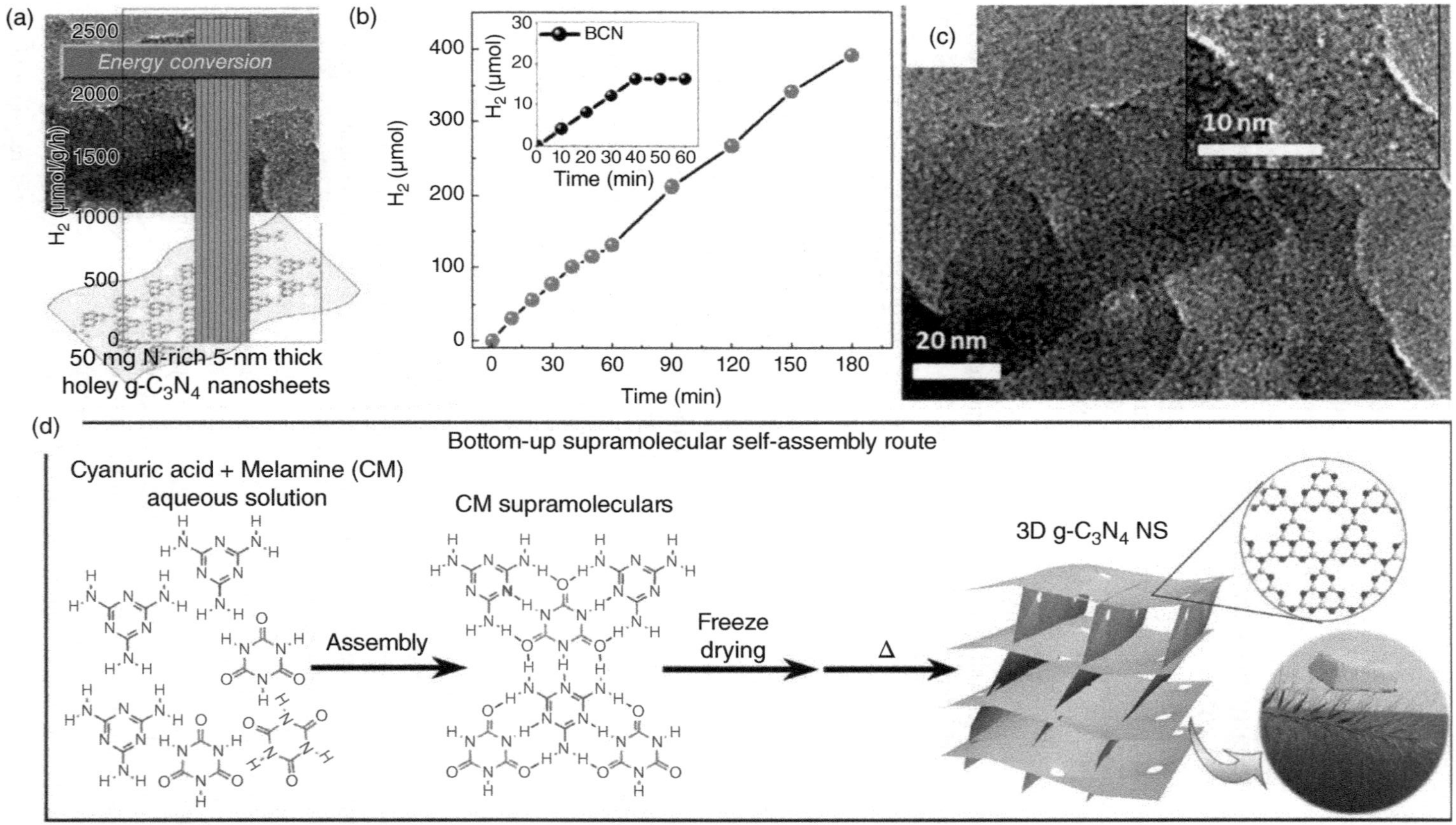

Figure 5.15 (a, b) Time course of H_2 evolution under the irradiation of visible light of holey g-$C_3N_{4.8}$ nanosheets (NCN) and bulk (BCN) photocatalyst. (c) High-resolution TEM image of holey g-$C_3N_{4.8}$ nanosheets (NCN). *Source:* Antil et al. [67] / with permission of American Chemical Society. (d) Schematic view for the development of 3D porous g-C_3N_4 nanosheets. *Source:* Chen et al. [95] / with permission of Elsevier.

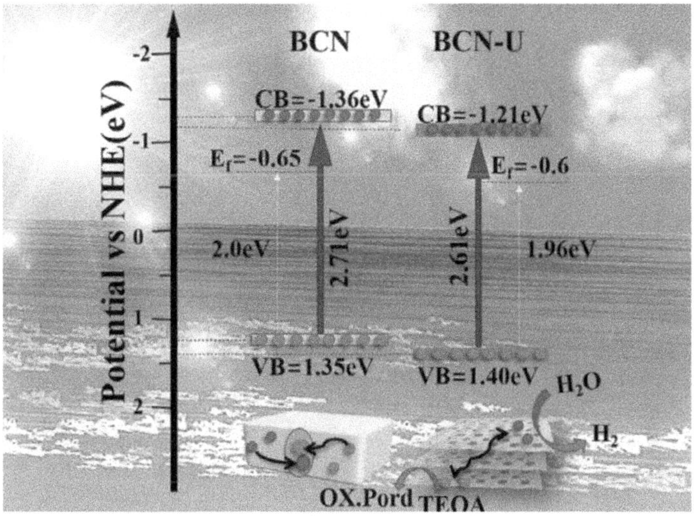

Figure 5.16 Diagrammatic representation of band structure and charge separation of BCN and BCN-U. *Source:* Bai et al. [96] / with permission of Elsevier.

Qian et al. [97] prepared 3D porous g-C_3N_4 using a salt-assisted ball-milling strategy where it presents several impressive properties vital for PHE reaction. The material exhibits a hydrogen evolution rate of 589 μmol/h/g and an apparent quantum yield of 3.31% at 420 nm. Recently, Huang et al. [98] introduced the design of defect-modified thin-layered and porous g-C_3N_4 for highly efficient solar-energy conversion. They used a new facile urea- and KOH-assisted thermal polymerization strategy where the thickness of the g-C_3N_4 was efficiently modulated, and the cyano groups, nitrogen vacancies, and mesopores were also incorporated. Due to these modifications, the material acquires a higher surface area, improved light-harvesting capability over a broad range, and efficient photogenerated charge separation. Also, the formation of in-plane mesopores induces a faster mass-transfer process helpful for rapid reaction kinetics. As a result of these synergistic effects, the obtained material demonstrates a significant improvement in the PHE activity (1.557 mmol/h/g) and an apparent quantum yield of 18.5% at 420 nm under visible-light irradiation which is approximately 48.5 times higher than that of pristine g-C_3N_4 prepared by direct thermal polymerization. Similarly, Deka et al. [99] in 2020 prepared highly ordered nitrogen-rich honey-comb-like mesoporous carbon nitride nanosheets following a liquid-crystal template/nanocasting approach using 3-amino-1,2,4-triazole precursor as shown in Figure 5.17a, b. The as-prepared material acquires a high surface area and uniformly ordered structure leading to an excellent PHE activity (8180 μmol/h/g) and an outstanding apparent quantum efficiency of 27.14% at 420 nm in the presence of Pt co-catalysts. This report not only essays a new synthetic protocol but also demonstrates a renewable green approach for the production of hydrogen fuel. This synergistic modulation strategy with tunable properties such as high surface area, pore size and volume, nitrogen-rich surface, reduced bandgap, and higher average lifetimes of charge carriers opens up new pathways for the modification of 2D materials toward advanced applications. Recently, Antil et al. have reported vermicular rope-like g-C_3N_4 nanostructure consisting of multichannel

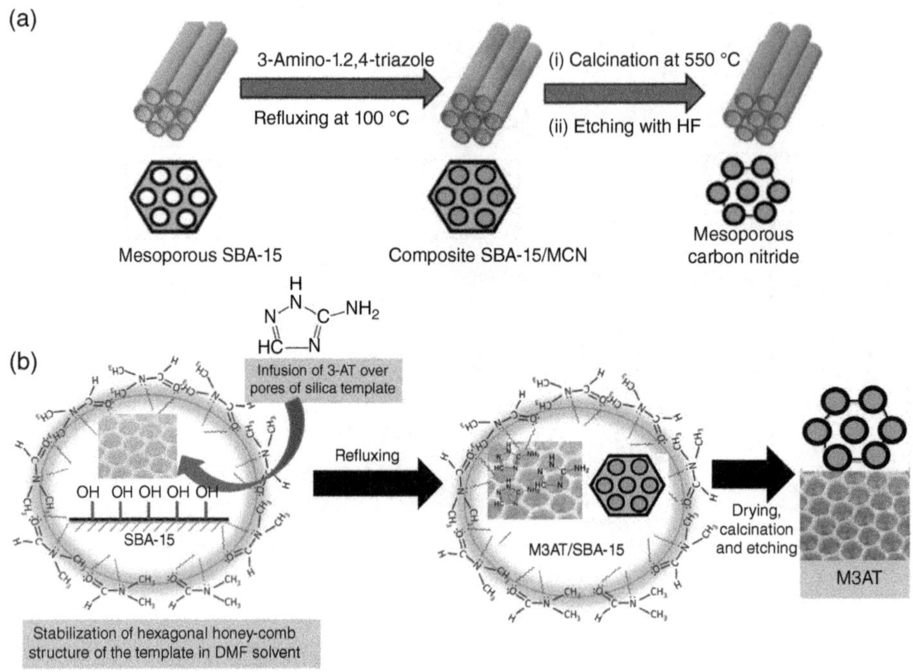

Figure 5.17 (a) Schematic representation of the preparation of the hexagonally ordered mesoporous carbon nitride material. (b) Study of the reaction mechanism for the formation of mesoporous carbon nitride material. *Source:* Antil et al. [99] / with permission of Royal Society of Chemistry.

tubular pores with excess nitrogen content, synthesized through a cost-effective approach featuring the final stoichiometry g-$C_3N_{4.7}$ for hydrogen evolution application [100]. This material possess abundant channels for the diffusion process, ample-electroactive sites from N-electron lone pairs, the immensely reduced recombination rate of photoexcited exciton pairs, and a high donor concentration. A visible-light-driven PHE rate for this photocatalyst is found to be as high as 4910 μmol/h/g with an apparent quantum yield (AQY) of 14.07% at bandgap absorption (2.59 eV, 479 nm) [100].

These examples reveal the possibility of efficient PHE activity of porous g-C_3N_4 photocatalyst under visible-light irradiation. However, a lot of studies and improvements in this area are still required and intensive investigations of the photocatalytic activity over various g-C_3N_4-based photocatalyst are highly encouraged.

5.9 Conclusion and Outlook

This chapter shows that porous nanostructured graphitic carbon nitrides are excellent visible-light hydrogen evolution photocatalysts due to their tunable electronic and optical properties. The carbon nitride material possesses a narrow bandgap and the appropriate positions of the CB and VB favor it as a visible-light-active photocatalyst for a wide range of reactions. Also, the high thermal and chemical stability (acids, bases, and organic solvents) enables the use of the material as a catalyst in various heterogeneous reactions. Carbon nitrides are usually prepared by simply thermal polymerizing the low-cost nitrogen-rich precursors; thus, it is cheap and environment friendly. The bulk graphitic carbon nitride usually obtained from the direct thermal polymerization technique. Despite having a stacked layered nanosheet structure, bulk g-C_3N_4 possesses low surface area. Therefore, to circumvent these advantages, the bulk sample is either liquid-exfoliated or thermally exfoliated to obtain a few-layered porous g-C_3N_4. The porosity is introduced into the carbon nitride nanostructure by using templating methods, supramolecular-chemistry approach and even template-free routes. The porous carbon nitride nanostructures usually gain a large surface area and improve light-harvesting ability and enhanced charge-carrier separation efficiency. Here, in this review, we have carefully summarized the current research trends in the area of porous carbon nitride materials and provided a brief discussion on the synthetic approach using various C and N feedstocks and the application of porous carbon nitrides in PHE under visible-light irradiation.

In the view of plenty of research articles, porous carbon nitrides demonstrate tremendous potential and offer exquisite performance in visible-light PHE reaction, but they still need expensive co-catalysts and sacrificial agents, making the whole process costly. Thus, it is highly desirable to design porous carbon nitride with an optimized bandgap structure acquainted with functional moieties into the mesoporous channels, which can effectively split water molecules without any kind of co-catalysts and sacrificial agents. To prepare carbon nitrides with a varied amount of C and N stoichiometric ratios is also a challenging task where the band edges of the materials can be carefully optimized, extending the visible-light absorption range. Porous carbon nitrides with different nitrogen contents are usually prepared by choosing suitable carbon and nitrogen precursors as well as apt carbonization temperature along with a strategy to induce porosity into the material. Porous carbon nitrides with interesting features, such as porosity, controlled morphology, and crystal structures, are highly aimed where the effects of controlled textural and structural parameters and optimized band structure will be highly advantageous for PHE reactions.

References

1 Asahi, R. (2001). Visible-light photocatalysis in nitrogen-doped titanium oxides. *Science* 293 (5528): 269–271.

2 Kudo, A. and Miseki, Y. (2009). Heterogeneous photocatalyst materials for water splitting. *Chem. Soc. Rev.* 38 (1): 253–278.

3 Wang, X., Maeda, K., Thomas, A. et al. (2009). A metal-free polymeric photocatalyst for hydrogen production from water under visible light. *Nat. Mater.* 8 (1): 76–80.

4 Fujishima, A. and Honda, K. (1972). Electrochemical photolysis of water at a semiconductor electrode. *Nature* 238 (5358): 37–38.

5 Inoue, T., Fujishima, A., Konishi, S., and Honda, K. (1979). Photoelectrocatalytic reduction of carbon dioxide in aqueous suspensions of semiconductor powders. *Nature* 277 (5698): 637–638.

6 Zhang, W., Peng, Q., Shi, L. et al. (2019). Merging single-atom-dispersed iron and graphitic carbon nitride to a joint electronic system for high-efficiency photocatalytic hydrogen evolution. *Small* 15 (50): 1905166.

7 Yu, J., Low, J., Xiao, W. et al. (2014). Enhanced photocatalytic CO_2 -reduction activity of anatase TiO_2 by coexposed {001} and {101} facets. *J. Am. Chem. Soc.* 136 (25): 8839–8842.

8 Jang, E.S., Won, J.-H., Hwang, S.-J., and Choy, J.-H. (2006). Fine tuning of the face orientation of ZnO crystals to optimize their photocatalytic activity. *Adv. Mater.* 18 (24): 3309–3312.

9 Kush, P., Ujjain, S.K., Mehra, N.C. et al. (2013). Development and properties of surfactant-free water-dispersible Cu_2ZnSnS_4 nanocrystals: a material for low-cost photovoltaics. *ChemPhysChem* 14 (12): 2793–2799.

10 Kush, P., Deori, K., Kumar, A., and Deka, S. (2015). Efficient hydrogen/oxygen evolution and photocatalytic dye degradation and reduction of aqueous Cr(vi) by surfactant free hydrophilic Cu_2ZnSnS_4 nanoparticles. *J. Mater. Chem. A* 3 (15): 8098–8106.

11 Cao, S.-W. and Zhu, Y.-J. (2008). Hierarchically nanostructured α-Fe_2O_3 hollow spheres: preparation, growth mechanism, photocatalytic property, and application in water treatment. *J. Phys. Chem. C* 112 (16): 6253–6257.

12 Chauhan, H., Soni, K., Kumar, M., and Deka, S. (2016). Tandem photocatalysis of graphene-stacked SnS_2 nanodiscs and nanosheets with efficient carrier separation. *ACS Omega* 1 (1): 127–137.

13 Liu, S., Yin, K., Ren, W. et al. (2012). Tandem photocatalytic oxidation of Rhodamine B over surface fluorinated bismuth vanadate crystals. *J. Mater. Chem.* 22 (34): 17759.

14 Huang, W.-C., Lyu, L.-M., Yang, Y.-C., and Huang, M.H. (2012). Synthesis of Cu_2O nanocrystals from cubic to rhombic dodecahedral structures and their comparative photocatalytic activity. *J. Am. Chem. Soc.* 134 (2): 1261–1267.

15 Chauhan, M., Soni, K., Karthik, P.E. et al. (2019). Promising visible-light driven hydrogen production from water on a highly efficient $CuCo_2S_4$ nanosheet photocatalyst. *J. Mater. Chem. A* 7 (12): 6985–6994.

16 Hu, Y., Gao, X., Yu, L. et al. (2013). Carbon-coated CdS petalous nanostructures with enhanced photostability and photocatalytic activity. *Angew. Chem. Int. Ed.* 52 (21): 5636–5639.

17 Tsuji, I., Kato, H., Kobayashi, H., and Kudo, A. (2004). Photocatalytic H_2 evolution reaction from aqueous solutions over band structure-controlled $(AgIn)_xZn_{2(1-x)}S_2$ solid solution photocatalysts with visible-light response and their surface nanostructures. *J. Am. Chem. Soc.* 126 (41): 13406–13413.

18 Tachibana, Y., Vayssieres, L., and Durrant, J.R. (2012). Artificial photosynthesis for solar water-splitting. *Nat. Photonics* 6 (8): 511–518.

19 Kim, Y.I., Salim, S., Huq, M.J., and Mallouk, T.E. (1991). Visible-light photolysis of hydrogen iodide using sensitized layered semiconductor particles. *J. Am. Chem. Soc.* 113 (25): 9561–9563.

20 Zou, Z., Ye, J., Sayama, K., and Arakawa, H. (2001). Direct splitting of water under visible light irradiation with an oxide semiconductor photocatalyst. *Nature* 414 (3): 625–627.

21 Liu, M., You, W., Lei, Z. et al. (2004). Water reduction and oxidation on Pt–Ru/$Y_2Ta_2O_5N_2$ catalyst under visible light irradiation. *Chem. Commun.* 19: 2192–2193.

22 Maeda, K., Teramura, K., Lu, D. et al. (2006). Photocatalyst releasing hydrogen from water. *Nature* 440 (7082): 295–295.

23 Wang, Y., Wang, Q., Zhan, X. et al. (2013). Visible light driven type II heterostructures and their enhanced photocatalysis properties: a review. *Nanoscale* 5 (18): 8326.

24 Yang, J., Wang, D., Han, H., and Li, C. (2013). Roles of cocatalysts in photocatalysis and photoelectrocatalysis. *Acc. Chem. Res.* 46 (8): 1900–1909.

25 Wang, Y., Wang, X., and Antonietti, M. (2012). Polymeric graphitic carbon nitride as a heterogeneous organocatalyst: from photochemistry to multipurpose catalysis to sustainable chemistry. *Angew. Chem. Int. Ed.* 51 (1): 68–89.

26 Lakhi, K.S., Park, D.-H., Al-Bahily, K. et al. (2017). Mesoporous carbon nitrides: synthesis, functionalization, and applications. *Chem. Soc. Rev.* 46 (1): 72–101.

27 Cohen, M.L. (1985). Calculation of bulk moduli of diamond and zinc-blende solids. *Phys. Rev. B* 32 (12): 7988–7991.

28 Liu, A.Y. and Cohen, M.L. (1989). Prediction of new low compressibility solids. *Science* 245 (4920): 841–842.

29 Cao, S., Low, J., Yu, J., and Jaroniec, M. (2015). Polymeric photocatalysts based on graphitic carbon nitride. *Adv. Mater.* 27 (13): 2150–2176.

30 Su, F., Antonietti, M., and Wang, X. (2012). mpg-C_3N_4 as a solid base catalyst for Knoevenagel condensations and transesterification reactions. *Catal. Sci. Technol.* 2 (5): 1005.

31 Liebig, J. (1834). Uber einige Stickstoff – Verbindungen. *Ann. Pharmacother.* 10 (1): 1–47.

32 Zhu, J., Xiao, P., Li, H., and Carabineiro, S.A.C. (2014). Graphitic carbon nitride: synthesis, properties, and applications in catalysis. *ACS Appl. Mater. Interfaces* 6 (19): 16449–16465.

33 Yang, S., Gong, Y., Zhang, J. et al. (2013). Exfoliated graphitic carbon nitride nanosheets as efficient catalysts for hydrogen evolution under visible light. *Adv. Mater.* 25 (17): 2452–2456.

34 Rono, N., Kibet, J.K., Martincigh, B.S., and Nyamori, V.O. (2020). A review of the current status of graphitic carbon nitride. *Crit. Rev. Solid State Mater. Sci.* 46 (3): 1–29.

35 Oh, T., Kim, M., Choi, J., and Kim, J. (2018). Design of graphitic carbon nitride nanowires with captured mesoporous carbon spheres for EDLC electrode materials. *Ionics* 24 (12): 3957–3965.

36 He, F., Chen, G., Miao, J. et al. (2016). Sulfur-mediated self-templating synthesis of tapered C-PAN/g-C_3N_4 composite nanotubes toward efficient photocatalytic H_2 evolution. *ACS Energy Lett.* 1 (5): 969–975.

37 Wang, X., Maeda, K., Chen, X. et al. (2009). Polymer semiconductors for artificial photosynthesis: hydrogen evolution by mesoporous graphitic carbon nitride with visible light. *J. Am. Chem. Soc.* 131 (5): 1680–1681.

38 Zhang, X., Xie, X., Wang, H. et al. (2013). Enhanced photoresponsive ultrathin graphitic-phase C_3N_4 nanosheets for bioimaging. *J. Am. Chem. Soc.* 135 (1): 18–21.

39 Zhu, J., Wei, Y., Chen, W. et al. (2010). Graphitic carbon nitride as a metal-free catalyst for NO decomposition. *Chem. Commun.* 46 (37): 6965.

40 Wirth, J., Neumann, R., Antonietti, M., and Saalfrank, P. (2014). Adsorption and photocatalytic splitting of water on graphitic carbon nitride: a combined first principles and semiempirical study. *Phys. Chem. Chem. Phys.* 16 (30): 15917–15926.

41 Dong, F., Zhao, Z., Xiong, T. et al. (2013). In situ construction of g-C_3N_4/g-C_3N_4 metal-free heterojunction for enhanced visible-light photocatalysis. *ACS Appl. Mater. Interfaces* 5 (21): 11392–11401.

42 Dai, K., Lv, J., Lu, L. et al. (2016). A facile fabrication of plasmonic g-C_3N_4/Ag_2WO_4/Ag ternary heterojunction visible-light photocatalyst. *Mater. Chem. Phys.* 177: 529–537.

43 Groenewolt, M. and Antonietti, M. (2005). Synthesis of g-C_3N_4 nanoparticles in mesoporous silica host matrices. *Adv. Mater.* 17 (14): 1789–1792.

44 Zhao, Z., Dai, Y., Lin, J., and Wang, G. (2014). Highly-ordered mesoporous carbon nitride with ultrahigh surface area and pore volume as a superior dehydrogenation catalyst. *Chem. Mater.* 26 (10): 3151–3161.

45 Wei, H., McMaster, W.A., Tan, J.Z.Y. et al. (2017). Mesoporous TiO_2/g-C_3N_4 microspheres with enhanced visible-light photocatalytic activity. *J. Phys. Chem. C* 121 (40): 22114–22122.

46 Hao, R., Wang, G., Tang, H. et al. (2016). Template-free preparation of macro/mesoporous g-C_3N_4/TiO_2 heterojunction photocatalysts with enhanced visible light photocatalytic activity. *Appl. Catal. B Environ.* 187: 47–58.

47 Zou, X., Yang, Y., Chen, H. et al. (2021). Hierarchical meso/macro-porous TiO_2/graphitic carbon nitride nanofibers with enhanced hydrogen evolution. *Mater. Des.* 202: 109542.

48 Liang, C., Hong, K., Guiochon, G.A. et al. (2004). Synthesis of a large-scale highly ordered porous carbon film by self-assembly of block copolymers. *Angew. Chem. Int. Ed.* 43 (43): 5785–5789.

49 Kosonen, H., Valkama, S., Nykänen, A. et al. (2006). Functional porous structures based on the pyrolysis of cured templates of block copolymer and phenolic resin. *Adv. Mater.* 18 (2): 201–205.

50 Meng, Y., Gu, D., Zhang, F. et al. (2006). A family of highly ordered mesoporous polymer resin and carbon structures from organic–organic self-assembly. *Chem. Mater.* 18 (18): 4447–4464.

51 Wang, Y., Wang, X., Antonietti, M., and Zhang, Y. (2010). Facile one-pot synthesis of nanoporous carbon nitride solids by using soft templates. *ChemSusChem* 3 (4): 435–439.

52 Peer, M., Lusardi, M., and Jensen, K.F. (2017). Facile soft-templated synthesis of high-surface area and highly porous carbon nitrides. *Chem. Mater.* 29 (4): 1496–1506.

53 Wang, Y., Zhang, J., Wang, X. et al. (2010). Boron- and fluorine-containing mesoporous carbon nitride polymers: metal-free catalysts for cyclohexane oxidation. *Angew. Chem. Int. Ed.* 49 (19): 3356–3359.

54 Yan, Q., Zhao, C., Zhang, L. et al. (2019). Facile two-step synthesis of porous carbon nitride with enhanced photocatalytic activity using a soft template. *ACS Sustain. Chem. Eng.* 7 (4): 3866–3874.

55 Arrachart, G., Carcel, C., Trens, P. et al. (2009). Silylated melamine and cyanuric acid as precursors for imprinted and hybrid silica materials with molecular recognition properties. *Chem. Eur. J.* 15 (25): 6279–6288.

56 Shalom, M., Guttentag, M., Fettkenhauer, C. et al. (2014). *In situ* formation of heterojunctions in modified graphitic carbon nitride: synthesis and noble metal free photocatalysis. *Chem. Mater.* 26 (19): 5812–5818.

57 Xiao, Y., Tian, G., Li, W. et al. (2019). Molecule self-assembly synthesis of porous few-layer carbon nitride for highly efficient photoredox catalysis. *J. Am. Chem. Soc.* 141 (6): 2508–2515.

58 Simanek, E.E., Mammen, M., Gordon, D.M. et al. (1995). Design and synthesis of hydrogen-bonded aggregates. Theory and computation applied to three systems based on the cyanuric acid-melamine lattice. *Tetrahedron* 51 (2): 607–619.

59 Li, Y., Li, X., Zhang, H., and Xiang, Q. (2020). Porous graphitic carbon nitride for solar photocatalytic applications. *Nanoscale Horiz.* 5 (5): 765–786.

60 Hummers, W.S. and Offeman, R.E. (1958). Preparation of graphitic oxide. *J. Am. Chem. Soc.* 80 (6): 1339–1339.

61 Niu, P., Zhang, L., Liu, G., and Cheng, H.-M. (2012). Graphene-like carbon nitride nanosheets for improved photocatalytic activities. *Adv. Funct. Mater.* 22 (22): 4763–4770.

62 Xu, J., Zhang, L., Shi, R., and Zhu, Y. (2013). Chemical exfoliation of graphitic carbon nitride for efficient heterogeneous photocatalysis. *J. Mater. Chem. A* 1 (46): 14766.

63 Challagulla, S., Payra, S., Chakraborty, C., and Roy, S. (2019). Determination of band edges and their influences on photocatalytic reduction of nitrobenzene by bulk and exfoliated g-C_3N_4. *Phys. Chem. Chem. Phys.* 21 (6): 3174–3183.

64 Sun, S., Li, J., Cui, J. et al. (2019). Simultaneously engineering K-doping and exfoliation into graphitic carbon nitride (g-C_3N_4) for enhanced photocatalytic hydrogen production. *Int. J. Hydrog. Energy* 44 (2): 778–787.

65 Yuan, Y.-J., Shen, Z., Wu, S. et al. (2019). Liquid exfoliation of g-C_3N_4 nanosheets to construct 2D-2D MoS_2/g-C_3N_4 photocatalyst for enhanced photocatalytic H_2 production activity. *Appl. Catal. B Environ.* 246: 120–128.

66 Hatamie, A., Jalilian, P., Rezvani, E. et al. (2019). Fast and ultra-sensitive voltammetric detection of lead ions by two-dimensional graphitic carbon nitride (g-C_3N_4) nanolayers as glassy carbon electrode modifier. *Measurement* 134: 679–687.

67 Antil, B., Kumar, L., Reddy, K.P. et al. (2019). Direct thermal polymerization approach to N-Rich holey carbon nitride nanosheets and their promising photocatalytic H_2 evolution and charge-storage activities. *ACS Sustain. Chem. Eng.* 7 (10): 9428–9438.

68 Talapaneni, S.N., Lee, J.H., Je, S.H. et al. (2017). Chemical blowing approach for ultramicroporous carbon nitride frameworks and their applications in gas and energy storage. *Adv. Funct. Mater.* 27 (1): 1604658.

69 Yan, J., Han, X., Zheng, X. et al. (2017). One-step template/chemical blowing route to synthesize flake-like porous carbon nitride photocatalyst. *Mater. Res. Bull.* 94: 423–427.

70 Liu, X., Pang, F., He, M., and Ge, J. (2017). Confined reaction inside nanotubes: new approach to mesoporous g-C_3N_4 photocatalysts. *Nano Res.* 10 (11): 3638–3647.

71 Osterloh, F.E. (2008). Inorganic materials as catalysts for photochemical splitting of water. *Chem. Mater.* 20 (1): 35–54.

72 Maeda, K., Wang, X., Nishihara, Y. et al. (2009). Photocatalytic activities of graphitic carbon nitride powder for water reduction and oxidation under visible light. *J. Phys. Chem. C* 113 (12): 4940–4947.

73 Lee, R.-L., Tran, P.D., Pramana, S.S. et al. (2013). Assembling graphitic-carbon-nitride with cobalt-oxide-phosphate to construct an efficient hybrid photocatalyst for water splitting application. *Catal. Sci. Technol.* 3 (7): 1694.

74 Zhang, G., Zhang, J., Zhang, M., and Wang, X. (2012). Polycondensation of thiourea into carbon nitride semiconductors as visible light photocatalysts. *J. Mater. Chem.* 22 (16): 8083.

75 Jorge, A.B., Martin, D.J., Dhanoa, M.T.S. et al. (2013). H_2 and O_2 evolution from water half-splitting reactions by graphitic carbon nitride materials. *J. Phys. Chem. C* 117 (14): 7178–7185.

76 Zhang, Y., Liu, J., Wu, G., and Chen, W. (2012). Porous graphitic carbon nitride synthesized via direct polymerization of urea for efficient sunlight-driven photocatalytic hydrogen production. *Nanoscale* 4 (17): 5300.

77 Niu, P., Liu, G., and Cheng, H.-M. (2012). Nitrogen vacancy-promoted photocatalytic activity of graphitic carbon nitride. *J. Phys. Chem. C* 116 (20): 11013–11018.

78 Wang, X.L., Fang, W.Q., Wang, H.F. et al. (2013). Surface hydrogen bonding can enhance photocatalytic H_2 evolution efficiency. *J. Mater. Chem. A* 1 (45): 14089.

79 Schwarze, M., Stellmach, D., Schröder, M. et al. (2013). Quantification of photocatalytic hydrogen evolution. *Phys. Chem. Chem. Phys.* 15 (10): 3466.

80 Liu, J., Zhang, Y., Lu, L. et al. (2012). Self-regenerated solar-driven photocatalytic water-splitting by urea derived graphitic carbon nitride with platinum nanoparticles. *Chem. Commun.* 48 (70): 8826.

81 Chen, X., Jun, Y.-S., Takanabe, K. et al. (2009). Ordered mesoporous SBA-15 type graphitic carbon nitride: a semiconductor host structure for photocatalytic hydrogen evolution with visible light. *Chem. Mater.* 21 (18): 4093–4095.

82 Tahir, M., Cao, C., Mahmood, N. et al. (2014). Multifunctional g-C_3N_4 nanofibers: a template-free fabrication and enhanced optical, electrochemical, and photocatalyst properties. *ACS Appl. Mater. Interfaces* 6 (2): 1258–1265.

83 Hong, J., Xia, X., Wang, Y., and Xu, R. (2012). Mesoporous carbon nitride with in situ sulfur doping for enhanced photocatalytic hydrogen evolution from water under visible light. *J. Mater. Chem.* 22 (30): 15006.

84 Yan, H. and Yang, H. (2011). TiO_2–g-C_3N_4 composite materials for photocatalytic H_2 evolution under visible light irradiation. *J. Alloys Compd.* 509 (4): L26–L29.

85 Chai, B., Peng, T., Mao, J. et al. (2012). Graphitic carbon nitride (g-C_3N_4)–Pt-TiO_2 nanocomposite as an efficient photocatalyst for hydrogen production under visible light irradiation. *Phys. Chem. Chem. Phys.* 14 (48): 16745.

86 Maeda, K., Kuriki, R., Zhang, M. et al. (2014). The effect of the pore-wall structure of carbon nitride on photocatalytic CO_2 reduction under visible light. *J. Mater. Chem. A* 2 (36): 15146–15151.

87 Yan, H. (2012). Soft-templating synthesis of mesoporous graphitic carbon nitride with enhanced photocatalytic H_2 evolution under visible light. *Chem. Commun.* 48 (28): 3430.

88 Cui, Y., Zhang, J., Zhang, G. et al. (2011). Synthesis of bulk and nanoporous carbon nitride polymers from ammonium thiocyanate for photocatalytic hydrogen evolution. *J. Mater. Chem.* 21 (34): 13032.

89 Han, Q., Wang, B., Gao, J. et al. (2016). Atomically thin mesoporous nanomesh of graphitic C_3N_4 for high-efficiency photocatalytic hydrogen evolution. *ACS Nano* 10 (2): 2745–2751.

90 Mane, G.P., Talapaneni, S.N., Lakhi, K.S. et al. (2017). Highly ordered nitrogen-rich mesoporous carbon nitrides and their superior performance for sensing and photocatalytic hydrogen generation. *Angew. Chem. Int. Ed.* 56 (29): 8481–8485.

91 Talapaneni, S.N., Mane, G.P., Park, D.-H. et al. (2017). Diaminotetrazine based mesoporous C_3N_6 with a well-ordered 3D cubic structure and its excellent photocatalytic performance for hydrogen evolution. *J. Mater. Chem. A* 5 (34): 18183–18192.

92 Cui, L., Song, J., McGuire, A.F. et al. (2018). Constructing highly uniform onion-ring-like graphitic carbon nitride for efficient visible-light-driven photocatalytic hydrogen evolution. *ACS Nano* 12 (6): 5551–5558.

93 Liu, Q., Wang, X., Yang, Q. et al. (2018). Mesoporous g-C_3N_4 nanosheets prepared by calcining a novel supramolecular precursor for high-efficiency photocatalytic hydrogen evolution. *Appl. Surf. Sci.* 450: 46–56.

94 Ren, H., Yang, D., Ding, F. et al. (2020). One-pot fabrication of porous nitrogen-deficient g-C_3N_4 with superior photocatalytic performance. *J. Photochem. Photobiol. A Chem.* 400: 112729.

95 Chen, X., Shi, R., Chen, Q. et al. (2019). Three-dimensional porous g-C_3N_4 for highly efficient photocatalytic overall water splitting. *Nano Energy* 59: 644–650.

96 Bai, J., Yin, C., Xu, H. et al. (2018). Facile urea-assisted precursor pre-treatment to fabricate porous g-C_3N_4 nanosheets for remarkably enhanced visible-light-driven hydrogen evolution. *J. Colloid Interface Sci.* 532: 280–286.

97 Qian, X., Meng, X., Sun, J. et al. (2019). Salt-assisted synthesis of 3D porous g-C_3N_4 as a bifunctional photo- and electrocatalyst. *ACS Appl. Mater. Interfaces* 11 (30): 27226–27232.

98 Huang, Y., Liu, J., Zhao, C. et al. (2020). Facile synthesis of defect-modified thin-layered and porous g-C_3N_4 with synergetic improvement for photocatalytic H_2 production. *ACS Appl. Mater. Interfaces* 12 (47): 52603–52614.

99 Antil, B., Ranjan, R., Gopinath, C.S., and Deka, S. (2020). Directed holey and ordered g-$C_3N_{4.5}$ nanosheets by a hard template nanocasting approach for sustainable visible-light hydrogen evolution with prominent quantum efficiency. *J. Mater. Chem. A* 8 (26): 13328–13339.

100 Antil, B., Kumar, L., Ranjan, R. et al. (2021). One-dimensional multichannel g-$C_3N_{4.7}$ nanostructure realizing an efficient photocatalytic hydrogen evolution reaction and its theoretical investigations. *ACS Appl. Energy Mater.* 4 (4): 3118–3129.

6

2D Transition Metal Carbides (MXenes) for Applications in Electrocatalysis

Devika Laishram, Divya Kumar, Kiran P. Shejale, Bhagirath Saini, Harikrishna, R. Krishnapriya, and Rakesh Kumar Sharma

Sustainable Materials and Catalysis Research Laboratory, Department of Chemistry, Indian Institute of Technology Jodhpur, Karwar, Rajasthan, India

6.1 Introduction on MXene

MXenes (pronounced as *Maxenes*) are a family of layered 2D transition metal (TM)-based carbides, nitrides, and carbon nitrides following the general form of $M_{n+1}X_nT_x$, where M is a TM such as Ti, V, Nb, and Mo. X is either a carbon or nitrogen and T_x represents various surface functional groups, including —O, —OH, and —F. Additionally, n = 1–3 is the number of layers of X [1]. Widely used typical MXenes include -$Ti_3C_2T_x$, $Nb_4C_3T_x$, and Mo_2CT_x. MXenes are synthesized from its MAX phase precursor where A is IIIA and IVA or groups 13 and 14, which are A-group elements. The A layers get selectively removed from the MAX to get MXenes. This 2D material is rapidly expanding because the variety of compositions can offer a large number of combinations not only with different M and X but also by varying the ratios of both. Therefore, limitless MXenes with different stoichiometry having unique and dissimilar properties can be formed. Figure 6.1a represents the elements that can be used for MXene formation and a representative three different typical layered structure of MXene is indicated next. Figure 6.1b is a table showing the different elements used for building MAX phase, MXenes, and other intercalated ions. Different elements with different color backgrounds indicate the formation possibility. The striped lanthanum series have a precursor MAX phase, but the MXenes are yet to be determined synthetically [1]. Figure 6.1c indicates the variations of MXene explored experimentally and computationally. It exhibits unique properties due to the solid solution mixing of different TMs forming 2D ordered structure via the formation of TM atomic sandwiches.

Owing to its exceptional characteristics, such as high specific surface area, good stability, high electrical conductivity, and hydrophilicity, the functionalization of MXenes is easier in bonding with other species. Additionally, MXene shows high negative zeta potential making colloidal solutions in water stable. Furthermore, they show electromagnetic wave absorption very efficiently. Consequently, they are used in various applications – e.g. energy storage in batteries and supercapacitors, electromagnetic

Heterogeneous Nanocatalysis for Energy and Environmental Sustainability: Energy Applications, First Edition. Edited by Putla Sudarsanam, Yusuke Yamauchi, and Pankaj Bharali.

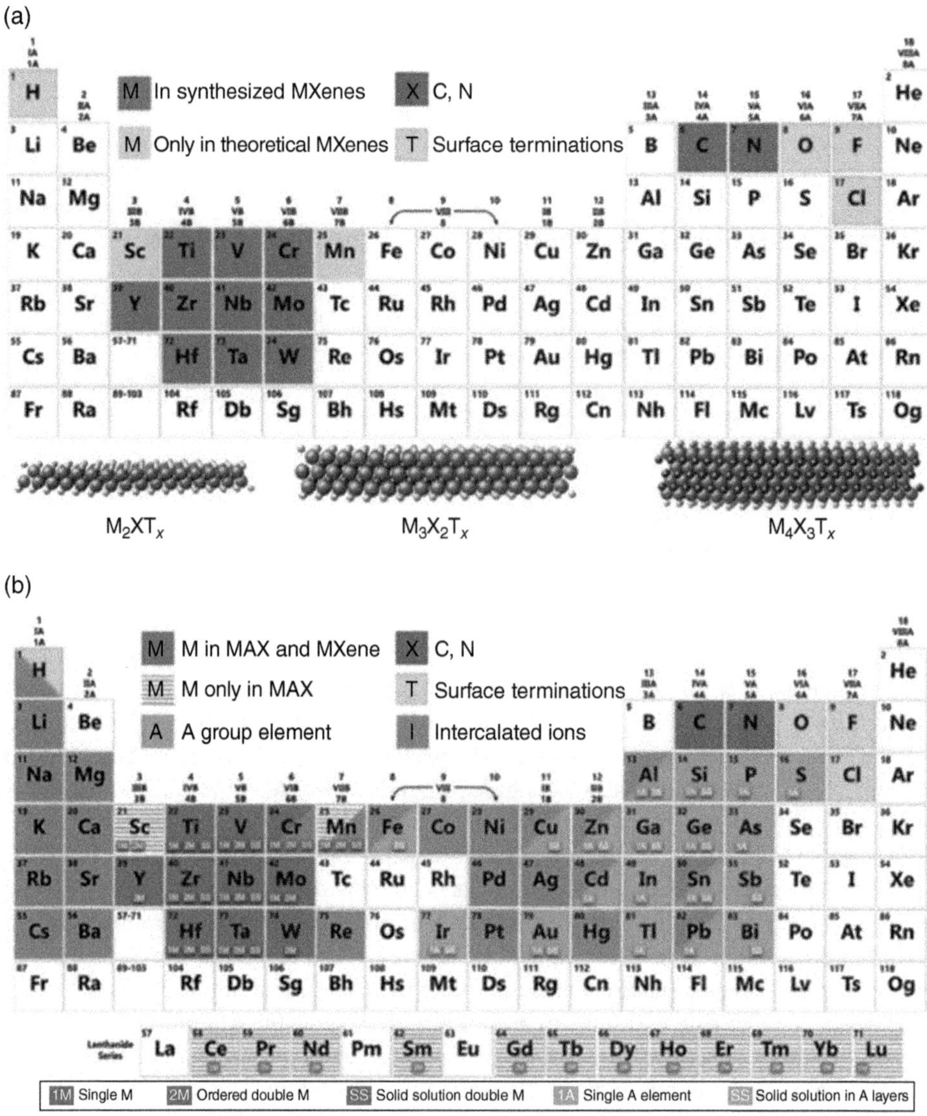

Figure 6.1 (a) Elements in the periodic table showing composition of MXenes with elements Sc and Mn show MXenes confirmed theoretically. The bottom three schemes are typical structures of MXenes. (b) MAX phases elements with intercalated ions. The striped elements are MAX phases for which MXenes have not yet been synthesized. The A elements in MAX phase, are later etched out to form MXene. The intercalated cations into MXenes are shown in green background. 1 M, 1A, SS, and 2 M represents the formation possibility of transitional metal, respectively, A element in the MAX phase, solid solution in A, and transition metal atom and an ordered double-transition metal MAX phase. *Source:* From [1] / with permission of American Chemical Society. (c) Variations in MXene reported with mono-M MXenes at the top row and double-M solid solutions (SS) at the second row, whereas the third row represents double-M MXenes and their compositions. Divacancy structure reported in M_2C MXenes with $M_{1.33}C$ composition as a result of ~33 atom % vacancies in the M layers represented by the elements in the fourth row. *Source:* From [2] / with permission of Springer Nature.

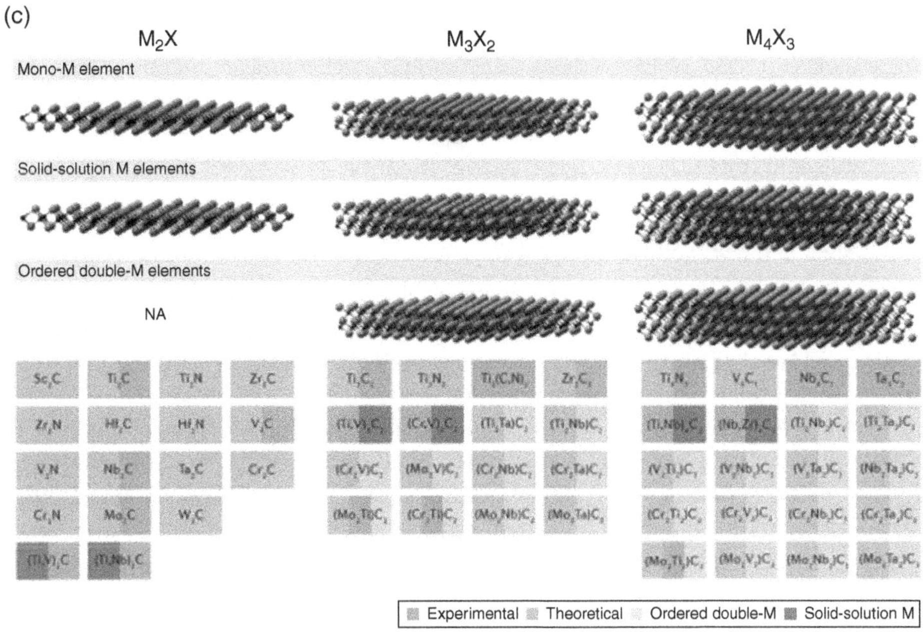

Figure 6.1 (Continued)

applications such as printable antennas and interference shielding, etc. [2–5] In the biomedical field, MXenes are used in photothermal therapy for cancer treatment, dialysis, biosensors, bio-implants, wound healing, drug and gene delivery, antibacterial, etc. [6–11] However, among the many applications of MXenes, they have attracted much attention as an effective and low-cost electrocatalyst substitute to platinum in applications such as HER, OER, NRR, MOR, CRR, and ORR [12]. These processes hold significant importance for the generation of clean energy conversion of atmospheric gases such as CO_2 and obtaining high-value products.

6.2 Synthetic Strategies

6.2.1 MAX Phase Synthesis

MAX phases can be synthesized by various synthetic processes. For bulk synthesis, MAX phase is mainly prepared by hot isotactic pressing (HIP) [13]. However, to suit for commercial and industrial applications, pressure-less methods of sintering are preferred [14]. Other methods such as solid–liquid reaction, 3D printing, spark plasma sintering, etc. have also been explored [15–17].

6.2.2 Methodologies for Thin-film Synthesis

Brief accounts for MAX phase synthesis using thin-film techniques are summarized, which are categorized into three approaches – chemical vapor deposition (CVD), physical vapor deposition (PVD), and solid-state reaction synthesis [18]. These techniques are atomistic techniques and not the methods used to grow thick coatings like spraying techniques [19–23].

6.2.3 Chemical Vapor Deposition (CVD)

CVD can be regarded as one of the bulk methods of synthesis, which utilizes a chemically based reaction. It generally operates at close to the thermodynamic equilibrium and high temperature. Nickl et al. made the first publication on MAX phase using CVD in 1972 [24]. They reported the deposition of Ti_3SiC_2 from a mixture of gases, $SiCl_4$, CCl_4, H_2, and $TiCl_4$. The formation of Ti_3SiC_2 by CVD method takes place at temperatures typically ranging from 1000 to 1300 °C, which is much higher compared to techniques such as magnetron sputtering. The formation of single phased Ti_3SiC_2 films was difficult to obtain by CVD in comparison to PVD. Basically, in CVD process, it is affected by the presence of species strongly adsorbed on the surface, which leads to the reduction of surface diffusion, and is required for the formation of complex nano-laminated structure of MAX phases. That is why high temperature is used in CVD process. Racault et al. [25], Pickering et al. [26], and Goto and Hirai [27] used this method and observed that simultaneous deposition of all elements and a reaction between gas with a solid phase, such as TiC can form Ti_3SiC_2. The term “reactive CVD” was coined by Jacques et al. and Faikh et al. who used it to synthesize Ti_3SiC_2/SiC multilayer coating [25, 28, 29]. In the future, there is still room for research in the synthesis of MAX phases by CVD and atomic layered deposition (ALD), a form of CVD for the synthesis of MAX phases.

6.2.4 Physical Vapor Deposition (PVD)

In PVD, the most common approaches used to synthesize MAX phases are sputtering and cathodic arc deposition techniques. It operates at a substrate temperature of around 700–1000 °C, which is relatively high. Therefore, the industrial application that is temperature sensitive cannot use these substrates. Ti-based MAX phases require high temperature; however, MXenes such as M_2AX phases can be vapor deposited at a temperature relatively lower (group 5 and 6 elements) [18], examples are V_2GeC, Cr_2GeC, etc. [30–32] Individual sources (targets) can be used for controlling the deposition of individual elements or from composite targets in sputter deposition [33–35]. MAX nitrides can be grown by using nitrogen gas that can act as a source of N_2 during reactive sputtering. However, reactive sputtering as a method for growing MAX carbide is limited as the growth of pure MAX phase film's process window is narrow. Interestingly, some research indicates broader potential usage for this technique [36, 37].

6.2.5 Cathodic Arc Deposition

MAX phase synthesis by cathodic arc deposition has more limited usage in comparison to sputtering [38]. The deposition flux in arc deposition has a comparatively higher degree of ionization. This compared to sputtering might suggest reducing temperature in the arc deposition method.

6.2.6 Solid-State Reaction Synthesis

Of the various thin-film synthesis techniques, there presents two main categories of solid-state reaction synthesis. The first technique is based on the film and the substrate, i.e. film–substrate, and the second one is based on the reaction between two films, i.e. film–film. Ti_3SiC_2 synthesized is one of the best examples of film–substrate-based reaction whereby annealing of Ti-based contacts is performed in SiC devices [39]. Furthermore, for film–film reaction, synthesis involves deposition in the right proportion on the film of the three precursor elements of MXene, namely, M, A, and X, for example, in a multilayer or in amorphous or in nanocrystalline form [40]. MAX phase transformation is initiated followed by annealing at a higher temperature. Examples where solid-state reaction has been utilized for making new MAX phases and other nano-laminated carbides include reactions of Au or other noble metal capping layers with nano-laminated carbide thin films such as Ti_3AuC_2, $Ti_3Au_2C_2$, Mo_2AuC, and Mo_2Au_2C [41–45].

6.2.7 Bulk Synthesis

Here, we discuss a brief description of the most important methods for bulk synthesis of MAX phases. Barsoum and El Raghy [13] in 1996 reported HIP to synthesize bulk Ti_3SiC_2 having greater than 95% phase purity. This remained an important method reported in various papers [46, 47]. Method such as pressure-less sintering is commercially viable [48]. Other procedures for bulk MAX phases synthesis include a variation of self-propagating high-temperature synthesis [49], solid–liquid reaction synthesis [15], 3D printing [16], using Al and Sn as a catalyst for the growth of Ti_3SiC_2 and Ti_3AlC_2 [50, 51], and spark plasma sintering process [52]. The preferred procedure for synthesis is to basically make a mixture of powders with accurate stoichiometry followed by heating them either in a vacuum or in an inert atmosphere (such as Ar), at a temperature ranging between 1350 and 1600 °C considering the requirement of MAX phase. Mostly, the MAX phases used for the synthesis of MXenes are Al based. There is a need for an excess of Al in the initial mix because Al tends to evaporate at high temperature [53].

6.2.8 MAX to MXenes Synthesis

Further synthesis of MXenes is commonly carried via etching selectively the specific "A" element from their layered precursor such as MAX phases (carbides and nitrides), having the general formula $M_{n+1}AX_n$ (where A is mostly an element from group 13 and 14). The selective etching of the A element is performed using acidic solutions of fluorides. On the other hand, the layers of M are terminated by groups such as —O, —OH, or —F, which are hydrophilic in nature. Till now, the reported MAX phases are 70 different types [54, 55]. Naguib et al. reported the synthesis of MXene (i.e. 2D Ti_3C_2) by extracting Al from Ti_3AlC_2 and replaced by -OH group by sonification in methanol following the reaction with HF (Figure 6.2) [56]. Another type of MXenes was synthesized by Naguib et al., where they have presented the preparation of MXenes by the exfoliation of the following MAX phases (Ti_2AlC, Ta_4ALC_3, $(Ti_{0.5}Nb_{0.5})_2AlC$, $(V_{0.5}Cr_{0.5})_3AlC_2$, Ti_3AlC_3) by simply immersion of their powders in HF at room temperature (Figure 6.3) [57].

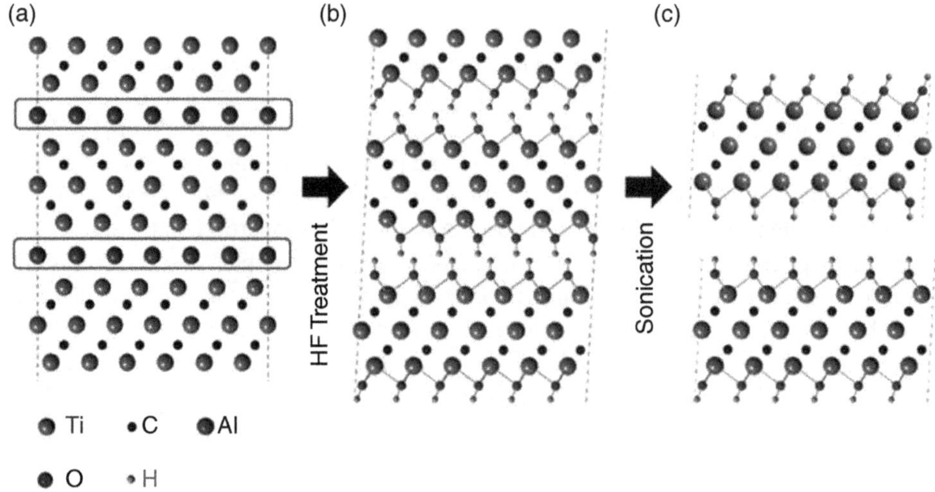

Figure 6.2 Schematic of the exfoliation process by HF etching method, (a) structure of Ti_3AlC_2, (b) replacement of Al by OH after reaction with HF, and (c) sonication in methanol resulting in separation of nanosheets. *Source:* From [56] / with permission of John Wiley & Sons.

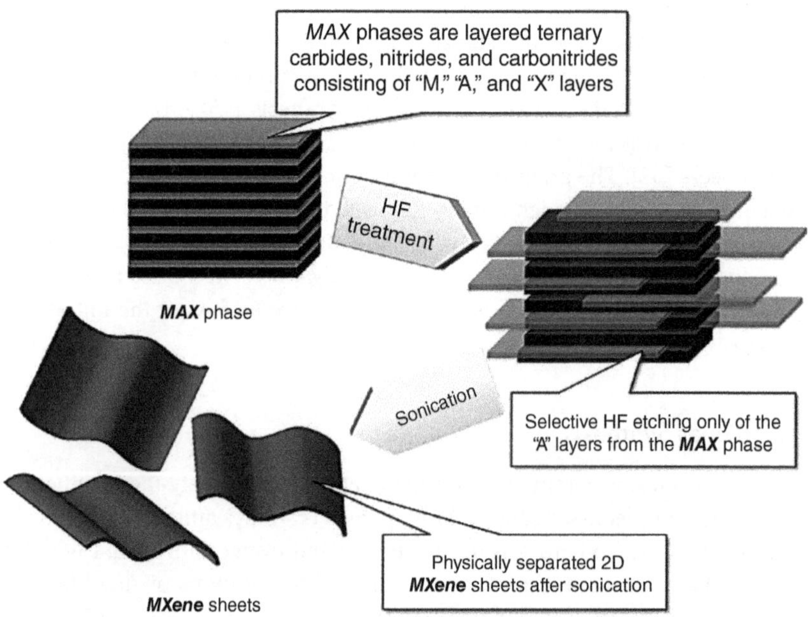

Figure 6.3 Schematic representation showing the formation of MXene from MAX phase. *Source:* From [57] / with permission of American Chemical Society.

However, there is a requirement of handling a high concentration of aqueous HF for etching and it can be a strenuous multistep procedure. Ghidiu et al. reported an easier and safe method of synthesizing MXenes using a solution of LiF and HCl for the in situ formation of HF. They showed that the resulting hydrophilic material on hydration swells

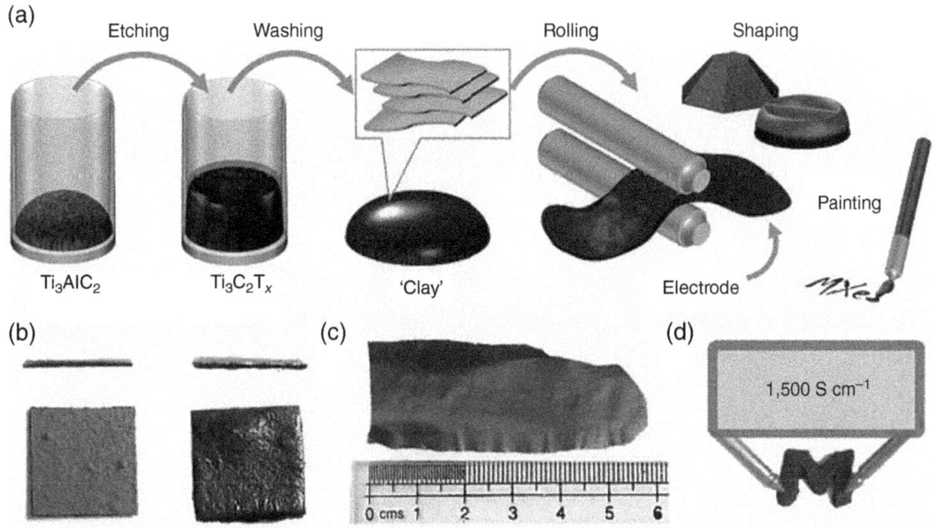

Figure 6.4 Schematic of MXene synthesis. (a) Etching of MAX phase, reaction product removed by washing with water, and pH raised toward neutral. Resulting in sediment that behaves like clay; (b) dried sample on hydration swell, (c) rolled film image, and (d) clay shaped into letter M. The etched material is $Ti_3C_2T_x$ [58]. *Source:* From [58] / with permission of Springer Nature.

in volume and can take up the shape of clay and dried into a highly conductive solid. This synthesis avoided the handling of hazardous concentrated HF (Figure 6.4) [58].

Due to the acute toxicity of HF, it was necessary to develop novel HF-free synthesis methods for MXenes. Li and coworkers demonstrated the synthesis of $Ti_3C_2(OH)_2$ by etching its precursor, i.e. Ti_3AlC_2 with KOH in the presence of water serving as etchant. They also observed that after drying the exfoliated $Ti_3C_2(OH)_2$, it could be restacked to form a new layered structured material (Figure 6.5a) [59]. Pang and coworkers proposed an electrochemical etching route to prepare MXene such as Ti_2CT_x, Cr_2CT_x, V_2CT_x, which is assisted thermally. This work showcased a nontoxic and HF-free route for synthesis [60]. Li et al. reported an elemental replacement approach for synthesis by reaction with Lewis acidic molten salts ($ZnCl_2$). In this work, Cl-terminated MXenes were prepared (Figure 6.5b) [61].

6.3 MXene Modifications to Enhance the Electrocatalyst Performance

6.3.1 Nanostructuring

Electrocatalysts offer surfaces that are actively involved in various electrochemical reactions. During electrolysis, the performance of HER is dependent on the approachability of active sites. During the fabrication of electrodes, 2D materials tend to restack, which act as a limitation for the utilization of active sites. Therefore, a 1D nanostructure is useful because abundant sites are available due to the presence of highly active surface area. Also, there is a good charge transfer that prevents the active components from restacking facilitated by

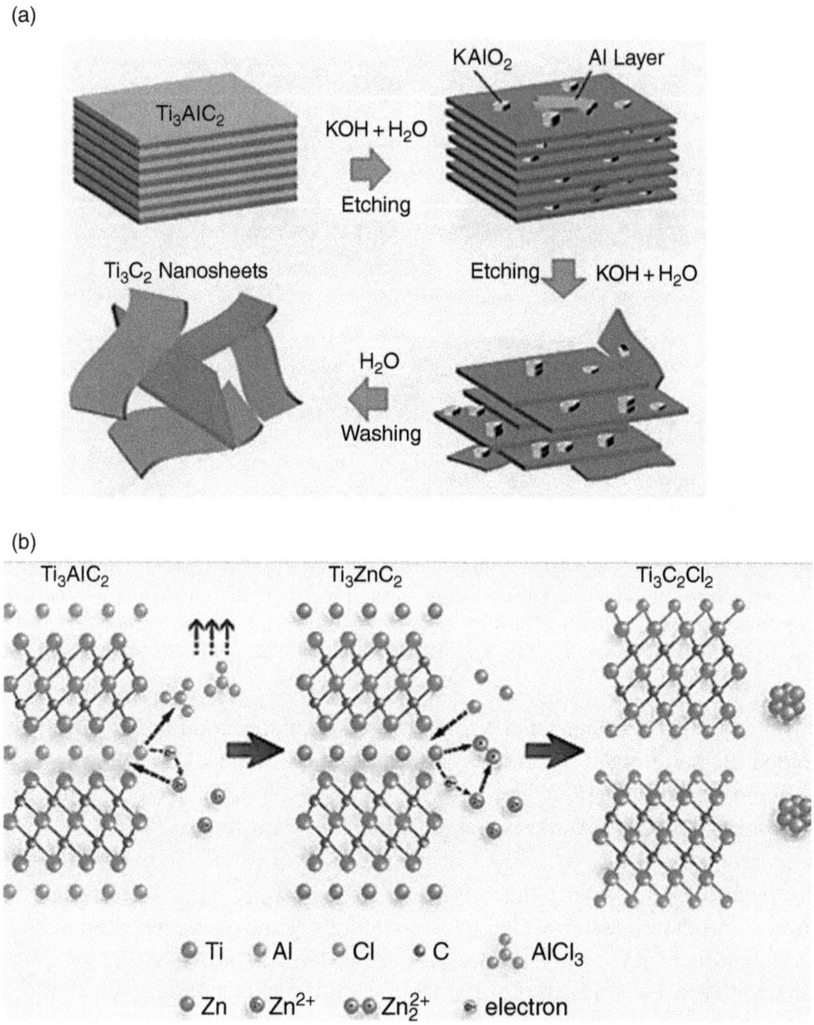

Figure 6.5 Schematic Illustration. (a) Etching process and exfoliation for Ti_3AlC_2. *Source:* From [59] / with permission of American Chemical Society, and (b) exfoliation process of Ti_3AlC_2 by molten $ZnCl_2$. *Source:* From [61] / with permission of American Chemical Society.

the axial dimension. In this case, the active sites in the MXene are well exposed and accessibility is highly improved [55]. Yang et al. in their work demonstrated the HER activity theoretically for MXene nanoribbons. They reported an increased efficiency in the case of MXene having nanoribbon morphology where the edges served as active sites compared to MXenes having O-terminated surface [62]. Yuan et al. designed highly active $Ti_3C_2T_x$ nanofibers by hydrolyzation of bulk MAX ceramic in KOH solution and followed by an HF etching process (Figure 6.6a) [63]. Ti—C bonds near defect sites are affected by the OH^- ions. The morphology of the nanofiber is retained after etching with a diameter of around 50 nm (Figure 6.6b). The newly synthesized MXene nanofibers showed enhancement of catalytic activity as it delivered a much higher specific surface area and exposed more active sites. As a result, at a current density of $10\,mA\,cm^{-2}$, the MXene nanofiber showed an

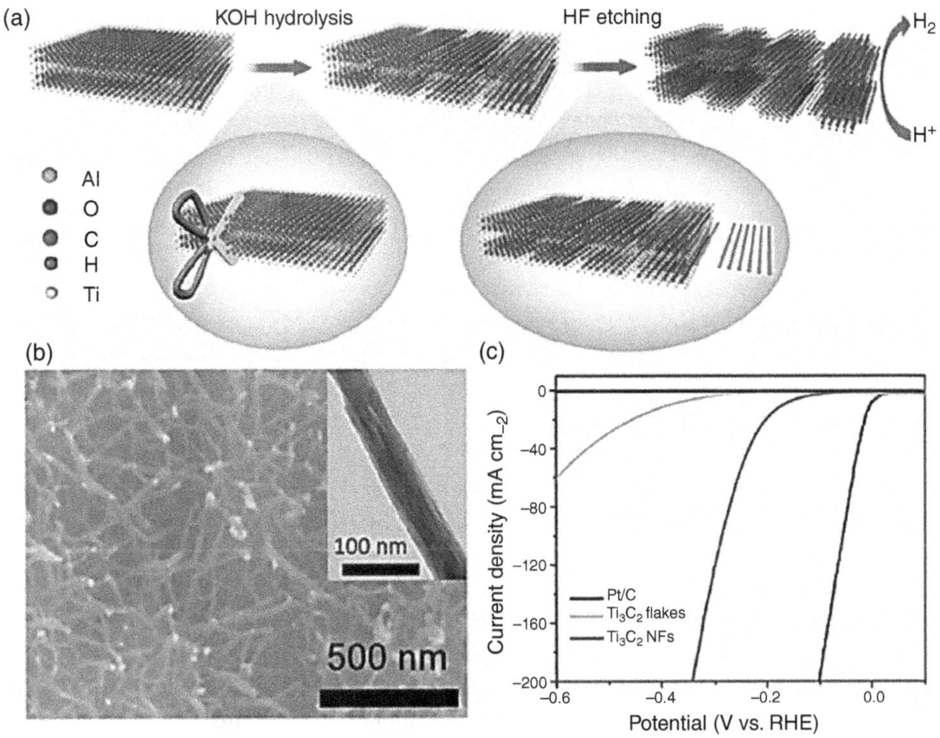

Figure 6.6 (a) Representation of the synthesis of $Ti_3C_2T_x$ nanofibers by hydrolyzation of bulk Max ceramic in KOH solution and followed by an HF etching process, (b) SEM image showing the diameter of the nanofiber, and (c) polarization curve for $Ti_3C_2T_x$ nanofibers [63]. *Source:* From [63] / with permission of American Chemical Society.

enhanced HER activity with a low overpotential of 169 mV, a depression in Tafel slope of 97 mV dec^{-1}, and lowering in electrochemical resistance (Figure 6.6c). Other researchers, such as Liu et al., developed a highly efficient noble-metal-free electrocatalyst [64]. Liquid nitrogen freezing followed by annealing is a synthesis method for making hierarchical "nanorolls" like $MoS_2/Ti_3C_2T_x$ hybrid. Using this material as an electrocatalyst, the exchange current density showed a 25 times increase with a small onset overpotential of 30 mV. This results in excellent catalytic activity for HER.

6.3.2 Hybridization

MXenes couple with other active materials to enhance the performance of HER, like chalcogenides [64, 65], phosphines [66], layered hydroxides [67], carbides [68], metal nanoparticles (NPs)/alloy [69, 70], and metal-free black phosphorous [71]. Wu and coworkers reported the synthesis of hierarchical $MoS_2/Ti_3C_2T_x$@C composites, in which structural degradation caused by spontaneous oxidation of MXenes was stabilized by the plated carbon layer [65]. Limited layered MoS_2 nanoplates were loosely packed on the surface of $Ti_3C_2T_x$ with 0.64 nm interlayer spacing. Meanwhile, around the $MoS_2/Ti_3C_2T_x$ an ultrathin carbon layer was formed. This unique assembly lowers the pathway for mass diffusion during the HER process and facilitates the charge transfer. The onset potential was −20 mV and the overpotential was 135 mV, which are much better than that of the standard MoS_2/rGO@C and MoS_2/oxidized MXenes. Many other TM chalcogenides are also investigated with MXenes to improve the HER activities, e.g. $NiSe_2$ [72], NiS_2 [73], and VS_2 [74]. Jiang et al. performed the one-pot hydrothermal synthesis of $NiSe_2/Ti_3C_2T_x$ hybrid. The synthesized material indicated strong interfacial and electrical properties due to the wrapping of $NiSe_2$ octahedral crystal and ultrathin $Ti_3C_2T_x$ MXene nanosheets (NSs) [72]. The interactions between the two significantly improve the property of MXene providing abundant active sites, fast charge transfer, and increased conductivity. MXenes supported by metal NPs and binary alloys have proven to be highly efficient HER electrocatalysts. Li and coworkers utilized in situ co-reduction for synthesizing Pt_3Ti NPs on $Ti_3C_2T_x$. The Pt_3Ti particles were distributed uniformly on the surface $Ti_3C_2T_x$ and were found to have a diameter of 6.6 ± 3.5 nm [75]. Compared to commercial Pt/C, it was observed that the synthesized $Pt_3Ti/Ti_3C_2T_x$ showed better HER activity. MXene supported NiCo nanoalloys [69], PtNi nanowires [70], and Pt NPs [76, 77] could enhance the HER electrocatalysts. Apart from this, crystalline CoNi nanodots dispersed homogeneously on the $Ti_{2.5}Nb_{0.5}C_2T_x$ NSs indicated performance with an overpotential of 43.4 mV at 10 mA cm^{-2}, which is equivalent to that of commercially available 10% Pt/C. Additionally, it has been observed that in an alkaline medium, the HER activity of NiFe layered double hydroxide (LDH) can be upgraded by using MXenes [67]. These ultrathin NiFe LDH grown on the surface of $Ti_3C_2T_x$ exposes more edge and defect sites. Additionally, it also decreases the charge-transfer distance. Therefore, composites such as these showed enhanced activity in comparison to commercial Pt/C at high current density. For promoting the HER performance of MXene-based composite, MXenes with designed nanostructure have been beneficial. For example, Liu et al. synthesized $MoS_2/Ti_3C_2T_x$ composite by a quick-freezing method with a nanoroll like hierarchical structure having diameter of 200 nm [64]. A low onset potential was shown by the nanoroll of around 30 mV and overpotential of 150 mV at 10 mA/cm^2. Few of the vertically aligned layered MoS_2 on the MXene possess exceedingly available edge sites.

6.3.3 Termination Modification

The electronic properties of MXenes can be tuned and enhanced by changing the terminal groups of MXene surfaces (—O, —OH, —F) [78, 79]. Hu and coworkers reported termination stabilized Ti_3C_2 monolayer matrix whereby the non-bonding valence electrons of surface Ti were saturated using the termination's low energy orbitals [80]. Schultz and coworkers investigated the electronic properties of $Ti_3C_2T_x$ for various surface terminations and identified the adsorption sites [81] and reported the change of work function with modification of terminations. It was observed that upon heating in a vacuum, there is an increase in work function from 3.9 to 4.8 eV. Additionally, change in work function can be due to water desorption, carbon-dominated contamination, and -OH species. Because of fluorine desorption, there is a decrease at high temperature to 4.1 eV. Hart and coworkers observed experimentally via surface defunctionalization an increased electronic conductivity of MXenes [82]. The HER activity of MXenes can also be altered by surface termination.

Handoko and coworkers demonstrated both through experimental and theoretical approaches to observe the effects of basal plane termination (T_x) on HER activity [83]. A study on HER activity as a result of the effects of F coverage of the basal plane was performed by making $Ti_3C_2T_x$ by means of using different etchants (HF, LiF-HCl). It was observed that the HER activity increases with the decrease of F coverages on the basal plane. Additionally, it was observed that ΔG_H increases monotonically with the F coverage, as in accordance with the DFT calculations. Furthermore, the investigation of other MXenes was also performed (Ti_2CT_x and Mo-based MXenes), which indicated similar trends. It was observed theoretically that Mo_2CT_x with O termination on the basal planes are more catalytically active toward the HER. MXenes with F terminations exhibit sluggish hydrogen adsorption kinetics, which attributes to a decrease in the HER activity. Whereas MXenes with O terminal groups enhance the HER performance greatly.

6.3.4 Heteroatom Doping

The chemical doping of MXenes with heteroatoms is one of the most viable strategies that have been established to enhance the performance of HER electrocatalysts by altering the elemental composition, surface chemistry, and modifying the electronic properties. MXenes doped with heteroatoms can be categorized into two types: metal-atom doping and non-metal-atom doping [55].

6.3.5 Metal-atom Doping

Metal-atom-doped catalysts utilize the active site of each metal atom and are known as single-atom catalysts [84]. The lattice or the coordination environment of these 2D materials is through strong covalent bond stabilization of single metal atoms and therefore acts as the active sites for HER. The strong electronic interaction develops between the single atoms and its 2D support, thus enriching the intrinsic activity at the active sites. Additionally, inspired by the other 2D single-atom catalysts, MXene confined single metal atoms are also studied to enhance HER activity [55]. Li et al. investigated by utilizing computational methods the optimization of MXenes of M_2XO_2 type by doping with several TM atoms [85]. $IrZr_2CO_2$, Ag-V_2CO_2, and Zn-Nb_2CO_2 show three kinds of O active sites toward HER. However, TM dopants such as Cu are not beneficial for HER activity. Figure 6.7a depicts the structural model of TM-doped M_2CO_2 MXenes and the potential H available adsorption sites. The total number of active sites is illustrated in Figure 6.7b for various TM-M_2CO_2. ΔG_H values of several TM-M_2CO_2 are given in Figure 6.7c with other possible modifications. The results showed that activation of M_2CO_2 sites could take place by suitable TM doping, except for Hf_2CO_2. It was observed that Mo-V_2CO_2 and Au-Sc_2CO_2 having various O sites with $|\Delta G_H| < 0.1$ are highly HER active. On the other hand, double TM MXenes such as $Mo_2TiC_2T_x$ immobilized by single Pt atoms are efficient HER electrocatalysts, as demonstrated by Zhang et al. [86]

MXene such as $Mo_2TiC_2T_x$-V_{Mo} with multiple Mo vacancies can have V_{Mo} immobilized single Pt atoms with the covalent Pt—C bonds simultaneously stabilizing it. The electrochemical exfoliation method was used to prepare $Mo_2TiC_2T_x$-V_{Mo}. The Pt atoms were immobilized at the Mo position on $Mo_2TiC_2T_x$. The Pt—Pt contribution observed at around

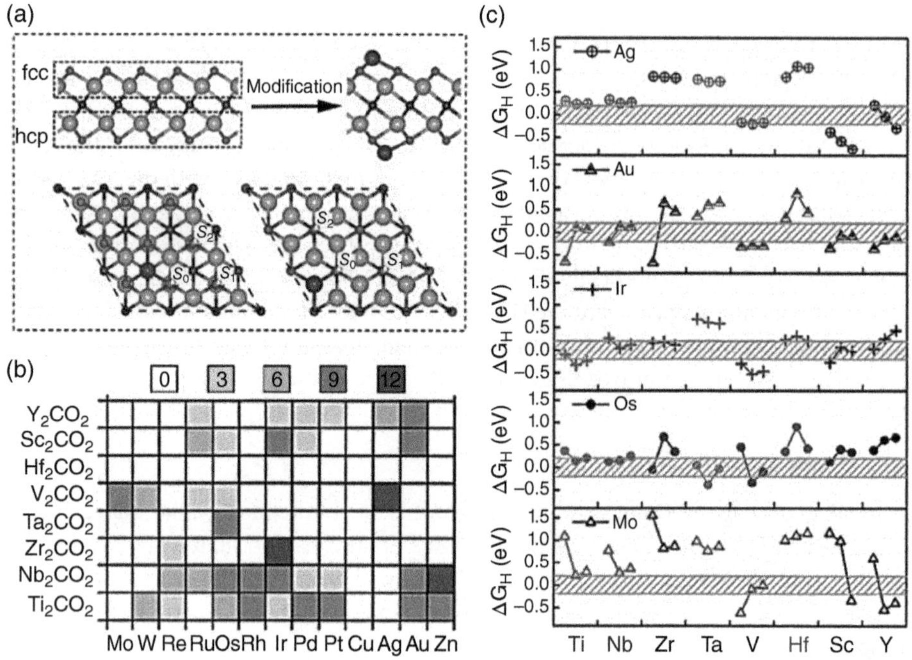

Figure 6.7 (a) Theoretical model of M_2CO_2 MXenes doped with transition metal and possible H adsorption sites are represented in the scheme; (b) tabular representation of various active sites in TM-M_2CO_2, where darker shade indicates increasing active sites and the lighter shades indicate the lesser number of the active site. White color represents no active site. (c) ΔG_H of various TM-M_2CO_2 MXenes with different TM on the X-axis. *Source:* From [85] / with permission of Royal Society of Chemistry.

2.7 A^0 for Pt foil was absent for $Mo_2TiC_2T_x$–Pt_{SA}. There is the presence of a shell at 1.9 A^0 corresponding to Pt—C bonds. Therefore, instead of Pt NPs or nanoclusters, the study strongly indicates the formation of single Pt atoms on MXene. Additionally, the result showed lower ΔG_H (−0.08 eV) than Pt (−0.1 eV) and $Mo_2TiC_2O_2$ (−0.19 eV). Also, $Mo_2TiC_2T_x$–Pt_{SA} exhibited mass activity of 8.3 A/mg at an overpotential of 77 mV and an overpotential of 30 mV at 10 mA/cm^2. Thus, it can be affirmed that quicker formation and release of H_2 on $Mo_2TiC_2O_2$–Pt_{SA} compared to Pt/C can be a result of the reduced strength with which the H_2 is adsorbed on the positively charged Pt atoms. Ramalingam et al. synthesized Ru_{SA}-N-S-$Ti_3C_2T_x$ via a pyrolysis strategy by developing N and S coordinated Ru single atoms on $Ti_3C_2T_x$. This resulted in higher HER activity in a wide pH range [87]. Additionally, there have been examples in which single metal-atom-doped MXenes were obtained from doped MAX precursors. Cobalt-doped Mo_2CT_x MXene (Mo_2CT_x:Co) was designed by Kuznetsov et al. from a Co-substituted Mo_2Ga_2C MAX phase [88]. The morphology was ultrathin hexagonal sheets, with Mo, C, and O uniformly distributed. However, due to the ultralow content (0.04 wt%), Co showed no detection. A lower overpotential of 180 mV at 10 mA/cm^2 was observed for Mo_2CT_x: Co than the pristine Mo_2CT_x (at 230 mV), leading to an enhancement of HER activity due to Co doping.

6.3.6 Non-metal-atom doping

In order to facilitate the HER performance, non-metals like N, S, P, etc., are also doped into 2D materials to tune the electronic structure. The functions of N dopant in MoS_2 as HER electrocatalysts were investigated by Xiao et al. [89] It was observed that it activated the S-edge sites and enhanced the conductivity of the basal plane. Interestingly, MXenes with non-metal doping were explored both theoretically and experimentally. Ding et al. reported DFT calculations that indicated the effect of heteroatom doping of N, S, B, and P on the HER performance of Ti_2CO_2 MXenes. They obtained ΔG_H comparable to Pt for N- and B-doped Ti_2CO_2 to be 0.087 eV and −0.097 eV, respectively [90]. The theoretical overpotential suggests an enhancement in the activity of HER with doping of N and B. The ΔG_H versus H coverage depicted an upward trend with H coverage increase and less H coverage has the potential to improve the HER performance of the X-Ti_2CO_2. This work opens the scope for optimizing the HER activity via non-metal doping of MXenes, theoretically. On the other hand, various strategies have been adopted to synthesize N- or P-doped MXenes. Yoon et al. by nitridation of Ti_2CT_x in sodium amide synthesized N-doped Ti_2CT_x [91]. They also performed phosphorization of V_2CT_x with triphenylphosphine to prepare P-doped V_2CT_x. DFT calculations revealed that the C—P bonds induced by P doping are stable composition [92].

6.3.7 MXene as Electrocatalyst

To meet the global demand for energy via a clean and carbon-free source of energy is going to be the goal in coming decades. The dependency on natural resources such as fossil fuels and rapid human growth has led to environmental problems and issues associated with it, such as global warming and climate change. In this regard, electrocatalysis will play a major role in providing clean energy by means of converting atmospheric gases such as H_2O, CO_2, and N_2 [12]. These processes will have the advantage of providing efficient clean fuel and high value-added products and a reliable option. It will be carried out through technologies such as batteries, supercapacitors, various fuel cells via processes such as HER, OER, CRR, and NRR [54]. Hitherto, electrocatalysis has been mostly based on noble, expensive, and less abundant metals such as Pt, Ir, Pd, and Ru. Therefore, it is vital to switch to an alternative without suffering much loss in efficiency and performance. In this aspect, MXene holds strong promise due to its unique and tunable properties. Incorporating heteroatoms, hybridizations, and changing the terminal groups can enhance the catalytic kinetics by optimizing the chemical and electronic configurations [55].

6.3.8 Hydrogen Evolution Reaction (HER)

HER is a half-reaction of H_2O electrolysis involving a two-electron transfer system with H*(adsorbed hydrogen) as a reaction intermediate. In this three-step reaction – Volmer, Heyrovsky, and Tafel, the H* formed during Volmer generates H_2 during Heyrovsky/Tafel. The rate at which HER occurs is determined by the factor ΔG_{H^*}, which is the free energy of H_2 adsorption [93]. The HER activity of an electrocatalyst is higher when ΔG_{H^*} value approaches zero and should have binding energy neither very high nor too low, according to the Sabatier's principle [94]. MXene exerts good electrocatalytic performance and hence is a good replacement for Pt as an electrocatalyst. However, replacing Pt totally has been a critical issue due to its high activity. In this regard, an excellent approach is the incorporation of a single atom on 2D materials [95]. Because MXene has excellent hydrophilicity and good conductivity, it acts as an anchor to the single-atom catalyst and enhances the metal–support interaction via coordination and also reduces Pt agglomeration [96]. For example, Yuan et al. reported Pt NPs on MXene ($Ti_3C_2T_x$) support as shown in Figure 6.8a exhibiting a current density of $10\,mA\,cm^{-2}$ and overpotential of 55 mV [76]. This was attributed to the fast charge transfer between the O-terminals in the MXene and the Pt catalyst. This is in fact due to the change in the electronic state of the Pt catalyst impacted by the O- and OH- terminals in MXene. Xiu et al. reported ultrafine Pt loaded 3D hollow nanostructure MXene that has 20% better performance than commercially available Pt/C [97]. As shown in Figure 6.8b Ti_3C_2 MXenes have been synthesized using ultrafast drying aerosols on the hard template to obtain engineered morphology, size, and chemical activity. Other combinations have been explored to compete with Pt catalyst such as bi-metal-based Co—Mo carbide, MXene, and carbon (Co_xMo_{x-2}/MXene/NC) as shown in Figure 6.8c [68]. Seawater electrolysis using this method has resulted in equivalent activity to Pt—C electrodes and was proven to have 64 times longer life expectancy than Pt—C. Xiu et al. also developed hierarchical 3D MXene architecture, which is resistant to aggregation using a capillary-forced assembly method as shown in Figure 6.9a and b [66]. Such an architecture will give easy accessibility for charge transfer as well as active sites. Figure 6.9c–e represent the scanning electron microscopy (SEM) images at different magnification and transmission electron microscopy (TEM) image indicating the uniform formation of a fluffy 3D architecture showing ridges of architecture CoP@3D Ti_3C_2-MXene [66]. This catalyst showed good HER performance in 1 M KOH solution. Figure 6.9f shows the polarization curves of CoP@3D Ti_3C_2-MXene against various other combinations along with a standard commercialized Pt/C for HER in 1 M KOH at 10 mVs^{-1}. The observed current density is $10\,mA/cm^2$ with 168 mV overpotential compared to 77 mV for Pt/C. Additionally, a smaller Tafel slope is observed for CoP@3D Ti_3C_2-MXene at 58 mV/dec compared to CoP@Ti_3C_2-MXene and MXene-free CoP, which suggests fast kinetics for the 3D MXene-based catalyst. Furthermore, compared to various other catalysts, this 3D architecture electrocatalyst showed a lower voltage in the steady-state polarization curve in Figure 6.9 h. A comparative histogram is shown in Figure 6.9i for the CoP@3D Ti_3C_2-MXene electrocatalyst, along with other bifunctional catalysts.

MXene and MXene-based hybrid catalysts have the potential for good performing HER by rationally enhancing the properties such as termination modification and subsequently the electronics configuration of the electrocatalyst. The synergistic coupling of various high-performing materials and MXenes has proven to increase the HER activity. For improving the HER performance using MXene-based catalyst to be comparable to the noble metal catalyst, various conditioning such as doping, modifying terminations, controlling the morphology, making hybrids, etc. are introduced.

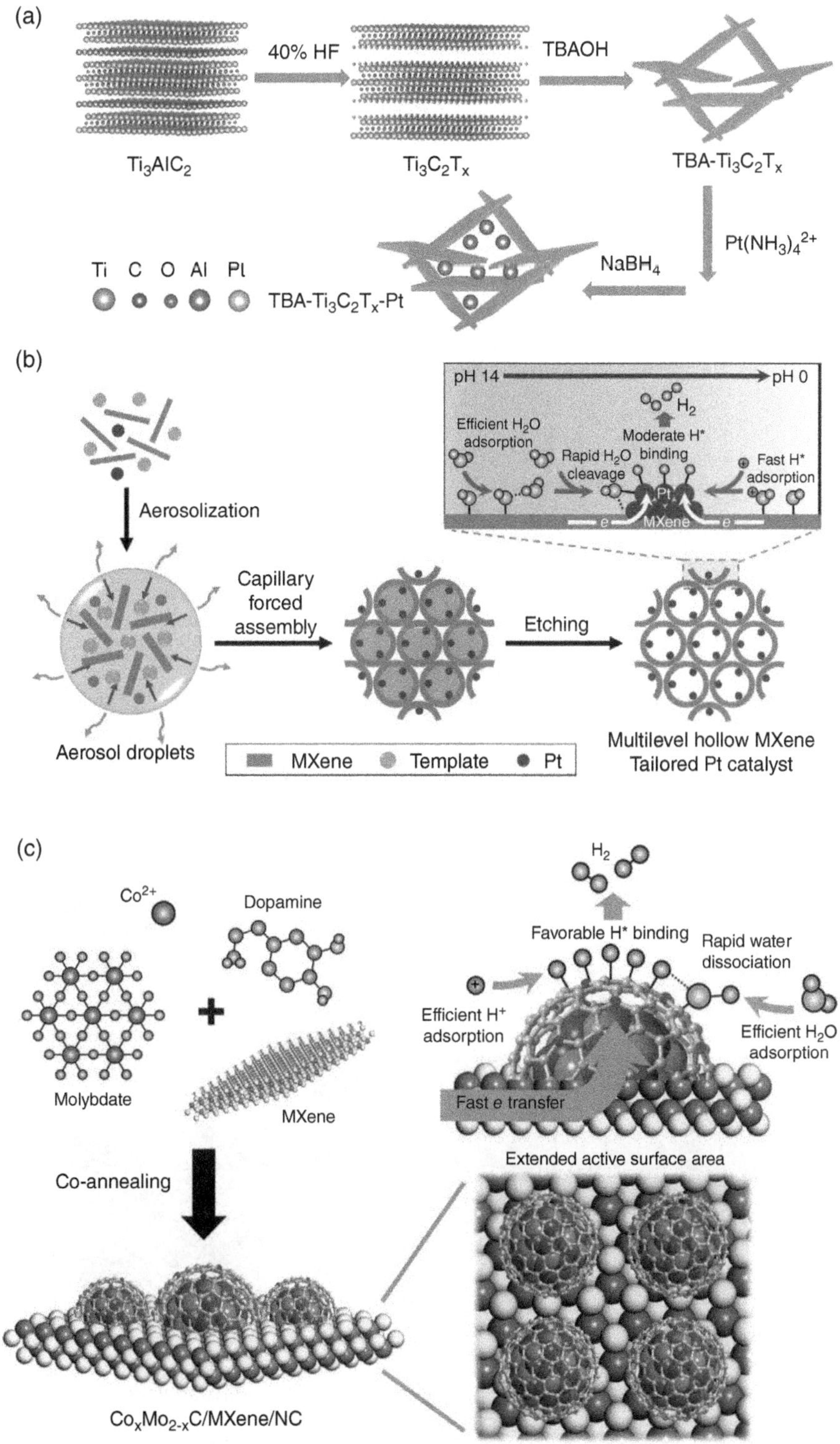

Figure 6.8 (a) Synthesis method of TBA-$Ti_3C_2T_x$-Pt-n. *Source:* From [76] / with permission of American Chemical Society, (b) low-Pt on hollow MXene having multifunctional catalytic interface for HER application at multiple pH synthesized via a template-engaged ultrafast aerosol drying strategy. *Source:* From [97] / with permission of John Wiley & Sons, and (c) synthetic strategy of the $Co_xMo_{2-x}C$/MXene/NC via co-annealing of the polymerized chelate of $Mo_7O_{24}^{6-}$ along with Co^{2+} and dopamine in the presence of Ti_3C_2 promoting HER at different pH due to collaborative interface in the catalyst.

6.4 Oxygen Evolution Reaction (OER)

The high conductivity and highly reachable electroactive sites are the keys to the remarkable electrocatalytic performance of catalysts. A strategy was reported to design a hierarchical FeOOH NSs/Ti_3C_2 by orienting the growth of FeOOH NSs on Ti_3C_2 NSs surface at room temperature. Such hierarchical framework proficiently reduces the NSs agglomeration and provides high exposure electroactive surface. The enhanced interface between FeOOH and Ti_3C_2 significantly increases charge transfer. Remarkably, FeOOH NSs/Ti_3C_2

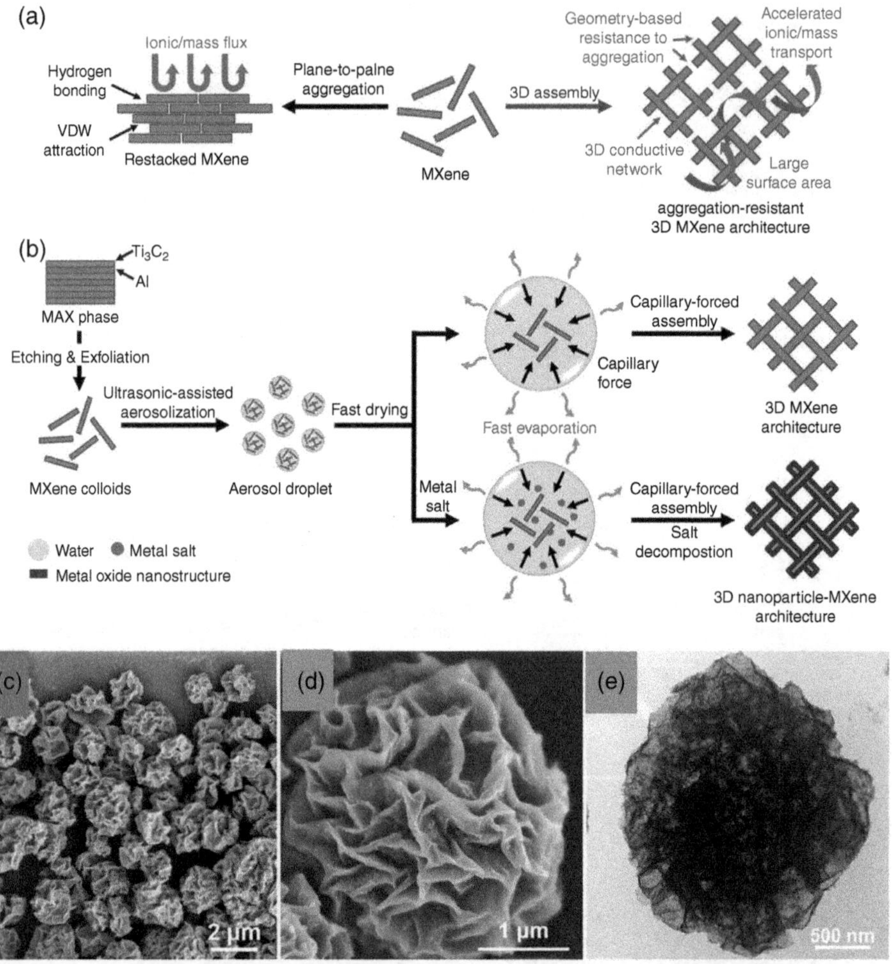

Figure 6.9 (a) Schematic representing the architecture of aggregation-resistant MXene. (b) 3D architecture and other hybrid systems of MXene assembled via capillary forced mechanism, (c) and (d) are the SEM and (e) TEM images showing the CoP@3D Ti_3C_2-MXene architecture indicating ridges with loose interior, (f) polarization curves of CoP-based MXene and without MXene electrocatalysts, (g) Tafel plots, (h) steady state for HER and OER activity, and (i) comparison of various bifunctional and CoP@3D Ti_3C_2-MXene catalyst. All the experiments are performed in 1 M KOH [66]. *Source:* From [66] / with permission of American Chemical Society.

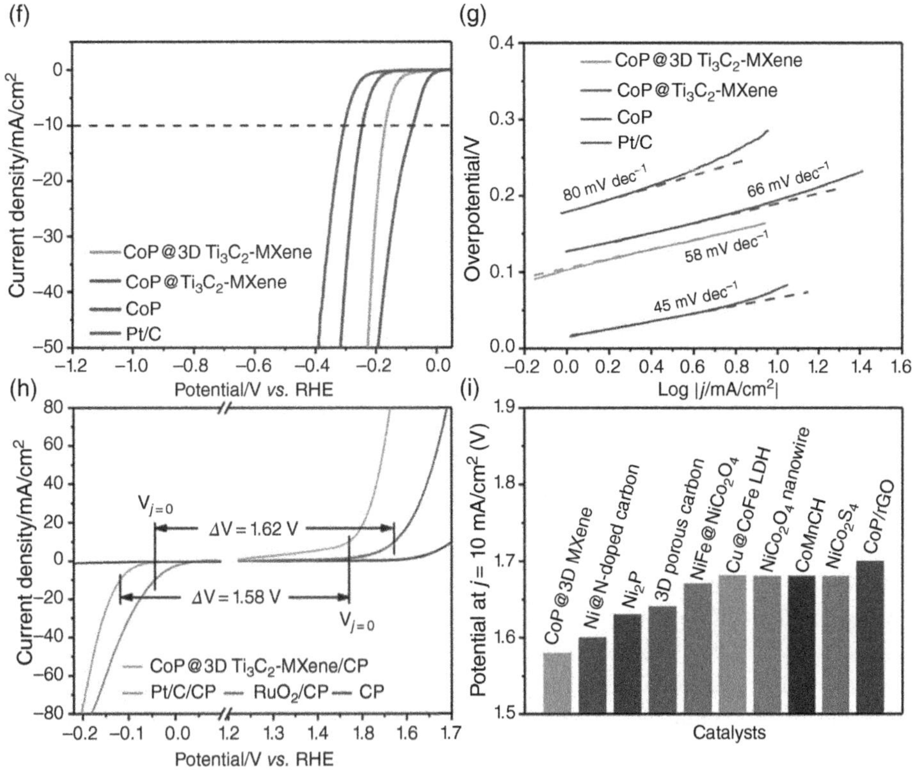

Figure 6.9 (Continued)

reported low overpotential (10 mA/cm^2 at 1.63 V) and long-term stability (no significant changes after 10 000 CV cycles) [98]. Controlled synthesis of rational design has gained tremendous attention and is the key for efficient electrocatalysts, specifically for promising energy conversion systems. The cobalt oxide nanostructures consisting of mixed valences evolved as a promising candidate for the OER, but it showed modest activity. The 0D Co_3O_4/2D Ti_3C_2 MXene demonstrated the effect of cubic Co_3O_4 electrocatalyst anchored on the 2D Ti_3C_2 MXene NSs. The strong interfacial electrostatic interactions between these two facilities outstanding catalytic activity. This heterojunction comprising intimate interfacial coupling exhibited stable and efficient electrocatalytic performance (300 mV at 10 mA/cm^2) due to abundant active sites, the reduced pathway for charge transfer, and superior electrical conductivity [99]. Being the cornerstone reaction in renewable energy technologies, OER based on the non-precious metal catalysts emerged as a cost-effective approach for the efficient electrocatalysts in alkaline electrolyte. The efficient non-precious metal catalysts have been developed by synergistic chemical and electronic coupling LDHs with 2D MXene. The hierarchical FeNi-LDH/Ti_3C_2-MXene nanohybrids provide strong electronic coupling and interfacial interaction between FeNi-LDH and Ti_3C_2 MXene, which facilities prominent charge-transfer, electrical conductivity, structural stability, and high redox process (Ni^{2+}/$Ni^{3+,4+}$). As a result, FeNi-LDH/Ti_3C_2-MXene electrocatalyst demonstrated excellent OER performance with superior durability [100].

The porous and free-standing TCCN film (an overlapped film of Ti_3C_2 nanosheets and graphitic carbon nitride) is prepared from the Ti–N_x motifs using 2D g-C_3N_4 and TiC (with MXene phase) NSs, which displayed outstanding OER activity and stability. A hierarchically porous structure with abundant Ti–N_x active sites and highly hydrophilic surface allows this electrocatalyst to remarkably promote activity, oxygen electrochemistry, and stability [101]. The 2D metal–organic framework (MOF) NSs have been used for combination due to their unique properties like fast mass and ion transfer, highly porous structure, and high exposure for active metal centers. The MXene/MOF hybrid (i.e. $Ti_3C_2T_x$–CoBDC) was formed via an interdiffusion reaction. The 2D cobalt 1,4-benzenedicarboxylate (CoBDC) layer hybridized over electronegative surfaces of $Ti_3C_2T_x$ NSs provides remarkable ion transfer and fast charge. The hybrid material was reported to achieve 10 mA/cm^2 at 1.64 V vs. RHE in 0.1 M KOH. Such outstanding performance outperforms the reported transition metal-based electrocatalysts due to less aggregation of 2D sheets, high accessibility for the electrolyte, large number of active sites, highly porous structure, and well-defined frameworks for rapid charge and ion transfer (see Figure 6.10a) [102]. In another case, hierarchical porous Ni-Co-mixed metal sulfide merged with $Ti_3C_2T_x$ MXene sheets was realized by MOF-based approach. The unique structure along with strong interfacial interaction between $Ti_3C_2T_x$ MXene sheets and NiCoS reported to generate a high active site and prominent charge-transfer conductivity. As a result, they showed excellent activity and stability as OER electrocatalysts (see Figure 6.10b) [103].

A MXene supported CoNi-ZIF-67 hybrid (CoNi-ZIF-67@$Ti_3C_2T_x$) was fabricated via in situ growth of bimetallic CoNi-ZIF-67 rhombic dodecahedrons on the $Ti_3C_2T_x$ matrix through coprecipitation reaction. The addition of cetyltrimethylammonium bromide (CTAB) during synthesis leads to intercalation of CoNi-ZIF-67 into $Ti_3C_2T_x$ layers and forms the two-phase comprised hybrid structure. It was observed that the introduction of the MXene matrix exclusively provides smaller size particles (CoNi-ZIF-67) and enhances the oxidation of Co/Ni elements. It endows the CoNi-ZIF-67@$Ti_3C_2T_x$ as a highly efficient non-precious metal OER catalyst, the effective synergy of enhanced conductivity, hybrid structure, increase in the Co/Ni oxidation state, and high electrochemically active surface area prompts the superior OER catalytic activity (275 mV vs. RHE) [104]. The Co(OH)F, CoP, and Co_7Se_8 decorated large-sized exfoliated 2D MXene sheets were prepared via phosphidation process and were reported with high electrocatalytic performances due to their specific conductivity and maximized surface area. CoP/MXene hybrid showed lower overpotential (230 mV vs. RHE) and enhanced OER activity. Hybrid swift charge-transfer kinetics showed remarkable bifunctional performance in alkaline medium. PO_x-enriched Co–OOH/CoP/MXene hybrid after CoP/MXene surface oxidation enables stable electrochemical activity. Such insights shed light on the MXene support's role in the hybridized electrocatalysts and their correlation with oxidized anion species and activity [105]. Combined with FeCo and supported with single-layer MXene ($Ti_3C_2T_x$) has been prepared and reported as a highly efficient OER electrocatalyst. The single-layer MXene with inherent advantages stabilizes the LDHs by preventing them from aggregation, and the strong synergistic interaction endows the high electron transfer and plenty of active sites. The FeCo-LDH/MXene hybrid exhibited electrocatalytic activity (268 mV at 10 mA/cm^2) and durability for OER and outperformed multiwalled carbon nanotubes (MWCNTs) and graphene supported FeCo-LDH activity. The 3D

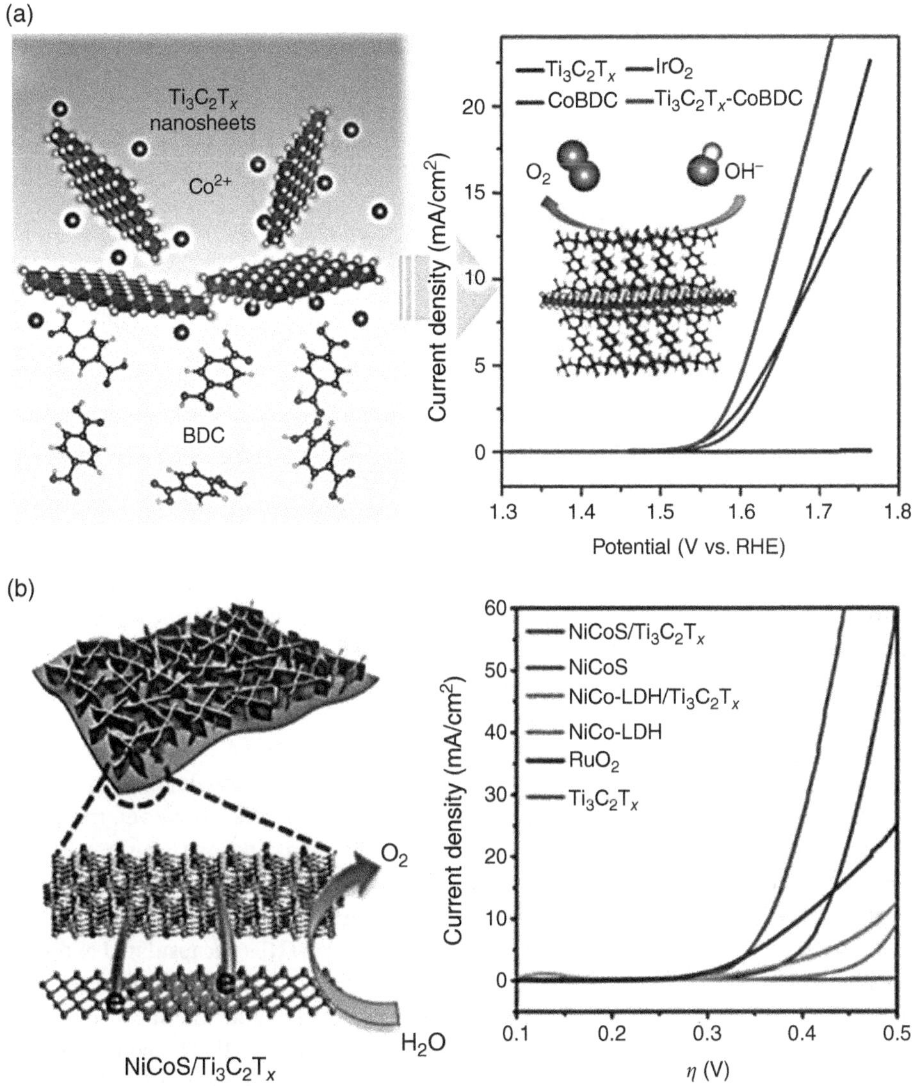

Figure 6.10 Electrocatalytic oxygen evolution from (a) 2D MOF and $Ti_3C_2T_x$ nanosheets hybridized by interdiffusion reaction. *Source:* From [102] / with permission of American Chemical Society, and (b) MOF developed Ni-CoS on ultrathin MXene nanosheets. *Source:* From [103] / with permission of American Chemical Society.

S-$NiFe_2O_4$@Ti_3C_2@NF hierarchical network structure has been successfully prepared by an in situ thiourea-assisted electrodeposition process followed by low-temperature calcination. It reported superior OER activity (1.50 V vs. RHE at 20 mA/cm^2) and remarkable stability, which are attributed synergistically to the unique and favorable 3D framework, few-layer Ti_3C_2 flakes (MXene) matrix enabled fast electron transport, and ultrasmall S-$NiFe_2O_4$ NSs derived abundant catalytic active sites [106]. A new type of NiFeCe-LDH/MXene hybrid structure was designed by in situ coprecipitation method by directly growing 2D Ce-NiFe-LDH nanoflakes over 2D $Ti_3C_2T_x$ MXene surface. Hierarchical nanoporous nature and strong interfacial junction between Ce doping and MXene coupling in this hybrid synergistically demonstrated an excellent OER activity (260 mV at 10 mA/cm^2) and durability in the alkaline medium [107].

6.5 Oxygen Reduction Reaction (ORR)

The non-Pt-based electrocatalysts with high activity become more vital due to their high potential for practical applications. The hybrid Co_3O_4/N-doped CNT catalyst supported on the Ti_3C_2 (Co_3O_4/NCNTs/Ti_3C_2) surface has been prepared by the in situ pyrolysis strategy. As-prepared Co_3O_4/NCNTs/Ti_3C_2 electrocatalyst demonstrated remarkable ORR activity owing to three active components viz., synergistic effects, improved electron transfers due to graphitized carbon, and abundant active site from the high Brunauer–Emmett–Teller (BET) surface area [108]. A novel FeNC/MXene hybrid NS was fabricated from the iron–ligand complex and highly dispersed MXene NSs as support via the pyrolysis approach. A thin and rugged FeNC coating was formed on the MXene surface, which creates an ideal path for high electron transport and many active sites for catalysis. FeNC/MXene hybrid NS showed excellent electrocatalytic performance (0.814 V versus RHE) and superb durability (only 2.6% decay after 20 000 seconds continuous cycles) (see Figure 6.11b) [109].

The 2D accordion-like $Ti_3C_2T_x$ NSs were prepared by the modified LiF/HCl solution-based strategy and uniformly loaded with the Pt particles. With good hydration properties, abundant active groups on the 2D layer surface, and high conductivity of MXenes, Pt easily seats uniformly all over the layer and exhibits remarkable activity and durability for ORR in both alkaline and acid medium by the quasi-four-electron process. However, the alkali condition favors the activity and durability due to the well-maintained 2D layer framework and low degree of Pt agglomeration that provides stable and uniform sites for ORR. Such phenomena also prevent the catalytic sites oxidative corrosion (see Figure 6.11a) [110]. A new type of composite (Co/N-CNTs@$Ti_3C_2T_x$) consisting Co NPs encapsulated N-doped CNTs supported on conductive $Ti_3C_2T_x$ MXene has been successfully designed as ORR and OER bifunctional electrocatalyst. The strong electron transfer and interfacial coupling were observed and endowed enhanced electrocatalytic activity [112]. The composites of cobalt-tipped CNT/Ti_3C_2 NS (Co-CNT/Ti_3C_2) were successfully prepared by the MOF-engaged strategy via in situ growing of ZIF-67 on Ti_3C_2 NSs followed by the pyrolysis that forms Co-tripped CNT. Being 2D conductive scaffolds for the growth, Ti_3C_2 NSs regulate the surface area and degree of graphitization. Co-CNT/Ti_3C_2 exhibited excellent activity as ORR electrocatalysts due to abundant Co–N/C sites at CNT tips, high degree of graphitized carbon, and large specific surface area from Ti_3C_2 incorporation. It significantly ensures the rapid electron transfer, large exposure of the active sites, and fast diffusion of O_2 [111]. A noble-metal-free composite material (N-rich carbon/MXene) consisting of porous N-doped carbon and MXene used polypyrrole and Ti_3C_2 MXene as C and N precursors. The highly conductive and reactive nature of these precursors, uniform distribution of the N, a high percentage of graphitic N, and abundant sites for the activity of N-rich carbon/MXene electrocatalyst showed remarkable stability and activity. Mostly, four-electron reduction pathway was followed during the ORR process [113].

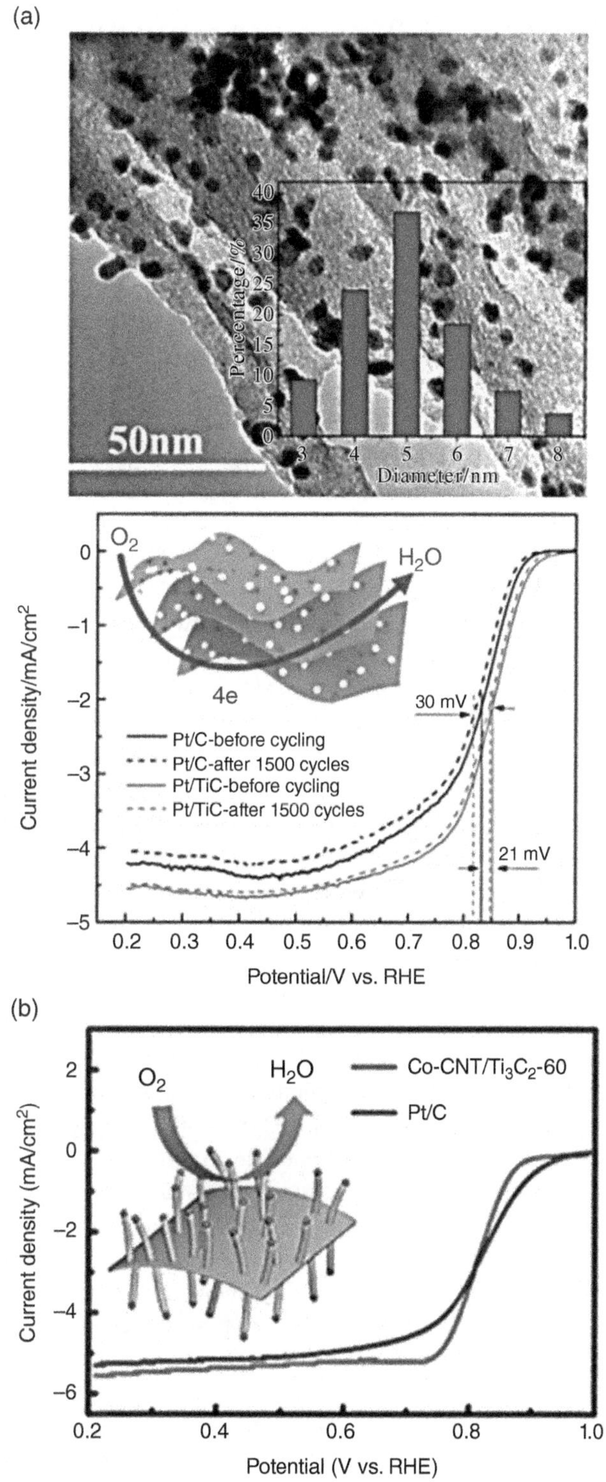

Figure 6.11 (a) Pt/MXene nanocatalysts. *Source:* From [110] / with permission of Elsevier and (b) Co-tipped CNT and MXene nanosheets-based electrocatalyst for an excellent ORR [111]. *Source:* From [111] / with permission of Royal Society of Chemistry.

6.6 Nitrogen Reduction Reaction (NRR)

NH_3 is used predominantly in fertilizers, textiles, dyes, and resins [114]. It is produced from high purity H_2 and N_2 mainly by an energy-intensive Haber–Bosch process in the presence of Fe as a catalyst at a reaction temperature and pressure of 400–500 °C and 20–40 MPa, respectively. Apart from accounting annually for over 1% of the world's energy supply, it also contributes 300 metric tons of CO_2, which is 1.4% of global emissions [115]. Therefore, electrocatalytic synthesis of NH_3 is appealing as an alternate route with lesser energy consumption and sustainable and milder ambient reaction conditions. Therefore, finding a low-cost, efficient electrocatalyst for N_2 fixation, which will utilize atmospheric N_2 as a source, is vital for advancing the NRR industry electrochemically. Selectivity issue due to the competition with HER is a major hurdle [116]. In this regard, MXene has been proposed by various groups as an electrocatalyst for NRR. Wang et al. performed DFT calculation and found high NRR activity for Mo_2TiC with an overpotential of 0.26 V and the competing HER activity curbed [117]. Zhao et al. found MXenes with different functional groups such as —F and —OH that catalyze N_2 to NH_3 under ambient atmosphere efficiently with 20.4 mg/h/mg_{cat} yield and 9.3% faradaic efficiency (FE) [118]. Azofra et al. identified five different pathways as a mechanism of electrochemical NRR on MXene, as shown in Figure 6.12a [119]. In this, additional two earlier postulates of Anderson et al. indicate three successive gains of H^+/e^- on the N centers to produce NH_3 molecule. The Azofra's group investigated that d2 M_3C_2 MXene preferably followed the fourth pathway while path 2 is preferred by d3 and d4 M_3C_2, which corresponds to the minimum energy path. For example, V_3C_2 and Nb_3C_2 MXenes (Figure 6.12b and c) are the predicted pathways calculated using DFT following minimum energy paths. The V_3C_2 MXene observed a lower activation energy barrier of 0.64 eV as the rate-determining step (first electron-proton transfer), whereas for Nb_3C_2 MXene, it is the second hydrogenation step with 0.9 eV. Apart from theoretical calculations, experimental verifications have been performed. For example, Zhang et al. hydrothermally prepared hybrid $TiO_2/Ti_3C_2T_x$, which yielded 23.62 μg/h/mg_{cat} at −0.6 V vs. RHE with 8.42% FE [120]. This indicates that MXene can synergistically combine with other materials and there is an enormous possibility of performance enhancement. Other metal oxides with excellent selectivity, such as hybrids of MnO_2 decorated over $Ti_3C_2T_x$ MXene showed brilliant durability yielding 34.12 μg/h/mg_{cat} with 11.39% FE vs. RHE [121]. This was attributed to the unsaturated Mn atoms that act as active sites for N_2 adsorption and activation. TiO_2 hydrothermally grown on conductive MXene substrates reported NH_3 yield and FE of 26.33 μg/h/mg_{cat} and 8.42%, respectively, at *0.6 V vs. RHE [120]. Further modification in TiO_2 such as oxygen vacant sites can act as main active sites for NH_3 synthesis, along with good electron transport of $Ti_3C_2T_x$ MXenes showed high activity with 32.17 μg/h/mg_{cat} and 16.07% FE at −0.55 V and −0.45 V vs. RHE [122]. Additionally, DFT calculations by Zheng and coworkers revealed that back-donation from B centers on B atoms decorated on MXenes such as Mo_2CO_2 and W_2CO_2 could enhance N_2 reduction [123].

Furthermore, experimental results suggest that more than the lateral Ti (−0.95 eV), C (−0.16 eV), and O (−1.21 eV), the central Ti atom in MXene $Ti_3C_2T_x$ is a more active site for N_2 adsorption. Additionally, the basal plane of MXene has low exposure Ti sites

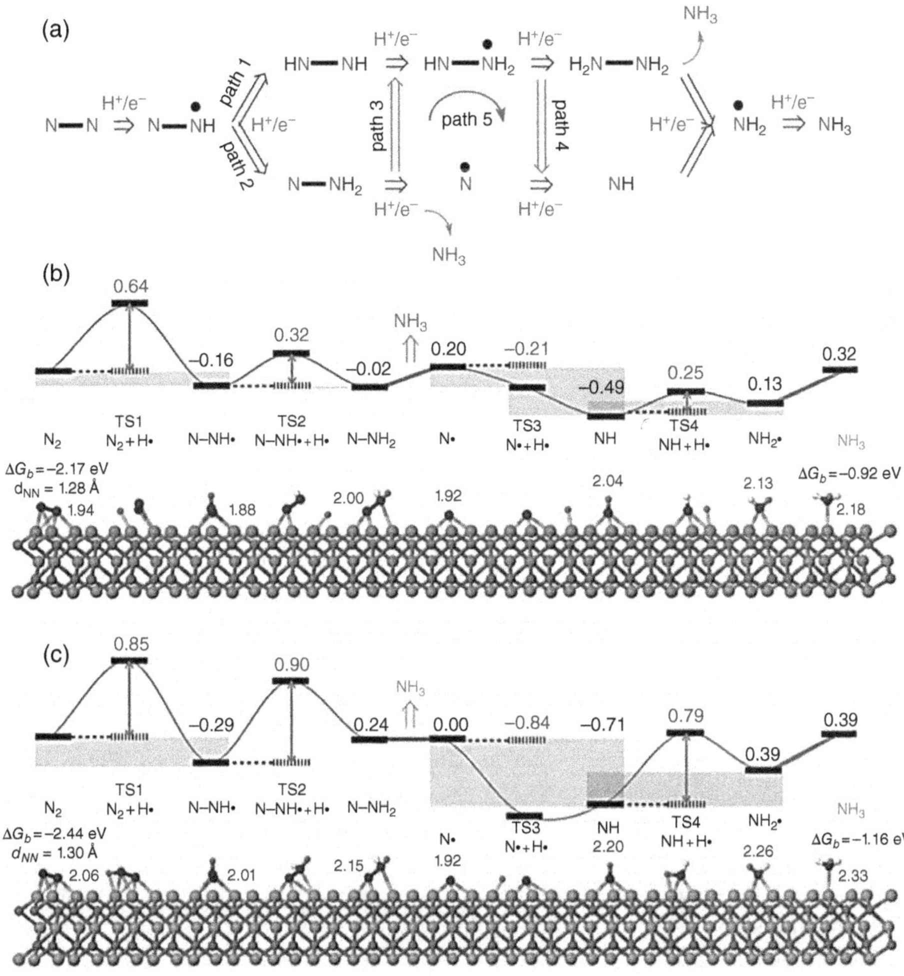

Figure 6.12 (a) Suggested possible pathways for the N_2 conversion mechanism, (b) and (c) represents the proposed routes for minimum energy path for NH_3 synthesis from N_2 catalyzed by V_3C_2 and Nb_3C_2, along with DFT+D3 computational calculation. TS represents the transition state. The text on the trough and crest respectively, are the Gibbs free energy and activation energy in eV vs. the standard hydrogen electrode at T = 298.15 K, pH = 0, f (H_2) = 101 325 Pa, and U = 0V. The texts in lighter shade at each step indicates the reactive hydrogen atom in the N_2 electroreduction. The two initial shaded regions are the spontaneous steps and the final shaded region is the non-spontaneous step, with dMN distances in Å. Also, the structures and energies of various intermediates formed are shown. *Source:* From [119] / with permission of Royal Society of Chemistry.

compared to edge planes, and therefore, the former is less active than the latter [124]. To develop an efficient MXene-based NRR electrocatalyst, active sites have to be exposed intentionally to achieve selective N_2 reduction as some MXenes have weaker binding to N_2. MXene and other MXene-based composites and hybrids can be used successfully designed as a low-cost electrocatalyst for electrochemical N_2 reduction to NH_3.

6.7 CO_2 Reduction Reaction (CRR)

The total utilization of natural resources, unchecked burning of fossil fuels, and other anthropogenic activities have resulted in large-scale emission and rise of CO_2. This has resulted in the greenhouse effect and an increase in global warming [125]. Therefore, it has become a worldwide challenge to mitigate and fix atmospheric CO_2. The electrochemical reduction can convert CO_2 into useful and high value-added products, such as CH_3OH, HCOOH, and higher hydrocarbons such as $CH_2 = CH_2$ (ethylene) [126]. MXenes with surface terminations such as —O and —OH have reduced energy input requirements for CO_2 to CH_4 conversion, which is enhanced compared to bare MXenes [12]. For example, DFT studies performed by Chen et al. revealed $Sc_2C(OH)_2$ MXenes to have promising activity for CRR to CH_4 [127]. This was attributed to the presence of highly reactive H in the -OH terminal that helps formation of stable intermediates and lowers the overpotential. Additionally, MXenes in the IV-VI series such as Cr_3C_2 and Mo_3C_2 were found to exhibit good conversion from CO_2 to CH_4 due to the nature of low charge migration in the potential-limiting step (Figure 6.13a, b). The overpotential required for Cr_3C_2 and Mo_3C_2 to convert CO_2 to CH_4 are 1.05 eV and 1.31 eV, respectively [128]. Li et al. found radicals such as OCHO• and HOCO• during the rate-determining step in the hydrogenation step [128]. Similarly, Handoko et al. reported high performance of W_2CO_2 and Ti_2CO_2, which they attributed to energetically favorable and highly accessible *HCOOH pathway [129]. This pathway is represented in the energy diagram in Figure 6.13c as the most favorable pathway for CO_2 reduction to methane for Ti_2CO_2 MXene. The CO_2 is adsorbed on the surface and the incoming H forms *COOH intermediates by attaching to the terminal O of CO_2. This intermediate is then bonded to Ti_2CO_2, as observed in Figure 6.13c. The good CRR effectiveness reported for MXene-based electrocatalyst can be due to the good selectivity and reported low overpotential. Apart from this, an energetically favorable pathway with an easily accessible *HCOOH pathway also boosts the performance. However, most studies in CRR are limited to DFT studies, and it is important to invest the theoretical data to perform CRR experimentally.

6.8 Methanol Oxidation Reaction (MOR)

Among the many means being explored for a sustainable, environmentally benign, and efficient energy source is the direct methanol fuel cells (DMFCs). However, sluggish kinetics and high overpotential of MOR are slowing the performance of DMFCs. MXene can be resourceful anchoring sites, with its high conductivity and abundant surface, functionalities can be a potent material for MOR. MWCNTs modified using MXene $Ti_3C_2T_x$ and MoS_2 quantum dots were used as electrocatalyst by Yang et al. for MOR [130]. The authors report a high current density of 160 A/g using this catalyst at a potential of 2.2 V in alkaline KOH. Additionally, MXenes can also be used in combination with noble metals, such as Pt, which diminishes the use of expensive metal without compromising the performance. For example, Wang et al. synthesized Pt/Ti_3C_2 and observed a current density of 1.137 mA/cm^2, which is higher than commercially available Pt/C [131]. Other compositions, such as Pd/MXene, were also synthesized, which showed a high MOR current density of 12.4 mA/cm^2.

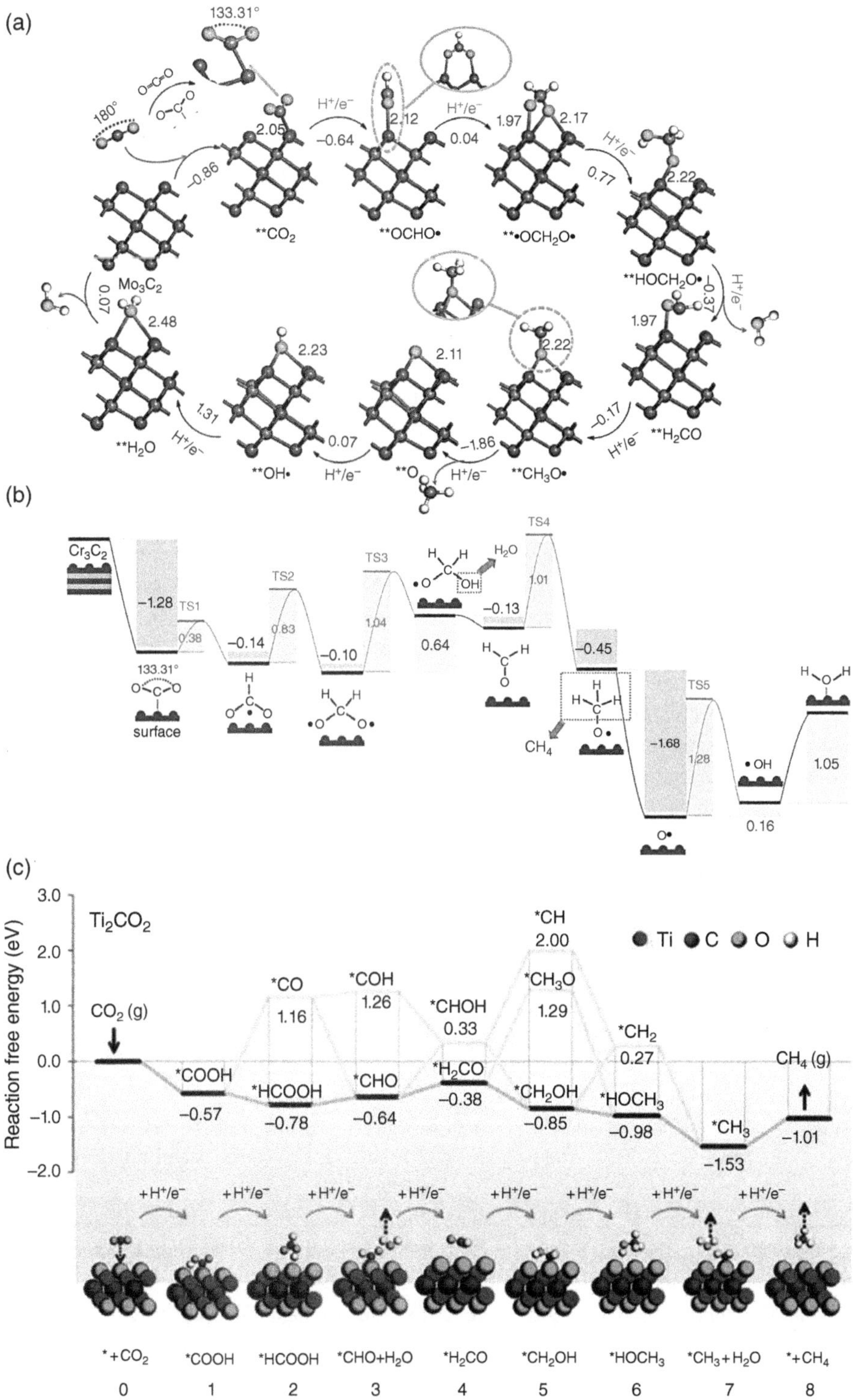

Figure 6.13 (a) and (b) represents the respective minimum energy path proposed for the catalyzed conversion of carbon dioxide into $^*CH_4$ and $^{**}H_2O$ using Mo_3C_2 and Cr_3C_2. *Source:* From [128] / with permission of American Chemical Society, and (c) free energy diagram and proposed routes of formation of CH_4 from CO_2 using O-terminated Ti_2CO_2 MXene represented by a ball-and-stick model. The route via *C was excluded as its unstable adsorption leads to the removal of O termination group. *Source:* From [129] / with permission of Royal Society of Chemistry.

The introduction of Pd onto MXene exposed that catalytically active Pd changes the electronic structure while improving the MOR activity with minimum Pd loading [132]. MXene and MXene hybrids are promising electrocatalyst for electro-oxidation for MOR due to their electronegative surface terminations inducing excellent support and interactions with metals that increases methanol adsorption.

6.9 Conclusion

A strong alternative to non-renewable fossil fuels is electrocatalysis-based technology that will curb CO_2 emission and other environmental challenges, such as the greenhouse effect and global warming. Briefly, MXene has been discussed and an attempt has been made to summarize state-of-the-art on the synthesis of MXenes from its MAX phase and various other modifications that can be made to enhance the performance when used as electrocatalyst. Additionally, the use of various MXenes as powerful electrocatalyst for HER, OER, ORR, NRR, CRR, and MOR has been comprehensively discussed. Designing MXene can be advantageous due to its characteristics such as large specific area, excellent mechanical and thermal stability, tunable morphological and bandgap, hydrophilicity and special carrier anisotropic mobility, etc. Moreover, to further enhance the electrocatalytic performance, various treatments and strategies can be adopted to further enhance the properties, such as single-atom loading, functional group, and defect state control and heterojunction formation, etc. Therefore, it is vital to focus on certain aspects of MXene development, such as controlling surface properties via regulation of morphology, size and surface properties, doping, or hybrids with metals. The metal-MXene hybrids will rearrange their electronic configuration and active sites and improve the charge-transfer kinetics as well as promote strong interfacial coupling. These modifications and the expanding family make MXene a strong candidate for future use as an electrocatalyst.

References

1 Gogotsi, Y. and Anasori, B. (2019). The rise of MXenes. *ACS Nano* 13 (8): 8491–8494.

2 Anasori, B., Lukatskaya, M.R., and Gogotsi, Y. (2017). 2D metal carbides and nitrides (MXenes) for energy storage. *Nature Reviews Materials* 2 (2): 16098.

3 Levi, M.D., Lukatskaya, M.R., Sigalov, S. et al. (2015). Solving the capacitive paradox of 2D MXene using electrochemical quartz-crystal admittance and in situ electronic conductance measurements. *Advanced Energy Materials* 5 (1): 1400815.

4 Shahzad, F., Alhabeb, M., Hatter, C.B. et al. (2016). Electromagnetic interference shielding with 2D transition metal carbides (MXenes). *Science* 353 (6304): 1137–1140.

5 Tong, Y., He, M., Zhou, Y. et al. (2018). Hybridizing polypyrrole chains with laminated and two-dimensional $Ti_3C_2T_x$ toward high-performance electromagnetic wave absorption. *Applied Surface Science* 434: 283–293.

6 Cheng, L., Wang, X., Gong, F. et al. (2020). 2D nanomaterials for cancer theranostic applications. *Advanced Materials* 32 (13): 1902333.

7 Mohammadpour, Z. and Majidzadeh-A, K. (2020). Applications of two-dimensional nanomaterials in breast cancer theranostics. *ACS Biomaterials Science and Engineering* 6 (4): 1852–1873.

8 Xu, B., Zhu, M., Zhang, W. et al. (2016). Ultrathin MXene-micropattern-based field-effect transistor for probing neural activity. *Advanced Materials* 28 (17): 3333–3339.

9 Cai, Y., Shen, J., Ge, G. et al. (2018). Stretchable $Ti_3C_2T_x$ MXene/carbon nanotube composite based strain sensor with ultrahigh sensitivity and tunable sensing range. *ACS Nano* 12: 56–62.

10 Han, X., Huang, J., Lin, H. et al. (2018). 2D ultrathin MXene-based drug-delivery nanoplatform for synergistic photothermal ablation and chemotherapy of cancer. *Advanced Healthcare Materials* 7 (9): 1701394.

11 Lin, H., Gao, S., Dai, C. et al. (2017). A two-dimensional biodegradable niobium carbide (MXene) for photothermal tumor eradication in NIR-I and NIR-II biowindows. *Journal of the American Chemical Society* 139 (45): 16235–16247.

12 Liu, A., Liang, X., Ren, X. et al. (2020). Recent progress in MXene-based materials: potential high-performance electrocatalysts. *Advanced Functional Materials* 30 (38): 2003437.

13 Barsoum, M.W. and El-Raghy, T. (1996). Synthesis and characterization of a remarkable ceramic: Ti_3SiC_2. *Journal of the American Ceramic Society* 79 (7): 1953–1956.

14 Amini, S., Barsoum, M.W., and El-Raghy, T. (2007). Synthesis and mechanical properties of fully dense Ti_2SC. *Journal of the American Ceramic Society* 90 (12): 3953–3958.

15 Luo, Y., Zheng, Z., Mei, X., and Xu, C. (2008). Growth mechanism of Ti_3SiC_2 single crystals by in-situ reaction of polycarbosilane and metal titanium with CaF_2 additive. *Journal of Crystal Growth* 310 (14): 3372–3375.

16 Yin, X., Travitzky, N., and Greil, P. (2007). Three-dimensional printing of nanolaminated Ti_3AlC_2 toughened $TiAl_3$–Al_2O_3 composites. *Journal of the American Ceramic Society* 90 (7): 2128–2134.

17 Tian, W., Vanmeensel, K., Wang, P. et al. (2007). Synthesis and characterization of Cr_2AlC ceramics prepared by spark plasma sintering. *Materials Letters* 61 (22): 4442–4445.

18 Eklund, P., Beckers, M., Jansson, U. et al. (2010). The $M_{n+1}AX_n$ phases: materials science and thin-film processing. *Thin Solid Films* 518 (8): 1851–1878.

19 Frodelius, J., Sonestedt, M., Björklund, S. et al. (2008). Ti_2AlC coatings deposited by high velocity oxy-fuel spraying. *Surface and Coatings Technology* 202 (24): 5976–5981.

20 Rech, S., Surpi, A., Vezzù, S. et al. (2013). Cold-spray deposition of Ti_2AlC coatings. *Vacuum* 94: 69–73.

21 Gutzmann, H., Gärtner, F., Höche, D. et al. (2013). Cold spraying of Ti_2AlC MAX-phase coatings. *Journal of Thermal Spray Technology* 22 (2): 406–412.

22 Pasumarthi, V., Chen, Y., Bakshi, S.R., and Agarwal, A. (2009). Reaction synthesis of Ti_3SiC_2 phase in plasma sprayed coating. *Journal of Alloys and Compounds* 484: 113–117.

23 Frodelius, J., Johansson, E.M., Córdoba, J.M. et al. (2011). Annealing of thermally sprayed Ti_2AlC coatings. *International Journal of Applied Ceramic Technology* 8 (1): 74–84.

24 Nickl, J.J., Schweitzer, K.K., and Luxenberg, P. (1972). Gasphasenabscheidung im system Ti_3Si_2C. *Journal of the Less Common Metals* 26 (3): 335–353.

25 Racault, C., Langlais, F., Naslain, R., and Kihn, Y. (1994). On the chemical vapour deposition of Ti_3SiC_2 from $TiCl_4$-$SiCl_4$-CH_4-H_2 gas mixtures. *Journal of Materials Science* 29 (15): 3941–3948.

26 Pickering, E., Lackey, W.J., and Crain, S. (2000). CVD of Ti_3SiC_2. *Chemical Vapor Deposition* 6 (6): 289–295.

27 Goto, T. and Hirai, T. (1987). Chemically vapor deposited Ti_3SiC_2. *Materials Research Bulletin* 22 (9): 1195–1201.

28 Fakih, H., Jacques, S., Berthet, M.P. et al. (2006). The growth of Ti_3SiC_2 coatings onto SiC by reactive chemical vapor deposition using H_2 and $TiCl_4$. *Surface and Coatings Technology* 201 (6): 3748–3755.

29 Fakih, H., Jacques, S., Dezellus, O. et al. (2008). Phase equilibria and reactive chemical vapor deposition (RCVD) of Ti_3SiC_2. *Journal of Phase Equilibria and Diffusion* 29 (3): 239–246.

30 Wilhelmsson, O., Eklund, P., Högberg, H. et al. (2008). Structural, electrical and mechanical characterization of magnetron-sputtered V–Ge–C thin films. *Acta Materialia* 56 (11): 2563–2569.

31 Eklund, P., Bugnet, M., Mauchamp, V. et al. (2011). Epitaxial growth and electrical transport properties of Cr_2GeC thin films. *Physical Review B* 84 (7): 075424.

32 Magnuson, M., Mattesini, M., Bugnet, M., and Eklund, P. (2015). The origin of anisotropy and high density of states in the electronic structure of Cr_2GeC by means of polarized soft X-ray spectroscopy and ab initio calculations. *Journal of Physics: Condensed Matter* 27 (415): 501.

33 Walter, C., Sigumonrong, D.P., El-Raghy, T., and Schneider, J.M. (2006). Towards large area deposition of Cr_2AlC on steel. *Thin Solid Films* 515 (2): 389–393.

34 Eklund, P., Beckers, M., Frodelius, J. et al. (2007). Magnetron sputtering of Ti_3SiC_2 thin films from a compound target. *Journal of Vacuum Science and Technology A* 25: 1381–1388.

35 Frodelius, J., Eklund, P., Beckers, M. et al. (2010). Sputter deposition from a Ti_2AlC target: process characterization and conditions for growth of Ti_2AlC. *Thin Solid Films* 518 (6): 1621–1626.

36 Scabarozi, T.H., Hettinger, J.D., Lofland, S.E. et al. (2011). Epitaxial growth and electrical-transport properties of $Ti_7Si_2C_5$ thin films synthesized by reactive sputter-deposition. *Scripta Materialia* 65 (9): 811–814.

37 Su, R., Zhang, H., O'Connor, D.J. et al. (2016). Deposition and characterization of Ti_2AlC MAX phase and Ti_3AlC thin films by magnetron sputtering. *Materials Letters* 179: 194–197.

38 Rosén, J., Ryves, L., Persson, P.O.Å., and Bilek, M.M.M. (2007). Deposition of epitaxial Ti_2AlC thin films by pulsed cathodic arc. *Journal of Applied Physics* 101 (5): 056101.

39 Wang, Z., Saito, M., Tsukimoto, S., and Ikuhara, Y. (2009). Interface atomic-scale structure and its impact on quantum electron transport. *Advanced Materials* 21 (48): 4966–4969.

40 Dolique, V., Jaouen, M., Cabioc'h, T. et al. (2008). Formation of (Ti,Al)N/Ti_2AlN multilayers after annealing of TiN/TiAl(N) multilayers deposited by ion beam sputtering. *Journal of Applied Physics* 103 (8): 083527.

41 Fashandi, H., Dahlqvist, M., Lu, J. et al. (2017). Synthesis of Ti_3AuC_2, $Ti_3Au_2C_2$ and Ti_3IrC_2 by noble metal substitution reaction in Ti_3SiC_2 for high-temperature-stable Ohmic contacts to SiC. *Nature Materials* 16 (8): 814–818.

42 Fashandi, H., Lai, C.C., Dahlqvist, M. et al. (2017). Ti_2Au_2C and $Ti_3Au_2C_2$ formed by solid state reaction of gold with Ti_2AlC and Ti_3AlC_2. *Chemical Communications* 53 (69): 9554–9557.

43 Lai, C.-C., Fashandi, H., Lu, J. et al. (2017). Phase formation of nanolaminated Mo_2AuC and $Mo_2(Au_{1-x}Ga_x)_2C$ by a substitutional reaction within Au-capped Mo_2GaC and Mo_2Ga_2C thin films. *Nanoscale* 9 (45): 17681–17687.

44 Lai, C.-C., Petruhins, A., Lu, J. et al. (2017). Thermally induced substitutional reaction of Fe into Mo_2GaC thin films. *Materials Research Letters* 5 (8): 533–539.

45 Lai, C.-C., Tao, Q., Fashandi, H. et al. (2018). Magnetic properties and structural characterization of layered $(Cr_{0.5}Mn_{0.5})_2AuC$ synthesized by thermally induced substitutional reaction in $(Cr_{0.5}Mn_{0.5})_2GaC$. *APL Materials* 6 (2): 026104.

46 Zhou, Y., Sun, Z., Chen, S., and Zhang, Y. (1998). In-situ hot pressing/solid-liquid reaction synthesis of dense titanium silicon carbide bulk ceramics. *Materials Research Innovations* 2 (3): 142–146.

47 Yongming, L., Wei, P., Shuqin, L. et al. (2002). Synthesis of high-purity Ti_3SiC_2 polycrystals by hot-pressing of the elemental powders. *Materials Letters* 52 (4): 245–247.

48 Hashimoto, S., Takeuchi, M., Inoue, K. et al. (2008). Pressureless sintering and mechanical properties of titanium aluminum carbide. *Materials Letters* 62 (10): 1480–1483.

49 Riley, D.P., Kisi, E.H., and Phelan, D. (2006). SHS of Ti_3SiC_2: ignition temperature depression by mechanical activation. *Journal of the European Ceramic Society* 26 (6): 1051–1058.

50 Dezellus, O., Gardiola, B., Andrieux, J., and Lay, S. (2015). Experimental evidence of copper insertion in a crystallographic structure of Ti_3SiC_2 MAX phase. *Scripta Materialia* 104: 17–20.

51 Mingxing, A., Hongxiang, Z., Yang, Z. et al. (2006). Synthesis of Ti_3AlC_2 powders using Sn as an additive. *Journal of the American Ceramic Society* 89 (3): 1114–1117.

52 Zhang, J., Wang, L., Jiang, W., and Chen, L. (2007). Fabrication of high purity Ti_3SiC_2 from Ti/Si/C with the aids of Al by spark plasma sintering. *Journal of Alloys and Compounds* 437 (1): 203–207.

53 Barsoum, M.W. and Eklund, P. (2019). The Mn + 1AXn phases: the precursors for MXenes. In: *2D Metal Carbides and Nitrides (MXenes): Structure, Properties and Applications* (ed. B. Anasori and Y. Gogotsi), 15–35. Cham: Springer International Publishing.

54 Liu, J., Peng, W., Li, Y. et al. (2020). 2D MXene-based materials for electrocatalysis. *Transactions of Tianjin University* 26 (3): 149–171.

55 Wang, H. and Lee, J.-M. (2020). Recent advances in structural engineering of MXene electrocatalysts. *Journal of Materials Chemistry A* 8 (21): 10604–10624.

56 Naguib, M., Kurtoglu, M., Presser, V. et al. (2011). Two-dimensional nanocrystals produced by exfoliation of Ti_3AlC_2. *Advanced Materials* 23 (37): 4248–4253.

57 Naguib, M., Mashtalir, O., Carle, J. et al. (2012). Two-dimensional transition metal carbides. *ACS Nano* 6 (2): 1322–1331.

58 Ghidiu, M., Lukatskaya, M.R., Zhao, M.-Q. et al. (2014). Conductive two-dimensional titanium carbide 'clay' with high volumetric capacitance. *Nature* 516 (7529): 78–81.

59 Li, G., Tan, L., Zhang, Y. et al. (2017). Highly efficiently delaminated single-layered MXene nanosheets with large lateral size. *Langmuir* 33 (36): 9000–9006.

60 Pang, S.-Y., Wong, Y.-T., Yuan, S. et al. (2019). Universal strategy for HF-free facile and rapid synthesis of two-dimensional MXenes as multifunctional energy materials. *Journal of the American Chemical Society* 141 (24): 9610–9616.

61 Li, M., Lu, J., Luo, K. et al. (2019). Element replacement approach by reaction with Lewis acidic molten salts to synthesize nanolaminated MAX phases and MXenes. *Journal of the American Chemical Society* 141 (11): 4730–4737.

62 Yang, X., Gao, N., Zhou, S., and Zhao, J. (2018). MXene nanoribbons as electrocatalysts for the hydrogen evolution reaction with fast kinetics. *Physical Chemistry Chemical Physics* 20 (29): 19390–19397.

63 Yuan, W., Cheng, L., An, Y. et al. (2018). MXene nanofibers as highly active catalysts for hydrogen evolution reaction. *ACS Sustainable Chemistry and Engineering* 6 (7): 8976–8982.

64 Liu, J., Liu, Y., Xu, D. et al. (2019). Hierarchical "nanoroll" like $MoS_2/Ti_3C_2T_x$ hybrid with high electrocatalytic hydrogen evolution activity. *Applied Catalysis B: Environmental* 241: 89–94.

65 Wu, X., Wang, Z., Yu, M. et al. (2017). Stabilizing the MXenes by carbon nanoplating for developing hierarchical nanohybrids with efficient lithium storage and hydrogen evolution capability. *Advanced Materials* 29 (24): 1607017.

66 Xiu, L., Wang, Z., Yu, M. et al. (2018). Aggregation-resistant 3D MXene-based architecture as efficient bifunctional electrocatalyst for overall water splitting. *ACS Nano* 12 (8): 8017–8028.

67 Yu, M., Wang, Z., Liu, J. et al. (2019). A hierarchically porous and hydrophilic 3D nickel–iron/MXene electrode for accelerating oxygen and hydrogen evolution at high current densities. *Nano Energy* 63: 103880.

68 Wu, X., Zhou, S., Wang, Z. et al. (2019). Engineering multifunctional collaborative catalytic interface enabling efficient hydrogen evolution in all pH range and seawater. *Advanced Energy Materials* 9 (34): 1901333.

69 Du, C.-F., Sun, X., Yu, H. et al. (2019). Synergy of Nb doping and surface alloy enhanced on water–alkali electrocatalytic hydrogen generation performance in Ti-based MXene. *Advanced Science* 6 (11): 1900116.

70 Jiang, Y., Wu, X., Yan, Y. et al. (2019). Coupling PtNi ultrathin nanowires with MXenes for boosting electrocatalytic hydrogen evolution in both acidic and alkaline solutions. *Small* 15 (12): 1805474.

71 Zhu, X.-D., Xie, Y., and Liu, Y.-T. (2018). Exploring the synergy of 2D MXene-supported black phosphorus quantum dots in hydrogen and oxygen evolution reactions. *Journal of Materials Chemistry A* 6 (43): 21255–21260.

72 Jiang, H., Wang, Z., Yang, Q. et al. (2019). Ultrathin $Ti_3C_2T_x$ (MXene) nanosheet-wrapped $NiSe_2$ octahedral crystal for enhanced supercapacitor performance and synergetic electrocatalytic water splitting. *Nano-Micro Letters* 11 (1): 31.

73 Kuang, P., He, M., Zhu, B. et al. (2019). 0D/2D NiS_2/V-MXene composite for electrocatalytic H_2 evolution. *Journal of Catalysis* 375: 8–20.

74 Wang, Z., Xu, W., Yu, K. et al. (2020). 2D heterogeneous vanadium compound interfacial modulation enhanced synergistic catalytic hydrogen evolution for full pH range seawater splitting. *Nanoscale* 12 (10): 6176–6187.

75 Li, Z., Qi, Z., Wang, S. et al. (2019). In situ formed Pt_3Ti nanoparticles on a two-dimensional transition metal carbide (MXene) used as efficient catalysts for hydrogen evolution reactions. *Nano Letters* 19 (8): 5102–5108.

76 Yuan, Y., Li, H., Wang, L. et al. (2019). Achieving highly efficient catalysts for hydrogen evolution reaction by electronic state modification of platinum on versatile $Ti_3C_2T_x$ (MXene). *ACS Sustainable Chemistry and Engineering* 7 (4): 4266–4273.

77 Zhang, X., Shao, B., Sun, Z. et al. (2020). Platinum nanoparticle-deposited $Ti_3C_2T_x$ MXene for hydrogen evolution reaction. *Industrial and Engineering Chemistry Research* 59 (5): 1822–1828.

78 Hantanasirisakul, K. and Gogotsi, Y. (2018). Electronic and optical properties of 2D transition metal carbides and nitrides (MXenes). *Advanced Materials* 30 (52): 1804779.

79 Magne, D., Mauchamp, V., Célérier, S. et al. (2016). Site-projected electronic structure of two-dimensional Ti_3C_2 MXene: the role of the surface functionalization groups. *Physical Chemistry Chemical Physics* 18 (45): 30946–30953.

80 Hu, T., Li, Z., Hu, M. et al. (2017). Chemical origin of termination-functionalized MXenes: $Ti_3C_2T_2$ as a case study. *The Journal of Physical Chemistry C* 121 (35): 19254–19261.

81 Schultz, T., Frey, N.C., Hantanasirisakul, K. et al. (2019). Surface termination dependent work function and electronic properties of $Ti_3C_2T_x$ MXene. *Chemistry of Materials* 31 (17): 6590–6597.

82 Hart, J.L., Hantanasirisakul, K., Lang, A.C. et al. (2019). Control of MXenes' electronic properties through termination and intercalation. *Nature Communications* 10 (1): 522.

83 Handoko, A.D., Fredrickson, K.D., Anasori, B. et al. (2018). Tuning the basal plane functionalization of two-dimensional metal carbides (MXenes) to control hydrogen evolution activity. *ACS Applied Energy Materials* 1 (1): 173–180.

84 Wang, Y., Mao, J., Meng, X. et al. (2019). Catalysis with two-dimensional materials confining single atoms: concept, design, and applications. *Chemical Reviews* 119 (3): 1806–1854.

85 Li, P., Zhu, J., Handoko, A.D. et al. (2018). High-throughput theoretical optimization of the hydrogen evolution reaction on MXenes by transition metal modification. *Journal of Materials Chemistry A* 6 (10): 4271–4278.

86 Zhang, J., Zhao, Y., Guo, X. et al. (2018). Single platinum atoms immobilized on an MXene as an efficient catalyst for the hydrogen evolution reaction. *Nature Catalysis* 1 (12): 985–992.

87 Ramalingam, V., Varadhan, P., Fu, H.-C. et al. (2019). Heteroatom-mediated interactions between ruthenium single atoms and an MXene support for efficient hydrogen evolution. *Advanced Materials* 31 (48): 1903841.

88 Kuznetsov, D.A., Chen, Z., Kumar, P.V. et al. (2019). Single site cobalt substitution in 2D molybdenum carbide (MXene) enhances catalytic activity in the hydrogen evolution reaction. *Journal of the American Chemical Society* 141 (44): 17809–17816.

89 Xiao, W., Liu, P., Zhang, J. et al. (2017). Dual-functional N dopants in edges and basal plane of MoS_2 nanosheets toward efficient and durable hydrogen evolution. *Advanced Energy Materials* 7 (7): 1602086.

90 Ding, B., Ong, W.-J., Jiang, J. et al. (2020). Uncovering the electrochemical mechanisms for hydrogen evolution reaction of heteroatom doped M2C MXene (M = Ti, Mo). *Applied Surface Science* 500: 143987.

91 Yoon, Y., Tiwari, A.P., Lee, M. et al. (2018). Enhanced electrocatalytic activity by chemical nitridation of two-dimensional titanium carbide MXene for hydrogen evolution. *Journal of Materials Chemistry A* 6 (42): 20869–20877.

92 Yoon, Y., Tiwari, A.P., Choi, M. et al. (2019). Precious-metal-free electrocatalysts for activation of hydrogen evolution with nonmetallic electron donor: chemical composition controllable phosphorous doped vanadium carbide MXene. *Advanced Functional Materials* 29 (30): 1903443.

93 Yang, C., Wang, H.-F., and Xu, Q. (2020). Recent advances in two-dimensional materials for electrochemical energy storage and conversion. *Chemical Research in Chinese Universities* 36 (1): 10–23.

94 Huang, Z.-F., Song, J., Dou, S. et al. (2019). Strategies to break the scaling relation toward enhanced oxygen electrocatalysis. *Matter* 1 (6): 1494–1518.

95 Yan, H., Lin, Y., Wu, H. et al. (2017). Bottom-up precise synthesis of stable platinum dimers on graphene. *Nature Communications* 8 (1): 1070.

96 Kang, Z., Khan, M.A., Gong, Y. et al. (2021). Recent progress of MXenes and MXene-based nanomaterials for the electrocatalytic hydrogen evolution reaction. *Journal of Materials Chemistry A* 9 (10): 6089–6108.

97 Xiu, L., Pei, W., Zhou, S. et al. (2020). Multilevel hollow MXene tailored low-Pt catalyst for efficient hydrogen evolution in full-pH range and seawater. *Advanced Functional Materials* 30 (47): 1910028.

98 Zhao, K., Ma, X., Lin, S. et al. (2020). Ambient growth of hierarchical FeOOH/MXene as enhanced electrocatalyst for oxygen evolution reaction. *ChemistrySelect* 5 (6): 1890–1895.

99 Lu, Y., Fan, D., Chen, Z. et al. (2020). Anchoring Co_3O_4 nanoparticles on MXene for efficient electrocatalytic oxygen evolution. *Science Bulletin* 65 (6): 460–466.

100 Yu, M., Zhou, S., Wang, Z. et al. (2018). Boosting electrocatalytic oxygen evolution by synergistically coupling layered double hydroxide with MXene. *Nano Energy* 44: 181–190.

101 Ma, T.Y., Cao, J.L., Jaroniec, M., and Qiao, S.Z. (2016). Interacting carbon nitride and titanium carbide nanosheets for high-performance oxygen evolution. *Angewandte Chemie International Edition* 55 (3): 1138–1142.

102 Zhao, L., Dong, B., Li, S. et al. (2017). Interdiffusion reaction-assisted hybridization of two-dimensional metal–organic frameworks and $Ti_3C_2T_x$ nanosheets for electrocatalytic oxygen evolution. *ACS Nano* 11 (6): 5800–5807.

103 Zou, H., He, B., Kuang, P. et al. (2018). Metal–organic framework-derived nickel–cobalt sulfide on ultrathin MXene nanosheets for electrocatalytic oxygen evolution. *ACS Applied Materials and Interfaces* 10 (26): 22311–22319.

104 Wen, Y., Wei, Z., Ma, C. et al. (2019). MXene boosted CoNi-ZIF-67 as highly efficient electrocatalysts for oxygen evolution. *Nanomaterials* 9 (5): 775.

105 Selvam, N.C.S., Lee, J., Choi, G.H. et al. (2019). MXene supported CoxAy (A = OH, P, Se) electrocatalysts for overall water splitting: unveiling the role of anions in intrinsic activity and stability. *Journal of Materials Chemistry A* 7 (48): 27383–27393.

106 Tang, Y., Yang, C., Yang, Y. et al. (2019). Three dimensional hierarchical network structure of S-$NiFe_2O_4$ modified few-layer titanium carbides (MXene) flakes on nickel foam as a high efficient electrocatalyst for oxygen evolution. *Electrochimica Acta* 296: 762–770.

107 Wen, Y., Wei, Z., Liu, J. et al. (2021). Synergistic cerium doping and MXene coupling in layered double hydroxides as efficient electrocatalysts for oxygen evolution. *Journal of Energy Chemistry* 52: 412–420.

108 Liu, J., Mi, L., Xing, Y. et al. (2020). Construction of Ti_3C_2 supported hybrid Co_3O_4/NCNTs composite as an efficient oxygen reduction electrocatalyst. *Renewable Energy* 160: 1168–1173.

109 Wen, Y., Ma, C., Wei, Z. et al. (2019). FeNC/MXene hybrid nanosheet as an efficient electrocatalyst for oxygen reduction reaction. *RSC Advances* 9 (24): 13424–13430.

110 Zhang, C., Ma, B., Zhou, Y., and Wang, C. (2020). Highly active and durable Pt/MXene nanocatalysts for ORR in both alkaline and acidic conditions. *Journal of Electroanalytical Chemistry* 865: 114142.

111 Chen, J., Yuan, X., Lyu, F. et al. (2019). Integrating MXene nanosheets with cobalt-tipped carbon nanotubes for an efficient oxygen reduction reaction. *Journal of Materials Chemistry A* 7 (3): 1281–1286.

112 Zhang, Y., Jiang, H., Lin, Y. et al. (2018). In situ growth of cobalt nanoparticles encapsulated nitrogen-doped carbon nanotubes among $Ti_3C_2T_x$ (MXene) matrix for oxygen reduction and evolution. *Advanced Materials Interfaces* 5 (16): 1800392.

113 Lei, Y., Tan, N., Zhu, Y. et al. (2020). Synthesis of porous N-rich carbon/MXene from MXene@polypyrrole hybrid nanosheets as oxygen reduction reaction electrocatalysts. *Journal of The Electrochemical Society* 167 (11): 116503.

114 Li, C., Mou, S., Zhu, X. et al. (2019). Dendritic Cu: a high-efficiency electrocatalyst for N_2 fixation to NH_3 under ambient conditions. *Chemical Communications* 55 (96): 14474–14477.

115 Capdevila-Cortada, M. (2019). Electrifying the Haber–Bosch. *Nature Catalysis* 2: 1055–1055.

116 Singh, A.R., Rohr, B.A., Schwalbe, J.A. et al. (2017). Electrochemical ammonia synthesis—the selectivity challenge. *ACS Catalysis* 7 (1): 706–709.

117 Gao, Y., Cao, Y., Zhuo, H. et al. (2020). Mo_2TiC_2 MXene: a promising catalyst for electrocatalytic ammonia synthesis. *Catalysis Today* 339: 120–126.

118 Zhao, J., Zhang, L., Xie, X.-Y. et al. (2018). Ti_3C_2Tx (T = F, OH) MXene nanosheets: conductive 2D catalysts for ambient electrohydrogenation of N_2 to NH_3. *Journal of Materials Chemistry A* 6 (47): 24031–24035.

119 Azofra, L.M., Li, N., MacFarlane, D.R., and Sun, C. (2016). Promising prospects for 2D d^2–d^4 M_3C_2 transition metal carbides (MXenes) in N_2 capture and conversion into ammonia. *Energy & Environmental Science* 9 (8): 2545–2549.

120 Zhang, J., Yang, L., Wang, H. et al. (2019). In situ hydrothermal growth of TiO_2 nanoparticles on a conductive $Ti_3C_2T_x$ MXene nanosheet: a synergistically active Ti-based nanohybrid electrocatalyst for enhanced N_2 reduction to NH_3 at ambient conditions. *Inorganic Chemistry* 58 (9): 5414–5418.

121 Kong, W., Gong, F., Zhou, Q. et al. (2019). An MnO_2–$Ti_3C_2T_x$ MXene nanohybrid: an efficient and durable electrocatalyst toward artificial N_2 fixation to NH_3 under ambient conditions. *Journal of Materials Chemistry A* 7 (32): 18823–18827.

122 Fang, Y., Liu, Z., Han, J. et al. (2019). High-performance electrocatalytic conversion of N_2 to NH_3 using oxygen-vacancy-rich TiO_2 in situ grown on $Ti_3C_2T_x$ MXene. *Advanced Energy Materials* 9 (16): 1803406.

123 Zheng, S., Li, S., Mei, Z. et al. (2019). Electrochemical nitrogen reduction reaction performance of single-boron catalysts tuned by MXene substrates. *The Journal of Physical Chemistry Letters* 10 (22): 6984–6989.

124 Luo, Y., Chen, G.-F., Ding, L. et al. (2019). Efficient electrocatalytic N_2 fixation with MXene under ambient conditions. *Joule* 3 (1): 279–289.

125 Li, Z. and Wu, Y. (2019). 2D early transition metal carbides (MXenes) for catalysis. *Small* 15 (29): 1804736.

126 Daiyan, R., Saputera, W.H., Masood, H. et al. (2020). A disquisition on the active sites of heterogeneous catalysts for electrochemical reduction of CO_2 to value-added chemicals and fuel. *Advanced Energy Materials* 10 (11): 1902106.

127 Chen, H., Handoko, A.D., Xiao, J. et al. (2019). Catalytic effect on CO_2 electroreduction by hydroxyl-terminated two-dimensional MXenes. *ACS Applied Materials and Interfaces* 11 (40): 36571–36579.

128 Li, N., Chen, X., Ong, W.-J. et al. (2017). Understanding of electrochemical mechanisms for CO_2 capture and conversion into hydrocarbon fuels in transition-metal carbides (MXenes). *ACS Nano* 11 (11): 10825–10833.

129 Handoko, A.D., Khoo, K.H., Tan, T.L. et al. (2018). Establishing new scaling relations on two-dimensional MXenes for CO_2 electroreduction. *Journal of Materials Chemistry A* 6 (44): 21885–21890.

130 Yang, X., Jia, Q., Duan, F. et al. (2019). Multiwall carbon nanotubes loaded with MoS_2 quantum dots and MXene quantum dots: non–Pt bifunctional catalyst for the methanol oxidation and oxygen reduction reactions in alkaline solution. *Applied Surface Science* 464: 78–87.

131 Wang, Y., Wang, J., Han, G. et al. (2019). Pt decorated Ti_3C_2 MXene for enhanced methanol oxidation reaction. *Ceramics International* 45 (2, Part A): 2411–2417.

132 Lang, Z., Zhuang, Z., Li, S. et al. (2020). MXene surface terminations enable strong metal–support interactions for efficient methanol oxidation on palladium. *ACS Applied Materials and Interfaces* 12 (2): 2400–2406.

7

Advances and Challenges in Pt-free Pd-based Catalysts for Oxygen Electro-Reduction in Alkaline Media

Lipipuspa Sahoo and Ujjal K. Gautam

Department of Chemical Sciences, Indian Institute of Science Education and Research, Mohali, Punjab, India

7.1 Introduction

After the first industrial revolution, advanced technologies and inventions have made human life immensely easier. Most of these advanced technologies heavily depend on fossil fuels as the source of energy, which led to its excessive demand and increasing emission of CO_2 [1]. According to the International Energy Agency, by 2035, 18 billion tons of oil equivalent energy will be the global annual demand to potentially produce 43 gigatons of CO_2 annually resulting in an energy crisis and severe global warming [2]. Therefore, researchers have focused on the development of clean energy technologies such as fuel cells having zero-emission while using H_2 and O_2 as fuel molecules and H_2O as the by-product [3–5]. Among various fuel cells, proton-exchange membrane fuel cells (PEMFCs) that operate in acidic media have been immensely investigated in the past two decades due to their fast refueling, high energy-conversion efficiency, and high power density [2–4, 6]. Pt is unanimously regarded as the most efficient catalyst for the cathodic oxygen reduction reaction (ORR) in PEMFCs [7–9]. Unfortunately, the exorbitant cost of Pt, difficulty in H_2 processing and handling, high cost of proton exchange membrane polymer (Nafion), and its instability at high temperature hinder the large-scale adoption and development of PEMFCs [10, 11]. Direct liquid feed fuel cells which use small organic molecules like methanol, ethanol, and ethylene glycol as fuel molecules in the anodic compartment have certain advantages over hydrogen gas feed fuel cells such as easy production, storage, and distribution of the liquid fuel molecules [12, 13]. However, fuel crossover in them from the anode to the cathode reduces the efficiency of the PEMFCs [13]. Therefore, as an alternative, scientists have devoted tremendous efforts toward the development of anion-exchange membrane fuel cells (AEMFCs) that provide a potential solution to the problems in PEMFCs (Figure 7.1) [14, 15]. AEMFCs, operating in alkaline conditions, are molecular structure insensitive toward these liquid fuels and open up the opportunity for the use of non-precious metal catalysts such as Pd, Ag, and Ni, instead of the Pt-based catalysts [14]. In the case of AEMFCs, the direction of the ionic flow of hydroxide ions is in the reverse direction of the crossover

Heterogeneous Nanocatalysis for Energy and Environmental Sustainability: Energy Applications, First Edition. Edited by Putla Sudarsanam, Yusuke Yamauchi, and Pankaj Bharali.

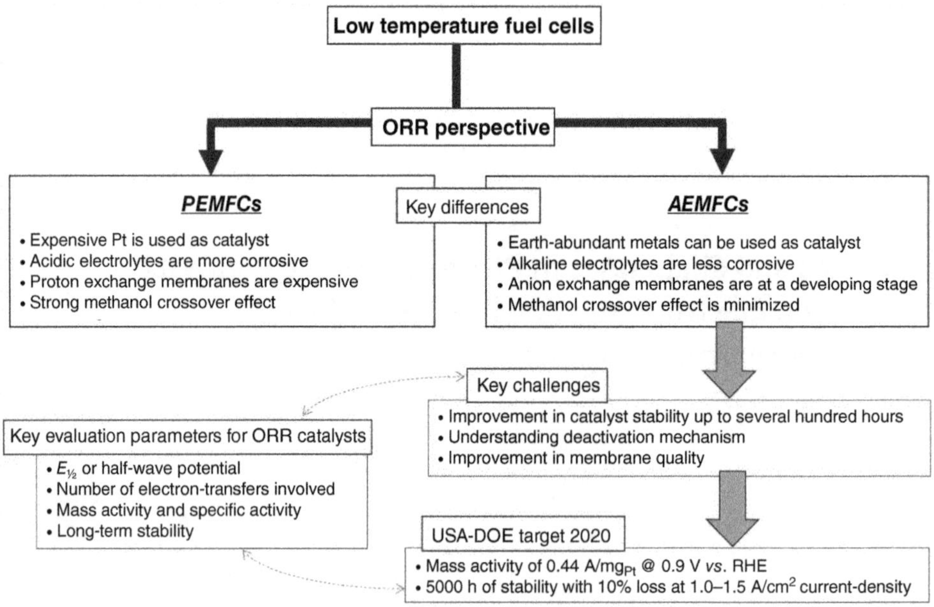

Figure 7.1 Key aspects of AEMFC research from an ORR perspective.

pathway, which minimizes the electro-osmotic drag of the fuel molecules, thereby reducing the fuel crossover [14–16].

In a typical AEMFC, fuel molecules (e.g. H_2 or CH_3OH) get oxidized at the anode, and O_2 gets reduced at the cathode in an alkaline medium [17]. Notably, the reduction of O_2 in alkaline media is also an important cathode reaction in alkaline metal-air batteries, alkaline hydrogen fuel cells, alkaline microbial fuel cells, and alkaline direct liquid fuel cells [18]. In these devices, the sluggish ORR kinetics (as it involves the step-wise transfer of four electrons) reduces the overall efficacy of the devices [19]. In order to achieve high energy efficiency, the oxygen electrodes of the electrochemical devices should trigger ORR with a minimum overpotential mimicking the reversible conditions to the extent possible. Therefore, the development of highly efficient ORR electrocatalysts, which could reduce the overpotential in alkaline media, is a stringent requirement. Notably, there are a few other advantages that can pave the way for a superior ORR. The adsorption energies of anions are lower in alkaline media, as desirable, as compared to acidic media because as per the Nernst equation, for every unit pH increase of reaction, the working potential of the electrochemical processes shifts by −59 mV in aqueous media, which is equivalent to a lowering of adsorption energy by 0.059 eV for a monovalent anion such as OH^- [20]. Furthermore, an alkaline electrochemical environment is less corrosive than an acidic one. Additionally, the initial ORR electron transfer processes are surface nonspecific in alkaline media which enables the use of a wide range of metal catalysts [20].

Even though Pt is considered the most efficient catalyst of ORR, its efficiency has appeared to be lower in alkaline media as compared to acidic media from many recent studies [21]. The properties of Pd are very similar to Pt as they belong to the same group in the periodic table, similar in atomic size, and have face-centered cubic (FCC) crystal structure, making it a suitable candidate for ORR. In addition, the cost of Pd is lower than the Pt due to its higher natural abundance [5, 10]. Therefore, Pd-based nanostructures have attracted significant attention as replacement cathode catalysts to Pt, and indeed many a time, their catalytic performances surpass the Pt-based catalysts in alkaline media [22]. Recent investigations have been centered on boosting the intrinsic catalytic activity of Pd-based nanocatalysts by adopting various strategies, including alloying, reducing the size of the nanocrystals, surface modification, morphology, composition control, and establishing desired metal-support interactions.

In this chapter, we first discuss the mechanism of ORR in alkaline media as established from experimental and theoretical studies. Then, we collate the recent developments in Pd-based ORR catalysts in alkaline media and describe the emerging strategies for improving their performances. The ORR electrocatalysts have been divided into four subsections: (i) shape-selective pure Pd nanostructures, (ii) alloys or intermetallics, (iii) core-shell nanostructures, and (iv) support-modified Pd nanostructures. Further, we concisely retrospect the long-term stability of Pd-based catalysts, which is deemed to be the next important challenge in the coming years. We end with the concluding remarks outlining the prospects and the emerging challenges of ORR in alkaline media.

7.2 ORR Mechanism in Alkaline Medium

The most commonly used alkaline electrolytes are NaOH and KOH for the electro-reduction of oxygen. Due to the difference in the size of the sodium and potassium cations, many of the physicochemical properties like oxygen solubility, oxygen diffusion coefficient, solution viscosity, and electrode interfacial phenomena, are very different in these two electrolytes. It has been observed that most of the catalysts demonstrate superior ORR performance in KOH as compared to NaOH as the former has a lower viscosity, high oxygen solubility, and weakly adsorbing species which improves the ORR activity [23]. Moreover, the concentration of the alkali also plays a crucial role in determining ORR performance [24]. With an increase in its concentration, the ORR performance reduces due to the increased surface coverage and bond strength of metal-OH groups, which blocks the active sites and alters the adsorption energy of O_2 and other intermediates. Therefore, most of the catalyst performances for ORR are evaluated in 0.1 M KOH solution [25]. Pd shows much higher ORR activity in the alkaline solutions than that in acidic solutions [26]. The ORR reaction mechanism on a Pd surface in alkaline solutions has not been investigated as thoroughly as that on Pt and Au surfaces. However, it should have many similarities to that on Pt surfaces.

In general, the ORR can occur via dissociative mechanism by either a direct four-electron transfer pathway to produce OH^- as given below:

$$O_2 + 2H_2O + 4e^- \rightarrow 4OH^-;\ E^0 = 0.401 \text{ vs SHE} \tag{7.1}$$

or via associative mechanism by a serial two-electron transfer pathway to generate peroxide species as intermediate as follows [27]:

$$O_2 + H_2O + 2e^- \rightarrow HO_2^- + OH^-;\ E^0 = -0.076 \text{ vs SHE} \tag{7.2}$$

which can further be reduced as

$$HO_2^- + H_2O + 2e^- \rightarrow 3OH^-;\ E^0 = 0.878 \text{ vs SHE} \tag{7.3}$$

Between these two mechanisms, the dissociative mechanism involving the four-electron ORR pathway is widely recognized as the most favorable pathway because the H_2O_2 formation via associative mechanism reduces energy-conversion efficiency [27, 28]. During ORR, several reaction intermediates could be generated, such as oxygenated (O*), hydroxyl (OH*), and hydroperoxyl (OOH*) species (* indicates the species attached to the catalyst surface) under the experimental conditions [29, 30]. However, various possible transformations among these intermediates make the overall process further complex [29, 30]. Within this complex scenario, the rate-determining step in ORR varies depending on the catalyst surface and environmental conditions, including temperature, pH of the reaction medium, and applied potential [31]. The dissociative ORR mechanism in alkaline media involving four single-electron transfers, one in each step, proceeds as follows (Eqs. 7.4–7.8): [32]

$$O_2 + e^- \rightarrow O_2^- \tag{7.4}$$

$$O_2^- + H_2O + e^- \rightarrow HO_2^- + OH^- \tag{7.5}$$

Note here that in Eq. 7.5, the O_2 is not yet dissociated. The O–O bond dissociation energy is higher than any nonmetal–O bond, and therefore the O–O dissociation becomes favorable on a catalyst surface that moderately adsorbs O_2 via several intermediates such as O, superoxide ion (O_2^-), and peroxide ions (HO_2^-) [33].

The first electron transfer occurs to the adsorbed O_2 species to form O_2^- which is widely regarded as the rate-determining step on Pt. [34, 35] The O_2^- intermediate has a sufficiently long lifetime as could be observed from a reverse sweep of cyclic voltammetry experiments [36]. Additionally, this intermediate was detected during ORR measurement at pH 11 on a Pt surface by using in-situ surface-enhanced infrared absorption spectroscopy [37]. Similarly, surface-enhanced Raman spectroscopy also evidenced the presence of O_2^- on polycrystalline Au surface during ORR measurement in 0.1 M NaOH [33]. The same reaction intermediate could not be detected in acidic media, possibly due to a much shorter lifetime of the intermediate as compared to alkaline media. The consequent second electron transfer results in the formation of the peroxide species (HO_2^-). The incomplete reduction occurring from the transfer of two electrons forming peroxide is undesirable as it reduces the efficiency of the catalyst. The peroxide species can further be reduced by either taking two more electrons as follows: [27]

$$HO_2^- + e^- \rightarrow O^- + OH^- \tag{7.6}$$

$$O^- + H_2O + e^- \rightarrow 2OH^- \tag{7.7}$$

or by disproportionation

$$2HO_2^- \rightarrow 2OH^- + O_2 \tag{7.8}$$

Thus, the adsorbed HO_2^- may further reduce to OH^- as per Eqs. (7.6) and (7.7) to complete the four-electron reduction or via disproportionation of itself (Eq. 7.8) to have $2e^- + 2e^-$ pathway. Although the ORR reactions can proceed either via a serial 2 + 2-electron pathway or a four-electron pathway, it is difficult to differentiate between a two-electron pathway followed by fast disproportionation of HO_2^- and a direct $4e^-$ pathway. Therefore, it is often called a "pseudo" $4e^-$ pathway instead of distinguishing it from a $2e^- + 2e^-$ pathway [38, 39].

It has emerged that the operational ORR mechanism depends on the adsorbed OH coverage on the catalyst surface, as inferred from several recent experimental as well as theoretical studies. The ORR mechanism can follow either $4e^-$ inner-sphere electron transfer pathway in low OH^- concentration or an outer-sphere $2e^-$ transfer in high OH^- concentration. In acidic media, the ORR follows an inner-sphere electron transfer mechanism due to the presence of low OH species (Figure 7.2a) [20]. On the other hand, more OH coverage in alkaline media induces competition between the inner-sphere and the outer-sphere electron transfer processes. The adsorbed OH species can interact with the solvated O_2 molecule to promote a two-electron reaction pathway to form HO_2^- intermediate [20]. This HO_2^- intermediate gets easily desorbed as the reaction product from the catalyst surfaces of the less-noble metals, resulting in a dominating outer-sphere mechanism. In the case of noble metals, the inner-sphere mechanism dominates as they can stabilize the HO_2^- intermediate, leading to the complete transfer of four electrons$^-$. Therefore, noble metals are considered the most facile catalysts in alkaline media as well. The high OH coverage on the catalyst surface in alkaline media hinders the direct adsorption of O_2 molecules by forming a compact H_2O layer by hydrogen bonding [20]. Molecular dynamics simulations suggest that adsorbed OH^- impedes the ORR kinetics on a Pt surface [42]. Density functional theory (DFT) calculations on Au (100) for ORR show that with increasing applied potential, the adsorption energies of the intermediates viz O and HO_2 increase while the same remains constant for OH (Figure 7.2b) [40]. The facile ORR kinetics on Au surface in alkaline media was attributed to the stronger adsorption of HO_2 intermediate at higher pH (Figure 7.2c) [40]. Unlike Pt surface, the adsorbed OH on Au surface can further stabilize the neighboring HO_2 intermediates, thereby facilitating the $4e^-$ reduction pathway [40].

The chemisorbed H_2O molecules also play a crucial role in determining the ORR mechanism, as observed in the case of Pt and Au. The ORR intermediates are stabilized to various extents depending on the interaction with neighboring H_2O molecules via hydrogen bonding. The chemisorbed H_2O can also act as the reducing agent for an adsorbed O_2 or O, and thus favoring the $4e^-$ transfer pathway on a Pt surface [20]. DFT calculation shows that the coadsorbed H_2O can promote O–O bond cleavage, which leads to $4e^-$ reduction on Au surface [43]. In addition, crystal imperfections or the creation of defects are highly beneficial for stabilizing these reaction intermediates. Figure 7.2d is a structure-dependent volcano plot on Au and Pt surfaces versus the general coordination number (GCN) [41]. It can be seen that the formation of *OOH and desorption of *OH are the rate-determining steps as shown in the right and the left side of the volcano. Pt exhibits the maximum activity at a GCN of 8.29, which is greater than the GCN of the (111) and (100) planes as it binds strongly with oxygenated species. On the other hand, Au has an optimal GCN of 5.75 at which it exhibits the maximum activity. Therefore the ORR kinetics can be strongly influenced by several factors such as pH, the concentration of adsorbed H_2O molecule, OH

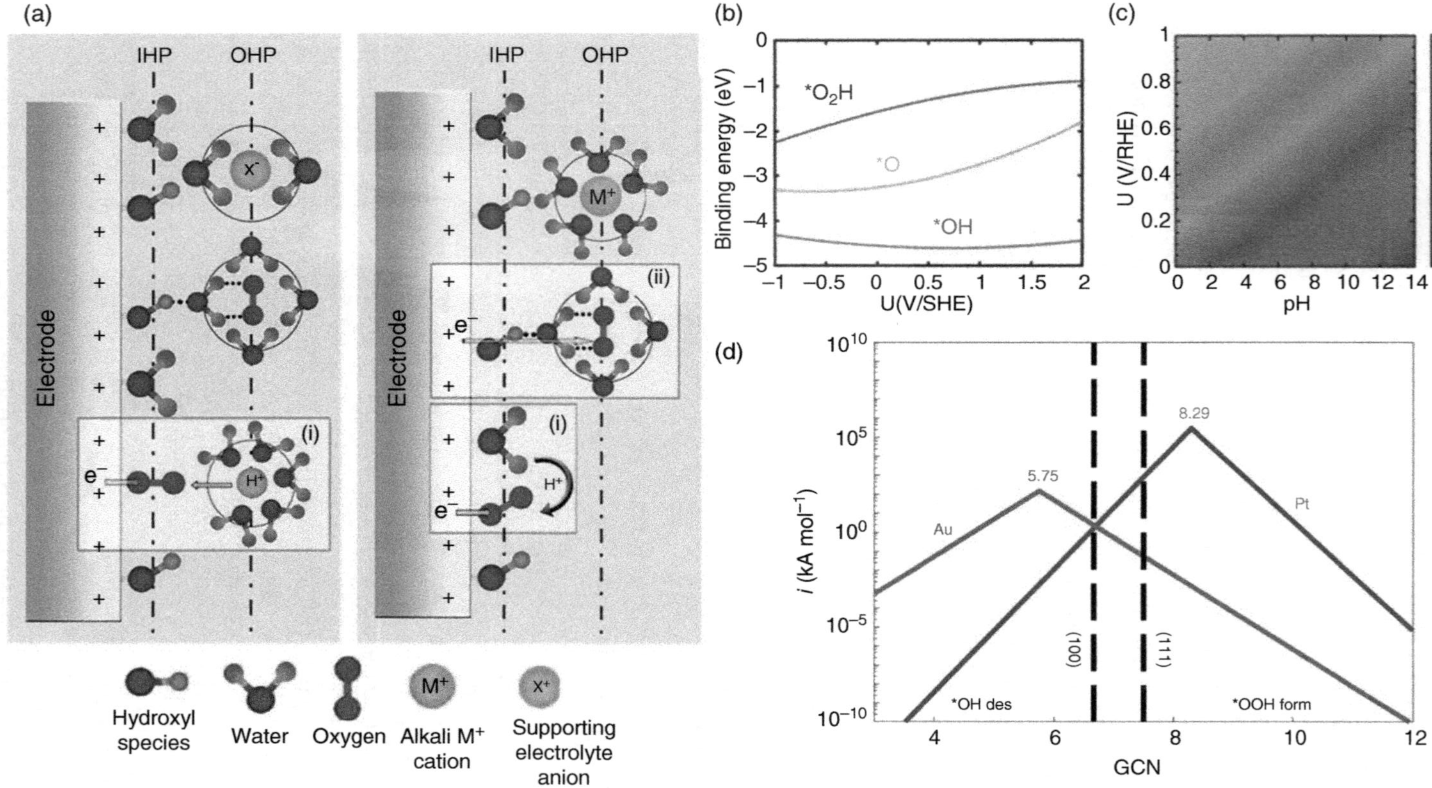

Figure 7.2 (a) Schematic representation of the double-layer structure during ORR in acidic media (left) and alkaline media (right). Insets (i) and (ii) represent the inner-sphere and the outer-sphere electron transfer processes. *Source:* From [20]/with permission of American Chemical Society. (b) Adsorption energies of *O, *OH, and $*O_2H$ as a function of U, (c) potential- and pH-dependent adsorption energy of $*O_2H$ on Au (001). *Source:* From [40]/with permission of American Chemical Society. (d) Volcano plot for ORR activities on Pt and Au vs general coordination number (GCN). The GCNs of the (111) and (100) planes are denoted as vertical dashed lines at 6.67 and 7.5, respectively. "*OOH form" and "*OH des" refer to areas of the volcano map corresponding to OOH formation and OH desorption, respectively, that are considered as the rate-determining steps. *Source:* From [41]/with permission of Springer Nature.

coverage density, and the coordination number of the surface atoms. Thus, to enhance the overall ORR activity, the surface of metals should be tuned in such a way that O, O_2, HO_2 intermediates binding will be suitable enough to allow the ORR via the $4e^-$ transfer mechanism.

To identify the electron transfer mechanism and associated intermediates, the rotating disk electrode (RDE) and rotating ring disk electrode (RRDE) techniques are used. The selectivity between the $2e^-$ and the $4e^-$ processes can be estimated from Koutecky–Levich (K-L) equation as follows: [27]

$$1/I = 1/I_k + 1/I_d \tag{7.9}$$

where I, I_k, and I_d are experimentally measured current, kinetic current, and diffusion-limited current, respectively. I_k is determined using the formula

$$I_k = nFkCO_2 \tag{7.10}$$

The I_d term can be obtained from the Levich equation:

$$I_d = 0.62nFD^{2/3}\upsilon^{-1/6}\omega^{1/2}CO_2 \tag{7.11}$$

where n is the number of electrons involved in the reaction, F is Faraday's constant (96485C/mol), D is the O_2 diffusion coefficient in 0.1 M KOH solution (1.9×10^{-5} cm^2/s), υ is the kinematic viscosity of the electrolyte (1.01×10^{-2} cm^2/s for 0.1 M KOH), ω is the angular frequency of rotation, CO_2 is the concentration of O_2 in the reaction medium (1.21×10^{-6} mol/cm^3), and k is electron transfer rate constant [44].

The percentage of peroxide formation is recorded using the RRDE ring electrode, and the corresponding electron transfer number, n, can be calculated as follows:

$$\%H_2O_2 = \frac{200\times(I_R/N)}{I_D + (I_R/N)} \tag{7.12}$$

$$n = 4\frac{I_D}{I_D + (I_R/N)} \tag{7.13}$$

where I_D and I_R are disk and ring currents, respectively, and N is the collection efficiency of H_2O_2 on the ring electrode. The formation of a high percentage of H_2O_2 during ORR measurement suggests that the ORR mechanism is dominated by the $2e^-$ transfer pathway.

The catalytic performance of a catalyst is determined by estimating the mass and specific activities. For this purpose, the ORR polarization curves are recorded at different rotation speeds and then using K-L equation (Eq. 7.9), 1/I vs $\omega^{1/2}$ can be plotted. From the plot of 1/I vs $\omega^{1/2}$, kinetic current can be determined from the intercept and the number of electrons transfer involved in the reaction can be estimated from the slope [45]. The mass activity is derived by estimating kinetic current and dividing it with metal mass loading on the electrode surface. The specific activity is determined by normalizing kinetic current with the corresponding electrochemically active surface area (ECSA). In general, the mass and specific activities of any catalyst are determined at an applied potential of 0.9 V vs RHE. Another important parameter, viz. the half-wave potential ($E_{1/2}$), is also widely used to evaluate the catalytic performance. $E_{1/2}$ is the potential required to achieve a current that is half that of I_d. A higher value of $E_{1/2}$ (potential more close to 1.23 V vs RHE) signifies a less overpotential needed for ORR and thus enhanced catalytic performance [45].

7.3 Recently Developed Pd-Based Catalysts for ORR in Alkaline Media

7.3.1 Shape-Controlled Pure Pd Catalysts

Controlling the shape and size of the nanocrystals has been widely adopted as a common strategy to improve the overall catalytic efficiency. The electro-reduction efficiencies of O_2 on different Pd surfaces follow the order Pd (110) < Pd (111) < Pd(100), which is in contrast to Pt facet-dependence, followed by Pt (100) << Pt (110) ~ Pt (111) [46, 47]. Pd nanocrystal enclosed with (100) facet can exhibit up to three times higher current density for ORR as compared to Pt (111) in a similar reaction condition. For example, Huang et al. recently prepared truncated Pd nanocubes enclosed with the (100) facets by varying the concentration of ascorbic acid in the reaction medium, which exhibited two times higher specific activity as compared to Pd nanoparticles (NPs) for ORR in 0.1 M NaOH [48]. The effect of size of the Pd nanocubes on ORR performance in alkaline media has been studied by Lusi et al. This revealed that Pd nanocubes' mass activity (at 0.9 V vs RHE) increases with decreasing the size from 30 to 7 nm [49]. However, the mass activity achieved for the 7-nm Pd cubes (~0.09 A/mg) was barely comparable to commercial Pd/C. The smaller the size of the NPs, the greater is the available surface area and low-coordination sites, such as edges, steps, and kinks, which are capable of boosting the catalytic performance due to the high surface energy [50, 51]. Therefore reducing the size of the NPs is broadly considered as an effective strategy to improve atom utilization efficiency and catalytic activity. It is also a great challenge to simultaneously control the morphology, facets, and size of Pd nanocrystals under a sub-10-nm regime.

Zhang et al. prepared monodisperse 4.9-nm Pd nanotetrahedra (NTd) enclosed by (111) facets using 1,1′-binaphthalene and NH_3 as additives (Figure 7.3a) [52]. These sub-5-nm Pd tetrahedrons demonstrated high ORR activity with an $E_{1/2}$ of 0.86 V and 4.4 times high specific activity as compared to commercial Pd black (Figure 7.3b). The dendritic Pd-lacinated nanourchins (LUs) were obtained by varying the reaction condition exhibited poor ORR activity as compared to Pd tetrahedrons. The enhanced activity was attributed to the ultra-small size of the Pd tetrahedrons, which helps in improving atomic utilization while the exposed (111) facets provide active sites for ORR. Very recently, 9.4-nm Pd nanocubes were synthesized by controlled acid etching (Figure 7.3c–f) [53]. These demonstrated very high $E_{1/2}$ of 0.95 V with ORR mass activity of 0.78 A/mg in 0.1 M KOH (Figure 7.3g, h), where the mass activity was 20 times higher than commercial Pd/C. These Pd nanocubes have a natural tendency to self-assemble to form a two-dimensional (2D) superlattice (Figure 7.3c, d). The excellent ORR activity of these Pd nanocubes has been attributed to the formation of uniform 2D superlattices on electrode surface, which can lower the

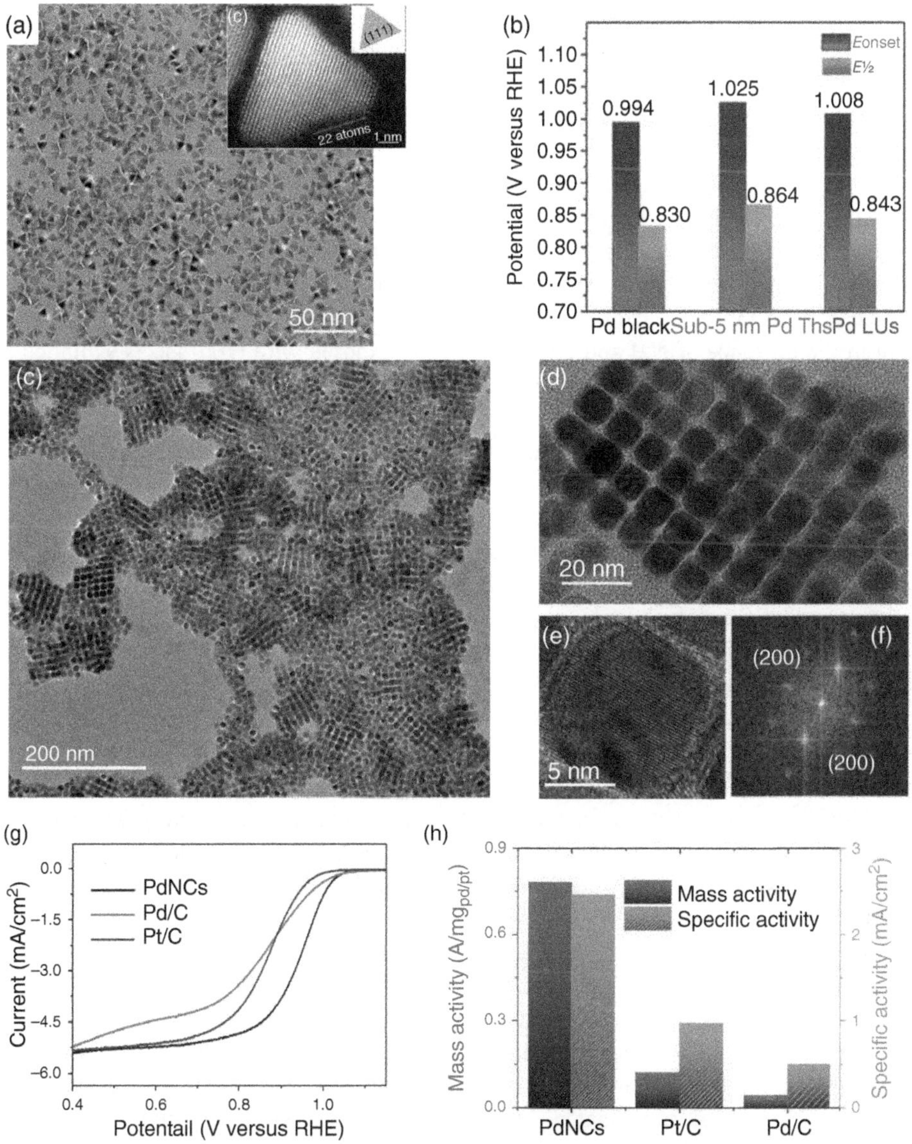

Figure 7.3 (a) TEM image of sub-5-nm Pd tetrahedra (inset: high-resolution HAADF-STEM image). *Source:* From [52] / with permission of John Wiley & Sons. (b) Comparison of the E_{onset} and $E_{1/2}$ for Pd nanotetrahedra, Pd-lacinated nanourchins (Pd Lus), and Pd black. *Source:* From [52] / with permission of John Wiley & Sons. (c) Low-magnification, and (d) high-magnification TEM images of Pd NCs having the tendency for of 2D superlattice formation due to edge-to-edge arrangement of the Pd NCs, (e) high-resolution TEM image of a Pd NC. (f) FFT pattern of a single PdNC *Source:* From [53] / with permission of American Chemical Society. (g) ORR polarization curves of PdNCs, commercial Pt/C and commercial Pd/C, (h) mass and specific activities of commercial Pt/C and commercial Pd/C. *Source:* From [53] / with permission of American Chemical Society.

activation barriers for charge transport by improving the electron conduction channels because of larger particle–particle or electrode–particle interface area compared to spherical particles [53]. The same superlattice array is also expected to impair the diffusion of particles on the electrode surface to arrest their leaching or agglomeration and provide improved stability.

Other than zero-dimensional (0D) Pd nanocrystals, one-dimensional (1D) nanostructures such as nanowires (NWs) and nanorods (NRs) have been shown to exhibit outstanding performance in oxygen electrocatalysis, benefiting from the high surface area and substantially better stability than nanocrystals [46, 54–56]. Xia et al. synthesized penta-twinned Pd NWs by controlled reduction of Pd precursors by ascorbic acid and diethylene glycol in the presence of NaI and HCl. These NWs exposed with (100) facets exhibited a mass activity of 0.22 A/mg, which is three times higher than commercial Pt/C. The NWs also demonstrated excellent durability by retaining 86% of their initial mass activity after 10 000 potential cycles. On the other hand, Pt/C suffered a 32% loss in mass activity in similar conditions [55]. The authors perceived higher activity and stability of Pd NWs to originate from surface strain and high structural stability. Similarly, the 1D Pd network synthesized by the wet chemical approach demonstrated ORR mass activity 2.1 times higher than commercial Pt/C with considerable tolerance toward methanol. The better ORR performance was attributed to the presence of a net-like porous structure providing a high surface area of 85.6 m^2/g [56]. Additionally, three-dimensional (3D) Pd nanodendrites prepared by using K_2PdCl_4 and *N,N′*-methylenebisacrylamide as the only additive also demonstrated excellent ORR performance with a specific activity of 3.1 mA/cm^2, which is 3.6 times better than commercial Pd black [57]. The outstanding performance of these nanodendrites was ascribed to the highly branched porous structure that ensures the presence of an abundant number of edges and corners as active sites, porous structure enabling easy accessibility of reactant molecules to the active sites, and long tentacles ensuring facile electron transfer, and low agglomeration tendency [57].

7.3.2 Pd-Based Alloys

A large body of recent works has established that alloying Pd with metals, such as Co, Fe, Ni, Mn, Sn, Cu, Ag, Ru, Au, and Bi, enhances the ORR kinetics further by several times as compared to pure Pd by improving the atomic and electronic structure of the catalyst [58–61]. Shao et al. investigated the origin of enhanced ORR activity in such alloys using DFT calculations [62]. The activities of the various Pd-M alloys were found to display a volcano-type dependence with the oxygen binding energy and the *d*-band center of the metals. The position of the *d*-band of a metal determines the adsorption strength of the oxygenated intermediates during catalysis, the lowering of which weakens the adsorption strength of the oxygenated intermediates (Figure 7.4a). As can be seen from Figure 7.4b, pure Pd binds oxygen too strongly, which hinders the facile removal of the reaction intermediate. On the other hand, PdFe alloy lies on the top of the volcano to exhibit the highest specific activity due to moderate binding with oxygen [62]. However, in the case of Pt-based alloy, Pt_3Ni was found to be the most active (Figure 7.4c) [63]. Motivated by such findings, researchers have devoted tremendous efforts to tuning diverse PdM alloys into better ORR catalysts by modulating their size, component, and shape.

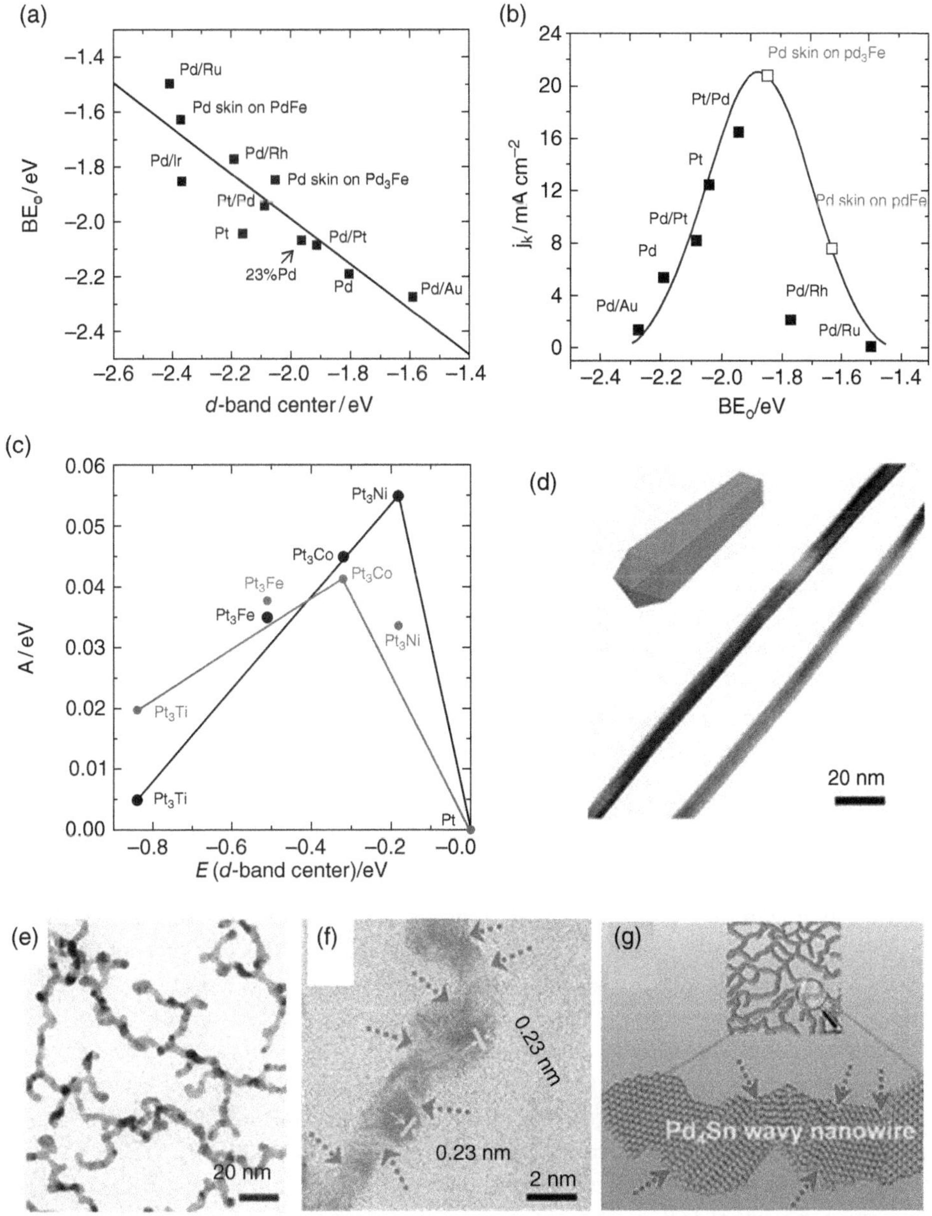

Figure 7.4 (a) Calculated oxygen-binding energies as a function of the Pd *d*-band center (relative to the Fermi level) of Pd– and Pt– overlayers on various substrates, (b) Volcano dependence of the ORR activity (expressed as the kinetic current density at 0.8 V vs RHE) on the calculated oxygen-binding energy. *Source:* From [62] / with permission of American Chemical Society. (c) The volcano plot of ORR activity vs the *d*-band center position of several Pt-based alloys measured experimentally. *Source:* From [63] / with permission of John Wiley & Sons. (d) TEM image of Pd_4Sn NWs, (e) TEM images of Pd_4Sn WNWs, (f) high-resolution TEM image and (g) schematic of Pd_4Sn WNWs showing defects present on the surface of WNWs, (h) ORR polarization curves of Pd_4Sn WNWs, Pd_4Sn NWs, commercial Pt/ and Pd/C, (i) mass activities histograms for of Pd_4Sn WNWs, Pd_4Sn NWs, commercial Pt/ and Pd/C before and after 10 000 cycles of ADTs. *Source:* From [64] / with permission of American Chemical Society.

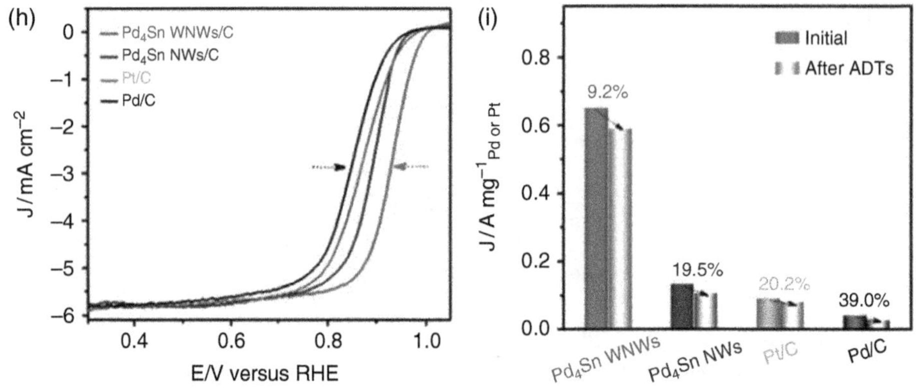

Figure 7.4 (Continued).

7.3.2.1 Shape-Controlled Pd Alloys

The composition and size of PdM alloys play a crucial role in determining the ORR activity. The composition ratio of Pd:M as well as the size of the alloy NPs are highly dependent on the synthesis method and alloying elements. Yang et al. synthesized PdCu NPs of four different sizes by varying the concentration of triphenylphosphine ligand in the reaction mixture [65]. They found out that the ORR performance decreases with increasing the size of the alloy nanostructures due to a decrease in the number of edges and corners, while the adsorption energy of O resembles that of the bulk phase with increasing particle size. Cabrera et al. demonstrated that PdAg alloy nanocrystals with a surface composition ratio of 2:1 (Ag:Pd) exhibit much higher ORR activity as compared to nanocrystals with 1 : 1 ratio [66]. The enhancement in catalytic performance was attributed to the synergistic interaction between Pd and Ag, which allows adsorption of O_2 on the Pd surface followed by its instant disproportionation. Another study by Jiao et al. investigated the effect of surface composition of PdAu alloy hollow nanospheres on ORR performance [67]. The activity of Pd-to-Au ratio of 3 : 1 exhibited a mass activity of 0.15 A/mg at 0.9 V vs RHE outperforms other compositions due to moderate strain on the surface. The effect of compositional variations on ORR activity was also studied with Pd alloys having dendritic structures [59, 67, 68]. $Pd_{59}Cu_{30}Co_{11}$ with dendritic morphology having 3.5% lattice strain improved the Pd mass activity to 0.38 A/mg at 0.9 V ($E_{1/2}$ = 0.91 V) vs RHE with pronounced methanol/ethanol tolerance behavior as compared to other PdCuCo compositions and commercial Pt/C [59]. $Pd_{59}Cu_{30}Co_{11}$ also demonstrated improved stability as only ~32% loss in mass activity occurred after 15 000 catalytic cycles. Another study showed the formation of tree-like PdAg nanocrystals with three different compositional ratios, i.e. Pd_3Ag, PdAg, and $PdAg_3$. The ORR activity follows the order $Pd_3Ag > PdAg_1 > PdAg_3$ [69]. The modified electronic structure in the different compositions led to variation in catalytic performances. Nanoporous alloys can demonstrate further improved ORR activity as porosity provides access to numerous active sites present throughout the interior as well as exterior surfaces of the nanostructures [70, 71]. Yamauchi et al. prepared PdCu mesoporous NPs which exhibited better ORR mass activity (0.19 A/mg at 0.85 V vs RHE) than Pd mesoporous NPs [70]. Another work showed the formation of nanoporous PdCe nanocubes with 5 nm pores by melt spinning combined with the chemical dealloying method [71]. The ORR mass activity of these PdCe nanocubes was much higher than Pd nanocube and commercial Pd/C due to weak binding with O intermediate in the PdCe due to its lowered *d*-band center than that of Pd as suggested from DFT calculation.

One-dimensional nanostructures such as NWs and nanochains, and 2D nanosheets (NSs) have attracted great attention in the recent past as their exterior surfaces consist of low-energy facets, which weaken the binding of oxygen than that of conventional NPs resulting in enhanced ORR activities [64, 72, 73]. Zhang et al. studied the defect-dependent ORR activity on PdSn alloy NWs. The Pd_4Sn wavy NWs contain numerous surface defects as compared to penta-twinned Pd_4Sn NWs (Figure 7.4d–g). Therefore, the defect-rich wavy Pd_4Sn NWs demonstrated much better mass activity (0.65 A/mg) as compared to penta-twinned Pd_4Sn NWs (0.13 A/mg) at 0.9 V vs RHE, which was further supported by the DFT studies revealing that surface defects create grain boundaries in the wavy NWs which helps in weakening the binding of O leading to high ORR activity (Figure 7.4h) [64]. In addition, these Pd_4Sn wavy NWs exhibited high stability up to 10 000 potential cycles with only 9% loss of the initial mass activity (Figure 7.4i). Similarly, Pd-Ag NWs rich in grain boundary defects showed 5.2 times higher mass activity (0.1 A/mg) as compared to PdAg NPs [72]. Jio et al. prepared PdAu hollow nanochain with concave hollow structure using cobalt nanochain as a sacrificial template via galvanic displacement method [73]. This unique nanostructure exhibited superior ORR activity and stability due to the presence of stronger strain and higher index facets on the concave corners than that of the arc edges. Wang et al. found that 2D porous PdAu NSs formed by the interconnection of NWs had 7.6 times higher specific activity (4.72 mA/cm^2) than commercial Pd black [68].

Remarkably, for the first time, Guo et al. synthesized PdMo bimetallene, which is 2D alloy NSs of Pd and Mo having structural morphology similar to graphene, i.e. sub-nanometer thickness and 100 nm lateral size (Figure 7.5a–c) [74]. These highly curved 2D PdMo NSs act as an efficient and stable electrocatalyst for ORR in alkaline media, demonstrating the highest mass activity of 16.37 A/mg at 0.9 V vs RHE (Figure 7.5d, e). This mass activity is exceptionally higher than those of commercial Pt/C (78 times) and Pd/C (327 times) in a similar reacting environment. The excellent performance was attributed to the optimized oxygen binding facilitated by changed electronic structure, causing a lowering of the *d*-band center and a high atom economy due to a small thickness. The changes in the electronic structure originate from several factors such as alloying of Pd with Mo, presence of strain on the curved geometry, and thinness of the sheets, which cause quantum size effect as revealed from the DFT studies. These PdMo NSs also demonstrated enhanced stability up to 30 000 potential cycles with only ~30% activity loss (Figure 7.5f).

From those findings, it is clear that alloying significantly improves the ORR activity by modifying the surface energy and strain in the nanostructures. However, due to slight variations in the experimental condition, these studies fail to provide a clear picture of best alloy. The inferences become more complicated due to the presence of different surface stabilizers on the nanocrystal surfaces, whose effect is yet to be quantified. Due to these limitations, the theoretical studies have remained the guiding light for the selection of the alloying elements.

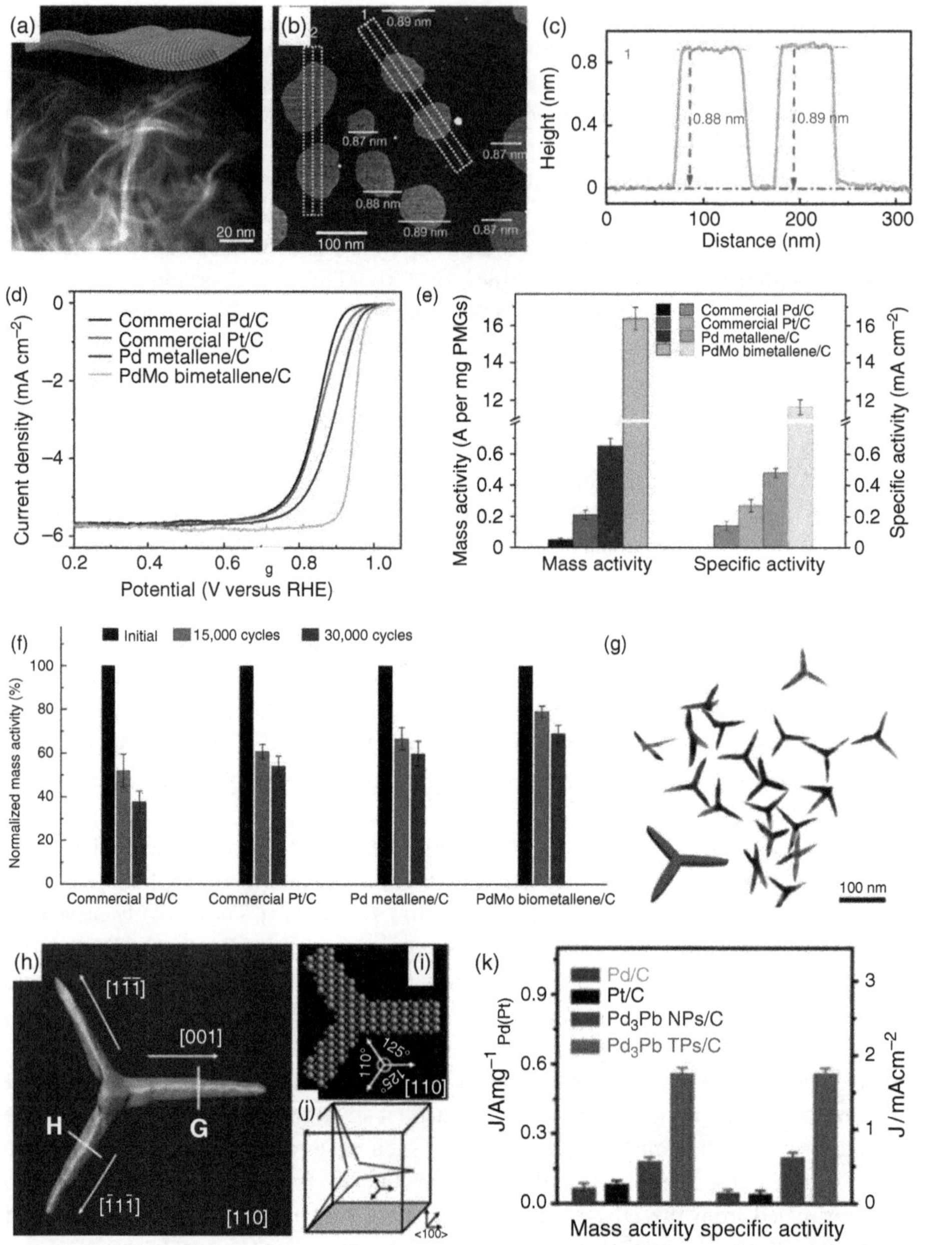

Figure 7.5 High-magnification (a) HAADF-STEM, (b) AFM image, (c) and the corresponding height profiles of PdMo bimetallene, (d) ORR polarization curves recorded in 0.1 M KOH and (e) the corresponading mass and specific activities of the PdMo bimetallene/C, Pd metallene/C, and the commercial Pt/C and Pd/C at 0.9 V vs RHE, (f) changes in normalized mass activities of the catalysts before and after 15 000 and 30 000 potential cycles. *Source:* From [74]/with permission of Springer Nature (g) TEM image of Pd_3Pb TPs, (h) reconstructed 3D tomogram and (i) atomic model along the [110] zone-axis, (j) 3D representation along the <100> zone-axis, (k) histogram of mass and specific activities of Pd_3Pb TPs/C, Pd_3Pb NPs/C, Pt/C, and Pd/C at 0.9 V vs RHE. *Source:* From [75]/with permission of Elsevier.

7.3.2.2 Ordered Pd Alloys

Further to the disordered alloys, recently, intermetallic Pd nanostructures have been considered by many to be better electrocatalysts for ORR, considering both the activity and stability aspects. The presence of periodic electronic configuration and uniform bond lengths in ordered intermetallic nanocrystals can potentially enhance ORR activities. Inspired by this, many ordered Pd alloys such as PdCu, PdCuNi, PdCuCo, PdBi, PdZn, AuPdZn, and PdPb. have been evaluated for ORR [76–79]. The ordered intermetallics are usually obtained using thermal annealing of the disordered alloy nanostructures. For example, Jiang et al. achieved ordered PdCuCo alloy by thermal annealing at 375 °C and the corresponding ORR mass activity (0.12 A/mg at 0.9 V vs RHE) exceeds that of commercial Pt/C [76]. DFT calculations reveal that improvement in ORR activity is contributed by compressive strain on Pd surface caused by the smaller atomic size of Co, Ni, and Cu. Similarly, Yang et al. obtained ordered PdZn nanocrystals by annealing treatment at 500 °C, which has an excessive compressive strain of 10% as compared to disordered ones [77]. The ordered PdZn exhibited 1.4 times higher mass activity as compared to the disorder nanocrystals. Ordered AuPdZn nanocrystals achieved by the galvanic displacement of Pd by gold demonstrated the highest mass activity of 0.105 A/mg (at 0.9 V vs RHE) and high stability with only 9.5% loss after 30 000 potential cycles. Along with thermal annealing, electrochemical techniques have been used for the synthesis of ordered intermetallic alloys [78]. Recently, Bi was shown to form ordered alloy nanocrystals with Pd by pulsed electrochemical deposition techniques having an $E_{1/2} = 0.92$ V vs RHE and very high mass activity of 0.95 A/mg. The high performance of these nanocrystals was attributed to the weakening of the surface Pd-O bonds due to alloying of Pd with Bi [78]. However, the stability of the PdBi alloy was poor as ~40% loss in initial mass activity occurred after 10 000 potential cycles.

Success in the synthesis of Pd-based intermetallic nanostructures simultaneously with shape control remains limited. Recently, Skrabalak et al. and Shi et al. demonstrated a low-temperature approach to synthesize Pd_3Pb nanocubes and Pd_3Pb NWs, respectively [61, 79]. The as-synthesized Pd_3Pb nanocubes exhibited a large enhancement in mass activity (0.15 A/mg at 0.9 V vs RHE) as compared to commercial Pt/C or Pd/C. The improved performance of Pd_3Pd is associated with the weak binding of OH intermediate on its surface when compared with Pd (100) surface, as suggested from the computational studies [61]. Likewise, The Pd_3Pb NWs exhibited three times higher mass activity as compared to commercial Pt/C [79]. Very recently, Bu et al. synthesized ordered Pd_3Pb tripods exposed predominantly with {110} facets, which exhibited high ORR performance as compared to Pd_3Pb NPs in alkaline media (Figure 7.5g–j) [75]. A wet-chemical synthesis approach was adopted to prepare the Pd_3Pb tripods with oleylamine and oleic acid as the shape directing agents. The Pd_3Pb tripods exhibited a mass activity of 0.56 A/mg at 0.9 V vs RHE, which is much higher than the Pd_3Pd NPs, Pt/C, and Pd/C (Figure 7.5k). The strong charge exchange between Pb-(sp) and Pd-4d orbitals on the {110} facet of Pd_3Pb tripod resulted in an orbital configuration similar to that of Pt, thereby exhibiting very high ORR performance. Another study demonstrated that 2% Au doping in ordered Pd_3Pb wavy NWs substantially improves the ORR performance [80]. These NWs also exhibited moderate durability up to 10 000 potential cycles with 23% loss in initial mass activity.

7.3.2.3 Pd-Nonmetal Alloys

Very recently, it has been discovered that alloying Pd with nonmetallic elements such as B, S, or P can alter the interatomic spacing of Pd more effectively to enhance the ORR performance by tuning the binding energies of oxygenated intermediates as compared to their multimetallic alloy counterparts [81–83]. For example, B doping in Pd crystal causes enlargements in Pd–Pd interatomic spacing, thereby affecting the surface electronic states of the Pd. [81, 84] The Pd–B nanoalloy with B doping up to 20% was achieved using $NaBH_4$ as reducing agent in *N,N*-dimethylformamide solvent by Li et al. This catalyst exhibited a mass activity (0.97 A/mg at 0.9 V vs RHE) that is 14 times and 8 higher than commercial Pt/C and Pd/C, respectively [81]. But the stability of Pd-B alloy was considerably poor with 60% loss in initial mass activity after 3000 cycles. DFT calculation suggested three primary active sites for ORR: (i) Pd-BO_2 assemblies, (ii) Pd atoms nearby these assemblies, and (iii) Pd atoms surrounded by B atoms at the surface. Pd_4S nanostructures also demonstrated superior ORR performance with $E_{1/2}$ of 0.88 V vs RHE than commercial Pt/C due to optimal oxygen adsorption free energy on different Pd_4S surface-sites [82]. Pd has also been co-alloyed with B and P by a soft-templating synthetic strategy to prepare PdBP mesoporous dendritic nanostructures [83]. The ORR mass activity of ternary PdBP alloy (1.45 A/mg at 0.9 V vs RHE) was 1.8 times and 3.3 times higher than those for the PdB and Pd mesoporous nanospheres, respectively. In addition, the PdBP demonstrated enhanced durability with only 32% loss in mass activity after 5000 potential cycles. The dendritic morphology and synergistic effects arising from co-alloying of Pd with B and P enable PdBP alloy to exhibit superior ORR performance. Nonmetal doping strategies are exciting because the nature of the bond length and electronic property variations is different from transition metal doping. However, investigations on such phases are still at an initial stage.

7.3.3 Core-Shell Pd Nanostructures

The utilization efficiency of the Pd atoms in NPs is low because only a fraction of the atoms exposed on the surface participates in the electrocatalytic reactions. In a 4-nm Pd particle, for example, only ~20% of the atoms are present on the surface, while the remaining ~80% are buried inside the particle. Constructing core-shell nanostructures, in which a thin Pd-based shell covers a less-expensive core, can be regarded as a promising strategy to improve the Pd atom utilization efficiency [58, 85, 86]. Therefore, tremendous efforts have been devoted toward designing Pd-based core-shell nanostructures for alkaline ORR in the recent past [87, 88]. The performance of the Pd-shell can be tuned by varying its composition, shape, and thickness of the intimate metal. For example, Jiang et al. varied the shell composition and content of monodisperse Ni@Pd (Ni-core and Pd-shell) core-shell NP [58]. They have observed that the Pd-enriched surface exhibited the highest ORR activity. The ORR performances of various core-shell compositions follow the order Ni@Pd_3 NPs > Ni@Pd_2 > Ni@Pd > Ni_3@Pd in 0.1 M KOH [58].

In a notable investigation by Huang et al., it was demonstrated that the ORR activity can be tuned by controlling the thickness of the Pd shell on an Au core. The Pd shell thickness systematically modifies the redox property of the surface Pd atoms, which in turn regulates the ORR performance [89]. Au core with 0.9 layer of Pd atom coverage exhibited the highest ORR mass activity of ~0.3 A/mg (at 0.9 V vs RHE) among the various submonolayer,

monolayer, or multilayer Pd-shells as the reduction peak potential of PdO shifted maximum for the same. The crystallinity of Au-core and the thickness of Pd-shell in Au@Pd NPs have been shown to regulate the ORR performance because these can create compressive strain on the Pd surface [86]. Electrochemical dealloying of alloy NPs also results in the formation of core-shell nanostructures [90, 91]. Abruna et al. have demonstrated that electrochemical dealloying strategy can be used to improve the ORR activity of PdNi and PdMn alloy nanocrystals where the improvement in ORR performance was attributed to the formation of Ni@Pd core-shell structure after electrochemical dealloying to expose electroactive Pd on the surface [91]. The dealloyed PdNi/C nanocrystals exhibited an initial mass activity of 0.37 A/mg at 0.85 V vs RHE, which decreases during the accelerated durability tests. Subsequently, they were able to restore the original catalytic activity after the durability test by repeating the dealloying procedure which removes the surface Ni–OH species and increases the surface concentration of Pd. Pd-shell on Au NW core also exhibits very high ORR activity due to the presence of high surface strain [92, 93].

Pd or Pd-based alloys have also been investigated as the core material in core-shell nanocrystals for ORR in alkaline media [94–96]. For example, Feng et al. synthesized Pd_6@Ni icosahedral core-shell nanostructures, which exhibited better ORR mass activity (0.22 A/mg at 0.9 V vs RHE) when compared with Pd_6Ni alloy icosahedral nanocrystal or Pd-rich surface [97]. The enhanced performance was ascribed to the presence of Ni on the surface, which lowers the Pd *d*-band center, reducing the adsorption energy of oxygen on Pd (111) surface. It was further observed that the presence of a PdFe shell on an icosahedral Pd core further improves the ORR mass activity up to 0.31 A/mg at 0.9 V vs RHE (Figure 7.6a–c) [98]. The DFT calculations revealed that the enhancement in ORR performance can be attributed to down filling of the *d*-band center of Pd due to alloy formation and also to strain arising due to core-shell configuration, which enables the optimum binding of the oxygenated intermediates (Figure 7.6d, e). Tang et al. studied the effect of core-shell morphology on ORR performance by designing Pd_3Pb@Pd core-shell nanostructures having two different morphologies, i.e. NS and nanocube (Figure 7.6f–i) [96]. The NSs outperform the nanocubes by exhibiting a mass activity of 0.57 A/mg. However, both the NSs and nanocubes exhibited 6.5 times and 5.3 times higher mass activity than commercial Pt/C, indicating that intermetallic Pd_3Pb core and few layers of Pd-shell is beneficial for achieving high catalytic activity. The DFT calculations suggest that Pd_3Pb/Pd (100) will possess higher ORR activity than that of Pd_3Pb (100) and Pd (100) due to the lowest overpotential and introduction of tensile strain effect causing favorable adsorption of oxygenated intermediates during ORR (Figure 7.6j).

While the atom economy is highly promising in core-shell nanostructures, generation of such structures with uniform shell thickness is a tough synthetic challenge. The regeneration of its surface after deactivation appears to be an even more significant challenge and would call for extensive research to evaluate their worthiness.

7.3.4 Support-Modified Pd Nanostructures

The low price and availability of carbon-based materials such as carbon black, graphene, and carbon nanotube (CNT) have made them desirable support materials for Pd-based electrocatalysts. They are endowed with the necessary characteristics for a suitable support

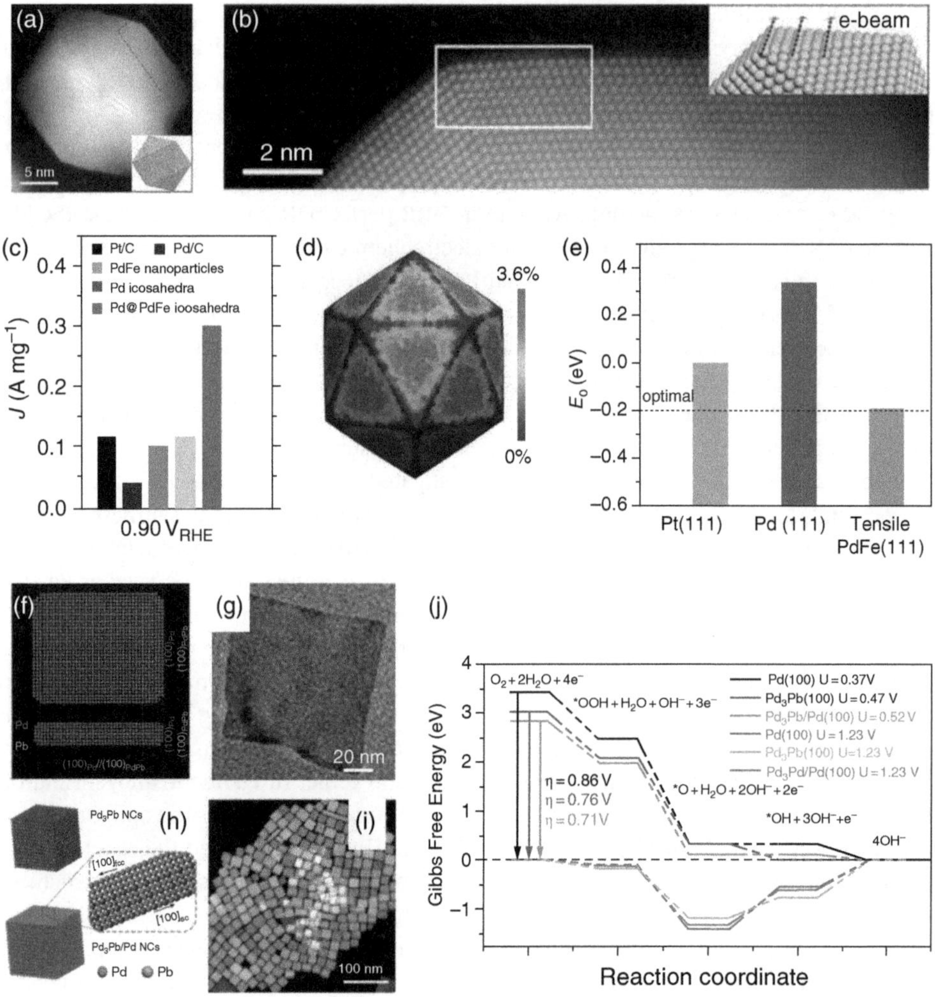

Figure 7.6 (a) HAADF-STEM image of a single Pd@PdFe icosahedron, (b) high-resolution HAADF-STEM image near edges marked in (a), (c) mass activity of Pt/C, Pd/C, PdFe nanoparticles, Pd icosahedra, and Pd@PdFe icosahedra at 0.9 V versus RHE, (d) surface-strain field in Pd@PdFe icosahedra and the strain percentage indicates in the gray scale with maximum strain in the center of the triangular face, (e) oxygen atom adsorption energy (E_O) on Pt (111) surface, Pd (111) surface, and PdFe (111) surface showing optimal value of −0.2 eV for PdFe (111). *Source:* From [98]/with permission of American Chemical Society. (f) schematic illustration of the Pd_3Pb nanosheets and Pd_3Pb/Pd nanosheets, (g) TEM image of Pd_3Pb/Pd nanosheets, (h) schematic illustration of the Pd_3Pb nanocubes and Pd_3Pb/Pd nanocubes, (i) TEM image of Pd_3Pb/Pd nanocubes, (j) Gibbs free energy changes in a $4e^-$ reaction process on Pd (100), Pd_3Pb (100), and $Pd_3Pb@Pd$ (100) surfaces. *Source:* From [96]/with permission of American Chemical Society.

material, such as high surface area and good electrical conductivity [74, 99–101]. A support provides for uniform distribution of Pd nanostructures and arrests their migration and non-homogeneous distribution on the electrode surface. Besides this, their interaction with the electrocatalysts can further be exploited to improve ORR efficiency. For example, graphene can be easily doped with other heteroatoms such as boron, nitrogen, sulfur, and phosphorous, which modifies its electronic properties [102, 103]. In particular, nitrogen doping has been widely studied as it can interact with the sp^2 carbons via π–π interactions and increases electrons density on graphene's carbon network, which in turn, tunes the electronic property of a Pd nanostructure [104]. The N atoms incorporated within a graphene sheet also activate the adjacent carbon atoms for facile oxygen adsorption and enhance ORR. Nitrogen dopants can further reduce the agglomeration and improve the distribution by introducing stronger binding sites for anchoring metal NPs on the carbon support [105]. For example, Zhang et al. synthesized ultrafine 4-nm PdCo bimetallic NPs encapsulated in N-doped porous carbon nanocapsules (PdCo@NPNCs), which demonstrated enhanced ORR mass activity of 0.45 A/mg at 0.9 V vs RHE as well high stability with only 8% loss after 20 000 seconds than commercial Pt/C [99]. Therein, the N-doped porous carbon shell could effectively prevent the detachment, agglomeration, and dissolution of NPs during fuel cell operation and also hindered their coalescence during the carbonization process. Similarly, Co@Pd core-shell NPs prepared by galvanic displacement and decorated on N-doped porous carbon exhibited enhanced ORR performance than that of commercial Pd/C [106].

N-doping creates primarily three different types of N, viz graphitic, pyridinic, and pyrrolic N. Among them, pyridinic N was found to be responsible for the enhancement in ORR activity [107, 108]. Liu et al. have demonstrated that graphene doped with pyridinic N would not only be beneficial for stabilizing Pd particles but also decrease the oxygenated intermediate binding energy resulting in enhanced ORR [107]. However, the influence of pyrrolic N for ORR has barely been investigated, and only very recently, its role in improving ORR activity as well as stability as found to be as important as pyrrolic N [44]. It has been observed that depositing Pd atom-by-atom on N-doped reduced graphene oxide (NRGO) support via an in-situ deposition technique, the ORR mass activity (0.59 A/mg at 0.9 V vs RHE) can be improved by two times as compared to Pd/NRGO composites obtained by ex-situ mixing of presynthesized components (Figure 7.7a, b). Besides, Pd NPs deposited on reduced graphene oxide (RGO) exhibited a poorer ORR performance as compared to both the in-situ and ex-situ Pd/NRGO samples, depicting the importance of N in improving ORR activity. Additionally, the Pd/NRGO in-situ sample exhibited excellent durability by maintaining a mass activity of 0.42 A/mg even after 15 000 potential cycles and better methanol tolerance ability as compared to commercial Pt/C (Figure 7.7c). Theoretical as well as X-ray photoelectron spectroscopy (XPS) studies revealed that substantial charge-transfer occurs from the Pd 3d-orbitals to the 2p-orbitals of pyrrolic N, thus lowering the Pd-3d binding-energy (Figure 7.7d), which results in a significant enhancement in the ORR activity as well as stability in the case of Pd/NRGO in-situ sample.

Some researchers have attempted to implement Pd NPs deposited on N-doped graphene support as cathode catalysts in AEMFCs. For instance, Atanassov and co-workers prepared Pd NPs on 3D nitrogen-doped graphene supports (Pd–N/3D-GNS) and implemented as cathode catalysts in H_2/O_2 fed AEMFC exhibiting a current density of ~250 mW/cm^2 due to their desirable chemical composition and unique structural features [103]. Another strategy to architect high-performance ORR catalysts is by tuning the interaction between Pd and graphene by boron doping [110]. B-doping suppresses the interaction between Pd and

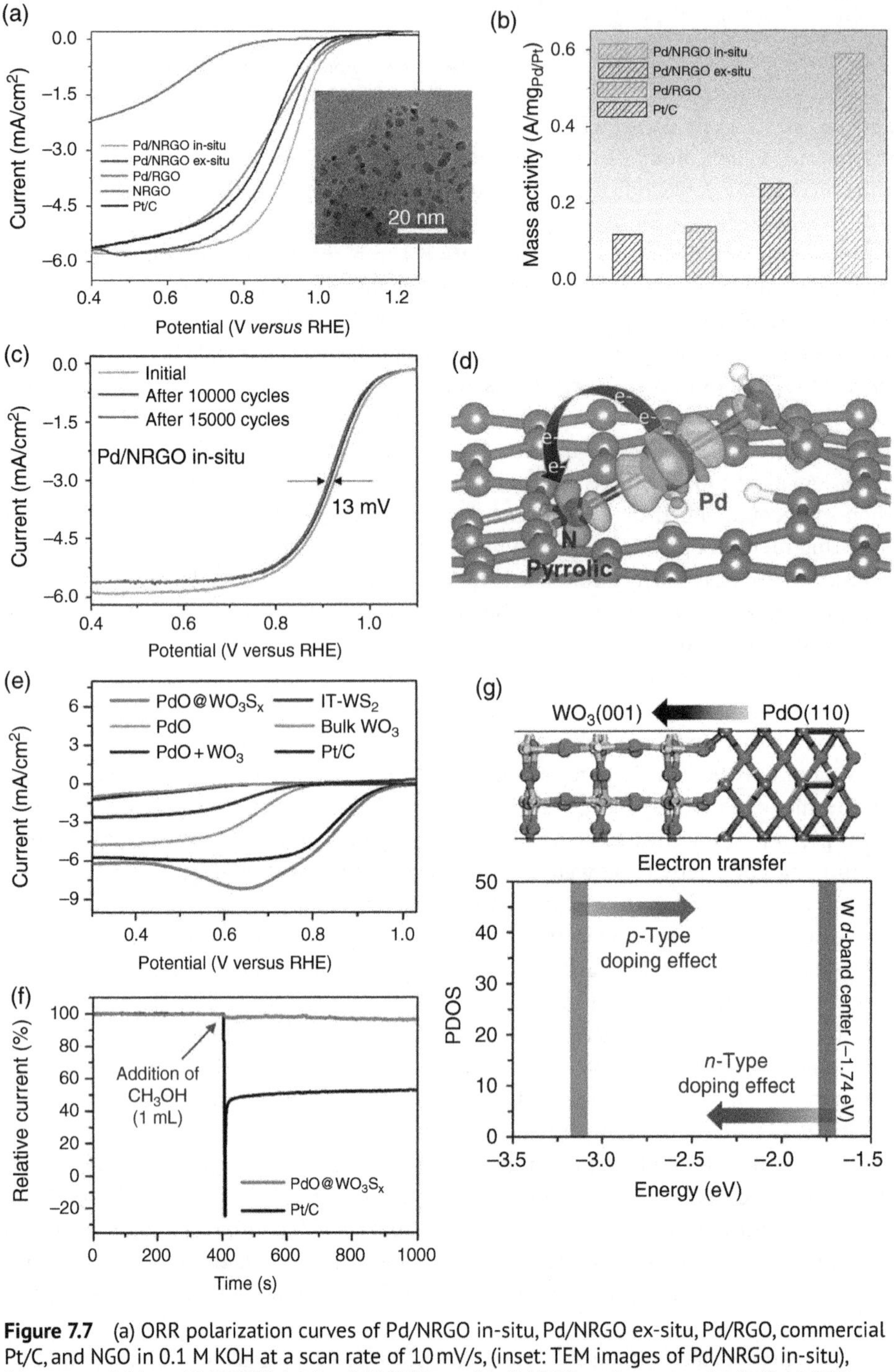

Figure 7.7 (a) ORR polarization curves of Pd/NRGO in-situ, Pd/NRGO ex-situ, Pd/RGO, commercial Pt/C, and NGO in 0.1 M KOH at a scan rate of 10 mV/s, (inset: TEM images of Pd/NRGO in-situ), (b) mass activities of Pd/NRGO in-situ, Pd/NRGO ex-situ, Pd/RGO, and commercial Pt/C, (c) ORR polarization curves of Pd/NRGO in-situ before and after stability tests in O_2 saturated 0.1 M KOH at 1600 rpm, (d) the electron charge density difference (0.008 e/Å 3) of the adsorbed Pd-adatom on the N Pyr-OH in which the light gray lobe depicts electron depletion and the dark gray lobes depict electron gain. *Source:* From [44]/with permission of Royal Society of Chemistry (e) ORR polarization curves of PdO@WO_3S_x hybrid, pure PdO, 1T-WS_2, bulk WO_3, a physical mixture of PdO and WO_3, and Pt/C recorded in O_2 saturated 0.1 M KOH, (f) relative current densities of PdO@WO_3S_x hybrid and Pt/C at 0.4 V (vs RHE) upon methanol addition at 400 s, (g) schematic representation of *d*-band center shifting of Pd and W in the hybrid structure due to electron transfer from PdO (110) to WO_3 (001). *Source:* From [109]/with permission of Royal Society of Chemistry.

C of graphene, which leads to enhancement in ECSA of Pd, and thus improves the ORR performance as established by the Raman and electron energy loss spectroscopic measurements. Moreover, DFT calculations suggested that electron transfer becomes more facile between Pd and graphene after B-doping [110]. The surface-functionalized carbon materials were shown to act as an excellent support material for stabilization of Pd nanostructures as they provide a confined environment for their growth and also improves sintering resistance [111, 112].

Non-carbon materials like TiO_2, CeO_2, WS_2, Ta_2O_3, Co_3O_4, WO_3, and $Co(OH)_2$, have also been investigated as support materials as they offer corrosion resistance under the extended operation periods [109, 113–116]. Lee et al. reported a $PdO@WO_3S_x$ composite which exhibited an enhanced half-wave potential (0.89 V vs RHE) than the bare components as well as commercial Pt/C (Figure 7.7e) [109]. In addition, it exhibited much higher methanol tolerance activity as compared to Pt/C (Figure 7.7f). Theoretical investigations have revealed that the *d*-band center of Pd in $PdO@WO_3S_x$ is upshifted to −2.57 eV from −3.19 eV than pure PdO, and that of W is downshifted to −2.61 eV from −1.74 eV than pure WO_3 so that it became similar to that of Pt (Figure 7.7g). The enhanced ORR activity in $PdO@WO_3S_x$ composite was attributed to its Pt-like *d*-band center position, which enabled facile ORR kinetics. Pd NPs loaded on TiO_2 possessing oxygen vacancies are shown to exhibit improved ORR activity as compared to TiO_2 without any oxygen vacancies [114]. The enhancement in ORR activity was attributed to the formation of electron-rich Pd surface in the case of oxygen vacancy possessing TiO_2 due to facile electron transfer as confirmed by XPS, ESR, and DFT studies. In addition, oxygen vacancies also led to improved stability as well as methanol tolerance ability in Pd/TiO_2.

It has also been observed that suitable inner structure tailoring can be highly beneficial. NiO_2/Pd nanostructures with local structure ordering exhibit superior ORR activity as compared to its counterpart without any order [117]. The local structure ordering causes severe lattice strain and electronegativity difference resulting in the relocation of electron density from Ni to Pd, which facilitates facile adsorption of the reaction intermediates. Pd NPs loaded transition metal dichalcogenides such as MoS_2 NSs have been investigated as oxygen reduction electrocatalysts. These covalently functionalized Pd/MoS_2 NSs outperform covalently functionalized Pd/graphene for ORR, as the latter is suggested to get damaged during the covalent functionalization [117].

With the advancements in electron microscopic techniques, it is now possible to observe single metal atoms dispersed on a support material. The investigations on single metal sites in metal-nitrogen-carbon composite with M–N coordination are growing as a promising class of electrocatalysts for ORR due to atom efficiency [118]. In one such important work, atomically dispersed Pd on the surface of nitrogen-doped hollow carbon cages (HNC) was prepared, having Pd–N coordination bonds [119]. The Pd-HNC nanocomposite ORR activity outperforms metallic Pd NPs, as well as commercial Pt/C. Xiang et al. have demonstrated that Pd atoms deposited on MnO_2 NWs and CNT exhibits two times higher ORR activity as compared to Pd/CNT [115]. Theoretical studies have revealed that the Pd single atoms homogeneously doped in MnO_2 synergistically activate the O_2 and moderately bind to the oxygenated intermediates due to a strong Pd and MnO_2 interaction. Single Pd atoms on various support are highly promising as the Pd load is significantly reduced, provided their long-term stability becomes well understood.

7.4 Conclusions and Perspective

AEMFCs using Pd-based catalysts and alkaline electrolytes are highly promising and emerging technology as compared to the PEMFCs. This chapter attempts to discuss the nanostructured, high-performance Pd-based electrocatalysts developed recently for ORR in alkaline media. In the last five years, there has been a rapid development on such electrocatalysts, which exhibit consistently improved activities and stabilities for the ORR that can surpass the performance of traditional Pt-based catalysts under normal operating conditions or even in the presence of methanol that swiftly poisons Pt. The activity and stability of these nanostructured electrocatalysts primarily depend on the morphology, composition, size, porosity, surface structure, and synthesis method. Various strategies have been adopted for improving their activities, such as (i) controlling the morphology of Pd nanostructures to maximize the exposer of the highly ORR active (100) facets, (ii) incorporating suitable non-noble transition metals to form alloy structures to increase the dispersion of Pd atoms and mass activity, (iii) forming a core-shell structure to maximize the Pd atoms utilization and improves the mass activity. The alloy or core-shell structures modify the electronic properties by inducing strain in the lattice and tailor the *d*-band center position, which in turn regulates the binding of the reaction intermediates on the surface of the catalyst. In addition, (iv) depositing Pd-based nanostructures on carbon or non-carbon support material prevents their agglomeration and facilitates the ORR kinetics by providing more active sites. The strong metal–support interactions cause substantial charge transfer between Pd and the supports and thereby can tune the *d*-band center position. However, a thorough understanding of the origin of the ORR activity in Pd-based nanostructures, as in the case of Pt, has remained a bottleneck for further developments of AEMFCs and will be crucial for their commercialization. In general, theoretical studies often focus only on a limited number of factors, while the actual operating conditions are much complicated. Hence, the development of more rigorous theories and their correlation with experiments can be highly beneficial to prepare high-performance electrocatalysts more efficiently. Moreover, catalyst activities must be investigated in real devices to determine the actual commercialization possibility. More importantly, the precise identification of catalytically active sites present on the electrocatalyst surface and the changes therein during the ORR is ought to be a key design consideration. For example, the activity of the catalysts is correlated with the coordination number of the surface atoms. A certain catalyst having an equal number of (111) terraces but more number of second nearest neighboring atoms can possess better activity toward ORR [126]. Therefore, defect engineering on the catalyst's surface can be highly beneficial to achieve significant enhancement in performance as active sites having the appropriate geometric motif can be created and stabilized.

In addition, other parameters like catalyst durability have been severely neglected as the primary focus was on improving their sustainability. As per DOE 2020 technical target, 5000 hours of cycling stability at 1.0–1.5 A/cm^2 current density with only 10% voltage decay in a membrane electrode assembly is necessary [127]. However, it can be seen from Table 7.1 that the stability of the catalysts investigated so far was only up to 220 hours, which is far below the target. The activity loss in metal-based catalysts during fuel cell operation can occur via either electrochemical Ostwald ripening of metal NPs or electrochemical metal dissolution and re-deposition. In the case of alloys, their compositional changes during operation are a subject of prime concern. When the metal NPs are loaded on a

Table 7.1 Summary of ORR activities of recently developed Pd-based catalysts in alkaline media.

S. No.	Materials	Electrolyte	$E_{1/2}$ (V vs RHE)	Mass activity @ 0.9 V (A/mg)	Specific activity @ 0.9 V (mA/cm^2)	Stability	Reference
Pure Pd catalysts							
1	Pd NCs	0.1 M KOH	0.95	0.78	2.46	20% loss in mass activity after 10 000 cycles (22 h)	[53]
2	Pd nanocube	1 M NaOH	~0.83	–	0.53	52.5% loss in current density after 1000 cycles	[48]
3	Pd cube/C	0.1 M KOH	~0.82	0.059	0.55	–	[49]
4	Pd nanowires/C	0.1 KOH	~0.87	0.22	0.83	13.8% loss in mass activity after 10 000 cycles	[55]
5	Pd-Net	0.1 M KOH	0.84	0.62 @0.85 V	1.88 @0.85 V	19.3% loss in ECSA after 3000 cycles	[56]
6	Pd HBND	0.1 M KOH	0.86	–	3.0	22.2% loss in current density after 20 000 s	[57]
7	Pd nanotetrahedra	0.1 M KOH	0.86	–	1.61	20.6% loss in current density after 12 000 s	[52]
Pd-based alloys							
8	PdCu/C	0.1 M KOH	0.82	0.3 @ 0.8 V	–	6.6% loss in mass activity after 1000 cycles	[65]
9	Pd@Ag/C	0.1 M KOH	0.83	0.88	–	10% loss in current density after 10 000 cycles	[66]
10	NP-PdCe	0.1 M KOH	~0.85	~0.03	–	15.3% loss in mass activity after 5000 cycles	[71]
11	PdCu MNs	0.1 M KOH	0.89	0.19 @0.85 V	–	4% loss in mass activity after 3000 cycles	[70]
12	Pd_2CoAg NCs	0.1 M KOH	~0.82	0.66	~0.7	Negligible loss after 5000 cycles	[120]
13	Pd_3Au/C	0.1 M KOH	0.88	0.15	0.13	14% loss in current density after 40 000 s	[67]
14	Pd_3Ag1 NTs	0.1 M KOH	0.84	–	1.19	14% loss in current density after 40 000 s	[69]
15	PdMo bimetallene/C	0.1 M KOH	0.95	16.3	11.6	~30% loss in mass activity after 30 000 cycles	[74]
16	Pd4Au HCs/C	0.1 M KOH	0.87	0.29	0.14	12% loss in current density after 20 000 s	[73]
17	2D PdAu PNSs	0.1 M KOH	0.932	–	4.72	1.8% loss in current density after 100 000 s	[68]
18	Pd1Ag1	0.1 M KOH	0.897	0.103	0.36	2 mV decrease in $E_{1/2}$ after 5000 cycles	[72]

(Continued)

Table 7.1 (Continued)

S. No.	Materials	Electrolyte	$E_{1/2}$ (V vs RHE)	Mass activity @ 0.9 V (A/mg)	Specific activity @ 0.9 V (mA/cm^2)	Stability	Reference
19	PdSn NWs	0.1 M KOH	0.93	0.65	1.51	10% loss after 10 000 cycles	[64]
20	PdCuCo/C nanoparticle	0.1 M KOH	0.872	0.13	–	4% loss in mass activity after 10 000 cycles	[76]
21	IM-Pd_3Pb NNs	0.1 M KOH	~0.9	0.61	15.7	23% loss in mass activity after 10 000 cycles	[79]
22	Pd_3Pb TPs/C	0.1 M KOH	~0.9	0.56	1.76	16.6% loss in mass activity after 20 000 cycles	[75]
23	PdBi	0.1 M KOH	0.92	0.95	2.42	40% loss in mass activity after 10 000 cycles	[78]
24	Au-O-PdZn	0.1 M KOH	~0.9	0.105	–	9.5% loss in mass activity after 10 000 cycles	[77]
25	2% Au-doped Pd_3Pb WNWs	0.1 M KOH	0.919	0.75	2.59	22.7% loss in mass activity after 10 000 cycles	[80]
26	Pd_3Pb NC/C	0.1 M KOH	0.88	0.15	–	12% loss in initial activity after 50 000 cycles	[61]
27	PdCu/C	1 M KOH	0.89	0.13	2.25	57% loss in mass activity 1 00 000 cycles (220 h)	[121]
28	$Pd_{59}Cu_{30}Co_{11}$		0.91	0.38		32% loss in mass activity after 15 000 cycles	[59]
29	B-Pd/C	0.1 M KOH	~0.94	0.97	1.27	60% loss in mass activity after 3000 cycles	[81]
30	Pd_4S	0.1 M KOH	0.877	–	–	8.6% loss in current density after 3000 cycles	[82]
31	PdBP MSs	0.1 M KOH	0.92	1.45	2.45	32% loss in mass activity after 5000 cycles	[83]
32	Pd S AuCu iNPs	1 M KOH	~0.93	1.78	1.08	Substantial decay after 10 000 cycles	[87]
Core-shell Pd nanostructures							
33	Ni@Pd_3 NPs	0.1 M KOH	0.86	0.035	0.08	5% loss in mass activity after 3000 cycles	[58]
34	Au-NWs@Pd 0.1 @PEI	0.1 M KOH	0.9	0.295	0.4	70% loss in mass activity after 10 000 cycles	[92]
35	Pd@NiO-0.3/C	0.1 M KOH	0.87	0.24	–	8.8% loss in mass activity after 10 000 cycles	[95]
36	Pd6Ni/C	1 M KOH	~0.87	0.22	0.66	5.6% loss in mass activity after 10 000 cycles	[97]
37	FFPC/NPCNFs-1	0.1 M KOH	0.81	0.48 @ 0.75 V	–	13% loss in current density after 20 000 s	[88]

38	Au dt –Pd NPs	0.1 M KOH	0.93	7.72	6.03	8 mV decrease in $E_{1/2}$ after 1000 cycles	[86]
39	Pd_3Pb/Pd NSs/C	0.1 M KOH	~0.9	0.574	1.31	8.1% loss in mass activity after 10 000 cycles	[96]
40	PdNi/C	0.1 M NaOH	0.87	0.37 @ 0.85 V	1.79 @0.85 V	14 mV decrease in $E_{1/2}$ after 16 000 cycles	[91]
41	PdCu@Pd/C	0.1 M KOH	0.901	~0.14	~1.0 @0.85V	5.8% loss in mass activity after 10 000 cycles	[94]
42	VW Au@Pd HNWs	0.1 M KOH	0.91	0.032	3.29	4.2% loss in current density after 12 000 s	[93]
43	Pd@PdFe icosahedra	0.1 M KOH	~0.88	0.31	1.56	15% loss in mass activity after 20 000 cycles	[98]
Support-modified Pd catalysts							
44	Pd/NRGO in-situ	0.1 M KOH	0.93	0.56	3.96	20% loss in mass activity 15 000 cycle (33 h)	[44]
45	Pd-Ru@NG	0.1 M KOH	0.8	–	–	20% loss after 2000 s	[85]
46	Pd-Co/N-doped porous C	0.1 M KOH	0.845	0.45	3.5	8% loss in current density after 20 000 s	[99]
47	PdNiCu/NG	0.1 M KOH	0.8	–	–	30 mV $E_{1/2}$ decrease after 5000 cycles	[122]
48	fct-PdFe@Pd@NG	0.1 M KOH	0.83	–	–	4% loss in current density after 10 000 s	[123]
49	Pd-HNC 10	0.1 M KOH	0.898	–	–	Negligible change in activity after 5000 cycles	[103]
50	PdO@WO_3S_x	0.1 M KOH	0.89	–	–	7.6% loss in current density after 36 000 s	[109]
51	Ni@Pd-70/CNT	0.1 M KOH	~0.88	0.23	0.42	25% loss in mass activity after 10 000 cycles	[117]
52	Pd/MnO_2–CNT	0.1 M KOH	~0.7	0.48	–	3% loss in current density after 18 000 s	[115]
53	EuPd/A	0.1 M KOH	0.88	0.35	–	~30% loss in mass activity after 5000 cycles	[124]
54	PdAu-GO	0.1 M KOH	0.9	0.53	1.57	12.9% loss in current density after 30 000 s	[125]

catalyst-support, migration of metal NPs on the support followed by their coalescence or detachment also causes loss in catalytic activity. Additionally, carbon corrosion (decomposition of carbon to CO_2) in the carbon-supported catalysts is another important factor for the degradation of catalyst performance over time. In order to alleviate such problems, the activity loss process is needed to be monitored using various state-of-the-art techniques such as identical location TEM, in-situ FT-IR, and in-situ XRD. Therefore, the stability of such catalysts still needs a great deal of attention. Future research should focus on the development of highly efficient and durable electrocatalysts for alkaline fuel cells in real operating conditions, assisted by a thorough understanding of their performance from fundamental studies.

References

1 Abas, N., Kalair, A., and Khan, N. (2015). Review of fossil fuels and future energy technologies. *Futures* 69: 31–49.

2 Wang, X., Li, Z., Qu, Y. et al. (2019). Review of metal catalysts for oxygen reduction reaction: from nanoscale engineering to atomic design. *Chem* 5: 1486–1511.

3 Daud, W.R.W., Rosli, R.E., Majlan, E.H. et al. (2017). PEM fuel cell system control: a review. *Renew. Energy* 113: 620–638.

4 Rosli, R.E., Sulong, A.B., Daud, W.R.W. et al. (2017). A review of high-temperature proton exchange membrane fuel cell (HT-PEMFC) system. *Int. J. Hydrogen Energy* 42: 9293–9314.

5 Gottesfeld, S., Dekel, D.R., Page, M. et al. (2018). Anion exchange membrane fuel cells: current status and remaining challenges. *J. Power Sources* 375: 170–184.

6 Sun, Z., Lin, B., and Yan, F. (2018). Anion-exchange membranes for alkaline fuel-cell applications: the effects of cations. *ChemSusChem* 11: 58–70.

7 Sui, S., Wang, X., Zhou, X. et al. (2017). A comprehensive review of Pt electrocatalysts for the oxygen reduction reaction: nanostructure, activity, mechanism and carbon support in PEM fuel cells. *J. Mater. Chem. A* 5: 1808–1825.

8 Chhetri, M., Rana, M., Loukya, B. et al. (2015). Mechanochemical synthesis of free-standing platinum nanosheets and their electrocatalytic properties. *Adv. Mater.* 27: 4430–4437.

9 Rana, M., Chhetri, M., Loukya, B. et al. (2015). High-yield synthesis of sub-10 nm Pt nanotetrahedra with bare ⟨111⟩ facets for efficient electrocatalytic applications. *ACS Appl. Mater. Interfaces* 7: 4998–5005.

10 Zhang, L., Chang, Q., Chen, H., and Shao, M. (2016). Recent advances in palladium-based electrocatalysts for fuel cell reactions and hydrogen evolution reaction. *Nano Energy* 29: 198–219.

11 Borup, R., Meyers, J., Pivovar, B. et al. (2007). Scientific aspects of polymer electrolyte fuel cell durability and degradation. *Chem. Rev.* 107: 3904–3951.

12 Demirci, U.B. (2007). Direct liquid-feed fuel cells: thermodynamic and environmental concerns. *J. Power Sources* 169: 239–246.

13 Li, X. and Faghri, A. (2013). Review and advances of direct methanol fuel cells (DMFCs) part I: design, fabrication, and testing with high concentration methanol solutions. *J. Power Sources* 226: 223–240.

14 McLean, G.F., Niet, T., Prince-Richard, S., and Djilali, N. (2002). An assessment of alkaline fuel cell technology. *Int. J. Hydrogen Energy* 27: 507–526.

15 Hren, M., Božič, M., Fakin, D. et al. (2021). Alkaline membrane fuel cells: anion exchange membranes and fuels. *Sustain. Energy Fuels* 5: 604–637.

16 Joghee, P., Malik, J.N., Pylypenko, S., and O'Hayre, R. (2015). A review on direct methanol fuel cells – in the perspective of energy and sustainability. *MRS Energy Sustain.* 2: E3.

17 Peng, X., Kulkarni, D., Huang, Y. et al. (2020). Using operando techniques to understand and design high performance and stable alkaline membrane fuel cells. *Nat. Commun.* 11: 3561.

18 Spendelow, J.S. and Wieckowski, A. (2007). Electrocatalysis of oxygen reduction and small alcohol oxidation in alkaline media. *Phys. Chem. Chem. Phys.* 9: 2654–2675.

19 Zhu, S., Hu, X., and Shao, M. (2017). Impacts of anions on the oxygen reduction reaction kinetics on platinum and palladium surfaces in alkaline solutions. *Phys. Chem. Chem. Phys.* 19: 7631–7641.

20 Ramaswamy, N. and Mukerjee, S. (2011). Influence of inner- and outer-sphere electron transfer mechanisms during electrocatalysis of oxygen reduction in alkaline media. *J. Phys. Chem. C* 115: 18015–18026.

21 Duong, H.T., Rigsby, M.A., Zhou, W.-P., and Wieckowski, A. (2007). Oxygen reduction catalysis of the Pt3Co alloy in alkaline and acidic media studied by X-ray photoelectron spectroscopy and electrochemical methods. *J. Phys. Chem. C* 111: 13460–13465.

22 Erikson, H., Sarapuu, A., Solla-Gullón, J., and Tammeveski, K. (2016). Recent progress in oxygen reduction electrocatalysis on Pd-based catalysts. *J. Electroanal. Chem.* 780: 327–336.

23 Jensen, K.D., Tymoczko, J., Rossmeisl, J. et al. (2018). Elucidation of the oxygen reduction volcano in alkaline media using a Copper–Platinum(111) alloy. *Angew. Chemie Int. Ed.* 57: 2800–2805.

24 Yan, W.-Y., Zheng, S.-L., Jin, W. et al. (2015). The influence of KOH concentration, oxygen partial pressure and temperature on the oxygen reduction reaction at Pt electrodes. *J. Electroanal. Chem.* 741: 100–108.

25 Jin, W., Du, H., Zheng, S. et al. (2010). Comparison of the oxygen reduction reaction between NaOH and KOH solutions on a Pt electrode: the electrolyte-dependent effect. *J. Phys. Chem. B* 114: 6542–6548.

26 Erikson, H., Sarapuu, A., Alexeyeva, N. et al. (2012). Electrochemical reduction of oxygen on palladium nanocubes in acid and alkaline solutions. *Electrochim. Acta* 59: 329–335.

27 Ge, X., Sumboja, A., Wuu, D. et al. (2015). Oxygen reduction in alkaline media: from mechanisms to recent advances of catalysts. *Acs Catal.* 5: 4643–4667.

28 Rana, M., Mondal, S., Sahoo, L. et al. (2018). Emerging materials in heterogeneous electrocatalysis involving oxygen for energy harvesting. *ACS Appl. Mater. Interfaces* 10: 33737–33767.

29 Awad, M.I. and Ohsaka, T. (2013). An electrocatalytic oxygen reduction by copper nanoparticles-modified Au(100)-rich polycrystalline gold electrode in 0.5 M KOH. *J. Power Sources* 226: 306–312.

30 Anastasijević, N.A., Vesović, V., and Adžić, R.R. (1987). Determination of the kinetic parameters of the oxygen reduction reaction using the rotating ring-disk electrode: part I. Theory. *J. Electroanal. Chem. interfacial Electrochem.* 229 (1–2): 305–316.

31 Mei, D., Da He, Z., Zheng, Y.L. et al. (2014). Mechanistic and kinetic implications on the ORR on a Au(100) electrode: pH, temperature and H–D kinetic isotope effects. *Phys. Chem. Chem. Phys.* 16 (27): 13762–13773.

32 Anastasijević, N.A., Vesović, V., and Adžić, R.R. (1987). Determination of the kinetic parameters of the oxygen reduction reaction using the rotating ring-disk electrode: part II applications. *J. Electroanal. Chem. Interfacial Electrochem.* 229 (1): 317–325.

33 Kim, J. and Gewirth, A.A. (2006). Mechanism of oxygen electroreduction on gold surfaces in basic media. *J. Phys. Chem. B* 110 (6): 2565–2571.

34 Hu, P., Song, Y., Chen, L., and Chen, S. (2015). Electrocatalytic activity of alkyne-functionalized AgAu alloy nanoparticles for oxygen reduction in alkaline media. *Nanoscale* 7 (21): 9627–9636.

35 W. Vielstich, A. Lamm and H. Gasteiger, 2003. Handbook of Fuel Cells. Fundamentals, Technology, Applications Wiley.

36 Divišek, J. and Kastening, B. (1975). Electrochemical generation and reactivity of the superoxide ion in aqueous solutions. *J. Electroanal. Chem. Interfacial Electrochem.* 65 (2): 603–621.

37 Shao, M., Liu, P., and Adzic, R.R. (2006). Superoxide anion is the intermediate in the oxygen reduction reaction on platinum electrodes. *J. Am. Chem. Soc.* 128 (23): 7408–7409.

38 Géniès, L., Bultel, Y., Faure, R., and Durand, R. (2003). Impedance study of the oxygen reduction reaction on platinum nanoparticles in alkaline media. *Electrochim. Acta* 48 (25–26): 3879–3890.

39 Wiggins-Camacho, J.D. and Stevenson, K.J. (2011). Mechanistic discussion of the oxygen reduction reaction at nitrogen-doped carbon nanotubes. *J. Phys. Chem. C* 115 (40): 20002–20010.

40 Duan, Z. and Henkelman, G. (2019). Theoretical resolution of the exceptional oxygen reduction activity of Au (100) in alkaline media. *ACS Catal.* 9 (6): 5567–5573.

41 Núñez, M., Lansford, J.L., and Vlachos, D.G. (2019). Optimization of the facet structure of transition-metal catalysts applied to the oxygen reduction reaction. *Nat. Chem.* 11 (5): 449–456.

42 Li, R., Li, H., Xu, S., and Liu, J. (2015). Theoretical investigation of the effect of OH− ions on O2 adsorption on low-index Pt surfaces in alkaline solution. *Appl. Surf. Sci.* 351: 853–861.

43 Lu, F., Zhang, Y., Liu, S. et al. (2017). Surface proton transfer promotes four-electron oxygen reduction on gold nanocrystal surfaces in alkaline solution. *J. Am. Chem. Soc.* 139 (21): 7310–7317.

44 Sahoo, L., Mondal, S., Gloskovskii, A. et al. (2021). Unravelling charge-transfer in Pd to pyrrolic-N bond for superior electrocatalytic performance. *J. Mater. Chem. A* 9 (17): 10966–10978.

45 Y. Garsany, O. A. Baturina, K. E. Swider-Lyons and S. S. Kocha, (2010). Experimental Methods for Quantifying the Activity of Platinum Electrocatalysts for the Oxygen Reduction Reaction. Anal. Chem. 2010, 82 (15): 6321–6328.

46 Zhang, H.-X., Wang, H., Re, Y.-S., and Cai, W.-B. (2012). Palladium nanocrystals bound by {110} or {100} facets: from one pot synthesis to electrochemistry. *Chem. Commun.* 48 (67): 8362–8364.

47 Shao, M., Chang, Q., Dodelet, J.-P., and Chenitz, R. (2016). Recent Advances in Electrocatalysts for Oxygen Reduction Reaction. *Chem. Rev.* 116 (6): 3594–3657.

48 Huang, K.-L., Liu, Z.-T., and Lee, C.-L. (2015). Truncated palladium nanocubes: synthesis and the effect of OH− concentration on their catalysis of the alkaline oxygen reduction reaction. *Electrochim. Acta* 157: 78–87.

49 Lüsi, M., Erikson, H., Sarapuu, A. et al. (2016). Oxygen reduction reaction on carbon-supported palladium nanocubes in alkaline media. *Electrochem. Commun.* 64: 9–13.

50 Nesselberger, M., Ashton, S., Meier, J.C. et al. (2011). The particle size effect on the oxygen reduction reaction activity of pt catalysts: influence of electrolyte and relation to single crystal models. *J. Am. Chem. Soc.* 133 (43): 17428–17433.

51 Sahoo, L., Dhindsa, P.K., Nihal, C.P., and Gautam, U.K. (2021). 'Pre-optimization' of the solvent of nanoparticle-synthesis for superior catalytic efficiency: a case study with Pd nanocrystals. *Nanoscale Adv.*, 3 (8): 2366–2376

52 Zhang, H., Qiu, X., Chen, Y. et al. (2020). Shape control of monodispersed sub-5 nm Pd tetrahedrons and laciniate Pd nanourchins by maneuvering the dispersed state of additives for boosting ORR performance. *Small* 16 (6): 1906026.

53 Sahoo, L. and Gautam, U.K. (2020). Boosting bifunctional oxygen reduction and methanol oxidation electrocatalytic activity with 2D superlattice-forming Pd nanocubes generated by precise acid etching. *ACS Appl. Nano Mater.* 3 (8): 8117–8125.

54 Zhang, Y.-L., Sui, X.-L., Zhao, L. et al. (2019). Controlling the surface roughness of chain-like Pd nanowires by pH values as excellent catalysts for oxygen reduction reaction. *Int. J. Hydrogen Energy* 44 (13): 6551–6559.

55 Huang, H., Ruditskiy, A., Choi, S.-I. et al. (2017). One-pot synthesis of penta-twinned palladium nanowires and their enhanced electrocatalytic properties. *ACS Appl. Mater. Interfaces* 9 (36): 31203–31212.

56 Begum, H., Ahmed, M.S., Cho, S., and Jeon, S. (2018). Freestanding palladium nanonetworks electrocatalyst for oxygen reduction reaction in fuel cells. *Int. J. Hydrogen Energy* 43 (1): 229–238.

57 Fan, C., Huang, Z., Wang, C. et al. (2019). Highly-branched palladium nanodandelions: simple, fast, and green fabrication with superior oxygen reduction property. *Chem. Eur. J.* 25 (19): 4920–4926.

58 Jiang, J., Gao, H., Lu, S. et al. (2017). Ni–Pd core–shell nanoparticles with Pt-like oxygen reduction electrocatalytic performance in both acidic and alkaline electrolytes. *J. Mater. Chem. A* 5 (19): 9233–9240.

59 Li, C., Yuan, Q., Ni, B. et al. (2018). Dendritic defect-rich palladium–copper–cobalt nanoalloys as robust multifunctional non-platinum electrocatalysts for fuel cells. *Nat. Commun.* 9 (1): 1–9.

60 Dai, S., Chou, J.P., Wang, K.W. et al. (2019). Platinum-trimer decorated cobalt-palladium core-shell nanocatalyst with promising performance for oxygen reduction reaction. *Nat. Commun.* 10 (1): 1–10.

61 Gamler, J.T.L., Shin, K., Ashberry, H.M. et al. (2020). Intermetallic Pd3Pb nanocubes with high selectivity for the 4-electron oxygen reduction reaction pathway. *Nanoscale* 12 (4): 2532–2541.

62 Shao, M., Liu, P., Zhang, J., and Adzic, R. (2007). Origin of enhanced activity in palladium alloy electrocatalysts for oxygen reduction reaction. *J. Phys. Chem. B* 111 (24): 6772–6775.

63 Stamenkovic, V., Mun, B.S., Mayrhofer, K.J.J. et al. (2006). Changing the activity of electrocatalysts for oxygen reduction by tuning the surface electronic structure. *Angew. Chemie Int. Ed.* 45 (18): 2897–2901.

64 Zhang, Y., Huang, B., Shao, Q. et al. (2019). Defect engineering of palladium-tin nanowires enables efficient electrocatalysts for fuel cell reactions. *Nano Lett.* 19 (10): 6894–6903.

65 Yang, Y., Dai, C., Wu, D. et al. (2018). The size effect of PdCu bimetallic nanoparticles on oxygen reduction reaction activity. *ChemElectroChem* 5 (18): 2571–2576.

66 Betancourt, L.E., Rojas-Pérez, A., Orozco, I. et al. (2020). Enhancing ORR performance of bimetallic PdAg electrocatalysts by designing interactions between Pd and Ag. *ACS Appl. Energy Mater.* 3 (3): 2342–2349.

67 Jiao, W., Chen, C., You, W. et al. (2020). Tuning strain effect and surface composition in PdAu hollow nanospheres as highly efficient ORR electrocatalysts and SERS substrates. *Appl. Catal. B Environ.* 262: 118298.

68 Wang, J., Li, M., Zhang, J. et al. (2020). Atom-ratio-conducted tailoring of PdAu bimetallic nanocrystals with distinctive shapes and dimensions for boosting the ORR performance. *Chem. – A Eur. J.* 26 (20): 4480–4488.

69 Jiang, X., Xiong, Y., Wang, Y. et al. (2019). Treelike two-level Pd x Ag y nanocrystals tailored for bifunctional fuel cell electrocatalysis. *J. Mater. Chem. A* 7 (10): 5248–5257.

70 Iqbal, M., Kim, Y., Li, C. et al. (2019). Tailored design of mesoporous PdCu nanospheres with different compositions using polymeric micelles. *ACS Appl. Mater. Interfaces* 11 (40): 36544–36552.

71 Chen, J., Li, Y., Lu, N. et al. (2018). Nanoporous PdCe bimetallic nanocubes with high catalytic activity towards ethanol electro-oxidation and the oxygen reduction reaction in alkaline media. *J. Mater. Chem. A* 6 (46): 23560–23568.

72 Liu, X.-J., Yin, X., Sun, Y.-D. et al. (2020). Interlaced Pd–Ag nanowires rich in grain boundary defects for boosting oxygen reduction electrocatalysis. *Nanoscale* 12 (9): 5368–5373.

73 Jiao, W., Chen, C., You, W. et al. (2020). Hollow nanochains: hollow palladium-gold nanochains with periodic concave structures as superior ORR electrocatalysts and highly efficient SERS substrates. *Adv. Energy Mater.* 10 (18): 2070082.

74 Luo, M., Zhao, Z., Zhang, Y. et al. (2019). PdMo bimetallene for oxygen reduction catalysis. *Nature* 574 (7776): 81–85.

75 Bu, L., Shao, Q., Pi, Y. et al. (2018). Coupled spd exchange in facet-controlled Pd3Pb tripods enhances oxygen reduction catalysis. *Chem* 4 (2): 359–371.

76 Jiang, K., Wang, P., Guo, S. et al. (2016). Ordered PdCu-based nanoparticles as bifunctional oxygen-reduction and ethanol-oxidation electrocatalysts. *Angew. Chemie Int. Ed.* 55 (31): 9030–9035.

77 Yang, Y., Xiao, W., Feng, X. et al. (2019). Golden palladium zinc ordered intermetallics as oxygen reduction electrocatalysts. *ACS Nano* 13 (5): 5968–5974.

78 Wang, Y. and Hall, A.S. (2020). Pulsed electrodeposition of metastable Pd31Bi12 nanoparticles for oxygen reduction electrocatalysis. *ACS Energy Lett.* 5 (1): 17–22.

79 Shi, Q., Zhu, C., Bi, C. et al. (2017). Intermetallic Pd 3 Pb nanowire networks boost ethanol oxidation and oxygen reduction reactions with significantly improved methanol tolerance. *J. Mater. Chem. A* 5 (45): 23952–23959.

80 Luo, S., Xu, D., Li, J. et al. (2020). Au-Doped intermetallic Pd 3 Pb wavy nanowires as highly efficient electrocatalysts toward the oxygen reduction reaction. *CrystEngComm* 22 (39): 6478–6484.

81 Li, J., Chen, J., Wang, Q. et al. (2017). Controllable increase of boron content in B-Pd interstitial nanoalloy to boost the oxygen reduction activity of palladium. *Chem. Mater.* 29 (23): 10060–10067.

82 Du, C., Li, P., Yang, F. et al. (2018). Monodisperse palladium sulfide as efficient electrocatalyst for oxygen reduction reaction. *ACS Appl. Mater. Interfaces* 10 (1): 753–761.

83 Lv, H., Xu, D., Sun, L. et al. (2019). Ternary palladium–boron–phosphorus alloy mesoporous nanospheres for highly efficient electrocatalysis. *ACS Nano* 13 (10): 12052–12061.

84 Vo Doan, T.T., Wang, J., Poon, K.C. et al. (2016). Theoretical modelling and facile synthesis of a highly active boron-doped palladium catalyst for the oxygen reduction reaction. *Angew. Chemie Int. Ed.* 55 (24): 6842–6847.

85 Barman, B.K., Sarkar, B., and Nanda, K.K. (2019). Pd-coated Ru nanocrystals supported on N-doped graphene as HER and ORR electrocatalysts. *Chem. Commun.* 55 (92): 13928–13931.

86 Wu, C., Li, H., He, H. et al. (2019). Compressive strain in core-shell Au-Pd nanoparticles introduced by lateral confinement of deformation twinnings to enhance the oxidation reduction reaction performance. *ACS Appl. Mater. Interfaces* 11 (50): 46902–46911.

87 Wang, G., Guan, J., Xiao, L. et al. (2016). Pd skin on AuCu intermetallic nanoparticles: a highly active electrocatalyst for oxygen reduction reaction in alkaline media. *Nano Energy* 29: 268–274.

88 Jiao, W., Chen, C., You, W. et al. (2019). Yolk–shell Fe/Fe4N@ Pd/C magnetic nanocomposite as an efficient recyclable ORR electrocatalyst and SERS substrate. *Small* 15 (7): 1805032.

89 Huang, X., Shumski, A.J., Zhang, X., and Li, C.W. (2018). Systematic control of redox properties and oxygen reduction reactivity through colloidal ligand-exchange deposition of Pd on Au. *J. Am. Chem. Soc.* 140 (28): 8918–8923.

90 Gunji, T., Wakabayashi, R.H., Noh, S.H. et al. (2018). The effect of alloying of transition metals (M = Fe, Co, Ni) with palladium catalysts on the electrocatalytic activity for the oxygen reduction reaction in alkaline media. *Electrochim. Acta* 283: 1045–1052.

91 Lu, X., Ahmadi, M., DiSalvo, F.J., and Abruña, H.D. (2020). Enhancing the electrocatalytic activity of Pd/M (M= Ni, Mn) nanoparticles for the oxygen reduction reaction in alkaline media through electrochemical dealloying. *ACS Catal.* 10 (10): 5891–5898.

92 Xue, Q., Bai, J., Han, C. et al. (2018). Au nanowires@ Pd-polyethylenimine nanohybrids as highly active and methanol-tolerant electrocatalysts toward oxygen reduction reaction in alkaline media. *ACS Catal.* 8 (12): 11287–11295.

93 Wang, J., Qiu, X., Su, K. et al. (2020). Breaking the lattice match of Pd on Au (111) nanowires: manipulating the island and epitaxial growth pathways to boost the oxygen reduction reactivity. *J. Mater. Chem. A* 8 (37): 19300–19308.

94 Huang, S., Lu, S., Hu, H. et al. (2020). A stable PdCu@Pd core-shell nanobranches with enhanced activity and methanol-tolerant for oxygen reduction reaction. *Electrochim. Acta* 354: 136680.

95 Feng, Y., Yang, C., Fang, W. et al. (2019). Anti-poisoned oxygen reduction by the interface modulated Pd@NiO core@shell. *Nano Energy* 58: 234–243.

96 Tang, C., Zhang, N., Ji, Y. et al. (2019). Fully tensile strained Pd3Pb/Pd tetragonal nanosheets enhance oxygen reduction catalysis. *Nano Lett.* 19 (2): 1336–1342.

97 Feng, Y., Shao, Q., Ji, Y. et al. (2018). Surface-modulated palladium-nickel icosahedra as high-performance non-platinum oxygen reduction electrocatalysts. *Sci. Adv.* 4 (7): eaap8817.

98 Li, X., Li, X., Liu, C. et al. (2020). Atomic-level construction of tensile-strained PdFe alloy surface toward highly efficient oxygen reduction electrocatalysis. *Nano Lett.* 20: 1403–1409.

99 Zhang, Z., Liu, S., Tian, X. et al. (2017). Facile synthesis of N-doped porous carbon encapsulated bimetallic PdCo as a highly active and durable electrocatalyst for oxygen reduction and ethanol oxidation. *J. Mater. Chem. A* 5 (22): 10876–10884.

100 Zhang, X., Zhu, J., Tiwary, C.S. et al. (2016). Palladium nanoparticles supported on nitrogen and sulfur dual-doped graphene as highly active electrocatalysts for formic acid and methanol oxidation. *ACS Appl. Mater. Interfaces* 8 (17): 10858–10865.

101 Perini, L., Durante, C., Favaro, M. et al. (2015). Metal–support interaction in platinum and palladium nanoparticles loaded on nitrogen-doped mesoporous carbon for oxygen reduction reaction. *ACS Appl. Mater. Interfaces* 7 (2): 1170–1179.

102 Razmjooei, F., Singh, K.P., Song, M.Y., and Yu, J.-S. (2014). Enhanced electrocatalytic activity due to additional phosphorous doping in nitrogen and sulfur-doped graphene: a comprehensive study. *Carbon N. Y.* 78: 257–267.

103 Kabir, S., Serov, A., Artyushkova, K., and Atanassov, P. (2017). Nitrogen-doped three-dimensional graphene-supported palladium nanocomposites: high-performance cathode catalysts for oxygen reduction reactions. *ACS Catal.* 7: 6609–6618.

104 Wang, H., Maiyalagan, T., and Wang, X. (2012). Review on recent progress in nitrogen-doped graphene: synthesis, characterization, and its potential applications. *Acs Catal.* 2: 781–794.

105 Jukk, K., Kongi, N., Matisen, L. et al. (2014). Electroreduction of oxygen on palladium nanoparticles supported on nitrogen-doped graphene nanosheets. *Electrochim. Acta* 137: 206–212.

106 Yang, H., Tang, Z., Wang, K. et al. (2018). Co@Pd core-shell nanoparticles embedded in nitrogen-doped porous carbon as dual functional electrocatalysts for both oxygen reduction and hydrogen evolution reactions. *J. Colloid Interface Sci.* 528: 18–26.

107 Liu, X., Li, L., Meng, C., and Han, Y. (2012). Palladium nanoparticles/defective graphene composites as oxygen reduction electrocatalysts: a first-principles study. *J. Phys. Chem. C* 116: 2710–2719.

108 Tian, Y., Liu, Y.J., Zhao, J.X., and Ding, Y.H. (2015). High stability and superior catalytic reactivity of nitrogen-doped graphene supporting Pt nanoparticles as a catalyst for the oxygen reduction reaction: a density functional theory study. *RSC Adv.* 5: 34070–34077.

109 Lee, J.-H., Yim, D., Park, J.H. et al. (2020). Tuning d-band centers by coupling PdO nanoclusters to WO 3 nanosheets to promote the oxygen reduction reaction. *J. Mater. Chem. A* 8 (27): 13490–13500.

110 Zhang, C., Yu, S., Xie, Y. et al. (2019). Suppressing the Pd-C interaction through B-doping for highly efficient oxygen reduction. *Carbon N. Y.* 149: 370–379.

111 Tamakloe, W., Agyeman, D.A., Park, M. et al. (2019). Polydopamine-induced surface functionalization of carbon nanofibers for Pd deposition enabling enhanced catalytic activity for the oxygen reduction and evolution reactions. *J. Mater. Chem. A* 7 (13): 7396–7405.

112 Jiang, X., Wang, J., Huang, T. et al. (2019). Sub-5 nm palladium nanoparticles in situ embedded in N-doped carbon nanoframes: facile synthesis, excellent sinter resistance and electrocatalytic properties. *J. Mater. Chem. A* 7 (46): 26243–26249.

113 Bao, Z., Zhou, H., Song, X. et al. (2019). Enhanced oxygen reduction activity on carbon supported pd nanoparticles via SiO 2. *ChemCatChem* 11: 1278–1285.

114 Li, J., Zhou, H., Zhuo, H. et al. (2018). Oxygen vacancies on TiO2 promoted the activity and stability of supported Pd nanoparticles for the oxygen reduction reaction. *J. Mater. Chem. A* 6: 2264–2272.

115 Xiang, W., Zhao, Y., Jiang, Z. et al. (2018). Palladium single atoms supported by interwoven carbon nanotube and manganese oxide nanowire networks for enhanced electrocatalysis. *J. Mater. Chem. A* 6 (46): 23366–23377.

116 Hyun, S., Saejio, A., and Shanmugam, S. (2020). Pd nanoparticles deposited on Co (OH) 2 nanoplatelets as a bifunctional electrocatalyst and their application in Zn–air and Li–O 2 batteries. *Nanoscale* 12 (34): 17858–17869.

117 Bhalothia, D., Chen, P.-C., Yan, C. et al. (2019). Heterogeneous NiO2-to-Pd epitaxial structure performs outstanding oxygen reduction reaction activity. *J. Phys. Chem. C* 124 (4): 2295–2306.

118 Liu, L. and Corma, A. (2018). Metal catalysts for heterogeneous catalysis: from single atoms to nanoclusters and nanoparticles. *Chem. Rev.* 118 (10): 4981–5079.

119 Liu, Q., Peng, Y., Li, Q. et al. (2020). Atomic dispersion and surface enrichment of palladium in nitrogen-doped porous carbon cages lead to high-performance electrocatalytic reduction of oxygen. *ACS Appl. Mater. Interfaces* 12 (15): 17641–17650.

120 Liu, S., Zhang, H., Mu, X., and Chen, C. (2019). Surface reconstruction engineering of twinned Pd2CoAg nanocrystals by atomic vacancy inducement for hydrogen evolution and oxygen reduction reactions. *Appl. Catal. B Environ.* 241: 424–429.

121 Yang, Y., Chen, G., Zeng, R. et al. (2020). Combinatorial studies of palladium-based oxygen reduction electrocatalysts for alkaline fuel cells. *J. Am. Chem. Soc.* 142 (8): 3980–3988.

122 Sun, L., Liao, B., Ren, X. et al. (2017). *Electrochim. Acta* 235: 543–552.

123 Maiti, K., Balamurugan, J., Peera, S.G. et al. (2018). Highly active and durable core–shell fct-PdFe@ Pd nanoparticles encapsulated NG as an efficient catalyst for oxygen reduction reaction. *ACS Appl. Mater. Interfaces* 10 (22): 18734–18745.

124 Sarma, S.C., Kaja, S.M., Mishra, V. et al. (2020). Tweaking palladium electronic structure to attain oxygen reduction activity superior to platinum/C. *ACS Appl. Energy Mater.* 3 (7): 6127–6132.

125 Zhu, W., Yuan, H., Liao, F. et al. (2020). Strain engineering for Janus palladium-gold bimetallic nanoparticles: enhanced electrocatalytic performance for oxygen reduction reaction and zinc-air battery. *Chem. Eng. J.* 389: 124240.

126 Calle-Vallejo, F., Tymoczko, J., Colic, V. et al. (2015). Finding optimal surface sites on heterogeneous catalysts by counting nearest neighbors. *Science (80-.).* 350 (6257): 185–189.

127 U. S. D. of Energy (2019). *DOE Technical Targets for Polymer Electrolyte Membrane Fuel Cell Components.*

8

Morphology- and Size-Selective Pd-Based Electrocatalyst for Fuel Cell Reactions

*Rashmi Chetry, Chiranjita Goswami, Biraj Jyoti Borah, and Pankaj Bharali**

Department of Chemical Sciences, Tezpur University, Nappam, Assam, India

8.1 General Introduction

The fast-growing population and global energy demands have caused the sharp depletion of natural energy sources. Burning nonrenewable energy sources, such as coal, natural gas, and fossil fuels, comprises a major portion of the formation of the total energy, whereas burning renewable energy sources such as solar, wind, water, and hydrothermal comprises minor portions [1]. The energy sources are depleting progressively, and the energy crisis will be inevitable if appropriate steps are not taken. Being alternative to the nonrenewable energy sources, the renewable sources mentioned here are earth-abundant and climate-friendly, but their erratic nature restrains their wide application. To overcome these problems, researchers are switching to highly efficient, high energy density, high power, more sustainable, and eco-friendly alternatives [2]. Thanks to advanced clean energy technologies that have risen in recent decades, including fuel cells (FCs), metal-air batteries, water electrolysis, and CO_2 valorizations. These aforementioned technologies contribute a lot to forming clean and green energy infrastructures with zero emissions of pollutants.

8.1.1 Brief Overview of Various Fuel Cells and Associated Reactions

FCs are electrochemical devices that convert chemical energy into electrical energy without fuel combustion [2–4]. They are open systems; cathode and anode act as the charge transfer media and entail a continuous supply of fuels outside the cell [5]. FCs are categorized mainly on the types of electrolytes employed. Based on this, they are classified into the following types: (i) phosphoric acid FCs (PAFCs) with an acidic electrolyte solution (e.g. phosphoric acid), (ii) alkaline FCs (AFCs) with alkaline electrolyte solution (e.g. potassium hydroxide), (iii) molten carbonate FCs (MCFCs) with molten carbonate salt as the electrolyte, (iv) proton exchange membrane FCs (PEMFCs), also known as polymer electrolyte membrane FCs with proton exchange membrane as the electrolyte, (v) direct

*Corresponding author

Heterogeneous Nanocatalysis for Energy and Environmental Sustainability: Energy Applications,
First Edition. Edited by Putla Sudarsanam, Yusuke Yamauchi, and Pankaj Bharali.

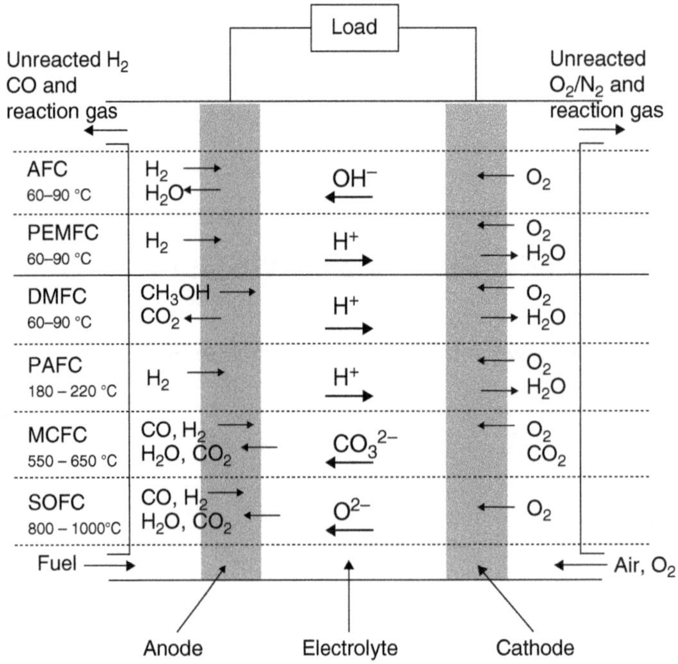

Figure 8.1 Summary of the reactions and processes that occur in the various fuel cell systems. *Source:* From [5] / with permission of American Chemical Society.

methanol FCs (DMFCs) with methanol as fuel, and (vi) solid oxide FCs (SOFCs) with ceramic ion-conducting electrolyte in solid oxide form [4]. Figure 8.1 represents the operation of different types of FCs [5]. Among various FCs, PEMFCs gained increasing demand due to their moderate temperature maintenance, high power density, good thermal/chemical stability, quick start-up, simple structure, and high energy conversion efficiency [4, 6, 7].

In PEMFC, at anode fuel (i.e. hydrogen H_2), an oxidation reaction occurs and hydrogen ion (H^+) is formed with the liberation of electrons whereas, at cathode oxygen (O_2), reduction occurs with the acceptance of those released electrons, and thus converts the chemically stored H_2 into electrical energy [8]. In such a device, the fuel can be H_2, C_2H_5OH, CH_3OH, or HCOOH. The primary reaction that occurs in the PEMFC is the oxygen reduction reaction (ORR). ORR is a multielectron reaction process and occurs in the cathode of the FCs in which O_2 leads to the formation of H_2O or OH^- based on the electrolyte employed [4].

Oxygen evolution reaction (OER) is the most important reaction that produces O_2 either in the unitized regenerative FCs (URFCs) or water electrolyzers. It is also the reversible process of ORR [5]. URFCs are compact electrochemical cells in which both FC and water electrolyzers are unitized into a single electrochemical system. Similarly, the hydrogen evolution reaction (HER) is another parallel reaction of the electrocatalytic H_2O-splitting process. It leads to the economic production of pure H_2 with zero greenhouse emissions [9–11]. H_2 is considered a promising fuel in PEMFC because of its ample natural richness and high energy density [12]. Thus, it contributes a major role in the production of clean energy and a green environment.

Direct fuel cells (DFCs) have also sparked FC technology. Different types of direct FCs considerably depend on the fuels employed (such as CH_3OH, C_2H_5OH, HCHO, HCOOH, etc.). Among all, direct formic acid FC (DFAFC), where liquid HCOOH is used as a fuel, is considered to be the most promising development because they are harmless compared to CH_3OH fuel. DFAFC possesses several advantages like fast oxidation kinetics, high power density, mild fuel crossover, ready availability, and high theoretical cell potential. The fast oxidation kinetics properties of DFAFC make them a highly efficient system. Because of all these advantages, DFAFC has wide application in portable electronic devices, electrical vehicles, etc. [13, 14]. Though FC technologies have great efficiency as an energy conversion device, they have some major limitations which restrain them from being commercialized. All these FCs need an electrocatalyst (EC) to acquire high voltage output, to reduce their electrochemical overpotential, to improve slow kinetics (in case of ORR and OER), and to reduce CO poisoning (in case of DFAFC). Platinum (in the form of Pt/C) or Pt-based materials have emanated as one of the most efficient ECs in all of the FC reactions mentioned here, but low durability, scarcity, low stability in acidic medium, and low OER efficiency of Pt have limited its commercialization on a larger scale [15–17]. Therefore, the development of alternative ECs based on earth-abundant metals or by lowering the amount of Pt content in ECs has attracted significant interest [18–20].

8.1.2 Current Development in Shape- and Size-Dependent Pd Electrocatalysts

A wide range of earth-abundant ECs has been growing up for either ORR, OER, HER, or formic acid oxidation reaction (FAOR). Pd being comparatively abundant has attracted attention due to its high electrocatalytic activity as high as Pt for all FC reactions [21, 22]. Pd-based materials by controlling size, shape, or morphology show significant improvement in their electrocatalytic activity. Different types of 1-dimensional (1-D) nanostructures (NSs) (e.g. nanotubes (NTs), nanorods (NRs), nanowires (NWs), and nanowire networks (NNWs)) are also synthesized over the past decades by various synthesis methods, for instance, selective surface passivation, templating or self-assembly methods. These synthesis methods of Pd nanoparticles (NPs) begin with nucleation, i.e. the development of small seeds or clusters with distinct crystallinity and stable structure. After nucleation, some of the single crystal seeds grow into different shapes such as octahedra, cubes, and cuboctahedra, depending on the passivation of specific crystalline facets. Single crystal seeds lead to the formation of multiple twinned structures to lower the surface energy. Geometrical structures such as decahedral or icosahedral can be adopted from fivefold twinned structures and may then lead to the formation of fivefold twinned palladium NRs and NWs. The existence of high index facets and twin defects in this 1-D-type structure shores up the activity toward electrochemical reactions [23].

Begum et al. synthesized Pd nanonetwork and studied their activity toward ORR [24]. They found that the synthesized EC provides higher current density as well as high electrochemically active surface area (ECSA). Similarly, 2-dimensional (2-D) (e.g. nanosheets) and 3-dimensional (3-D) (e.g. nano-buds or nano-dendrites) ECs were also fabricated in recent years to improve the electrocatalytic activity. He et al. and Yang et al. reported Pd nanosheets and nanodendrites, respectively, and investigated their ORR activity [25, 26]. These reported ECs showed higher electrocatalytic activity and longer durability. The classification of different Pd NSs is shown in Figure 8.2.

Owing to the high cost of Pd, it is alloyed with other transition metals to form bimetallics or trimetallics. It is observed that these stoichiometric compositions showed enhancement in certain properties relative to pure Pd. To date, several works have been done on Pd bimetallic and trimetallic compositions and desired results were obtained. Zhang et al. [27] reported that trimetallic PdCuW dramatically enhances electrocatalytic activity as well as current density toward FAO. We reported bimetallic carbon-supported PdNi ECs and studied their activity toward ORR and FAO. Electrochemical measurements revealed that these ECs exhibited high current density and longer durability [28].

Several studies have shown that maintaining different experimental conditions, low index facet, and high index facet ECs could be achieved, and these ECs can boost up the specific activity (SA) and the current density [22, 27]. Myriads of synthetic approaches have been performed to synthesize the desired facets and crystals of Pd with diverse morphologies. Recent studies have proved that the thermodynamic and kinetic controlled methods lead to forming the most coveted structure of ECs. Figure 8.3 represents the schematic illustrations of different experimental conditions that lead to the formation of Pd with disparate morphologies. From this scheme, it is observed that the morphology of the products is mostly dependent on the crystallinity of Pd seeds. Shapes such as the cube, octahedron, or cuboctahedron develop gradually from single-crystal seeds and are dependent on the growth ratio rates along (100) and (111) directions. For instance, truncated octahedral and multiple twinned icosahedral formed by polyol synthesis methods. Similarly, nanosheets and nanodendrite structures can be formed through microemulsion and hydrothermal methods, respectively [23, 29, 30].

8.2 Oxygen Reduction Reaction

ORR is a multielectron reaction that occurs on the cathode surface of FCs. It proceeds through many elementary steps via the formation of different intermediates depending on the electrolyte medium used. In the alkaline medium, O_2 reduction proceeds via two different pathways.

i) Partial $2e^-$ pathway: In this pathway, O_2 forms HO_2^- as intermediate and then OH^- as shown here:

$$O_2 + H_2O + 2e^- \rightarrow HO_2^- + OH^-$$

$$HO_2^- + H_2O + 2e^- \rightarrow 3OH^-$$

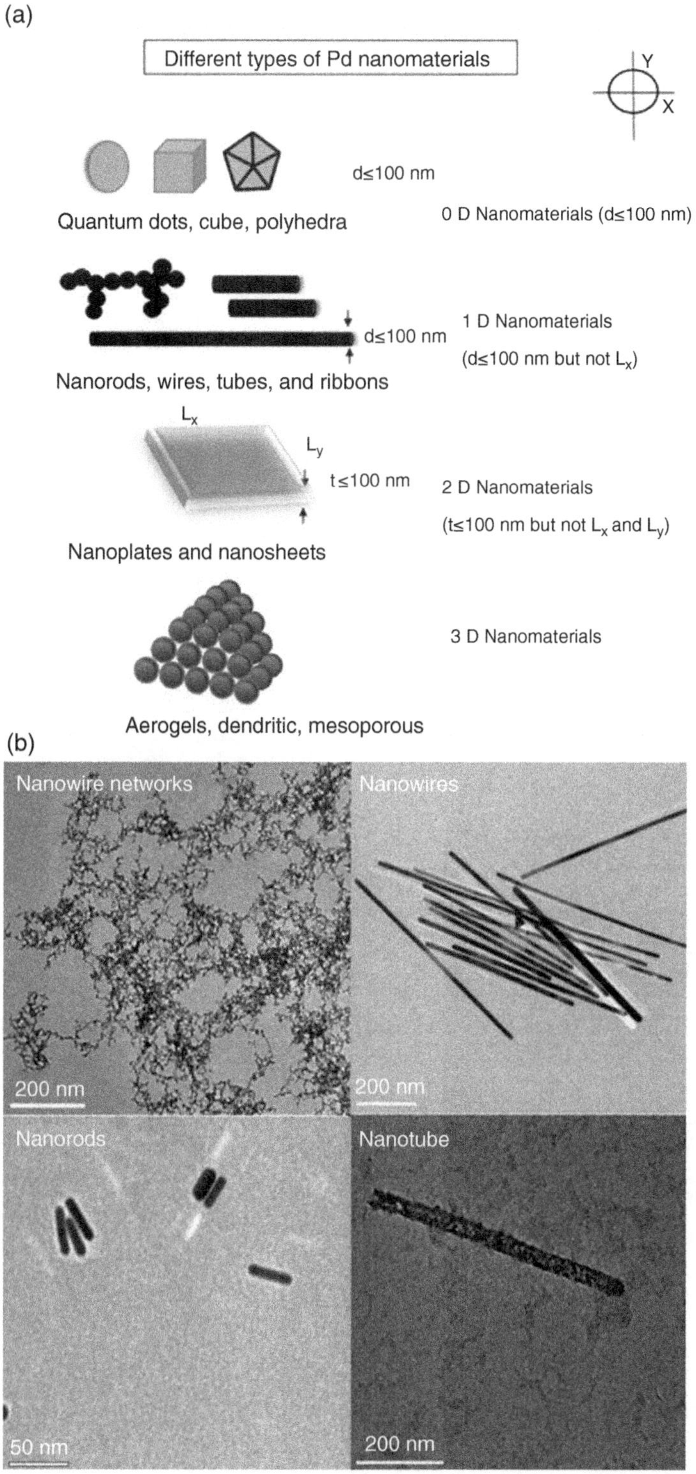

Figure 8.2 (a) Classification of Pd nanostructures, (b) representative TEM images of different types of Pd nanostructures. *Source:* From [23] / with permission of Royal Society of Chemistry.

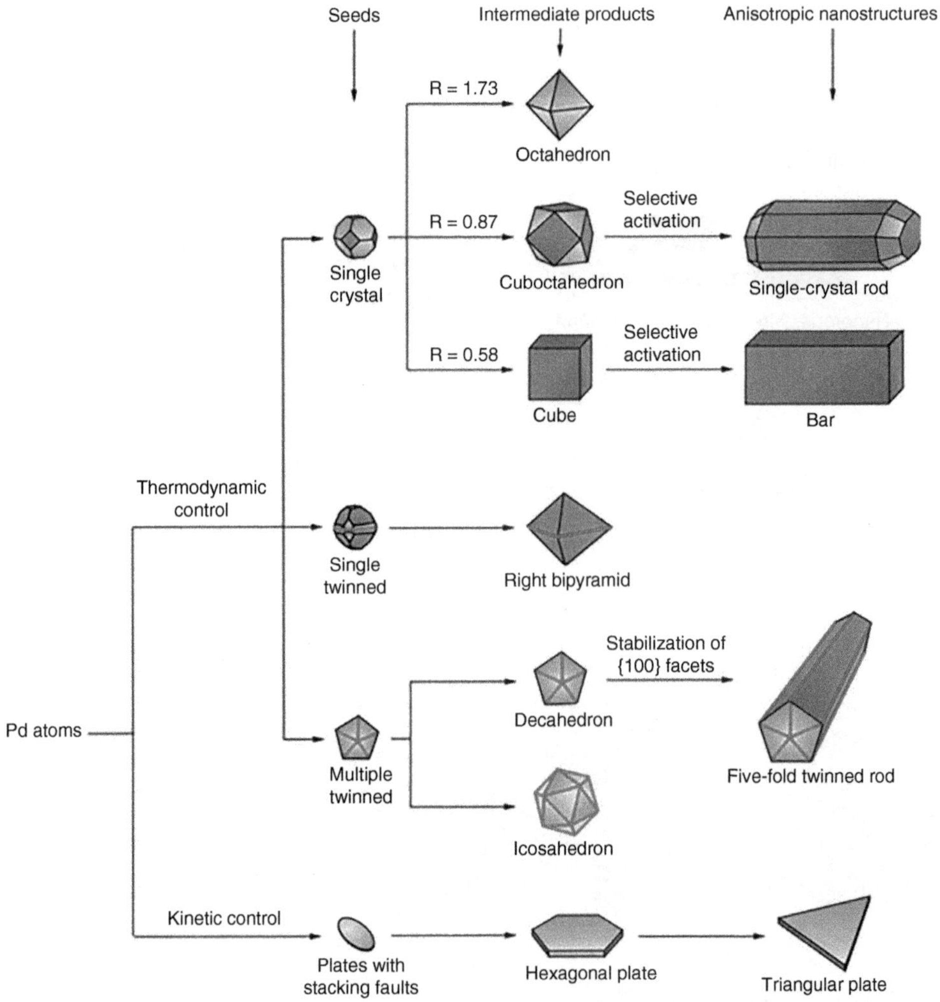

Figure 8.3 Schematic illustration of the reaction pathways that lead to Pd nanostructures with different morphologies showing {100}, {111}, and {110} facets, and twin planes. The parameter R is defined as the ratio between the growth rates along the ⟨100⟩ and ⟨111⟩ axes. *Source:* From [23] / with permission of Royal Society of Chemistry.

ii) Direct $4e^-$ pathway: Here O_2 is converted into OH^-:

$$O_2 + 2H_2O + 4e^- \rightarrow 4\ OH^-$$

Similarly, in an acidic medium, it follows two pathways:

i) Direct $4e^-$ electron pathway: Here O_2 is converted into H_2O

$$O_2 + 4H^+ + 4e^- \rightarrow 2H_2O$$

ii) Partial $2e^-$ pathway: In this pathway, O_2 forms H_2O_2 as intermediate and then H_2O as shown here:

$$O_2 + 2H^+ + 2e^- \rightarrow H_2O_2$$
$$H_2O_2 + 2H^+ + 2e^- \rightarrow 2H_2O$$

The most restraining aspect in the energy conversion efficiency of FCs is the sluggish kinetics of ORR at the cathode. To overcome this difficulty, Pt-based ECs have been used broadly for the desired ORR [31]. Though Pt shows a high catalytic property, its high cost and scarcity limit its demand in FC's application [32, 33]. Furthermore, the kinetics of ORR becomes slow while Pt is used as the cathode EC. Pt also possesses low alcohol tolerance in DAFC, which is a significant obstruction in the practical applications of low-temperature FCs [34]. For that reason, it is vital to develop non-Pt-based ORR ECs with budget-friendly, higher efficiency, and superior alcohol tolerance ability. Consequently, Pd and Pd-based ECs have gained immense interest in the FC industry [35]. The use of Pd and Pd-based ECs alternative for ORR is due to its low cost, earth-abundant, and analogous properties with Pt, including the close lattice parameters and identical crystalline structure [36]. However, the intrinsic activity of Pd for ORR is still lesser than that of Pt [37]. Recent research suggests that Pd-based hybrids with different sizes and shapes are promising due to their exceptional functional properties like large surface area, low cost, and high conductivity. Several reports have been made on the synthesis of ECs with definite size and shape such as nanothorn [38], 3-D nanosheet [39], hollow echinus [40], concave decahedra [41], nanobranches [42], branched alloy networks [43], hollow sphere [44], octahedron [45, 46], nanosphere [47], and fivefold twinned [48], which are very effectively used for the enhancement of electrocatalytic activity.

Lian et al. synthesized hierarchical Pd_4Fe flowerlike nanomaterials prepared by lowering palladium acetylacetonate with oleylamine and $Fe(CO)_5$. The specific hierarchical structure of the synthesized EC showed superior activity toward ORR in both the acidic and alkaline media. The nanoflowers (NFs) exhibited SA of $1.57\,mA/cm^2$ in both acidic and alkaline media. The EC also exhibited the highest mass activity (MA) of $4.07\ A/mg_{Pd}$ at 0.85 V vs. reversible hydrogen electrode (RHE) [49]. Different images of the synthesized Pd_4Fe are presented in Figure 8.4. The author reported that flowerlike structures were assembled from a number of uniform 2D nanosheets with a thickness of around 2–3 nm. From EDS-mapping images, it was found that Pd and Fe were homogeneously dispersed in the entire Pd_4Fe NFs (Figure 8.4d–g). The HRTEM image (Figure 8.4 h) gives detailed information about the Pd_4Fe NFs. The interfringe distance of the nanosheets in Pd_4Fe NFs was about 0.23 nm, which was relatively smaller than that of 1/3(422) fringes of face-centered cubic (fcc) Pd (0.24 nm) due to the formation of alloy NS with Fe. The appearance of 1/3(422) reflections recommended that the nanosheets in the hierarchical Pd_4Fe NFs have (111) basal planes. So, from this observation, the authors conclude that the EC synthesized was a hierarchical flowerlike structure developed from Pd_4Fe alloy nanosheets with the highly exposed (111) crystal planes.

Figure 8.4 SEM image (a), TEM image (b), HAADF-STEM images (c, d), EDS elemental mapping images (e–g), and HRTEM image (h) of the prepared Pd_4Fe NFs. *Source:* From [49] / with permission of Royal Society of Chemistry.

In another report, a facile one-pot synthesis method was employed for the synthesis of 3-D bimetallic PdAu nanodendrite, which contains numerous low and high index facets at room temperature. The unique hierarchical dendritic architecture and high index facets of the EC provided more active sites and exhibited improvement in ORR under alkaline media. The electrochemical investigation also showed that the synthesized EC manifested more positive half-wave potential (0.890 V) and 8.3-fold increment in MA compared to commercial Pd black (0.808 V) [50]. A simple approach was adopted to synthesize intermetallic Pd_3Pb NFs, nanocubes, and nanodendrites using $Pd(acac)_2$, $Pb(acac)_2$ as the salt precursors. The NSs have various advantages, such as highly opened ordered phase and low-coordinated edge sites that significantly improve the electrocatalytic ORR activity. Out of the three hierarchical NSs, the intermetallic Pd_3Pb NFs demonstrated MA (1.14 $mA/\mu g_{Pd}$) in an alkaline solution at 0.9 V vs. RHE with outstanding durability compared to commercial Pt/C, owing to their chemically stable intermetallic structures [51].

Pt—Pd nanodendrite alloys possessing a 3-D porous structure consisting of interconnected branches of highly alloyed Pt—Pd NRs are also reported. The carbon-supported Pt—Pd nanodendrites possess three times more ORR MA than Pt/C. Moreover, the EC showed considerably enhanced stability due to the unique branched porous structure. The 3-D interconnected dendrite structures made the EC notably beneficial over the Pt/C in reducing the mass transport and ohmic polarization, which would become important at high current density in the membrane electrode assembly [52].

Recently, our group reported a simplistic one-step effective strategy for the fabrication of unique half-embedded and half-exposed interfacial PdFeCu (~ 6–7 nm) without using any surfactant and structure-directing organic solvent at a relatively low temperature [54]. The electronic modification with the optimized composition of PdFe/C using Cu incorporation benefiting from the synergistic effect and their unique structures makes the trimetallic nanoalloy network high-performing ORR EC in both alkaline and acidic media. The synthesized EC exhibited higher ECSA than 20 wt% Pt/C and Pd/C. As a result, the synthesized EC offers a higher number of ORR-active sites and eases the electron transport during the ORR process. For example, it exhibited 1.5- and 2.4-fold higher MA, low onset potential, and longer stability compared to 20 wt% Pt/C and Pd/C and other compositions of PdFe/C. Thus, this work offers a new-angled strategy for the design and synthesis of Pt-free-efficient EC with exceptional ORR activity and stability. Huang et al. developed a single-step, surfactant-assisted synthetic method to prepare highly branched mesoporous Au–Pd–Pt trimetallic NFs blooming on the reduced graphene oxide (rGO) with the enhanced ORR. The synergetic effects and highly branched mesoporous structures between the Au–Pd–Pt alloy compositions are the major reasons behind this catalytic enhancement. Moreover, the distinctive NS, incorporation of Au in Pt—Pd atoms, and the rGO support promote electrocatalytic activity and durability. Figure 8.5 represents the ORR polarization curves, CV plots, mass, and specific activities of Au–Pd–Pt/NFs rGO and Pt/C in 0.1 M $HClO_4$ aqueous solution [53].

Trimetallic CoAuPd nanoclusters are reported through classical successive reduction methods by using $NaBH_4$ as a reductant and P123 as a protecting agent. It has two kinds of super 3D structures: the nano-thornbush and nanocluster structure. It exhibits superior ORR activity after the dealloying process. The unusually high electrocatalytic activity of the dealloyed CoAuPd is developed from the special epitaxial multilevel structure like

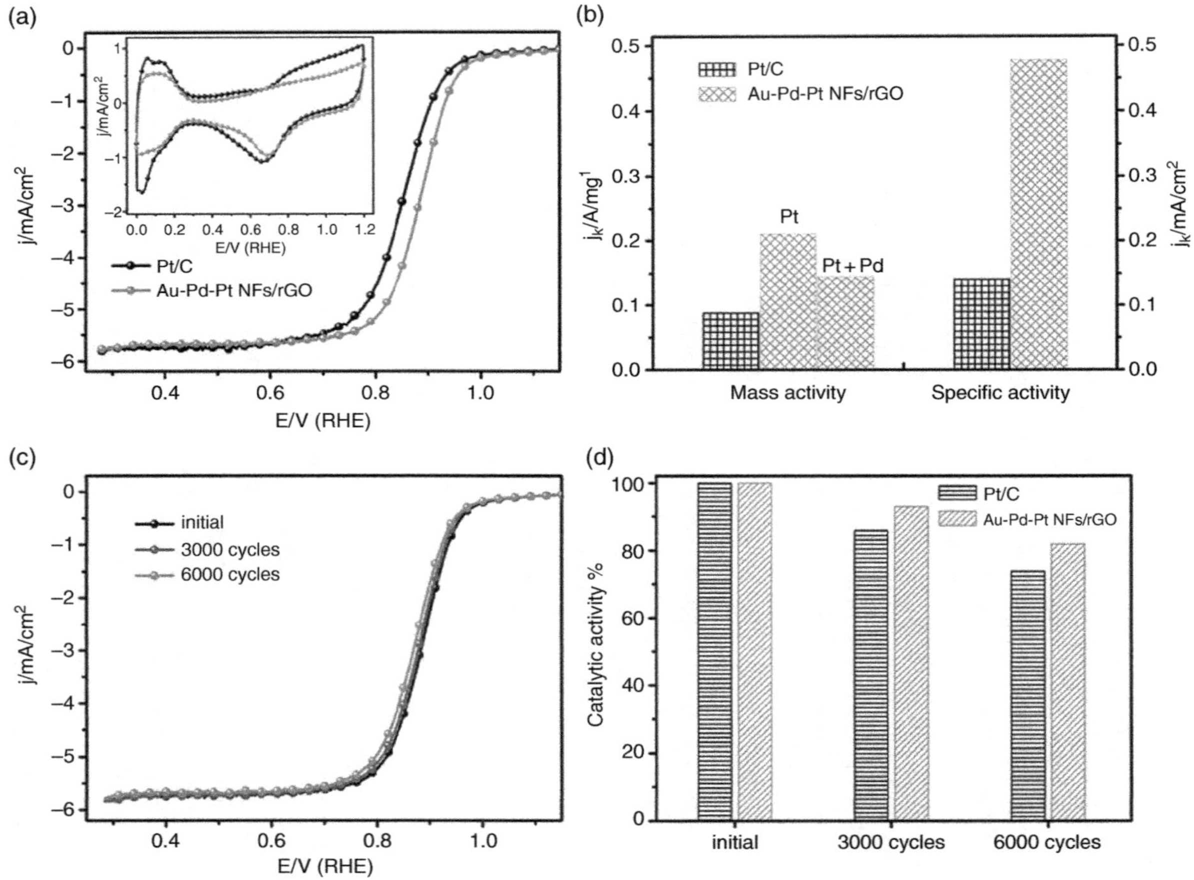

Figure 8.5 Electrocatalytic properties of Au–Pd–Pt NFs/rGO and a Pt/C catalyst (E-TEK, 20 wt% Pt). (a) The ORR polarization curves were recorded at room temperature in an O_2-saturated 0.1 M $HClO_4$ aqueous solution with a rotation rate of 1600 rpm and a sweep rate of 10 mV/s. Cyclic voltammograms (inset of (a)) were recorded at room temperature in N_2-saturated 0.1 M $HClO_4$ solution with a sweep rate of 50 mV/s. (b) MA and SA at 0.9 V vs. RHE for the catalysts, which are given as kinetic current densities normalized to the loading amount of Pt (Pt + Pd) and the ECSA, respectively.
(c) ORR polarization curves for Au–Pd–Pt NFs/rGO after prolonged CV cycles. (d) Loss of catalytic activity of the catalysts after potential sweep cycles. *Source:* From [53] / with permission of Royal Society of Chemistry.

nanocluster and nano-thornbush structure, the accumulated component of featherness NPs and hollow spherical NPs, as well as numerous filamentous nanocrystals. CoAuPd NPs also have exposed active sites and can form the CoAuPd@AuPd core-shell structure [55].

In addition, Pd alloys in the form of NWs and NRs have been used as a substitute for Pt. A surfactant-less simple and efficient technique was employed to prepare $Pd_{1-x}Au_x$ and $Pd_{1-x}Pt_x$ NWs with various compositions and sizes [56]. The improved activity of NWs has been credited to the size-induced contraction of the NW surface. As a result, it is helpful to change the electronic properties of the NWs [57–59]. PtPd NRs were prepared in the form of porous and alloy structures. Due to the exceptional morphology, larger specific surface area, and alloy structure, the porous PtPd NRs showed better electrocatalytic ORR activity via an efficient $4e^-$ pathway [60]. Moreover, there are several reports on Pt-free EC for ORR [28, 61, 62]. For example, Liu et al. [63] synthesized an efficient surfactant-based synthesis of ultrathin PdNi NSs, which followed a volcano-type relationship between ORR activity and chemical composition. The composition $Pd_{0.90}Ni_{0.10}$ exhibited the highest activity (i.e. 1.96 mA/cm^2 at 0.8 V) among the various NWs tested. 1-D Pd – Ni NSs represent a promising platform for designing ORR with a highly regarded activity. This is mainly due to the surface structure and strain of the NWs. As a result of this observation, the author tried to convey that these NWs act as an economical, high-performance, and attractive EC alternative to conventional Pt for ORR.

8.3 Oxygen Evolution Reaction

Another type of FC technology that has gained tremendous interest is the URFC. As we have already discussed, URFCs can function as FCs as well as water electrolyzers at the same time. Here comes the significance of OER that occurs during water electrolysis mode in the URFCs. Apart from OER, HER also takes place concurrently. Thus, the generated O_2 and H_2 molecules eventually combine to produce H_2O (or generate electricity) in the FC mode [64, 65]. This is considered as one of the best approaches to produce high-purity H_2 at an economic interest. Thanks to the high energy density and insignificant carbon emission of H_2 for which it can be considered as the clean and green energy carrier. H_2 also plays a dynamic part in the industries where the manufacture of NH_3 and petrochemicals are the key areas [66]. Hence, OER is an essential half-reaction of H_2O splitting and is highly explored in recent decades [67–69]. OER proceeds through 4 H^+/e^--coupled processes and occurs both in acidic and alkaline media. It also involves the formation of O—O bond for which the reaction is kinetically unfavorable and requires external ECs to boost the reaction. In neutral and acidic media, H_2O molecules are converted into $4H^+$ and O_2 by releasing a total of $4e^-$. In an alkaline medium, OH^- is oxidized into H_2O and O_2. The electrode half-reactions in 1 atm and 25 °C are summarized here [70]:

$$\text{Acidic medium}: 2H_2O \leftrightarrow 4H^+ + O_2 + 4e^- \quad E_a^\circ = 1.23\,V\left(vs.\ \text{NHE}\right)$$

$$\text{Alkaline medium}: \ 4OH^- \leftrightarrow 2H_2O + O_2 + 4e^- \quad E_a^\circ = 0.404\,V\left(vs.\ \text{NHE}\right)$$

where NHE stands for normal hydrogen electrode

To generate a potential difference of 1.23 V (vs. NHE) at pH = 0, it is necessary to use an external circuit. In such a condition, a potential shift (~59 mV per unit pH) occurs according to the Nernst equation. To minimize the influence of pH on the applied potential, RHE is generally used to maintain the working voltage at ~1.23 V (vs. RHE) [67]. The two most commonly used ECs for OER are the Ru and IrO_2, which have optimum binding characteristics with the reactive intermediates. However, the pitfalls like high price and instability in all electrolytes deprived these benchmark oxides of commercial applications. Thus, the design of inexpensive ECs for OER with low overpotential is highly demanding. The amount of Ru or Ir is reduced by incorporating them into binary/ternary metal oxides or by replacing them with various non-precious metals-based ECs with equal or even better activity. The morphology of the EC also plays a critical role in the overall electrocatalytic performance as it is a surface phenomenon. Hence, finding the best morphological structure is equally significant besides finding the appropriate material [71]. Zhang and co-workers [66] synthesized Ir—Pd nanoalloys with improved surface-microstructure-sensitive electrocatalytic activity toward OER in acidic and alkaline media. Various shapes of Ir—Pd nanoalloys include nanotetrahedrons, NWs, and nano-hollow-spheres (NHSs), which were synthesized via a simple one-pot solvothermal method. The MA of Ir—Pd NWs and NTs were found to be five times higher in comparison to that of standard Ir/C catalyst in acidic and alkaline media at an overpotential of 0.25 V (vs. RHE). Pd over Co-based metal–organic framework-derived carbon (MOFDC) was designed to study bifunctional OER and HER. The EC was found to be active in both reactions in an alkaline medium [72]. Joya et al. [73] reported nanoporous Pd-derived EC for OER at a very low overpotential in an alkaline medium. H_2O oxidation occurred at an onset potential of 1.23 V (vs. RHE). The EC also exhibited long-term stability as well as high electrocatalytic activity. This Pd-based EC has a unique single metal structure that gets converted into thin films in a single step and can be directly employed for water oxidation.

Xia and co-workers [74] adopted a facile strategy to develop a low-cost Ir-based EC with a core–shell structure where most reactive Ir atoms are on the surface. The resultant Pd@Ir_{nL} (n: number of atomic layers of Ir) core–shell nanocubes exhibited a volcano-type plot toward OER corresponding to the number of atomic layers of Ir. Pd@Ir_{3L} (i.e. for n = 3) shows the best OER performance compared to the other variations and commercial Ir/C catalyst, which may be due to good intermixing of Pd and Ir and ligand/strain effects.

Different electron microscopy images of the synthesized Pd@Ir_{nL} are presented in Figure 8.6 along with elemental mapping to establish cubic morphology with different shell thicknesses.

Datta et al. [75] deposited Pd NPs on Mg thin films that act as nanoportals to investigate the activity of individual NPs. The Pd NPs could transfer electrons to the Mg thin films in spite of the surface oxidation of Mg atoms. Magnetron sputtering inert-gas condensation method was used to form an assembly of active uncoupled single nanoparticle electrodes (SNPEs). The SNPE assemblies exhibited good OER in an alkaline medium. DFT studies also supported experimental findings. Jiang's group [76] employed first-principle simulations to a hybrid structure of the type TM@CN (TM = Co, Ni, Cu, Pd, and Pt) supported on graphitic carbon nitride monolayer. DFT calculations suggested that TM@CN and TM-O@CN acted as high-performance and high-durability ECs for OER in a water-alkali

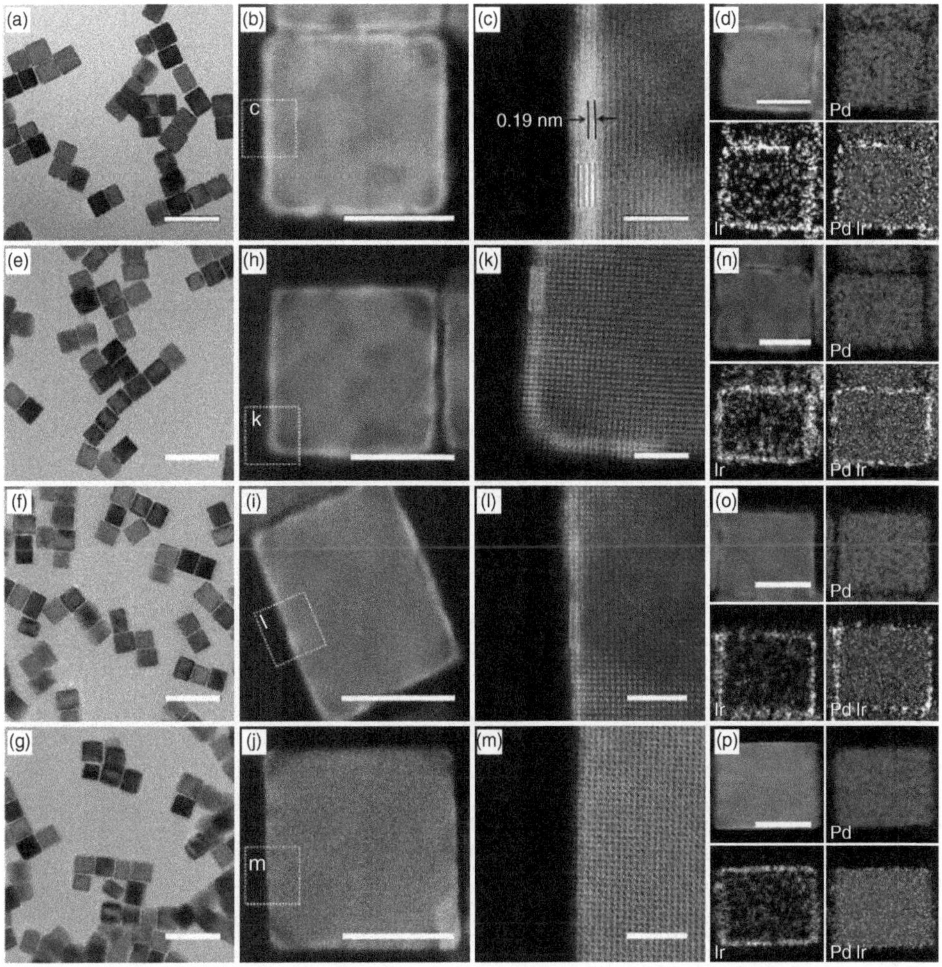

Figure 8.6 (a) TEM image of Pd@IrL$_{4L}$ nanocubes (scale bar: 50 nm). (b) HAADF–STEM image of Pd@Ir$_{4L}$ nanocubes (scale bar: 10 nm). (c) High-resolution HAADF–STEM image taken from the region marked in (b) (scale bar: 2 nm). (d) EDX mapping of a Pd@Ir$_{4L}$ nanocube (scale bar: 10 nm). (e–g) TEM images of (e) Pd@Ir$_{3L}$, (f) Pd@Ir$_{2L}$, and (g) Pd@Ir$_{1L}$ nanocubes (scale bar: 50 nm). (h–j) HAADF–STEM images of (h) Pd@Ir$_{3L}$, (i) Pd@Ir$_{2L}$, and (j) Pd@Ir$_{1L}$ nanocubes (scale bar: 10 nm). (k–m) High-resolution HAADF–STEM images taken from the regions marked in (h–j) (scale bar: 2 nm). (n–p) EDX mappings of (n) Pd@Ir$_{3L}$, (o) Pd@Ir$_{2L}$, and (p) Pd@Ir$_{1L}$ nanocubes (scale bar: 10 nm). *Source:* From [74] / with permission of American Chemical Society.

environment. PdNi alloy NPs supported on carbon nanofiber (PdNi/CNF) were reported as highly effective ECs with a bifunctional activity toward OER and HER. PdNi/CNF was synthesized via electrospinning and carbonization methods where the size and homogeneity of the Pd—Ni alloy NPs were maintained by controlling the appropriate molar feed ratio of the metal salts [77].

A cubic-core hexagonal-branch route was adopted to synthesize bimetallic NPs having *fcc*-Pd as the core and *hcp*-Ru as the branches using oleylamine as surfactant. Low index (10–11) and {0001} facets were formed due to the use of oleylamine, which gives not only

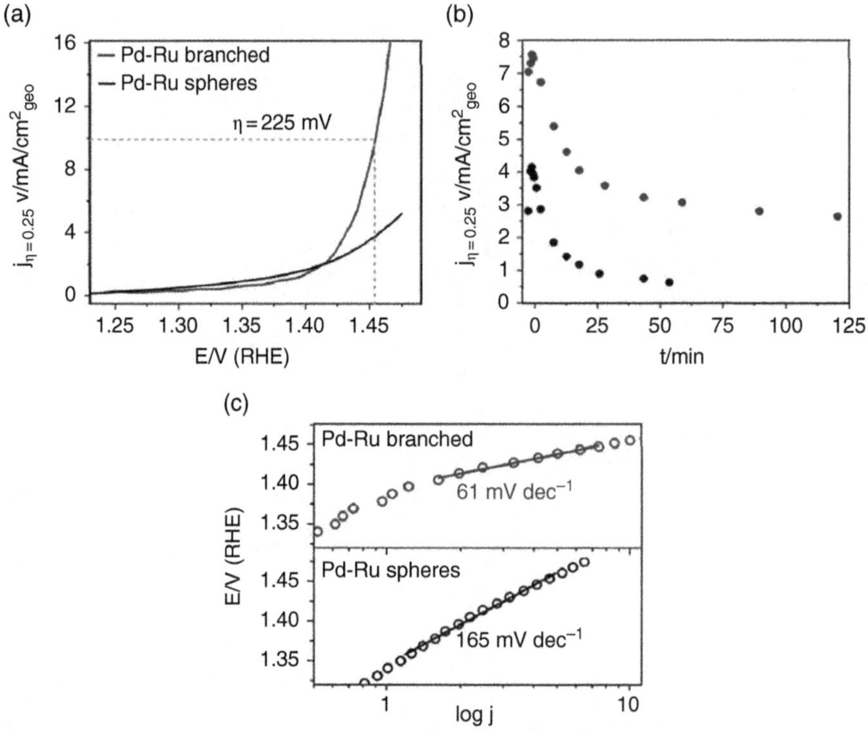

Figure 8.7 (a) iR and capacitive current-corrected oxidation scans of cyclic voltammetry of Pd – Ru-branched NPs and Pd–Ru spheres at maximum current density. The dashed line shows the overpotential required to reach 10 mA/cm$_{geo}$. (b) Chronoamperometry curves at a constant overpotential of 0.25 V. (c) Tafel plots for Pd – Ru-branched NPs and Pd–Ru spheres shown in panel a. *Source:* From [78] / with permission of American Chemical Society.

high OER activity but also provides long-term stability. Figure 8.7 represents the results of electrocatalytic analyses of Pd – Ru-branched NPs and Pd – Ru spheres toward OER [78].

Bifunctional nanocatalysts (NCs) like Pd overlayers (1–5 atomic layers) covered nanoporous gold (NPG) and carbon-supported Ir—Pd bimetallic alloy networks reported by Wang et al. [79] and Nguyen et al. [80], respectively, to study ORR and OER in alkaline medium. NPG-Pdx NCs were synthesized via Cu underpotential developed (UPD-Cu) followed by galvanic displacement method. The enhanced bifunctional activity was attributed to the electronic alteration in Pd, unique 3-D NS of NPG, and proper utilization of Pd atoms in the Pd overlayers. Likewise, the 3-D architecture and the rough surface modify the surface electronic structure of the nanoporous Ir—Pd alloy nanocomposites on carbon support, which are responsible for its superior activity toward oxygen electrocatalysis. Among various Ir:Pd ratios, $Ir_{23}Pd_{77}/C$ revealed the best ORR and OER performance. Cubic $Pt_{39}Ir_{10}Pd_{11}$ trimetallic nanocages with excellent activity and durability toward bifunctional ORR and OER were reported by Xia and co-workers [81] in an acidic medium. The resultant EC has prominent (100) facets, thin porous walls with a thickness of about five atomic layers and an average edge length of ~12.3 nm. The nanocages were synthesized in two steps: co-deposition of Pt and Ir atoms over Pd nanocubes followed by removal of Pd core via chemical etching. The remarkable activity of this dual EC was attributed to the unique open arrangement of the nanocages and the promising electronic coupling Pt and Ir atoms in the lattice. Thus, it can be understood that Pd-based ECs of various morphologies played a very crucial role for OER in the few decades.

8.4 Formic Acid Oxidation Reaction

As already stated earlier, DFAFCs have significant importance in modern-day energy technologies [82]. It is mainly because of the product formed during FAOR, i.e. H_2, which is a potential energy carrier [83]. Basically, FAOR occurs via two different pathways, viz. direct or dehydrogenation pathway and indirect or dehydration pathway. The corresponding mechanisms are summarized as follows: [84, 85]

1) The dehydrogenation pathway is a two-step H^+/e^- transfer process that supports the formation of an active intermediate, i.e. HCOO*, on the catalyst surface that eventually produces CO_2.

$$\mathrm{HCOOH} \rightarrow \mathrm{H}^+(\mathrm{aq}) + \mathrm{e}^- + \mathrm{HCOO}*$$

$$\mathrm{HCOO}* \rightarrow \mathrm{CO_2} + \mathrm{H}^+(\mathrm{aq}) + \mathrm{e}^-$$

2) The dehydration process involves the adsorption of CO species on the catalyst surface via COOH* intermediate formation, which, in turn, oxidizes to CO_2.

$$\mathrm{HCOOH} \rightarrow \mathrm{H}^+(\mathrm{aq}) + \mathrm{e}^- + \mathrm{COOH}*$$

$$\mathrm{COOH}* \rightarrow \mathrm{CO}* + \mathrm{OH}*$$

$$\mathrm{CO}* + \mathrm{OH}* \rightarrow \mathrm{CO_2} + \mathrm{H}^+ + \mathrm{e}^-$$

The commercialization of DFAFCs is still restricted due to certain challenges like deactivation of the EC, poisoning via intermediates, H_2O–CO_2 management in the stacks of FCs, and poor electrocatalytic activity [13]. Generally, Pt surface is more vulnerable to CO poisoning effects compared to that of Pd [86]. Moreover, it has been observed that the formation of CO intermediate leads to the degradation of the EC surface as well as its dissolution, particularly in the acidic medium after a few electrocatalytic cycles [87]. Commonly, Pd-based ECs are used as anode materials in FAOR due to their high power densities and superior electrocatalytic activity. But, their surface is very prone to poisoning via CO species that affects the overall durability of the ECs. Researchers have adopted various approaches, like adding second metals/nonmetals/ metal oxides and the use of various supports (like graphene, rGO, carbon nanotubes, etc.) to increase the efficiency of the Pd-based ECs [88]. Despite these approaches, the development of morphology- and size-dependent Pd-based ECs has been highly prioritized in recent years [89–91]. This is because the surface of the morphology-controlled ECs is accustomed to specific planes depending on their shapes which enhance the catalytic activity and selectivity. Peter's group [92] designed flower-like ordered Pd_3Pb nanocrystals endowed with connecting networks to employ it as ECs in FAOR as well as C_2H_5OH oxidation reaction. The nanocrystals synthesized via the polyol method in 10s unveiled the best electrocatalytic results among those nanocrystals synthesized at a longer time. This is mainly attributed to the morphological variation of Pd_3Pb and the promotional effect of Pb that greatly amplify the ECSA of the resultant EC. Further, to effectively utilize and minimize the amount of Pd, a core–shell and alloy nanostructure is a significant choice [93].

Metal catalysts in the nano-range have drawn huge attention of researchers owing to their structure–activity relationship [94]. Wang et al. [95] developed ultrathin Pd NWs of 2–3 nm in diameter that displayed a decent anodic current density which is 2.5 times higher than that of conventional Pd/C. The unique wavy structure of the nanowires with twin defects is mainly responsible for the enhanced electrocatalytic activity. High-quality Pd nanocubes were obtained via a one-pot method in an aqueous phase, where NaI was used as a capping agent, $PdCl_2$ as the precursor, and poly(vinylpyrrolidone) as a surfactant and reducing agent. Replacing iodide with other halogens does not lead to cubic shapes of the corresponding nanocrystals. Thus, the use of Pd in an appropriate amount is very significant to attain the Pd nanocubes. The Pd nanocubes synthesized via this procedure was found to be highly effective as well as stable toward FAOR [96].

Li and co-workers [97] developed ultrathin ternary PdIrCu alloy nanosheets having flower-like morphology (PdIrCu/C NCF) via a one-step solvothermal reduction method. Figure 8.8 displays the corresponding FESEM, TEM, and HRTEM images of PdIrCu nanosheets along with the elemental mapping of the elements present. The as-synthesized material exhibited excellent stability and high MA toward FAO reaction compared to that of commercially accessible Pd/C. Figure 8.9 represents the corresponding electrocatalytic properties of the as-synthesized Pd-based ECs and commercial Pd/C. The ECSA of PdIrCu/C NCF ($51.67\,m^2/g_{Pd}$) was found to be much higher than that of PdCu/C ($45.46\,m^2/g_{Pd}$), PdIrCu/C NPs ($40.25\,m^2/g_{Pd}$), and commercial Pd/C ($23.65\,m^2/g_{Pd}$).

Numerous studies on FAOR have also been accomplished using 2-D and 3-D structures of EC. For instance, 3-D-rGO hollow nanospheres on Pd nanohybrids with network structure were prepared to investigate the electrocatalytic activity and durability for FAOR [98]. The unique properties of 3-D self-supported NSs like low density, high porosity, exceptional charge, and mass transfer improve catalytic activity. Yazdan-Abad and co-workers [89] investigated the electrocatalytic activity and stability of 3-D and 2-D Pd NSs toward FAOR and ethanol oxidation. They had converted the 2-D Pd nanosheets into 3-D Pd NSs by raising the temperature from 30 to 60 °C. The results of electrochemical analyses showed that the 3-D Pd NS exhibited excellent electrocatalytic activity toward both FAOR and ethanol oxidation reactions in comparison to that of the corresponding 2-D structures. Thin-layered tremella-like Pd NS consisting of 4–5 atoms (ULT—Pd) with uniform distribution were reported [99]. They have adopted an ultrafast (1-minute) surfactant-free one-pot route to synthesize the NSs. The ULT-Pd EC delivered a more negative peak potential and very high cathodic current density toward FAOR compared to that of commercial Pd/C. Recently, 2-D self-assembled Pd nanosheets

Figure 8.8 Morphology images of PdIrCu NCF (a) FESEM, (b) TEM, and (c) HRTEM, inset: the thickness of PdIrCu NCF, (d) further large HRTEM of the box from (c), and (e–h) elemental mapping. *Source:* From [97] / with permission of American Chemical Society.

were prepared at room temperature in the presence and absence of surfactant to investigate the effect of morphology on the activity [100]. The unsupported ECs were found to exhibit better electrocatalytic activity toward FAOR than that of carbon-supported Pd. The improved electrocatalytic activity was attributed to the high surface area and self-support features. They have also observed that Pd nanosheets synthesized without surfactants are superior to the nanosheets synthesized using surfactants. Similarly, other structures such as Pd NNWs [101], twisted Pd—Cu nanochains [102], Pd—Cu alloy with hierarchical network structure [10], highly dispersed palladium/carbon nanofiber [103], PdAg alloy NWs [104], and ultrathin freestanding porous Pd nanosheets [105] were also designed by various renowned research groups to explore their electrocatalytic behavior for FAOR.

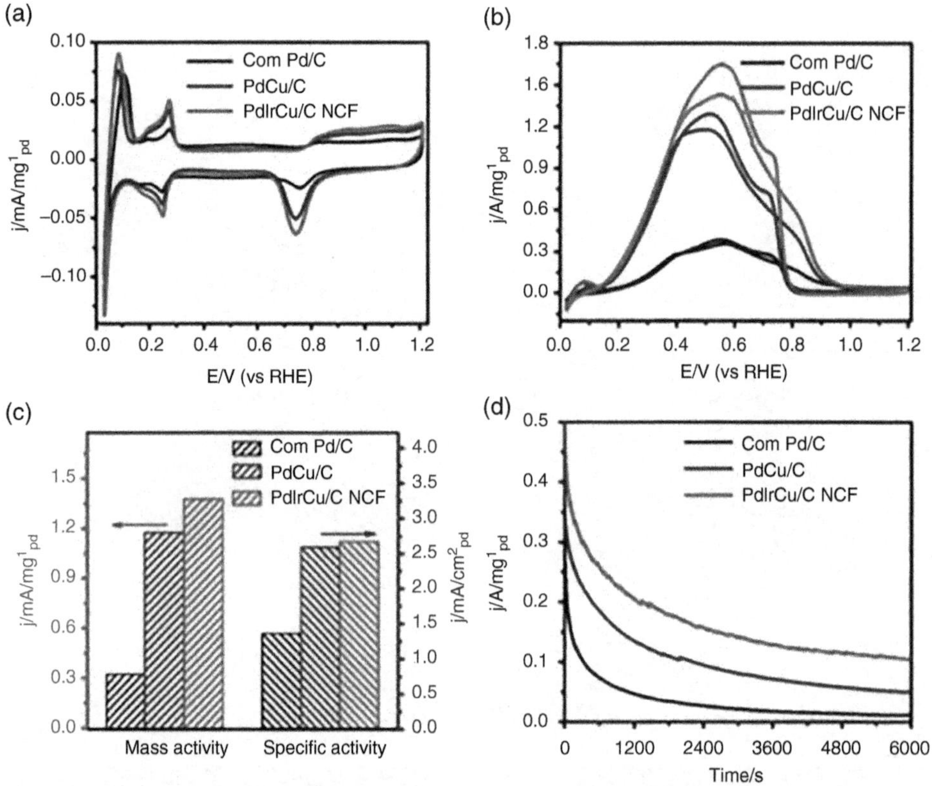

Figure 8.9 CV curves of different Pd-based catalysts in N_2-saturated (a) 0.5 M H_2SO_4 solution and (b) 0.5 M H_2SO_4 + 0.5 M HCOOH solutions and (c) MA and SA values at 0.5 V, and (d) chronopotentiometry curves of Pd-based catalysts at 0.35 V in N_2-saturated 0.5 M H_2SO_4 + 0.5 M HCOOH solutions. *Source:* From [97] / with permission of American Chemical Society.

8.5 Conclusions and Future Perspectives

Over the past few decades, extensive research efforts have been devoted to synthesizing commercially viable ECs for FC applications. Due to the wide abundance of Pd as well as some inveterate properties of it being similar to Pt has fascinated the attention of researchers toward different FC reactions. With the increasing research intrigue so far, various physical and chemical fabrication methods have been developed to synthesize required Pd ECs. These methods include hydrothermal and solvothermal methods, electrochemical deposition, microemulsions, photochemically assisted synthesis, and electroless deposition. Apart from FC reactions, Pd and Pd-based nanomaterials have extended applications toward carbon dioxide reduction reaction (CO_2RR), purification and storage of hydrogen, biosensors, gas sensors, capacitors, and pollutant degradation. It is expected that in the coming decades, newer methods will be developed to design high-performance Pd-based ECs. With the unconstrained consumption of fossil fuels and increasing environmental concerns, there is an urgent requirement to find sustainable and climate-friendly energy storage and conversion devices. Even though Pd-based nanomaterials behave as ideal ECs, there are certain issues that should be resolved. The major drawback of it is the soaring price of Pd. The price of Pd has been going up in the past few years and is anticipated to hike up again in the forthcoming days as the research interest in Pd nanomaterial expands. This leads us to a massive demand for the formation of state-of-the-art ECs while minimizing the amount of noble metals, at the same time increasing their electrochemical activity and stability. In recent years, various methods have been used to achieve cost-effective ECs, including non-noble metal ECs. Such ECs consist of various transition metals and their alloys, transition metal oxides (TMOs), carbon materials, doped carbonaceous materials, transition metal carbides, mixed transition metal oxides (MTMOs) with different structures, such as spinel oxide, perovskites, and metal-N complex on carbon matrices. Different bimetallic and trimetallic compositions of transition metals, TMOs, and MTMOs with various compositions have recently enticed increasing research interest globally due to their impressive electrochemical and synergistic properties. These non-noble metals play an important role in forming low-cost, green, and clean technologies. Therefore, in the coming future, the research field will be highly focused on developing economic transition metal-based ECs so that they act as great environmental remediation.

Acknowledgments

The authors thank Tezpur University (TU) and the Department of Science & Technology (DST No: INT/RUS/RFBR/P-322), New Delhi, for generous financial support. CG thanks DST, New Delhi for the research fellowship. RC gratefully acknowledges TU for institutional fellowship and "Research and Innovation Grant."

References

1 Guo, S. and Wang, E. (**2011**). Noble metal nanomaterials: controllable synthesis and application in fuel cells and analytical sensors. *Nano Today* 6 (3): 240–264.
2 Lokanathan, M., Patil, I.M., and Kakade, B. (2018). Trimetallic PtNiCo nanoflowers as efficient electrocatalysts towards oxygen reduction reaction. *Int. J. Hydrogen Energy* 43 (18): 8983–8990.
3 Gong, Y., Liu, X., Gong, Y. et al. (2018). Synthesis of defect-rich palladium-tin alloy nanochain networks for formic acid oxidation. *J. Colloid Interface Sci.* 530: 189–195.
4 Peighambardoust, S.J., Rowshanzamir, S., and Amjadi, M. (2010). Review of the proton exchange membranes for fuel cell applications. *Int. J. Hydrogen Energy* 35 (17): 9349–9384.
5 Winter, M. and Brodd, R.J. (2004). What are batteries, fuel cells, and supercapacitors? *Chem. Rev.* 104 (10): 4245–4270.
6 Vielstich, W., Lamm, A., and Gasteiger, H. (2003). *Handbook of Fuel Cells. Fundamentals, Technology and Applications*, 1e. Wiley.
7 Shao, M., Chang, Q., Dodelet, J.P., and Chenitz, R. (2016). Recent advances in electrocatalysts for oxygen reduction reaction. *Chem. Rev.* 116 (6): 3594–3657.
8 Guo, S., Zhang, S., and Sun, S. (2013). Tuning nanoparticle catalysis for the oxygen reduction reaction. *Angew. Chem. Int. Ed.* 52s (33): 8526–8544.

9 Jiao, Y., Zheng, Y., Jaroniec, M., and Qiao, S.Z. (2015). Design of electrocatalysts for oxygen-and hydrogen-involving energy conversion reactions. *Chem. Soc. Rev.* 44 (8): 2060–2086.

10 Wang, J., Cui, W., Liu, Q. et al. (2016). Recent progress in cobalt-based heterogeneous catalysts for electrochemical water splitting. *Adv. Mater.* 28 (2): 215–230.

11 Guo, Y., Tang, J., Wang, Z. et al. (2018). Elaborately assembled metal sulfides as a bifunctional core-shell structured catalyst for highly efficient electrochemical overall water splitting. *Nano Energy* 47: 494–502.

12 Ma, J., Wei, H., Liu, Y. et al. (2020). Application of Co_3O_4-based materials in electrocatalytic hydrogen evolution reaction: a review. *Int. J. Hydrogen Energy* 45 (41): 21205–21220.

13 Jiang, K., Zhang, H.X., Zou, S., and Cai, W.B. (2014). Electrocatalysis of formic acid on palladium and platinum surfaces: from fundamental mechanisms to fuel cell applications. *Phys. Chem. Chem. Phys.* 16 (38): 20360–20376.

14 Ong, B.C., Kamarudin, S.K., and Basri, S. (2017). Direct liquid fuel cells: a review. *Int. J. Hydrogen Energy* 42 (15): 10142–10157.

15 Wang, Y., Li, J., and Wei, Z. (2018). Transition-metal-oxide-based catalysts for the oxygen reduction reaction. *J. Mater. Chem. A* 6 (18): 8194–8209.

16 Morales-Guio, C.G., Stern, L.A., and Hu, X. (2014). Nanostructured hydrotreating catalysts for electrochemical hydrogen evolution. *Chem. Soc. Rev.* 43 (18): 6555–6569.

17 Suen, N.T., Hung, S.F., Quan, Q. et al. (2017). Electrocatalysis for the oxygen evolution reaction: recent development and future perspectives. *Chem. Soc. Rev.* 46 (2): 337–365.

18 Li, Y., Li, F.M., Meng, X.Y. et al. (2018). Ultrathin Co_3O_4 nanomeshes for the oxygen evolution reaction. *ACS Catal.* 8 (3): 1913–1920.

19 Chaudhari, N.K., Jin, H., Kim, B., and Lee, K. (2017). Nanostructured materials on 3D nickel foam as electrocatalysts for water splitting. *Nanoscale* 9 (34): 12231–12247.

20 Zhang, J., Jiang, Y., Wang, Y. et al. (2019). Ultrathin carbon coated mesoporous Ni-$NiFe_2O_4$ nanosheet arrays for efficient overall water splitting. *Electrochim. Acta* 321 (134): 652.

21 Pires, F.I. and Villullas, H.M. (2012). Pd-based catalysts: influence of the second metal on their stability and oxygen reduction activity. *Int. J. Hydrogen Energy* 37 (22): 17052–17059.

22 Kondo, S., Nakamura, M., Maki, N., and Hoshi, N. (2009). Active sites for the oxygen reduction reaction on the low and high index planes of palladium. *J. Phys. Chem. C* 113 (29): 12625–12628.

23 Kumar, A., Mohammadi, M.M., and Swihart, M.T. (2019). Synthesis, growth mechanisms, and applications of palladium-based nanowires and other one-dimensional nanostructures. *Nanoscale* 11 (41): 19058–19085.

24 Begum, H., Ahmed, M.S., Cho, S., and Jeon, S. (2018). Freestanding palladium nanonetworks electrocatalyst for oxygen reduction reaction in fuel cells. *Int. J. Hydrogen Energy* 43 (1): 229–238.

25 He, C., Tao, J., and Shen, P.K. (2018). Solid synthesis of ultrathin palladium and its alloys' nanosheets on RGO with high catalytic activity for oxygen reduction reaction. *ACS Catal.* 8 (2): 910–919.

26 Yang, G.M., Chen, X., Li, J. et al. (2011). Bubble dynamic templated deposition of three-dimensional palladium nanostructure catalysts: approach to oxygen reduction using macro-, micro-, and nano-architectures on electrode surfaces. *Electrochim. Acta* 56 (19): 6771–6778.

27 Zhang, L.Y., Meng, X., Wu, H. et al. (2020). Tungsten-induced synthesis of defective palladium–copper–tungsten trimetallic nanochains to highly enhance activity for formic acid electrooxidation. *Mater. Today Energy* 18: 100558.

28 Goswami, C., Saikia, H., Tada, K. et al. (2020). Bimetallic palladium–nickel nanoparticles anchored on carbon as high-performance electrocatalysts for oxygen reduction and formic acid oxidation reactions. *ACS Appl. Energy Mater.* 3 (9): 9285–9295.

29 Shao, M., Liu, P., Zhang, J., and Adzic, R. (2007). Origin of enhanced activity in palladium alloy electrocatalysts for oxygen reduction reaction. *J. Phys. Chem. B* 111 (24): 6772–6775.

30 Chen, A. and Ostrom, C. (2015). Palladium-based nanomaterials: synthesis and electrochemical applications. *Chem. Rev.* 115 (21): 11999–12044.

31 Meku, E., Du, C., Wang, Y. et al. (2016). Concentration gradient Pd-Ir-Ni/C electrocatalyst with enhanced activity and methanol tolerance for oxygen reduction reaction in acidic medium. *Electrochim. Acta* 192: 177–187.

32 Stauffer, D. B., Hirschenhofer, J. H., Klett, M. G., and Engleman, R. R. (1998). *Fuel cell handbook* (No. DOE/FETC-99/1076). Federal Energy Technology Center (FETC), Morgantown, WV, and Pittsburgh, PA (date of access 2 February 2021).

33 Brouzgou, A., Song, S.Q., and Tsiakaras, P. (2012). Low and non-platinum electrocatalysts for PEMFCs: current status, challenges and prospects. *Appl. Catal., B* 127: 371–388.

34 Antolini, E. (2017). Alloy vs. intermetallic compounds: effect of the ordering on the electrocatalytic activity for oxygen reduction and the stability of low temperature fuel cell catalysts. *Appl. Catal., B* 217: 201–213.

35 Bianchini, C. and Shen, P.K. (2009). Palladium-based electrocatalysts for alcohol oxidation in half cells and in direct alcohol fuel cells. *Chem. Rev.* 109 (9): 4183–4206.

36 Antolini, E. (2009). Palladium in fuel cell catalysis. *Energy Environ. Sci.* 2 (9): 915–931.

37 Shao, M. (2013). Palladium-based electrocatalysts for oxygen reduction reaction. In: *Electrocatalysis in Fuel Cells, A Non- and Low- Platinum Approach*, 513–531. London: Springer.

38 Wang, H., Yin, S., Li, Y. et al. (2018). One-step fabrication of tri-metallic PdCuAu nanothorn assemblies as an efficient catalyst for oxygen reduction reaction. *J. Mater. Chem. A* 6 (8): 3642–3648.

39 Bu, L., Tang, C., Shao, Q. et al. (2018). Three-dimensional Pd_3Pb nanosheet assemblies: high-performance non-Pt electrocatalysts for bifunctional fuel cell reactions. *ACS Catal.* 8 (5): 4569–4575.

40 Shu, Y., Shi, X., Ji, Y. et al. (2018). Hollow echinus-like PdCuCo alloy for superior efficient catalysis of ethanol. *ACS Appl. Mater. Interfaces* 10 (5): 4743–4749.

41 Wang, X., Vara, M., Luo, M. et al. (2015). Pd@ Pt core–shell concave decahedra: a class of catalysts for the oxygen reduction reaction with enhanced activity and durability. *J. Am. Chem. Soc.* 137 (47): 15036–15042.

42 Wang, Y., Wang, W., Xue, F. et al. (2018). One-pot synthesis of Pd@ Pt_3Ni core–shell nanobranches with ultrathin Pt_3Ni {111} skins for efficient ethanol electrooxidation. *Chem. Commun.* 54 (41): 5185–5188.

43 Cui, X., Xiao, P., Wang, J. et al. (2017). Highly branched metal alloy networks with superior activities for the methanol oxidation reaction. *Angew. Chem. Int. Ed.* 56 (16): 4488–4493.

44 Jiang, Z., Jiang, Z.J., Maiyalagan, T., and Manthiram, A. (2016). Cobalt oxide-coated N-and B-doped graphene hollow spheres as bifunctional electrocatalysts for oxygen reduction and oxygen evolution reactions. *J. Mater. Chem. A* 4 (16): 5877–5889.

45 Xiong, Y., Xin, P., Chen, W. et al. (2018). PtAl truncated octahedron nanocrystals for improved formic acid electrooxidation. *Chem. Commun.* 54 (32): 3951–3954.

46 Lim, J., Shin, H., Kim, M. et al. (2018). Ga–doped Pt–Ni octahedral nanoparticles as a highly active and durable electrocatalyst for oxygen reduction reaction. *Nano Lett.* 18 (4): 2450–2458.

47 Xia, Y., Hu, H., Shen, T. et al. (2016). Hierarchically porous Pd nanospheres: facile synthesis and their application in HCOOH electrooxidation. *Chem. Commun.* 52 (65): 10064–10067.

48 Liu, S., Zhang, Q., Li, Y. et al. (2015). Five-fold twinned Pd_2NiAg nanocrystals with increased surface Ni site availability to improve oxygen reduction activity. *J. Am. Chem. Soc.* 137 (8): 2820–2823.

49 Lian, C., Cheng, Y., Chen, L. et al. (2018). Synthesis and electrocatalytic properties for oxygen reduction of Pd_4Fe nanoflowers. *Chem. Commun.* 54 (51): 7058–7061.

50 Huang, X.Y., Wang, A.J., Zhang, L. et al. (2019). A simple wet-chemical strategy for facile fabrication of hierarchical PdAu nanodentrites as excellent electrocatalyst for oxygen reduction reaction. *J. Colloid Interface Sci.* 552: 51–58.

51 Luo, S., Ou, Y., Li, L. et al. (2019). Intermetallic Pd_3Pb ultrathin nanoplate-constructed flowers with low-coordinated edge sites boost oxygen reduction performance. *Nanoscale* 11 (37): 17301–17307.

52 Xiong, X., Chen, W., Wang, W. et al. (2017). Pt-Pd nanodendrites as oxygen reduction catalyst in polymer-electrolyte-membrane fuel cell. *Int. J. Hydrogen Energy* 42 (40): 25234–25243.

53 Borah, B.J., Saikia, H., Goswami, C. et al. (2019). Unique half embedded/exposed PdFeCu/C interfacial nanoalloy as high-performance electrocatalyst for oxygen reduction reaction. *ChemCatChem* 11: 3522–3529.

54 Huang, L., Han, Y., and Dong, S. (2016). Highly-branched mesoporous Au–Pd–Pt trimetallic nanoflowers blooming on reduced graphene oxide as an oxygen reduction electrocatalyst. *Chem. Commun.* 52 (56): 8659–8662.

55 Luo, L.M., Zhang, R.H., Chen, D. et al. (2018). Synthesis of 3D thornbush-like trimetallic CoAuPd nanocatalysts and electrochemical dealloying for methanol oxidation and oxygen reduction reaction. *ACS Appl. Energy Mater.* 1 (6): 2619–2629.

56 Koenigsmann, C., Sutter, E., Chiesa, T.A. et al. (2012). Highly enhanced electrocatalytic oxygen reduction performance observed in bimetallic palladium-based nanowires prepared under ambient, surfactantless conditions. *Nano Lett.* 12 (4): 2013–2020.

57 Koenigsmann, C. and Wong, S.S. (2011). One-dimensional noble metal electrocatalysts: a promising structural paradigm for direct methanol fuel cells. *Energy Environ. Sci.* 4 (4): 1161–1176.

58 Koenigsmann, C., Zhou, W.P., Adzic, R.R. et al. (2010). Size-dependent enhancement of electrocatalytic performance in relatively defect-free, processed ultrathin platinum nanowires. *Nano Lett.* 10 (8): 2806–2811.

59 Zhou, W.P., Li, M., Koenigsmann, C. et al. (2011). Morphology-dependent activity of Pt nanocatalysts for ethanol oxidation in acidic media: nanowires versus nanoparticles. *Electrochim. Acta* 56 (27): 9824–9830.

60 Lu, Y., Jiang, Y., and Chen, W. (2013). PtPd porous nanorods with enhanced electrocatalytic activity and durability for oxygen reduction reaction. *Nano Energy* 2 (5): 836–844.

61 Yang, T., Ma, Y., Huang, Q., and Cao, G. (2016). Palladium–iridium nanocrystals for enhancement of electrocatalytic activity toward oxygen reduction reaction. *Nano Energy* 19: 257–268.

62 Goswami, C., Saikia, H., Borah, B.J. et al. (2021). Boosting the electrocatalytic activity of Pd/C by Cu alloying: insight on Pd/Cu composition and reaction pathway. *J. Colloid Interface Sci.* 587: 446–456.

63 Liu, H., Koenigsmann, C., Adzic, R.R., and Wong, S.S. (2014). Probing ultrathin one-dimensional Pd–Ni nanostructures as oxygen reduction reaction catalysts. *ACS Catal.* 4 (8): 2544–2555.

64 Zhang, Y., Zhang, H., Ma, Y. et al. (2010). A novel bifunctional electrocatalyst for unitized regenerative fuel cell. *J. Power Sources* 195 (1): 142–145.

65 Wang, Y., Leung, D.Y., Xuan, J., and Wang, H. (2016). A review on unitized regenerative fuel cell technologies, part-A: unitized regenerative proton exchange membrane fuel cells. *Renewable Sustainable Energy Rev.* 65: 961–977.

66 Zhang, T., Liao, S.A., Dai, L.X. et al. (2018). Ir-Pd nanoalloys with enhanced surface-microstructure-sensitive catalytic activity for oxygen evolution reaction in acidic and alkaline media. *Sci. China Mater.* 61 (7): 926–938.

67 Gong, M. and Dai, H. (2015). A mini review of NiFe-based materials as highly active oxygen evolution reaction electrocatalysts. *Nano Res.* 8 (1): 23–39.

68 Dau, H., Limberg, C., Reier, T. et al. (2010). The mechanism of water oxidation: from electrolysis via homogeneous to biological catalysis. *ChemCatChem* 2 (7): 724–761.

69 Carmo, M., Fritz, D.L., Mergel, J., and Stolten, D. (2013). A comprehensive review on PEM water electrolysis. *Int. J. Hydrogen Energy* 38 (12): 4901–4934.

70 Tahir, M., Pan, L., Idrees, F. et al. (2017). Electrocatalytic oxygen evolution reaction for energy conversion and storage: a comprehensive review. *Nano Energy* 37: 136–157.

71 Eftekhari, A. (2017). Tuning the electrocatalysts for oxygen evolution reaction. *Mater. Today Energy* 5: 37–57.

72 Ipadeola, A.K. and Ozoemena, K.I. (2020). Alkaline water-splitting reactions over Pd/Co-MOF-derived carbon obtained via microwave-assisted synthesis. *RSC Adv.* 10 (29): 17359–17368.

73 Joya, K.S., Ehsan, M.A., Sohail, M., and Yamani, Z.H. (2019). Nanoscale palladium as a new benchmark electrocatalyst for water oxidation at low overpotential. *J. Mater. Chem. A* 7 (15): 9137–9144.

74 Zhu, J., Lyu, Z., Chen, Z. et al. (2019). Facile synthesis and characterization of Pd@ Ir_nL (n= 1–4) core–shell nanocubes for highly efficient oxygen evolution in acidic media. *Chem. Mater.* 31 (15): 5867–5875.

75 Datta, A., Porkovich, A.J., Kumar, P. et al. (2019). Single nanoparticle activities in ensemble: a study on Pd cluster nanoportals for electrochemical oxygen evolution reaction. *J. Phys. Chem. C* 123 (43): 26124–26135.

76 Li, X., Cui, P., Zhong, W. et al. (2016). Graphitic carbon nitride supported single-atom catalysts for efficient oxygen evolution reaction. *Chem. Commun.* 52 (90): 13233–13236.

77 Chen, J., Chen, J., Yu, D. et al. (2017). Carbon nanofiber-supported PdNi alloy nanoparticles as highly efficient bifunctional catalysts for hydrogen and oxygen evolution reactions. *Electrochim. Acta* 246: 17–26.

78 Gloag, L., Benedetti, T.M., Cheong, S. et al. (2018). Cubic-core hexagonal-branch mechanism to synthesize bimetallic branched and faceted Pd–Ru nanoparticles for oxygen evolution reaction electrocatalysis. *J. Am. Chem. Soc.* 140 (40): 12760–12764.

79 Wang, Y., Huang, W., Si, C. et al. (2016). Self-supporting nanoporous gold-palladium overlayer bifunctional catalysts toward oxygen reduction and evolution reaction. *Nano Res.* 9 (12): 3781–3794.

80 Nguyen, A.T.N. and Shim, J.H. (2018). Facile one-step synthesis of Ir-Pd bimetallic alloy networks as efficient bifunctional catalysts for oxygen reduction and oxygen evolution reactions. *J. Electroanal. Chem.* 827: 120–127.

81 Zhu, J., Xie, M., Chen, Z. et al. (2020). Pt-Ir-Pd trimetallic nanocages as a dual catalyst for efficient oxygen reduction and evolution reactions in acidic media. *Adv. Energy Mater.* 10 (16): 1904114.

82 Grasemann, M. and Laurenczy, G. (2012). Formic acid as a hydrogen source–recent developments and future trends. *Energy Environ. Sci.* 5 (8): 8171–8181.

83 Zhang, Z., Luo, Y., Liu, S. et al. (2019). A PdAg-CeO_2 nanocomposite anchored on mesoporous carbon: a highly efficient catalyst for hydrogen production from formic acid at room temperature. *J. Mater. Chem. A* 7 (37): 21438–21446.

84 Wang, L., Zhai, J.J., Jiang, K. et al. (2015). Pd–Cu/C electrocatalysts synthesized by one-pot polyol reduction toward formic acid oxidation: structural characterization and electrocatalytic performance. *Int. J. Hydrogen Energy* 40 (4): 1726–1734.

85 Klinkova, A., De Luna, P., Sargent, E.H. et al. (2017). Enhanced electrocatalytic performance of palladium nanoparticles with high energy surfaces in formic acid oxidation. *J. Mater. Chem. A* 5 (23): 11582–11585.

86 Zhou, Z.Y., Kang, X., Song, Y., and Chen, S. (2011). Butylphenyl-functionalized palladium nanoparticles as effective catalysts for the electrooxidation of formic acid. *Chem. Commun.* 47 (21): 6075–6077.

87 Orilall, M.C., Matsumoto, F., Zhou, Q. et al. (2009). One-pot synthesis of platinum-based nanoparticles incorporated into mesoporous niobium oxide carbon composites for fuel cell electrodes. *J. Am. Chem. Soc.* 131 (26): 9389–9395.

88 El-Nagar, G.A., Darweesh, A.F., and Sadiek, I. (2016). A novel nano-palladium complex anode for formic acid electro-oxidation. *Electrochim. Acta* 215: 334–338.

89 Yazdan-Abad, M.Z., Alfi, N., Farsadrooh, M. et al. (2019). Deposition of palladium-copper nanostructure on reduced graphene oxide by a simple method toward formic acid oxidation. *J. Electroanal. Chem.* 848 (113): 299.

90 Yang, F., Zhang, Y., Liu, P.F. et al. (2016). Pd–Cu alloy with hierarchical network structure as enhanced electrocatalysts for formic acid oxidation. *Int. J. Hydrogen Energy* 41 (16): 6773–6780.

91 Zhou, Y., Hu, X.C., Fan, Q., and Wen, H.R. (2016). Three-dimensional crumpled graphene as an electro-catalyst support for formic acid electro-oxidation. *J. Mater. Chem. A* 4 (12): 4587–4591.

92 Jana, R., Subbarao, U., and Peter, S.C. (2016). Ultrafast synthesis of flower-like ordered Pd_3Pb nanocrystals with superior electrocatalytic activities towards oxidation of formic acid and ethanol. *J. Power Sources* 301: 160–169.

93 Hong, J.W., Kim, D., Lee, Y.W. et al. (2011). Atomic-distribution-dependent electrocatalytic activity of Au–Pd bimetallic nanocrystals. *Angew. Chem. Int. Ed.* 50 (38): 8876–8880.

94 Ma, T., Li, C., Liu, T., and Yuan, Q. (2018). Size-controllable synthesis of dendritic Pd nanocrystals as improved electrocatalysts for formic acid fuel cells' application. *J. Saudi Chem. Soc.* 22 (7): 846–854.

95 Wang, Y., Choi, S.I., Zhao, X. et al. (2014). Polyol synthesis of ultrathin Pd nanowires via attachment-based growth and their enhanced activity towards formic acid oxidation. *Adv. Funct. Mater.* 24 (1): 131–139.

96 Liu, X., Li, Z., Wang, K. et al. (2019). Facile synthesis of Pd nanocubes with assistant of iodide and investigation of their electrocatalytic performances towards formic acid oxidation. *Nanomater.* 9 (3): 375.

97 An, H.M., Zhao, Z.L., Zhang, L.Y. et al. (2018). Ir-alloyed ultrathin ternary PdIrCu nanosheet-constructed flower with greatly enhanced catalytic performance toward formic acid electrooxidation. *ACS Appl. Mater. Interfaces* 10 (48): 41293–41298.

98 Qiu, X., Wu, P., Xu, L. et al. (2015). 3D graphene hollow nanospheres@ palladium-networks as an efficient electrocatalyst for formic acid oxidation. *Adv. Mater. Interfaces* 2 (18): 1500321.

99 Zhang, L.Y., Ouyang, Y., Wang, S. et al. (2020). Ultrafast synthesis of uniform 4–5 atoms-thin layered tremella-like Pd nanostructure with extremely large electrochemically active surface area for formic acid oxidation. *J. Power Sources* 447 (227): 248.

100 Saravani, H., Farsadrooh, M., Mollashahi, M.S. et al. (2020). Two-dimensional engineering of Pd nanosheets as advanced electrocatalysts toward formic acid oxidation. *Int. J. Hydrogen Energy* 45 (41): 21232–21240.

101 Wang, J., Chen, Y., Liu, H. et al. (2010). Synthesis of Pd nanowire networks by a simple template-free and surfactant-free method and their application in formic acid electrooxidation. *Electrochem. Commun.* 12 (2): 219–222.

102 Zhang, L.Y., Gong, Y., Wu, D. et al. (2019). Twisted palladium-copper nanochains toward efficient electrocatalytic oxidation of formic acid. *J. Colloid Interface Sci.* 537: 366–374.

103 Qin, Y.H., Yang, H.H., Zhang, X.S. et al. (2011). Synthesis of highly dispersed and active palladium/carbon nanofiber catalyst for formic acid electrooxidation. *J. Power Sources* 196 (10): 4609–4612.

104 Lu, Y. and Chen, W. (2012). PdAg alloy nanowires: facile one-step synthesis and high electrocatalytic activity for formic acid oxidation. *ACS Catal.* 2 (1): 84–90.

105 Qiu, X., Zhang, H., Wu, P. et al. (2017). One-pot synthesis of freestanding porous palladium nanosheets as highly efficient electrocatalysts for formic acid oxidation. *Adv. Funct. Mater.* 27 (1): 1603852.

9

Nanocatalysis of Prussian Blue Analogues Related to H_2O_2 Production and Utilization

Yusuke Yamada[1,2] *and Hiroyasu Tabe*[1,2]

[1] *Department of Chemistry and Bioengineering, Graduate School of Engineering, Osaka Metropolitan University, Osaka, Japan*
[2] *Research Center for Artificial Photosynthesis (ReCAP), Osaka Metropolitan University, Osaka, Japan*

9.1 Introduction

Hydrogen peroxide (H_2O_2) is a versatile and valuable chemical used not only for industries but also for our ordinary life. Currently, H_2O_2 is mainly used for pulp- and paper-bleaching in industries [1]. In the near future, H_2O_2 is expected to be used as a strong but selective oxidant to produce various oxygenated products from hydrocarbons such as alkanes and alkenes to alcohols, epoxides [2, 3]. For the household purpose, H_2O_2 adducts, such as sodium perborate and sodium percarbonate, are typically used as mild bleaches in laundry detergents. Additionally, H_2O_2 is used for disinfection and sanitary purposes. Thus, the global production amount of H_2O_2 exceeded 4.1 million tons in 2019 and is anticipated to grow at the compound average growth rate of 5.1% between 2021 and 2030 [4].

Today, H_2O_2 is industrially produced by the anthraquinone process, in which anthraquinone derivative is reduced to the corresponding anthrahydroquinone by hydrogenation in the presence of Pd or Ni catalyst, and then, the anthrahydroquinone reduces molecular oxygen in the air to produce H_2O_2 [5, 6]. The process was developed by BASF Aktiengesellschaft and patented in 1939 [7]. This old process accompanies with nuisances in which the catalyst regeneration step is necessary, and expensive anthraquinone can be over reduced during the regeneration. Thus, the direct catalytic process of reducing molecular oxygen with hydrogen has been investigated mainly with Au or Pd nanoparticles (NPs) catalysts [8]. Addition to this, photocatalytic H_2O_2 production has been thoroughly investigated with various catalytic systems [9].

Such photocatalytic H_2O_2 production paves the way for the utilization of H_2O_2 as an energy carrier for next-generation because it can be used for electrical power generation in H_2O_2 fuel cells (FCs) [10, 11]. The hallmark of H_2O_2 acting as both reductant and oxidant allows to construct a simple FC with one-compartment or membrane-less structure by employing selective electrodes for both the cathode and anode [11]. The products after the

Heterogeneous Nanocatalysis for Energy and Environmental Sustainability: Energy Applications, First Edition. Edited by Putla Sudarsanam, Yusuke Yamauchi, and Pankaj Bharali.

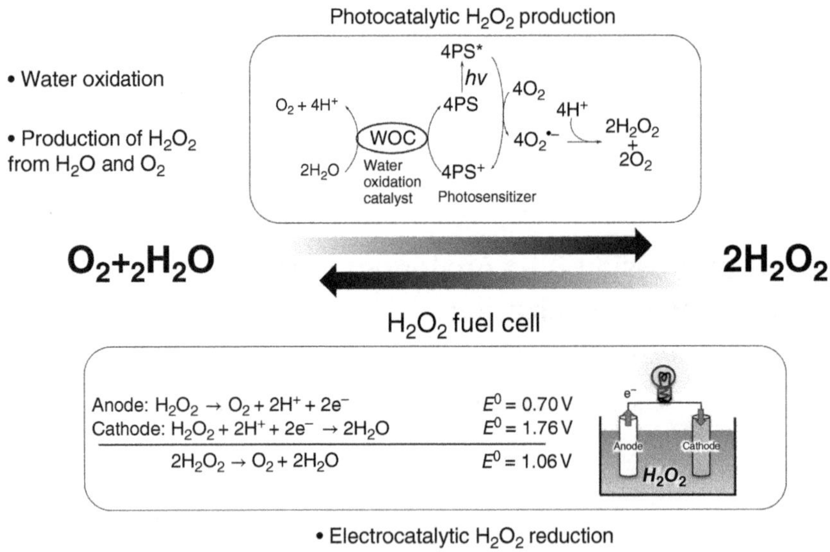

Figure 9.1 H_2O_2 as the energy carrier of next generation.

FC reactions are O_2 and H_2O. Thus, photocatalytic H_2O_2 production from O_2 and H_2O allows to realize ideal recycling society based on O_2 or H_2O_2 (Figure 9.1), which was first proposed in 2010 [12].

Superior catalysts should be developed for both photocatalytic production of H_2O_2 and electrical power generation with H_2O_2 FCs. Photocatalytic H_2O_2 production from H_2O and O_2 demands water oxidation catalysts and selective O_2 reduction catalysts. Recently, polynuclear cyano-bridged metal complexes, often called as Prussian blue analogues (PBAs), have shown high catalytic activity for water oxidation [13–16]. Also, H_2O_2 FCs require electrode catalysts for both oxidation and reduction of H_2O_2. Prussian blue (PB) has been reported as a highly active electrocatalyst for H_2O_2 reduction [17]. PBAs can be utilized as promising catalytic materials to realize O_2- or H_2O_2-based society.

9.2 Atomic and Nanoarchitecture Control of PBAs

PB is the oldest porous coordination polymer with a long history of more than 300 years [18]. Typical PBAs consist of metal ions coordinated to C or N atoms, noted as M^C and M^N, with orthogonal octahedral structures, resulting in the cubic jungle-gym framework. However, the jungle-gym framework is not always perfect because the number of M^C and M^N is not the same due to charge compensation (Figure 9.2). For example, the chemical formula of PB expressed as $Fe^{III}_4[Fe^{II}(CN)_6]_3$ indicates that the number of Fe^{II}(3) is smaller than Fe^{III}(4), resulting in 1/4 positions for $[Fe^{II}(CN)_6]$ units are absent to produce defects in the framework [19]. Also, excess Fe^{III} ions surrounding the defect site are coordinated by water or solvent molecules to fulfill the octahedral coordination [20]. Another crystal structure of PBAs is also suggested for $Co^{II}_3[Co^{III}(CN)_6]_2$, in which excess

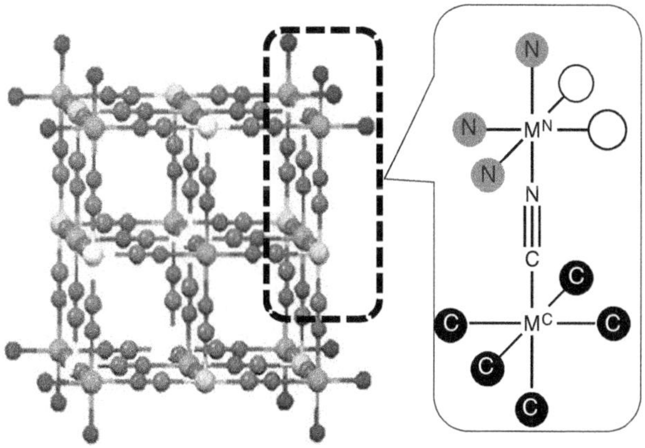

Figure 9.2 A typical structure of Prussian blue analogues (PBAs).

amount of Co^{II} ions occupy interstitial sites, out of the framework [21]. The real structures of imperfect frameworks have been thoroughly pursued by both single crystal and powder X-ray diffractions. Recently, X-ray diffuse scattering of single crystals of PBAs has been used for investigating vacancy networks [22].

Such defect-rich structures are highly attractive for researchers interested in heterogeneous catalysis because the defect formation results in the increase of potential active sites with coordinatively unsaturated structures after removal of coordinating water or solvent molecules. Also, the catalysis of PBAs can be modulated by innocent framework metal ions with a coordinatively saturated structure through metal–metal interaction [15].

PBAs have a porous structure; however, the aperture size of nanopores is too small. Thus, the catalysis of PBAs can be also modulated by particle sizes and morphologies. Nanostructured PBA particles have been synthesized by various methods. Mann et al. have synthesized PB nanocubes by using reverse microemulsions [23]. The synthesized PB nanocubes are almost uniform in size which assembled to superlattices [23]. This microemulsion method is applicable to synthesize PBAs nanocubes, such as cobalt hexacyanoferrate, cobalt pentacyanonitrosylferrate, and chromium hexacyanochromate [24]. Shen et al. have successfully synthesized PB nanocubes and spheres separately under hydrothermal synthesis in the presence of polyvinylpyrrolidone (PVP) and HCl at various temperatures (Figure 9.3) [25]. Yamauchi and Wu have reported both the shape and the size control of PB NPs covered with PVP [26]. More precise shape control was possible for $Co^{II}_3[Co^{III}(CN)_6]_2$ [27]. Cubic, rod, and truncated cube (polyhedral) nanocrystals were separately synthesized by changing the precursor ($K_3[Co^{III}(CN)_6]$) concentrations and using cetyltrimethylammonium bromide (CTAB) as a surfactant in a mixed solvent of cyclohexane and *n*-pentanol (Figure 9.4) [27].

One-dimensional (1D) channel structured PBAs were also prepared by using pentacyanoferrate with alkyl pyridinium surfactants as a precursor [28]. The formamide solution of the precursors was mixed with the formamide solution of octadecypyradinium to form the 1D channel structure (Figure 9.5) after removal of the octadecypyradinium by washing. The mesoporous structure of the obtained material was directly observed by TEM measurements [28].

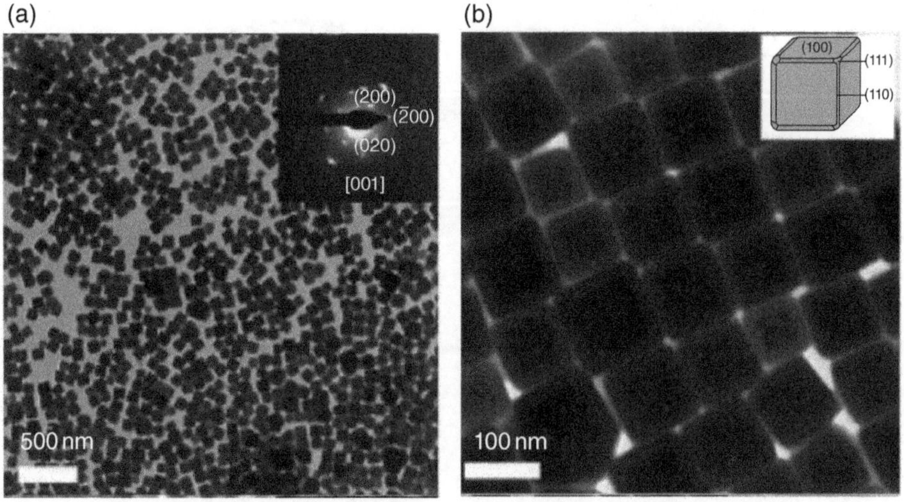

Figure 9.3 TEM images of (a) cubic PB nanoparticles and (b) superlattices. Scale bars 100 nm. *Source:* From [25] / with permission of Elsevier.

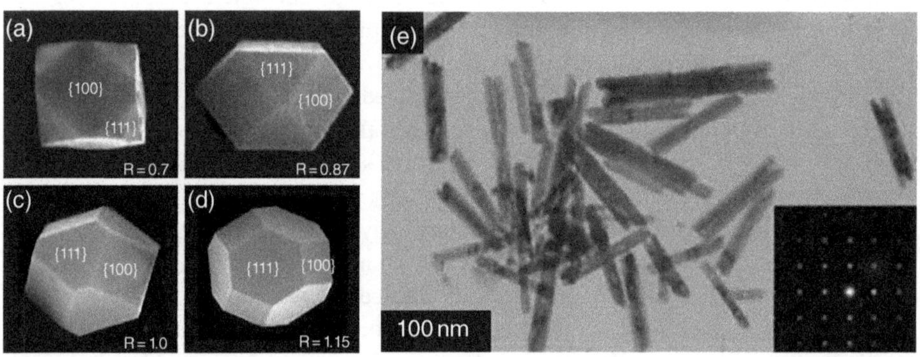

Figure 9.4 (a–d) SEM images of the four $Co_3[Co(CN)_6]_2$ polyhedra. (e) TEM image of $Co_3[Co(CN)_6]_2$ nanorods, the inset is a selected area electron diffraction (SAED) image of a single nanorod [27]. *Source:* From [27] / with permission of American Chemical Society.

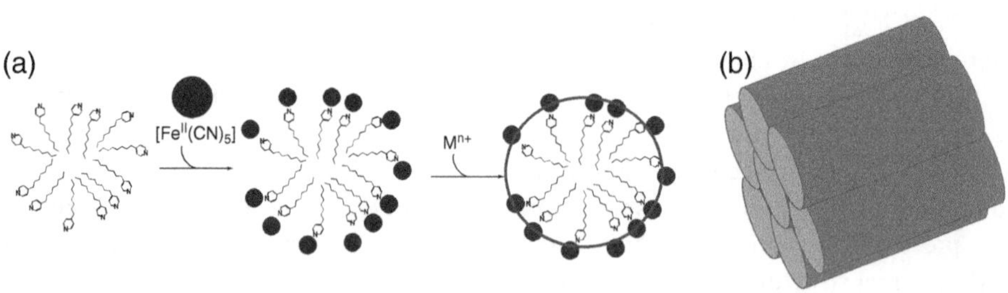

Figure 9.5 (a) Synthetic approach and (b) idealized structure of a hexagonal mesostructured PB analogue. *Source:* From [28] / with permission of John Wiley & Sons.

9.3 Photocatalytic Water Oxidation by PBAs

Water oxidation is a difficult process because it involves four electrons and four protons transfer with O–O bond formation. However, water oxidation is the most crucial reaction step in artificial photosynthesis as well as photosynthesis. Thus, various metal oxides and metal complexes have been examined for water oxidation [29, 30]. Recently, metal-organic frameworks (MOFs) have been regarded as a new class of water oxidation catalysts since 2013, when Galán-Mascarós et al. have reported the water oxidation catalysis of PBAs [14]. The turnover frequency based on the number of Co ions in $[Co^{II}(H_2O)_2]_{1.5}[Co^{III}(CN)_6]$ (CoCo), which is the most active catalyst in the report, for photocatalytic water oxidation was 3.0×10^{-4}/s under photoirradiation of the reaction solution containing $[Ru^{II}$(2,2′-bipyridine)$_3]^{2+}$ and $S_2O_8^{2-}$ as a photosensitizer and a sacrificial electron acceptor, respectively (Figure 9.6a) [31]. Water oxidation catalysis of CoCo was concisely summarized in a recent review article [16]. A plausible reason for the high catalytic activity compared with other MOFs can be ascribed to the porous structure resulted from many defects in CoCo. The enlarged pore window at a defect site allows the photosensitizer to access Co^{II} ions regarded as active species located at subsurface (Figure 9.6b) [32].

The catalytic efficiency of the subsurface Co^{II} ions was evidenced by catalysis examination of core-shell NPs with inactive core and CoCo as the shell [32]. Shell thickness effect on CoCo catalysis was examined with $[Cu^{II}(H_2O)_2]_{1.5}[Fe^{III}(CN)_6]$ (CuFe) core and CoCo shell NPs, in which CuFe core is inactive but isostructural to CoCo. The TEM images of CuFe@CoCo with elemental mapping images indicate the formation of the core-shell structure (Figure 9.7) [32].

The shell thickness effect on the catalytic activity of a series of core-shell **CuFe@CoCo** NPs was investigated for photocatalytic water oxidation under visible-light irradiation in the presence of $[Ru(bpy)_3]SO_4$ and $Na_2S_2O_8$ (Figure 9.6a). Expected relationships between catalytic activity and shell thickness are indicated in Figure 9.8 in the cases of subsurface

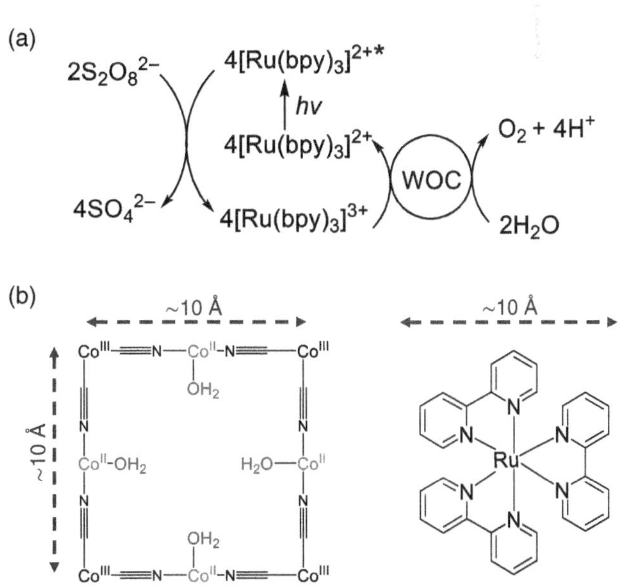

Figure 9.6 (a) Overall catalytic cycle of visible-light-driven photocatalytic water oxidation. WOC: water oxidation catalyst ($[Co^{II}(H_2O)_2]_{1.5}[Co^{III}(CN)_6]$); $[Ru(bpy)_3]$: [Ru(2,2′-bipyridine)$_3$]. (b) The sizes of an enlarged micropore of $[Co^{II}(H_2O)_2]_{1.5}[Co^{III}(CN)_6]$ and $[Ru(bpy)_3]$. *Source:* From [32] / with permission of Elsevier.

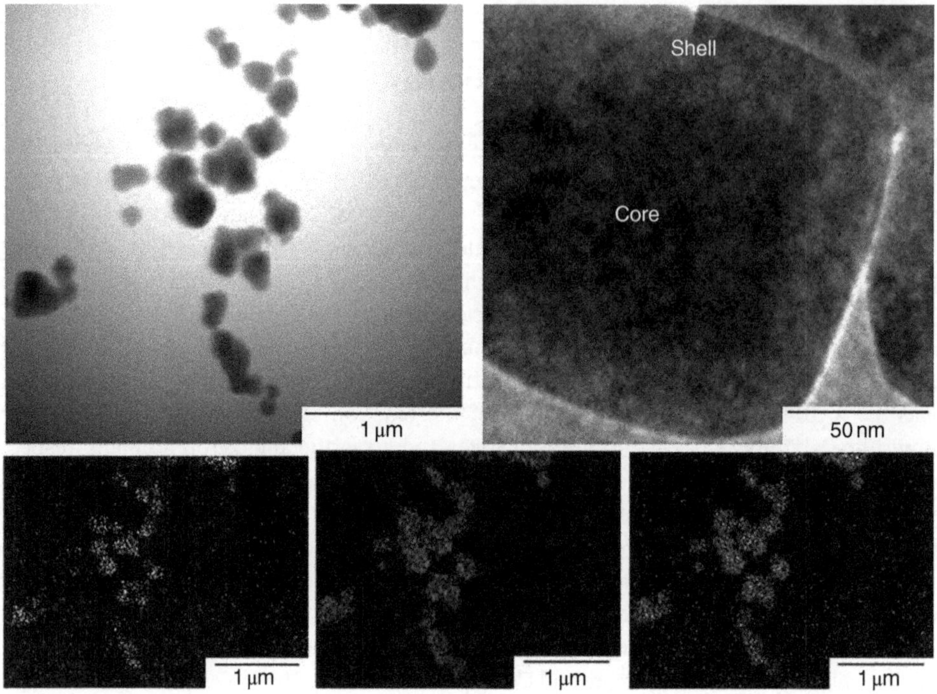

Figure 9.7 Transmission electron microscope (TEM) images (upper left and upper right) and elemental mapping images of Fe–K (bottom left), Co–K (bottom center), and their overlap (bottom right) of $[Cu^{II}(H_2O)_2]_{1.5}[Fe^{III}(CN)_6]@[Co^{II}(H_2O)_2]_{1.5}[Co^{III}(CN)_6]$. *Source:* From [32]/with permission of Elsevier.

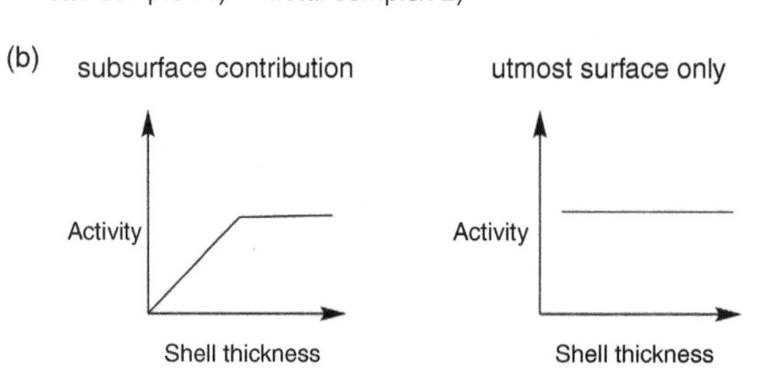

Figure 9.8 (a) Nanoparticles of polynuclear cyano-bridged metal complexes with inactive core and active shell. (b) Expected relationship between catalytic activity and shell thickness depending on subsurface contribution.

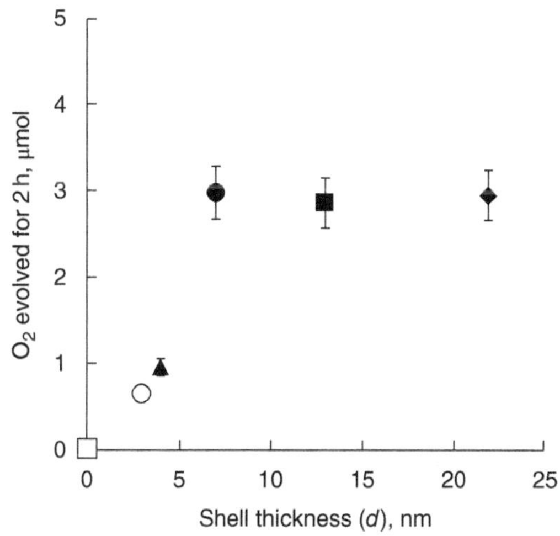

Figure 9.9 The amount of O_2 evolved for two hours by water oxidation performed by photoirradiation of an aqueous solution containing the core-shell nanoparticles of CuFe@CoCo with various CoCo shell thickness. *Source:* From [32]/with permission of Elsevier.

and only the utmost surface involved. When the subsurface is contributed to the catalysis, the catalytic activity increases up to a certain thickness and then becomes constant. On the other hand, when only the utmost surface is involved in the reaction, no clear dependence of shell thickness is expected. The O_2 yields for two hours reactions highly depend on the shell thickness thinner than 7 nm; however, the O_2 yields were independent of the shell thickness thicker than 7 nm. These results indicate the subsurface up to 7 nm is involved in the catalytic reaction (Figure 9.9) [32].

The catalytic activity of CoCo can be improved by replacing C-bound Co^{III} ions to high-valent Pt^{IV} ions because of the electronic interaction between the N-bound Co^{II} ion and the adjacent Pt^{IV} ions [15]. The more positive charge of Pt^{IV} ion than that of Co^{III} ion increases the oxidizing ability of the N-bound Co^{II} ion. The catalytic activity of CoCo can be also improved by partial replacement of the octahedral $Co^{III}(CN)_6$ unit with the tetrahedral $Pt^{II}(CN)_4$ unit because of increasing the number of defect sites. Furthermore, loading of Lewis acids such as Ca^{2+} or Sc^{3+} to the CoCo framework increases the oxidizing ability of Co^{II} ions as well as Ca^{2+} ion in the oxygen-evolving center of photosynthesis [33].

9.4 Photocatalytic H_2O_2 Production

The water oxidation catalysis of PBAs was combined with O_2 reduction catalysis of a photosensitizer, $[Ru^{II}(4,7\text{-dimethyl-1,10-phenanthroline})_3]^{2+}$, in an aqueous solution for the photocatalytic production of H_2O_2 (Figure 9.10a) [34]. Photocatalytic H_2O_2 production was examined for the reaction solutions containing 10 different PBAs in the presence of Sc^{3+} ion. The Sc^{3+} ion was added to the reaction solution for stabilizing $O_2^{\bullet -}$ formed by one-electron reduction of O_2 to be converted into H_2O_2 and O_2 by disproportionation. The highest H_2O_2 concentration recorded for the reaction solution containing $[Fe^{II}(H_2O)_2]_{1.5}[Co^{III}(CN)_6]$ was as low as 350 µM (Figure 9.10b).

The low H_2O_2 concentration may be due to the separation of a photosensitizer and a water oxidation catalyst. Thus, a photosensitizer and a water oxidation catalyst were integrated into one PBA complex. In $[Fe^{II}(H_2O)_3]_2[Ru^{II}(CN)_6]$ complex, the Fe^{II} and Ru^{II} moieties are in charge of water oxidation and photosensitizing with O_2 reduction, respectively [35].

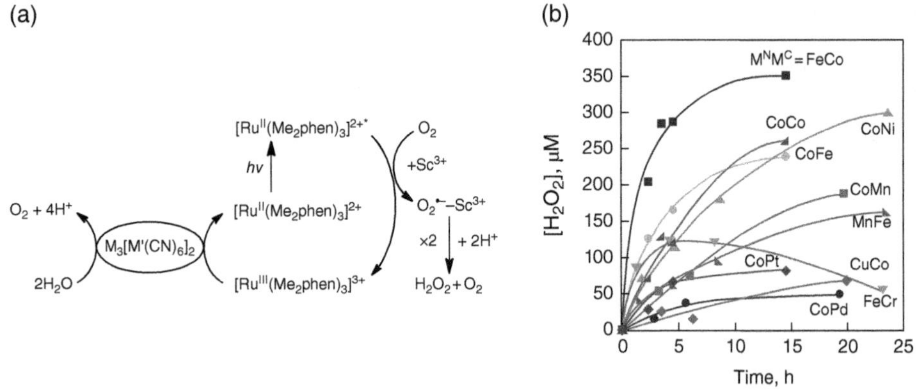

Figure 9.10 (a) Overall catalytic cycle for photocatalytic H_2O_2 production from water and molecular oxygen. (b) Time courses of production of H_2O_2 from H_2O and O_2 in an O_2-saturated aqueous solution (2.0 ml) of $[Ru(Me_2phen)_3]^{2+}$ (100 mM), $Sc(NO_3)_3$ (100 mM) and a heteropolynuclear cyanide metal complex (1.0 mg) under photoirradiation of visible light ($\lambda > 420$ nm) with a xenon lamp using a UV light cut filter at room temperature. The employed heteropolynuclear cyanide complexes are $Fe_3[Co(CN)_6]_2$, $Co_3[Co(CN)_6]_2$, $Cu_3[Co(CN)_6]_2$, $Co[Ni(CN)_4]$, $Fe_3[Cr(CN)_6]_2$, $Mn_3[Fe(CN)_6]_2$, $Co_3[Mn(CN)_6]_2$, $Co_3[Fe(CN)_6]_2$, $Co[Pd(CN)_4]$ and $Co[Pt(CN)_4]$. *Source:* From [34] / Royal Society of Chemistry / CC BY 3.0.

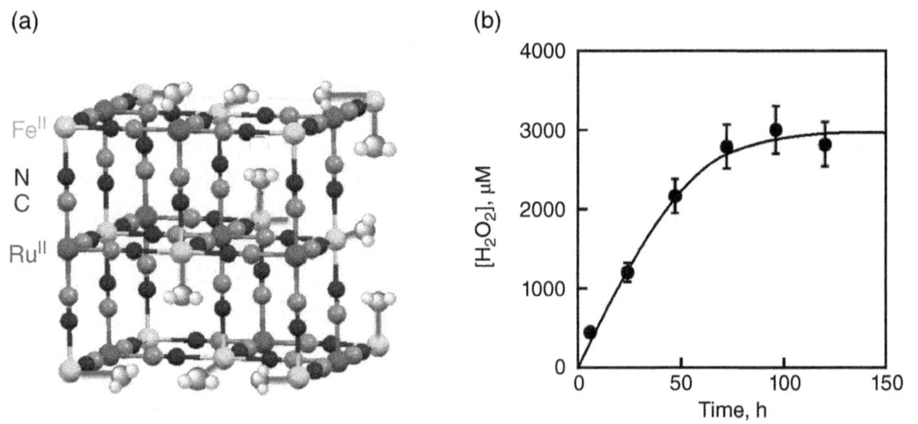

Figure 9.11 (a) Partial structure of $[Fe(H_2O)_3]_2[Ru(CN)_6]$ and (b) Time course of production of H_2O_2 from H_2O and O_2 in an O_2-saturated MeCN (2.9 ml) containing $[Fe(H_2O)_3]_2[Ru(CN)_6]$@sAl-MCM-41 (1.0 mg, 17 μM of $[Fe(H_2O)_3]_2[Ru(CN)_6]$), water (0.40 ml), and $Sc(NO_3)_3$ (0.10 M) under photoirradiation with a xenon lamp. *Source:* From [35] / with permission of American Chemical Society.

Photocatalytic production of H_2O_2 from H_2O and O_2 was performed by photoirradiation of an O_2-saturated aqueous dispersion containing $[Fe(H_2O)_3]_2[Ru(CN)_6]$@sAl-MCM-41 and $Sc(NO_3)_3$ as shown in Figure 9.11. The $[Fe(H_2O)_3]_2[Ru(CN)_6]$ was loaded on sAl-MCM-41 to increase the effective surface area. The maximum H_2O_2 concentration in this photocatalytic system reached ~3.0 mM [35], which is more than eight times higher than that obtained for the above-mentioned separated system. However, the photocatalytic reaction system was not responsible to visible light.

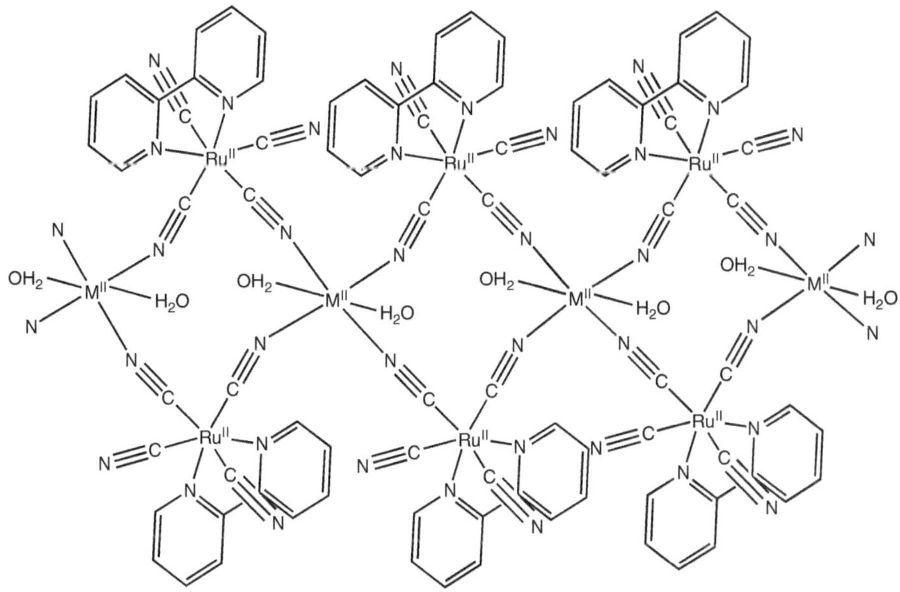

Figure 9.12 Schematic drawing of 4,2-ribbon-like chain structure. *Source:* From [36]/Royal Society of Chemistry.

The responsibility of ruthenium cyano complexes to visible light can be improved by partial replacement of the cyano ligands to 2,2′-bipyridine (bpy) or phenanthroline (phen) ligand because of strong metal-to-ligand charge transfer (MLCT). The MLCT bands of Ru complexes enable their wide uses as photosensitizers in visible-light responsive photocatalytic systems. Thus, $[Ru^{II}(bpy)(CN)_4]^{2+}$ was used as a building unit together with divalent metal ions such as Fe^{II}, Mn^{II}, Ni^{II}, and Cu^{II} ions to construct polynuclear cyano-bridged metal complexes for photocatalytic H_2O_2 production. The complexes are expected to have 4,2-ribbon-like chain structure not the jungle-gym structure (Figure 9.12) [36].

Photocatalytic H_2O_2 production from H_2O and O_2 was performed by visible light irradiation of an O_2-saturated dispersion containing Sc^{3+} ions and a Ru-based polynuclear cyano-bridged metal complex chosen from $Fe^{II}[Ru^{II}(CN)_4(bpy)]$, $Fe^{II}[Ru^{II}(CN)_4(Me_2phen)]$, and $Fe^{II}[Ru^{II}(CN)_4\{(MeO)_2bpy\}]$ (Figure 9.13) [36]. Among the complexes examined, $Fe^{II}[Ru^{II}(CN)_4(bpy)]$ showed the highest catalytic activity where the maximum concentration of H_2O_2 reached 1 mM [36]. Then, the activity of $M^{II}[Ru^{II}(CN)_4(bpy)]$ ($M^{II} = Ni^{II}$, Co^{II} or Cu^{II}) for photocatalytic H_2O_2 production was investigated under the same reaction conditions [36]. The time profiles indicated in Figure 9.13b showed that a complex with $M^{II} = Ni^{II}$ exhibited the highest catalytic activity. The replacement of reaction solvent from water to a mixture of MeOH and water resulted in the improvement in the H_2O_2 yield of the photocatalytic reaction. As a bit surprise in this reaction system, the electron donor is water, not MeOH, as evidenced by O_2 evolution by separated experiments of photocatalytic water oxidation. Thus, the reaction mechanism of photocatalytic H_2O_2 production is depicted in Figure 9.14. Ni^{II} ions are the most active species for photocatalytic water oxidation in $M^{II}[Ru^{II}(CN)_4(bpy)]$ complexes, although Co^{II} ions are the most active species in PBAs. Perhaps electronic interaction with Ru^{II} ions enhances the catalytic activity of Ni^{II} ions.

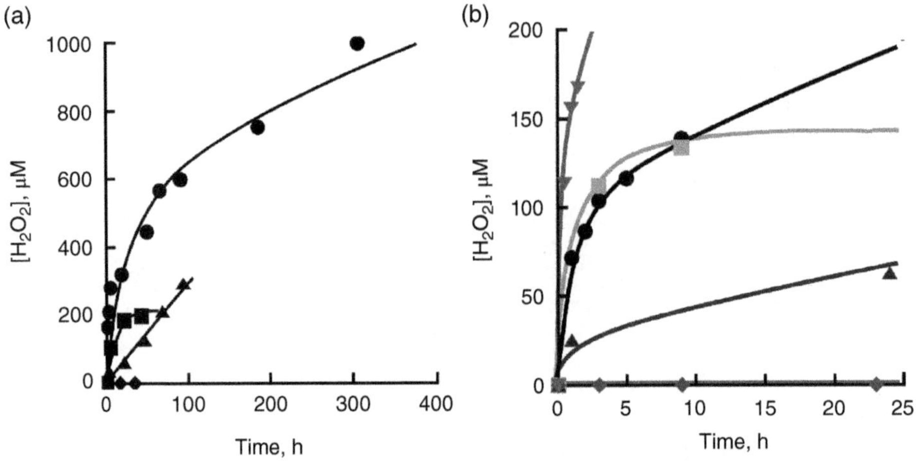

Figure 9.13 (a) Time profiles of production of H_2O_2 from H_2O and O_2 with (a) $Fe^{II}[Ru^{II}(CN)_4(bpy)]$ (1.2 mM, ●), $Fe^{II}[Ru^{II}(CN)_4(Me_2phen)]$ (1.1 mM, ■), $Fe^{II}[Ru^{II}(CN)_4\{(MeO)_2bpy\}]$ (1.0 mM, ▲), and $Fe^{II}_2[Ru^{II}(CN)_6]$ (1.4 mM, ♦), and (b) $M^{II}[Ru^{II}(CN)_4(bpy)]$ (0.12 mM; M = Ni (▼), Mn (■), Fe (●), Co (▲), and Cu (♦)), in the presence of $Sc(NO_3)_3$ (0.10 M) in O_2-saturated H_2O (2.0 ml) under visible-light irradiation with a xenon lamp using a UV cut-off filter ($\lambda > 390$ nm) at 298 K. *Source:* From [36] / with permission of Royal Society of Chemistry.

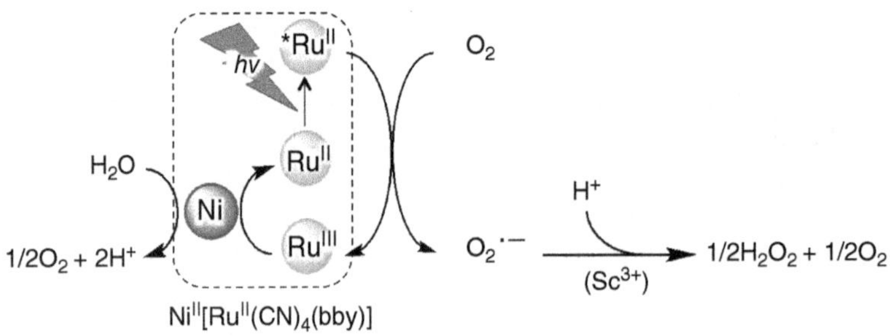

Figure 9.14 Catalytic cycle of visible-light-driven water oxidation by O_2 with $Ni^{II}[Ru^{II}(CN)_4(bpy)]$. *Source:* From [36] / with permission of Royal Society of Chemistry.

9.5 Electroreduction of H_2O_2 for Electrical Power Generation

Electrical power generation by using H_2O_2 as both oxidant and fuel was first reported by Hasegawa et al. in 2005 [10]. They designed a microflow reactor that provides the maximum power density of 23 mW/cm^2 with a cell voltage of 300 mV and current density of 76 mA/cm^2. After a while, Yamazaki et al. have reported that simple one-compartment H_2O_2 FCs without the microflow reactor can be constructed by employing Au and Ag metal electrodes as anode and cathode, respectively, in 2008 [11]. A serious problem of the one-compartment FCs was low power density because of large overpotential at the cathode. Various cathode materials such as Ag-Pb alloy and iron phthalocyanate complex were effective for reducing overpotential at H_2O_2 reduction; however, more active electrodes for H_2O_2 reduction were strongly demanded for improving the cell performance [37].

The excellent catalysis of PB for electroreduction of H_2O_2 was reported by Karyakin et al. to be used as a highly sensitive and stable electrode for biological H_2O_2 detection [38]. Thus, Shaegh et al. have used PB loaded on a carbon paper as the cathode of one-compartment H_2O_2 FC using Ag or Ni as an anode [39]. When Ag was used as an anode, the open circuit potential was 0.53 V with a maximum power density of ca. 0.8 mW/cm^2 at 0.3 V. Slightly higher open circuit potential of 0.6 V with a maximum power density of 1.55 mW/cm^2 at 0.3 V was obtained for the H_2O_2 FC with Ni anode in an aqueous H_2O_2 solution (pH 1.0) [39].

The electrocatalytic activity of nine PBAs for H_2O_2 reduction was examined for PBAs containing Fe, Co, Mn, and Cr ions [40]. Among the PBAs examined, PBAs containing Fe^{II} ions as M^N ions showed the highest activity. The performance of H_2O_2 FC using $Fe^{II}_3[\{Co^{III}(CN)_6\}_2]$ showing the highest activity as the cathode and Ni mesh as an anode was compared with that using PB as cathode. The power density of the H_2O_2 (pH 3) FC using $Fe^{II}_3[Co^{III}(CN)_6]_2$ was 110 μW/cm^2, which was almost double that with PB (~60 μW/cm^2) [40]. The power density of the H_2O_2 FC using $Fe^{II}_3[Co^{III}(CN)_6]_2$ was improved to 1.2 mW/cm^2 by decreasing the pH of the H_2O_2 solution to 1.0 from 3.0 [40]. The power density of 1.2 mW/cm^2 was achieved by using the small amount of the catalyst (2.0 mg/cm^2) (Figure 9.15) [40]. The obtained power density was slightly small but comparable to that reported for the H_2O_2 FC using PB as cathode (1.55 mW/cm^2).

The power density of H_2O_2 FCs was improved by using pillared 2D-polynuclear cyano-bridged metal complexes as cathodic materials [41]. The micropore window size of conventional PBAs is too small for penetration of H_2O_2. On the other hand, the window size of pillared 2D-polynuclear cyano-bridged metal complexes can be controlled by an appropriate choice of a pillar molecule. When pyrazine was chosen as the pillar molecule to connect the 2D complex of $Fe[M^C(CN)_4]$ (M^C = Pt or Pd), the window size was enlarged to 2.8 × 7.43 Å (Figure 9.16) [41]. The power density of H_2O_2 FCs constructed with pyrazine-bridged $Fe[M^C(CN)_4]$ (M^C = Pt or Pd) as cathodes was as high as 4.2 mW/cm^2, whereas the

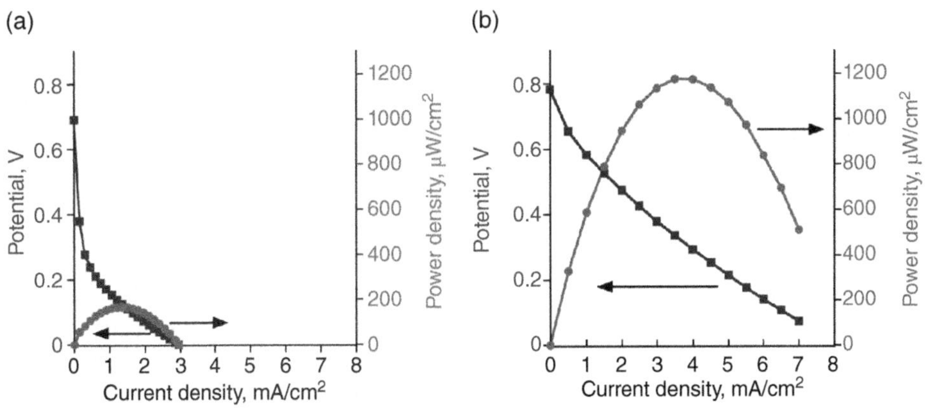

Figure 9.15 *I*–*V* (blue) and *I*–*P* (red) curves of a one-compartment H_2O_2 fuel cell with a Ni anode and a carbon-cloth electrode that was modified with a PBA: (a) $Fe^{III}_4[Fe^{II}(CN)_6]_3$ and (b) $Fe^{II}_3[Co^{III}(CN)_6]_2$. Performance tests were conducted in an aqueous solution of $HClO_4$ (pH 1) that contained H_2O_2 (0.30 M) and NaCl (1.0 M). *Source:* From [40] / with permission of John Wiley & Sons.

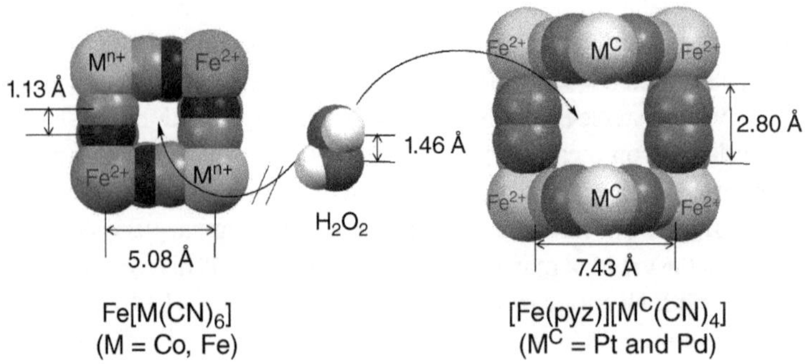

Figure 9.16 Comparison of aperture sizes of conventional PBAs and pillared 2D cyano-bridged metal complex. *Source:* From [41] / with permission of American Chemical Society.

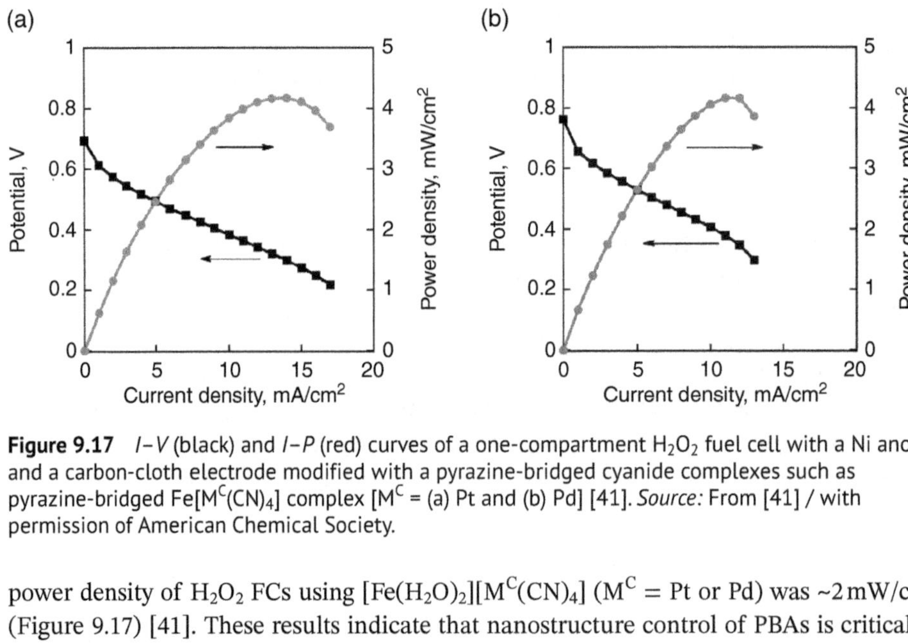

Figure 9.17 *I–V* (black) and *I–P* (red) curves of a one-compartment H_2O_2 fuel cell with a Ni anode and a carbon-cloth electrode modified with a pyrazine-bridged cyanide complexes such as pyrazine-bridged $Fe[M^C(CN)_4]$ complex [M^C = (a) Pt and (b) Pd] [41]. *Source:* From [41] / with permission of American Chemical Society.

power density of H_2O_2 FCs using $[Fe(H_2O)_2][M^C(CN)_4]$ (M^C = Pt or Pd) was ~2 mW/cm^2 (Figure 9.17) [41]. These results indicate that nanostructure control of PBAs is critical to achieve the high catalytic activity.

Summary

Hydrogen peroxide is a potential energy carrier of next generation. To be a real energy carrier, highly active catalysts should be developed for both production and electric power generation. Polynuclear cyano-bridged metal complexes, so-called PBAs, are attractive materials as heterogeneous catalysts because not only the coordination structure but also the electronic structure of metal ions with coordinatively unsaturated sites can be rationally controlled under the consideration of innocent metal ions in the frameworks. Micro- to nano-structure modification of PBAs will be profitable to improve the catalysis, resulting in opening the door to the energy society based on O_2 and H_2O_2.

References

1 Hage, R. and Lienke, A. (2006). Applications of transition-metal catalysts to textile and wood-pulp bleaching. *Angew. Chem. Int. Ed.* 45: 206.

2 Sanderson, W.R. (2000). Cleaner industrial processes using hydrogen peroxide. *Pure Appl. Chem.* 72: 1289.

3 Burek, B.O., Bormann, S., Hollmann, F. et al. (2019). Hydrogen peroxide driven biocatalysis. *Green Chem.* 21: 3232.

4 Global Hydrogen Peroxide Market (2020). Global Hydrogen Peroxide Market: Plant Capacity, Production, Demand & Supply, Concentration, Grade, End Use, Sales Channel, Competition, Trade, Customer & Price Intelligence Market Analysis, 2030. ChemAnalyst.

5 Santacesaria, E., Serio, M.D., Veloti, R., and Leone, U. (1994). Kinetics, mass transfer, and palladium catalyst deactivation in the hydrogenation step of the hydrogen peroxide synthesis via anthraquinone. *Ind. Eng. Chem. Res.* 33: 277.

6 Hâncu, D. and Beckman, E.J. (1999). Production of hydrogen peroxide in liquid CO_2. 2. catalytic hydrogenation of CO_2-philic anthraquinones. *Ind. Eng. Chem. Res.* 38: 2833.

7 Riedl, H. and Pfleiderer, G. (1936 and 1935). U.S. Patent 2,158,525 (2 October 1936 in USA, and 10 October 1935 in Germany) to I. G. Farbenindustrie, Germany.

8 Nomura, Y., Ishihara, T., Hata, Y. et al. (2008). Nanocolloidal Pd-Au as catalyst for the direct synthesis of hydrogen peroxide from H_2 and O_2. *ChemSusChem* 1: 619.

9 Centi, G., Perathoner, S., and Abate, S. (2009). Direct synthesis of hydrogen peroxide: recent advances. In: *Modern Heterogeneous Oxidation Catalysis: Design, Reactions and Characterization* (ed. N. Mizuno). Wiley-VCH.

10 Hasegawa, S., Shimotani, K., Kishi, K., and Watanabe, H. (2005). Electricity generation from decomposition of hydrogen peroxide. *Electrochem. Solid State Lett.* 8: A119.

11 Yamazaki, S., Siroma, Z., Senoh, H. et al. (2008). A fuel cell with selective electrocatalysts using hydrogen peroxide as both an electron acceptor and a fuel. *J. Power Sources* 178: 20.

12 Yamada, Y., Fukunishi, Y., Yamazaki, S., and Fukuzumi, S. (2010). Hydrogen peroxide as sustainable fuel: electrocatalysts for production with a solar cell and decomposition with a fuel cell. *Chem. Commun.* 46: 7334.

13 Han, L., Tang, P., Reyes-Carmona, A. et al. (2016). Enhanced activity and acid pH stability of Prussian Blue-type oxygen evolution electrocatalysts processed by chemical etching. *J. Am. Chem. Soc.* 138: 16037.

14 Pintado, S., Goberna-Ferron, S., Escudero-Adan, E.C., and Galan-Mascaros, J.R. (2013). Fast and persistent electrocatalytic water oxidation by Co–Fe Prussian Blue coordination polymers. *J. Am. Chem. Soc.* 135: 13270.

15 Yamada, Y., Oyama, K., Gates, R., and Fukuzumi, S. (2015). High catalytic activity of heteropolynuclear cyanide complexes containing cobalt and platinum ions: visible-light driven water oxidation. *Angew. Chem. Int. Ed.* 54: 5613.

16 Ghobadi, T.G.U., Ozbay, E., and Karadas, F. (2021). How to build Prussian Blue based water oxidation catalytic assemblies: common trends and strategies. *Chem. Eur. J.* 27: 3638.

17 Neff, V.D. (1978). Electrochemical oxidation and reduction of thin films of Prussian Blue. *J. Electrochem. Soc.* 128: 886.

18 Brown, D.J. (1724). observations and experiments upon the foregoing preparation. *Philos. Trans.* 33: 17.

19 Buser, H.J., Schwarzenbach, D., Petter, W., and Ludi, A. (1977). The crystal structure of Prussian Blue: $Fe_4[Fe(CN)_6]_3.xH_2O$. *Inorg. Chem.* 16: 2704.

20 Ludi, A., Güdel, H.-U., and Ruegg, M. (1970). Structural chemistry of Prussian blue analogs. Single-crystal study of manganese(II) hexacyanocobaltate(III), $Mn_3[Co(DcN)_6]_2$. xH_2O. *Inorg. Chem.* 9: 2224.

21 Shriver, D.F. and Brown, D.B. (1969). Environment of interstitial ions in a Prussian blue analog, $Co_3[Co(CN)_6]_2$. *Inorg. Chem.* 8: 42.

22 Simonov, A., De Baerdemaeker, T., Boström, H.L.B. et al. (2020). Hidden diversity of vacancy networks in Prussian blue analogues. *Nature* 578: 256.

23 Vaucher, S., Li, M., and Mann, S. (2000). Synthesis of Prussian Blue nanoparticles and nanocrystal superlattices in reverse microemulsions. *Angew. Chem. Int. Ed.* 39: 1793.

24 Vaucher, S., Fielden, J., Li, M. et al. (2002). Molecule-based magnetic nanoparticles: synthesis of cobalt hexacyanoferrate, cobalt pentacyanonitrosylferrate, and chromium hexacyanochromate coordination polymers in water-in-oil microemulsions. *Nano Lett.* 2: 225.

25 Shen, X., Wu, S., Liu, Y. et al. (2009). Morphology syntheses and properties of well-defined Prussian Blue nanocrystals by a facile solution approach. *J. Colloid Interface Sci.* 329: 188.

26 Ming, H., Torad, N.L.K., Chiang, Y.-D. et al. (2012). Size- and shape-controlled synthesis of Prussian Blue nanoparticles by a polyvinylpyrrolidone-assisted crystallization process. *CrystEngComm* 14: 3387.

27 Cao, M., Wu, X., He, X. et al. (2005). Shape-controlled synthesis of Prussian blue analogue $Co_3[Co(CN)_6]_2$ nanocrystals. *Chem. Commun.* 2005: 2241.

28 Roy, X., Thompson, L.K., Coombs, N., and MacLachlan, M.J. (2008). Mesostructured Prussian Blue analogues. *Angew. Chem. Int. Ed.* 47: 511.

29 Harriman, A., Pickering, I.J., Thomas, J.M., and Christensen, P.A. (1988). Metal oxides as heterogeneous catalysts for oxygen evolution under photochemical conditions. *J. Chem. Soc. Faraday Trans. 1* 84: 2795.

30 Li, S., Gao, Y., Li, N. et al. (2021). Transition metal-based bimetallic MOFs and MOF-derived catalysts for electrochemical oxygen evolution reaction. *Energy Environ. Sci.* https://doi.org/10.1039/D0EE03697H.

31 Goberna-Ferrón, S., Hernández, W.Y., Rodríguez-García, B., and Galán-Mascarós, J.R. (2014). Light-driven water oxidation with metal hexacyanometallate heterogeneous catalysts. *ACS Catal.* 4: 1637.

32 Tabe, H., Kitase, A., and Yamada, Y. (2020). Utilization of core-shell nanoparticles to evaluate subsurface contribution to water oxidation catalysis of $[Co^{II}(H_2O)_2]_{1.5}[Co^{III}(CN)_6]$ nanoparticles. *Appl. Catal. B Environ.* 262: 118101.

33 Yamada, Y., Oyama, K., Suenobu, T., and Fukuzumi, S. (2017). Photocatalytic water oxidation by persulphate with a Ca^{2+} ion-incorporated polymeric cobalt cyanide complex affording O_2 with 200% quantum efficiency. *Chem. Commun.* 53: 3418.

34 Isaka, Y., Oyama, K., Yamada, Y. et al. (2016). Photocatalytic production of hydrogen peroxide from water and dioxygen using cyano-bridged polynuclear transition metal complexes as water oxidation catalysts. *Catal. Sci. Technol.* 6: 681.

35 Aratani, Y., Oyama, K., Suenobu, T. et al. (2016). Photocatalytic hydroxylation of benzene by dioxygen to phenol with a cyano-bridged complex containing Fe^{II} and Ru^{II} incorporated in mesoporous silica–alumina. *Inorg. Chem.* 55: 5780.

36 Aratani, Y., Suenobu, T., Ohkubo, K. et al. (2017). Dual function photocatalysis of cyano-bridged heteronuclear metal complexes for water oxidation and two-electron reduction of dioxygen to produce hydrogen peroxide as a solar fuel. *Chem. Commun.* 53: 3473.

37 Yamada, Y. (2018). Electrocatalyis for hydrogen peroxide reduction used in fuel cells. In: *Anion Exchange Membrane Fuel Cells* (ed. L. An and T.S. Zhao). Springer-Nature.

38 Karyakin, A.A. (2001). Prussian Blue and its analogues: electrochemistry and analytical applications. *Electroanalysis* 13: 813.

39 Shaegh, S.A.M., Nguyen, N.T., Ehteshami, S.M.M., and Chan, S.H. (2012). A membraneless hydrogen peroxide fuel cell using Prussian Blue as cathode material. *Energy Environ. Sci.* 5: 8225.

40 Yamada, Y., Yoneda, M., and Fukuzumi, S. (2013). A robust one-compartment fuel cell with a polynuclear cyanide complex as a cathode for utilizing H_2O_2 as a sustainable fuel at ambient conditions. *Chem. Eur. J.* 19: 11733.

41 Yamada, Y., Yoneda, M., and Fukuzumi, S. (2014). High power density of one-compartment H_2O_2 fuel cells using pyrazine-bridged $Fe[M^C(CN)_4]$ ($M^C = Pt^{2+}$ and Pd^{2+}) complexes as the cathode. *Inorg. Chem.* 53: 1272.

10

Nanostructured Graphene Oxide-Based Catalysts for Fischer–Tropsch Synthesis

Swagotom Sarkar[1], Nittan Singh[2,3], Mala Khan[1], and Putla Sudarsanam[4]

[1] *Bangladesh Reference Institute for Chemical Measurements (BRiCM), Dhaka, Bangladesh*
[2] *Catalysis and Inorganic Chemistry Division, CSIR-National Chemical Laboratory, Pune, Maharashtra, India*
[3] *Academy of Scientific and Innovative Research (AcSIR), Ghaziabad, Uttar Pradesh, India*
[4] *Department of Chemistry, Indian Institute of Technology Hyderabad, Kandi, Telangana, India*

10.1 Introduction

The primary sources of energy, fuels, and chemicals currently used in different sectors are mainly derived from fossil fuels [1]. The US Energy Information Administration (EIA) assessed that around 94.3 million barrels of petroleum and liquid fuels per day were used globally in August 2020 [2]. As estimated by the Organization of the Petroleum Exporting Countries (OPEC), the consumption of the Organization for Economic Co-operation and Development (OECD) countries per capita intake is going to be 1.05 barrels in 2030 [1]. In addition, the countries that are not under OECD but developing economies of Asia, Africa, and Latin America, such as China, India, and Brazil, are about to increase their petrochemicals consumption in the future because of the rapid growth of industrialization [1]. Therefore, it is vital to find out the potential ways to produce alternative fuels and chemicals instead of entirely depending on fossil fuels to meet the increasing energy demand and the needs of future generations.

Fischer–Tropsch synthesis (FTS) is a widely used industrial process to produce liquid hydrocarbons. In FTS, syngas ($CO + H_2$) is transformed into hydrocarbons using heterogeneous catalysts, consisting of transition metals, such as Co, Fe, or Ru [3, 4]. Liquid fuels, lubricants, and other valuable chemicals are produced from FTS-synthesized hydrocarbons [3]. The catalysts' selectivity in FTS is controlled by various factors, such as catalyst's particle size, surface crystallography, promoter, support and its porosity, solvent, chemical state, and metal localization on the support surface [5–8]. In the case of bulk catalyst (i.e. Fe), sintering can be occurred during the reaction due to the application of elevated reaction temperatures, resulting in poor catalytic efficacy [9, 10]. Hence, Fe metal nanoparticles (NPs) were deposited on different supporting materials like silica (SiO_2) and alumina (Al_2O_3) to achieve improved FTS activity [11]. However, SiO_2 and Al_2O_3 supports have some limitations. For example, these supports form iron silicates and aluminates, which

Heterogeneous Nanocatalysis for Energy and Environmental Sustainability: Energy Applications, First Edition. Edited by Putla Sudarsanam, Yusuke Yamauchi, and Pankaj Bharali.

Table 10.1 Properties of graphene and its related materials.

Characteristics	Graphene	Graphene oxide	Reduced graphene oxide
Synthesis	Chemical vapor deposition	Graphite's oxidation followed by exfoliation	Thermal or chemical reduction of graphene oxide
C:O ratio	No oxygen	2–3	8–246
Young's modulus (TPa)	1–2	2–4	2–4
Electron mobility ($cm^2/V/S$)	10 000–50 000	Insulator	0.05–200
Thermal conductivity (W/mK)	$4.84–5.30 \times 10^{-3}$	1500–5800	0.1–2
Electrical conductivity (S/m)	7200	$1–5 \times 10^{-5}$	$2–7 \times 10^{-3}$
Production cost	High	Low	Low

Source: Adapted from [17].

are unlikely to be reduced and not very active catalytically. Carbon nanotubes (CNTs) are also widely used as supports for FTS catalysts [12–16]. These CNTs also have several drawbacks, associated with the complex synthesis process, high price, and low catalytic activity in FTS.

Recently, graphitic nanomaterials (Table 10.1), such as graphene (G), graphene oxide (GO), and reduced graphene oxide (rGO) having remarkable physical and chemical properties are considered as alternative supporting materials to the conventional oxide supports for FTS catalysts [18–20]. For example, these G nanomaterials exhibit high surface area, well-defined porosity, excellent electrical conductivity, remarkable thermal stability, and flexible surface chemistry, giving promising results in catalysis [21]. Then, these G-related materials can be used as carbon sources to produce active iron carbides, which are necessary for high FTS activity [11]. Besides, due to the presence of the extended π-system and the high electron density in G, reduction of metal-oxide NPs (Fe or Co oxides) may be easier on the G's surface, hence creating a large number of catalytically active metal centers for FTS [17]. Several studies have reported the significance of GO-based nanomaterials as support in FTS catalysts for achieving superior performances [22–30]. Recently, Zhao et al. investigated the influence of oxygenated species on GO, on which the pre-synthesized Fe-oxide NPs were deposited [31]. They found that the surface property of Fe-oxide NPs may be changed by using oxygenated groups or oxygen functional groups. Hence, their performance in FTS was influenced. Wei et al. reported mesoporous Fe-oxide NP-decorated GO catalysts for efficient FTS [11]. The hierarchically mesoporous structure of the catalysts exhibits a high surface area, porosity, and facile reduction/carburization properties, providing more surface-exposed active sites for syngas adsorption/activation and enhancing the access/release of the reactant and products to/from the active sites. Thus, the hierarchically mesoporous Fe-oxide/GO catalysts showed excellent FTS activity, with high selectivity to heavier hydrocarbons and olefins. Then, the surface of rGO contains oxygen-comprising groups and more atomic defects, making it suitable for anchoring metal oxide NPs. The defective nature of rGO might retard the coalescence of different metals (i.e. iron NPs). As well, the presence of acidic oxygen-comprising groups on the rGO plays a favorable role in

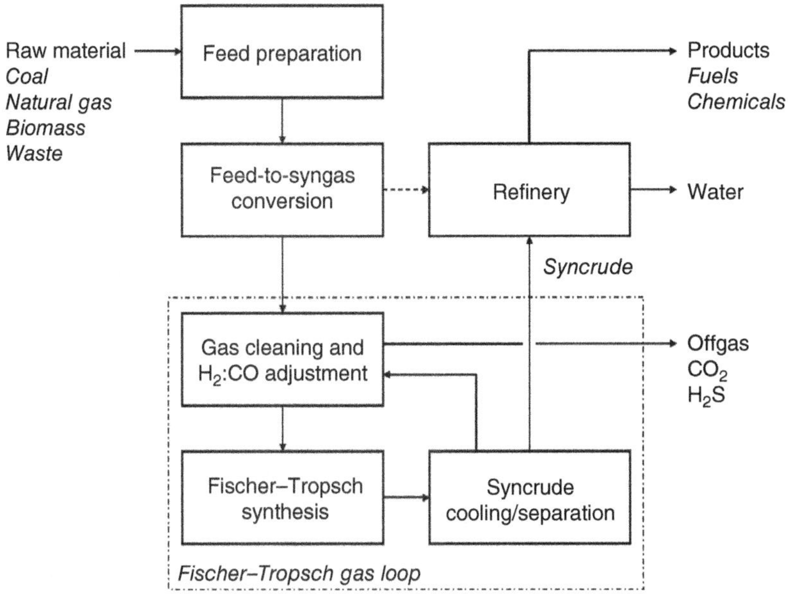

Figure 10.1 Simplified diagram of FTS facility. *Source:* Adapted from [34].

cleaving heavier hydrocarbons to the gasoline fraction [17]. So, in this chapter, we will discuss GO or related supported nanocatalysts for efficient FTS.

Franz Fischer and Hans Tropsch developed the FTS process in the 1920s [32, 33]. This process, after refining, can produce olefins, wax, alcohols, gasoline, diesel fuel, and jet fuel. The composition of the final product depends on the catalyst properties and operating conditions. Figure 10.1 shows the simplified diagram of the FTS facility [34].

At first, feed material, i.e. coal, natural gas, and biomass, is converted into syngas. For example, steam reforming of CH_4 produces syngas (1 mol CO and 3 mol H_2) [34].

$$CH_4 + H_2O \rightarrow CO + 3H_2 \tag{10.1}$$

To achieve promising results in FTS reaction, some amount of feed material is required to burn out to produce necessary heat energy, which will help to induce the reforming reaction [34]. After that, syngas is purified by removing sulfur and its related compounds, as these compounds act as the catalyst poisons for Fe- and Co-based FTS catalysts. Feed material, such as natural gas, can be freed from sulfur by applying hydro-treating and ZnO adsorption. Synthesis or syngas can also be purified by washing with cold CH_3OH. For efficient FTS, the molar ratio of H_2 and CO is adjusted after syngas purification. Under stoichiometric and appropriate reaction conditions like catalyst, pressure, temperature, and the molar ratio of the feedstock, syngas is converted into the following products:

$$\text{Alkane production: } (2n+1)H_2 + nCO \rightarrow C_nH_{2n+2} + nH_2O \tag{10.2}$$

$$\text{Alkene production: } 2nH_2 + nCO \rightarrow C_nH_{2n} + nH_2O \quad (10.3)$$

$$\text{Alcohol and ether production: } 2nH_2 + nCO \rightarrow C_nH_{2n+2}O + (n-1)H_2O \quad (10.4)$$

$$\text{Aldehyde and ketone production: } (2n-1)H_2 + nCO \rightarrow C_nH_{2n}O + (n-1)H_2O \quad (10.5)$$

$$\text{Carboxylic acid and ester production: } (2n-2)H_2 + nCO \rightarrow C_nH_{2n}O_2 + (n-2)H_2O \quad (10.6)$$

The FTS reaction is highly exothermic (the heat of the reaction is around 10 kJ/g). Finally, the produced syncrude is separated and refined for use in different sectors [34, 35].

10.2 Synthesis Methods of Various Graphene-Based Materials

Brodie first developed the GO synthesis process in 1859 [36–38]. This process (Figure 10.2), in general, uses an oxidation reaction to oxidize graphite to graphite oxide, which is exfoliated to produce GO [36, 37]. Later, this process has been modified by different strategies (Staudenmaier in 1898, Hofmann in 1937, Hummers in 1958) so that it could be

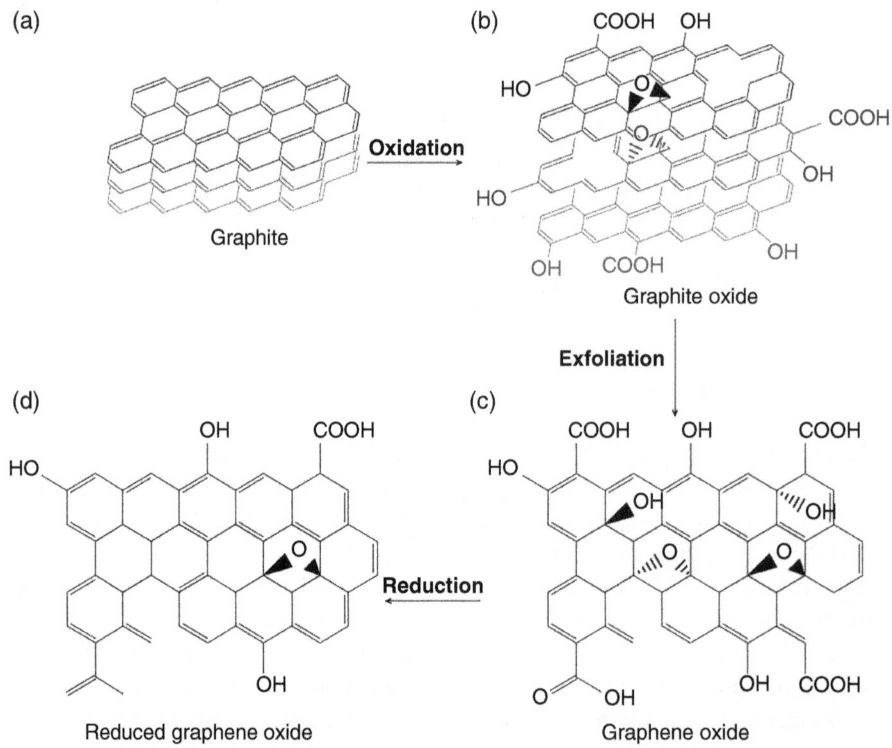

Figure 10.2 General presentation of graphite (a), its oxidation to graphite oxide (b), exfoliation of graphene oxide (c), and reduced graphene oxide (d) synthesis from graphite oxide. *Source:* Cronqvist/KTH Royal Institute of Technology [36].

cost-effective and efficient [37]. But these methods give low yields of GO and release gases like NO_2/N_2O_4 and acid vapors, which are very toxic, as well as ClO_2 gas, which is highly explosive on mixing with strong acids. Next, for the Staudenmaier method, the synthesis involves fuming HNO_3, concentrated H_2SO_4, and $KClO_3$. So, this method also produces explosive ClO_2 gas. In the Hofmann method, the synthesis also involves concentrated HNO_3, concentrated H_2SO_4, and $KClO_3$. Again, there is the evolution of toxic gases like NO_x, ClO^-, and ClO_2. The Hummer's method is quite environmentally friendly because it can be done without $KClO_3$ and concentrated HNO_3. Instead, they used H_2SO_4, $NaNO_3$, and $KMnO_4$. Nevertheless, to guarantee enough oxidation, there is a need for a large amount of concentrated acid and $KMnO_4$, which produce sewage water containing heavy metal ions and strong acids. This method is particularly time-consuming because it may require more than 100 hours for oxidation. Furthermore, the removal of Na^+ and NO_3^- residual ions is complicated after the synthesis. Due to these drawbacks, this method has also been modified by several researchers. For example, Marcano et al. [38] synthesized GO without using carcinogenic $NaNO_3$. In the following sections, several synthesis protocols developed for G, GO, and rGO are discussed.

Tour method: The synthesis procedure for GO is as follows: 96 and 10.7 ml of H_2SO_4 and H_3PO_4, are mixed maintaining a volume ratio of 9 : 1. Another mixture was prepared by mixing graphite powder and $KMnO_4$ with a molar ratio of 1 : 8 (0.6 g : 4.8 g). $KMnO_4$ plays an essential role in oxidizing graphite to GO [39]. The first mixture was slowly added into the second mixture with constant stirring for 12 hours at 40 °C. The obtained mixture was taken in 250 ml deionized water, followed by the addition of 15 ml H_2O_2, which will result in yellow coloration, suggesting overoxidation. The mixture was purified to remove metal ions, giving a yellowish paste. The resulting paste was washed with 5% HCl using the centrifugation step and continued to remove SO_4^{2-} completely, which can be examined by adding barium chloride into the supernatant. If there is no SO_4^{2-} in the supernatant, no white precipitate will be produced. The collected solid material was washed again with deionized water several times until the pH of the supernatant reaches 7, followed by drying at 50 °C for 24 hours, producing graphite oxide [38, 39].

Preparation of GO: In a typical procedure, 0.5 g of graphite oxide, ground with mortar and pestle, was dispersed in deionized water and then stirred at mild temperatures (60 °C) for 12 hours to obtain GO [38, 39].

Preparation of rGO: GO can be reduced by different methods, such as hydrothermal and solvothermal reduction [40], electrochemical reduction [41], UV and solar irradiation [42], supercritical fluids-based reduction [43], and photothermal reduction [44]. These methods use various reducing agents, such as H_2S [45], $NaBH_4$ [46], N_2H_4 [47], alkaline solution [48, 49], hydrohalic acid [50], sodium citrate [51, 52], amino acids [53–55], sugar [56, 57], ascorbic acid [58–60], plant extract [61–68], and micro-organisms [69–71]. Out of these reducing agents, N_2H_4 is the cost-effective reducing agent [47]. But, it is unstable and toxic, so it is not environmentally friendly [56, 60]. Alternatively, ascorbic acid is one of the promising green reducing agents, which can effectively reduce GO to rGO [58–60, 72].

Advantages of ascorbic acid-reducing agents: First, it is environmentally friendly and cheap [72, 73]. Second, it gives stable rGO products by inhibiting agglomeration due to the disruption of pi–pi stacking between rGO sheets [58, 73]. Stable rGO products can be synthesized in common organic solvents, such as *N,N*-dimethylformamide (DMF) or *N*-methyl-2-pyrrolidone (NMP) [46]. Third, this process prevents the presence of heteroatom in the reduced product [46]. Finally, this process is efficient and scalable for industrial production [74]. About 5 g of ascorbic acid needs to be added to the GO solution to initiate the reduction. The obtained GO-ascorbic acid solution was stirred at 60 °C for 0.5 hours. The supernatant was separated by centrifugation. It follows the addition of the excess amount of H_2O_2 and then stirring at 60 °C for 0.5 hours. The resulted product was washed with ethanol and water, followed by drying at 100 °C for 24 hours [39]. The mechanism of GO's reduction by ascorbic acid can be explained as follows: nucleophilic reagents attack the epoxy groups of GO, resulting in the epoxy ring-opening due to nucleophilic substitution reaction. It is assumed that protons and electrons are donated to GO by ascorbic acid. Then epoxide rings open, which get converted to easily leaving hydroxyl groups (–OH) and finally dehydration, giving a G sheet [39].

Synthesis of scalable rGO: About 50 ml of concentrated H_2SO_4 was added to 1 g of graphite flakes in a beaker, which is placed in an ice-water bath under mild stirring conditions [75]. To this mixture, 3 g $KMnO_4$ was added slowly while keeping the temperature under 10 °C. The suspension's temperature was increased to 25 °C and the stirring was continued for 25 minutes. Afterward, the suspension was sonicated for five minutes using an ultrasonic bath. The stirring-sonication steps were repeated 12 times, followed by the addition of 200 ml distilled water. The ultrasonication was again performed for 2 hours. The sample was divided into two equal parts to prepare rGO and oxidized graphite. For the rGO synthesis, the pH of the solution was adjusted using NaOH. The sonication was done for 1 hours, followed by the addition of 10% L-ascorbic acid solution at room temperature. The reduction reaction was conducted at 95 °C for 1 hours. The obtained black precipitates were collected by filtration using cellulose filter paper and washed using 1M HCl solution and distilled water until the pH of the solution reaches seven. Finally, the sample was dried to obtain rGO powder.

Preparation of GO using sonication method (improved Hummer's method): An improved Hummer's method was developed to prepare GO because it is eco-friendly and does not need toxic chemicals like $NaNO_3$ [76]. In brief, about 5 g of graphite was ball milled for several hours. The obtained graphite powder was added to 200 ml H_2SO_4 in a beaker placed in an ice bath, and the stirring was continued for 1 hours. Then, about 30 g of $KMnO_4$ was added slowly to the above mixture and kept it for 24 hours, resulting in a dark brown solution. Before sonication of the solution for 5 hours at 30 °C, about 100 ml distilled water was added to the above solution. The mixture was heated at 90 °C for 1 hours, and then an appropriate amount of distilled water and 30 ml H_2O_2 was added under stirring conditions. For GO, the suspension was washed with HCl and distilled water alternatively until the pH of the solution reaches seven. The sample was collected by centrifugation, and then, the sample was dried in an oven [77].

Preparation of rGO: Here, L-ascorbic acid was used to reduce GO to rGO. The procedure involves the addition of 100 ml distilled water to 10 g L-ascorbic acid. This solution was added to GO-s (prepared by sonication) and GO-ws (prepared without sonication). The obtained mixture was heated at 95 °C for 1 hours, followed by sonication immediately for 15 minutes. The obtained black precipitates were filtered using a vacuum pump, followed by washing with 1M HCl and distilled water until it becomes neutral. Finally, the purified product was dried at oven temperature overnight [77].

Synthesis of rGO by modified Hummer's method: About 108 ml H_2SO_4 and 12 ml H_3PO_4 were added to a mixture of 5 g graphite and 2.5 g $NaNO_3$. The obtained mixture was placed in an ice bath for 10 minutes under stirring conditions. Then, 15 g of $KMnO_4$ was added to maintain the temperature below 5 °C. The suspension was placed in an ice bath for 2 hours, followed by stirring for 60 minutes. The temperature of the suspension was increased to 40 °C in a water bath and maintained it for 60 minutes under stirring conditions. Then, the temperature was increased to 98 °C and kept it for 60 minutes. During this time, deionized water was added to make the suspension volume of 400 ml. After five minutes, 15 ml H_2O_2 was added, followed by centrifugation with deionized water and 5% HCl simultaneously, and finally, dried the product at 60 °C [78].

Purification of graphene and synthesis of functionalized graphene support: The impurities in graphene need to be removed before the experiments as they can adversely affect the products' properties. The purification of graphene involves the refluxing of the mixture of pristine graphene with 30% HNO_3 at 120 °C overnight, followed by washing with distilled water, and then drying at 120 °C for 6 hours. To remove amorphous carbons, the material is treated at 400 °C in air flow, with a ramping rate of 10 °C/min for 20 minutes, obtaining graphene nanosheet (GNS) [79]. To prepare nitrogen-functionalized graphene nanosheet (N-GNS), the purified graphene is treated at 400 °C for 6 hours in ammonia (10 vol.% NH_3 in He) with a flow rate of 25 sccm. After that, ammonia was neutralized with an acetic acid solution [80].

10.3 Activity of Mono- and Bimetallic Catalysts Dispersed on GO and rGO Supports

A one-pot aqueous hydrolysis-decrease system was used for synthesizing rGO, which is used as a support for Fe_2O_3 NPs. The developed Fe-rGO catalyst is very stable and selective for FTS reaction. The defective nature of rGO can inhibit the aggregation of Fe NPs. The rGO material, with acidic hydrogen species, may help in the cracking of heavier hydrocarbons [8]. Karimi et al. synthesized G- and CNT-supported catalysts with 15 wt% Co loading using the wet impregnation method for FTS reaction [81]. The G-supported catalyst contains a high dispersion of Co clusters with about 6 nm of average particle size, as noticed by TEM analysis. Then, a decrease in the reduction temperature of Co-oxide species was observed in the case of G support. Interestingly, the CO conversion rate increased from 61 to 74%. Besides, about 22% increase was observed in the rate of FTS; the product distribution moved to the higher atomic weight hydrocarbons, in contrast with CNTs-supported catalysts. On the G-supported Co catalysts, about 22% of the primary CO conversion was decreased during FTS over 480 hours time-on-stream. Similarly, for CNTs-supported catalyst, the CO conversion was decreased to 34%.

The FTS performance of GO-supported Fe catalysts can be tuned efficiently by simply changing the Fe precursor [79]. The Fe/G-C catalyst showed the highest FTS activity and C5+ selectivity. High dispersion of NPs, significantly increased BET surface area, and superior reduction/carburization properties were the reasons for the observed catalytic activity and selectivity. DFT calculations estimated the highest binding energy in between Fe precursors and GO in Fe/G-C. The advantage of using Fe NPs is that active sites do not aggregate by combining Fe and oxygen-containing functional groups. So, iron NPs and oxygen-containing functional groups also help the catalysts to perform exceptionally stable. Since the catalyst possesses higher reduction/carburization properties, it gave higher selectivity toward heavier hydrocarbons. Wei et al. synthesized hierarchically mesoporous Fe-oxide/GO composites for FTS using a one-step hydrothermal method [11]. The catalysts provide the following properties: hierarchical porous geometry, superior surface area, weak Fe-GO interaction, and facile reduction/carburization properties. They reported that with the increase in the number of hierarchical pore structures, there is an increase in the catalytic active sites. Consequently, CO conversion rate reaches around 83–90%, C5+ selectivity reaches about 67–71%, and olefin to paraffin ratio of the catalyst becomes 4.9–5.5. Thus, hierarchical pore structure and weaker Fe-GO interaction along with other properties contributed to higher FTS activity.

The performance of FeO/GO catalysts for FTS toward C5+ hydrocarbon was increased due to the pyrolytic treatment of the support at 600 °C [31]. It was found that catalytic properties are related to the presence of oxygenated functional groups on the carbon support. They suggested that the number of surface oxygen groups of carbon support might change the properties of the iron phase. As a result, FTS activity might also change. Torshizi et al. synthesized a Co/rGO catalyst using the impregnation method with Co loading of 5, 15, and 30 wt% [82]. The H_2-TPR results revealed that if the Co loading is increased, the carbon defects are also increased, confirmed by the Raman study. These defects increase the Co–support interaction and C5+ selectivity and decrease CH_4 selectivity. The higher C5+ selectivity by the catalyst was due to chain propagation in the primary chain growth reactions. Nanosized Pd catalysts, dispersed on GO and GO-TiO_2, were prepared using $PdCl_2(MeCN)_2$ and $Pd(OAc)_2$ precursors [83]. The Pd precursor and the synthetic method affect the catalyst's structure. By introducing Pd into the GO matrix increases the number of sp^2 centers, which ultimately increase photoemission. With the application of 1 atm. CO, 0.2 mol% catalyst, the activity and selectivity of the catalyst were increased toward di-aryl ketones formation. The best results were achieved (95% ketone selectivity) over Pd/GO-TiO_2 with good recyclability of the catalyst. Nasser et al. synthesized an Fe NPs/rGO catalyst with Mn as a promoter, and this catalyst is used to produce light olefins from FTS reaction [84]. The catalyst structure contains Mn as shell and Fe as core. It was found that Mn shell thickness controls the catalyst's efficacy and olefin selectivity. For example, with the increase in Mn shell thickness, both the rates of water-gas-shift reaction and CO conversion were decreased. About 19% olefin yield was achieved at the optimum conditions: 2 MPa, 300 °C, and 4.2 l/g/h of 1 : 1 H_2 : CO flow.

Worm-like Ru nanostructures dispersed on G sheets were developed for photocatalytic FTS reaction [85]. The reaction conditions used are 14.4 mol CO mol/Ru/h, 150 °C, 2.0 MPa H_2, 1 MPa CO, and visible light. When the irradiation intensity was increased or irradiation wavelength was decreased, the FTS reaction rate was found to be improved. There is a work function difference between G and Ru. So, hot electrons move from G to Ru enhance the catalytic activity. The Ru/G catalyst was stable for over 30 hours period of time with a slight decrease of catalytic activity after 10 hours. This work provides a green and coherent photocatalytic route for FTS. Hajjar et al. synthesized GO and nanoporous G (NPG) for Co catalysts with high BET surface area of 290 and 700 m^2/g, respectively [86]. The Co/NPG catalysts produced heavier hydrocarbons due to their higher surface area and efficient hydrogen carrier properties. In contrast, Co/Al_2O_3 catalyst showed higher selectivity toward lower hydrocarbon (C_1–C_4) than Co/NPG catalyst.

Li et al. developed an Fe-based catalyst supported on graphite. Results indicated that adding the right amount of potassium (K) to the catalyst improved the reductive nature and significantly increased the selectivity to liquid hydrocarbons [87]. The presence of K promoter makes graphite as an electron donor to Fe. As a result, the catalyst becomes more active, suppressing lower hydrocarbon (i.e. methane) selectivity and increasing heavier hydrocarbon (C_4–C_{10}) productivity. On the other hand, due to the absence of K promoter, graphite withdraws electrons from Fe; thus higher reaction temperature is needed to activate the catalyst, resulting in increased lower hydrocarbon selectivity. Bimetallic Co–Mn catalysts supported on G and rGO were developed for FTS reaction. The catalyst showed better performance when the Co-Mn ratio was 1. These catalysts showed excellent stability. The higher thermal conductivity of the G support causes higher stability to the catalyst during the high-temperature reaction. Functional groups and defect sites present on rGO, confirmed by XPS and Raman study, were contributed to increased conversion and selectivity to light olefins compared to G-supported catalyst [88].

G-supported Fe-K and Fe-Mn catalysts exhibited excellent FTS activity for heavier hydrocarbons [22]. Fe_{15}-G catalyst showed 51.5% CO conversion, which was increased to 73.5 and 92% over $Fe_{15}K_5$-G and $Fe_{15}Mn_5$-G catalysts, respectively. The selectivity for heavier hydrocarbons (C8+) increased from 66.5% for Fe_{15}-G to 86.7% for $Fe_{15}K_5$-G catalysts. The high efficacy of these catalysts was explained by the presence of well-scattered NPs, the large surface area of G, high defects site on G formation of active Fe_5C_2, and well-balanced interaction between NPs and GO support. Deerattrakula et al. synthesized rGO nanosheets and used as a support for bimetallic Cu-Zn catalysts for CH_3OH production from FTS reaction [89]. They demonstrated that 250 °C reaction temperature and 10 wt% Cu and Zn metal loading could give excellent catalytic properties. When the reaction temperature is above 250 °C, the space-time yield of CH_3OH was decreased. Besides, if the Cu and Zn loading is more than 10 wt%, aggregation might happen, dispersion and specific surface area might decrease, which results in insufficient catalytic activity. This catalyst produced CH_3OH of around 424 $mg_{MeOH}\ g_{cat}^{-1}\ h^{-1}$, which can be scalable.

10.4 Research Gaps and Conclusions

Although considerable research has been done so far using GO support, we need to give extensive focus on G-supported FTS research for future cost-effective and reliable sources of synthetic fuel. From the above studies, we found that the properties of G-based supported catalysts, such as defective nature, acidic oxygen-containing functional groups, hierarchical pore structure, high BET surface area, facile reduction/carburization properties, highest binding energy in between NPs and G support, doping, recyclability, promoter thickness, photocatalytic behavior of graphene, electron-donating properties of graphite in the presence of K promoter, the higher thermal conductivity of G, bimetallic NP ratio, formation of active iron carbide, and many other properties have contributed to the enhancement of FTS performance. The development of GO-supported FTS catalyst is still at its primary stage. So, we need to overcome the following drawbacks to produce an optimized G-supported FTS catalyst.

- Few research studies have been done in developing bimetallic catalysts using GO support. So, this gap can be filled by doing extensive research toward developing bimetallic catalysts on G-based supports.
- In-depth understanding of the properties of the catalysts using advanced analytic techniques to derive better structure-activity properties toward improved FTS performance, which will guide us for future research directions.
- Cost-effective GO support needs to be synthesized. Currently used GO materials are expensive.
- Synthesis methodology has to be reproducible and should produce desired properties in FTS catalysts.
- Photocatalytic FTS routes have been explored as well. But only a few works have been done using GO support. So, these gaps need to be fulfilled by doing further research as well.
- It is difficult sometimes to deposit a few metal oxides on G due to the trivial interaction between G and metal oxide NPs. So, by developing a cost-effective and efficient metal oxide depositing method, FTS performance can be improved.
- Tuning the surface interactions between bimetallic NPs and functional groups on rGO needs to be done to achieve improved FTS performances toward heavier hydrocarbons.
- Green synthesis methods for producing G, GO, and rGO, including the use of green reducing agents need to be explored.
- The mechanism of incorporating mono- or bimetallic NPs on GO-based support needs to be investigated more deeply.
- The influence of the iron carbide phase, which is generated by Fe-GO catalyst, has not been investigated very well. So, the influence of iron carbide, support on iron carbides formation, and the configuration of iron carbides has to be investigated to understand the performance of FTS.

Acknowledgments

Nittan Singh acknowledges the Council of Scientific and Industrial Research (CSIR), New Delhi, for a research fellowship.

References

1 Santos, R. and Alencar, A. (2020). Biomass-derived syngas production via gasification process and its catalytic conversion into fuels by Fischer Tropsch synthesis: a review. *Int. J. Hydrog. Energy.* 45 (36): 18114–18132.

2 Short term energy outlook. https://www.eia.gov/outlooks/steo/report/global_oil.php (accessed September 2020).

3 Chernyak, S., Strokova, N., Fedorova, E. et al. (2019). Adsorption of water and n-hexane on pristine and oxidized carbon nanotube supports of cobalt-based Fischer–Tropsch catalysts. *Phys. Chem. Chem. Phys.* 21 (24): 13234–13240.
4 Gao, W., Zhu, Q., and Ma, D. (2018). Nanostructured catalyst for Fischer–Tropsch synthesis. *Chin. J. Chem.* 36 (9): 798–808.
5 Bezemer, G., Bitter, J., Kuipers, H. et al. (2006). Cobalt particle size effects in the Fischer–Tropsch reaction studied with carbon nanofiber supported catalysts. *J. Am. Chem. Soc.* 128 (12): 3956–3964.
6 Khodakov, A., Chu, W., and Fongarland, P. (2007). Advances in the development of novel cobalt Fischer–Tropsch catalysts for synthesis of long-chain hydrocarbons and clean fuels. *ChemInform.* 107 (5): 1692–1744.
7 Zhu, Y., Ye, Y., Zhang, S. et al. (2012). Synthesis and catalysis of location-specific cobalt nanoparticles supported by multiwall carbon nanotubes for Fischer–Tropsch synthesis. *Langmuir.* 28: 8275–8280.
8 Zhai, P., Sun, G., Zhu, Q., and Ma, D. (2013). Fischer-Tropsch synthesis nanostructured catalysts: understanding structural characteristics and catalytic reaction. *Nanotechnol. Rev.* 2 (5): 547–576.
9 Shroff, M., Kalakkad, D., Coulter, K. et al. (1995). Activation of precipitated iron Fischer-Tropsch synthesis catalysts. *J. Catal.* 156 (2): 185–207.
10 Torres Galvis, H. and de Jong, K. (2013). Catalysts for production of lower olefins from a synthesis gas: a review. *ACS Catal.* 3 (9): 2130–2149.
11 Wei, Y., Yan, L., Ma, C. et al. (2020). Mesoporous iron oxide nanoparticle-decorated graphene oxide catalysts for Fischer-Tropsch synthesis. *ACS Appl. Nano Mater.* 3 (7): 7182–7191.
12 Schulte, H., Graf, B., Xia, W., and Muhler, M. (2011). Nitrogen- and oxygen-functionalized multiwalled carbon nanotubes used as support in iron-catalyzed, high-temperature Fischer–Tropsch synthesis. *ChemCatChem.* 4 (3): 350–355.
13 van Steen, E. and Prinsloo, F. (2002). Comparison of preparation methods for carbon nanotubes supported iron Fischer–Tropsch catalysts. *Catalysis Today* 71 (3–4): 327–334.
14 Guczi, L., Stefler, G., Geszti, O. et al. (2006). CO hydrogenation over cobalt and iron catalysts supported over multiwall carbon nanotubes: effect of preparation. *J. Catal.* 244: 24–32.
15 Chen, W., Pan, X., and Bao, X. (2007). Tuning of redox properties of iron and iron oxides via encapsulation within carbon nanotubes. *J. Am. Chem. Soc.* 129 (23): 7421–7426.
16 Chen, W., Fan, Z., Pan, X., and Bao, X. (2008). Effect of confinement in carbon nanotubes on the activity of Fischer-Tropsch iron catalyst. *J. Am. Chem. Soc.* 130 (29): 9414–9419.
17 Sharma, B., Shekhar, S., Malik, P., and Jain, P. (2018). Study of mechanism involved in synthesis of graphene oxide and reduced graphene oxide from graphene nanoplatelets. *Mater. Res. Express.* 5 (6): 065012.
18 Serp, P. and Figueiredo, J. (2009). *Carbon Materials For Catalysis*. Hoboken, NJ: Wiley.
19 Wei, Y., Zhang, C., Liu, X. et al. (2018). Enhanced Fischer–Tropsch performances of graphene oxide-supported iron catalysts via argon pretreatment. *Catal. Sci. Technol.* 8 (4): 1113–1125.

20 Xiong, H., Jewell, L., and Coville, N. (2015). Shaped carbons as supports for the catalytic conversion of syngas to clean fuels. *ACS Catal.* 5 (4): 2640–2658.

21 Sun, B., Xu, K., Nguyen, L. et al. (2012). preparation and catalysis of carbon-supported iron catalysts for Fischer-Tropsch synthesis. *ChemCatChem* 4 (10): 1498–1511.

22 Moussa, S., Panchakarla, L., Ho, M., and El-Shall, M. (2014). Graphene supported iron-based nanoparticles for catalytic production of liquid hydrocarbons from synthesis gas. *ACS Catal.* 4 (2): 535–545.

23 Geim, A. and Novoselov, K. (2007). The rise of graphene. *Nat. Mater.* 6 (3): 183–191.

24 Dreyer, D., Ruoff, R., and Bielawski, C. (2010). From conception to realization: an historial account of graphene and some perspectives for its future. *Angew. Chem. Int. Ed.* 49 (49): 9336–9344.

25 Zhang, Y., Tan, Y., Stormer, H., and Kim, P. (2005). Experimental observation of the quantum Hall effect and Berry's phase in graphene. *Nature.* 438: 201–204.

26 Zhu, Y., Murali, S., Cai, W. et al. (2010). Graphene and graphene oxide: synthesis, properties, and applications. *Adv. Mater.* 22 (35): 3906–3924.

27 Su, D., Perathoner, S., and Centi, G. (2013). Nanocarbons for the development of advanced catalysts. *Chem. Rev.* 113 (8): 5782–5816.

28 Jasuja, K., Linn, J., Melton, S., and Berry, V. (2010). Microwave-reduced uncapped metal nanoparticles on graphene: tuning catalytic, electrical, and raman properties. *J. Phys. Chem.* 1 (12): 1853–1860.

29 Moussa, S., Abdelsayed, V., and Samy El-Shall, M. (2011). Laser synthesis of Pt, Pd, CoO and Pd–CoO nanoparticle catalysts supported on graphene. *Chem. Phys. Lett.* 510 (4-6): 179–184.

30 Moussa, S., Siamaki, A., Gupton, B., and El-Shall, M. (2011). Pd-partially reduced graphene oxide catalysts (Pd/PRGO): laser synthesis of Pd nanoparticles supported on PRGO nanosheets for carbon—carbon cross coupling reactions. *ACS Catal.* 2 (1): 145–154.

31 Zhao, H., Zhu, Q., Gao, Y. et al. (2013). Iron oxide nanoparticles supported on pyrolytic graphene oxide as model catalysts for Fischer Tropsch synthesis. *Appl. Catal. A. Gen.* 456: 233–239.

32 Hubbard, W., Biles, L., Mayfield, C., and Ashton, S. (2007). *Sustainable Forestry for Bioenergy and Bio-based Products: Trainers Curriculum Notebook*. Athens, GA: Southern Forest Research Partnership, Inc.

33 Steynberg, A.P. and Dry, M.E. (ed.) (2004). *Fischer-Tropsch Technology*. Elsevier.

34 Klerk, A.D. and Furimsky, E. (2010). *Catalysis in the Refining of Fischer–Tropsch Syncrude*. Royal Society of Chemistry.

35 https://www.sigmaaldrich.com/materials-science/material-science-products.html?TablePage=112007852 (accessed October 2020)

36 Cronqvist, P. (2019). Chemistry of ascorbic acid reduction of graphene oxide: reduction of graphene oxide in solution and film. Master's thesis. KTH Royal Institute of Technology, School of Information and Communication Technology, SE-100 44 Stockholm, Sweden.

37 De Silva, K., Huang, H., Joshi, R., and Yoshimura, M. (2017). Chemical reduction of graphene oxide using green reductants. *Carbon.* 119: 190–199.

38 Marcano, D.C., Kosynkin, D.V., and Berlin, J.M. (2010). Improved synthesis of graphene oxide. *ACS Nano.* 4 (8): 4806–4814.

39 Emiru, T. and Ayele, D. (2017). Controlled synthesis, characterization and reduction of graphene oxide: a convenient method for large scale production. *Egypt. J. Basic Appl. Sci.* 4 (1): 74–79.
40 Zhou, Y., Bao, Q., Tang, L. et al. (2009). Hydrothermal dehydration for the "green" reduction of exfoliated graphene oxide to graphene and demonstration of tunable optical limiting properties. *Chem. Mater.* 21 (13): 2950–2956.
41 Zhang, X., Zhang, D., Chen, Y. et al. (2012). Electrochemical reduction of graphene oxide films: preparation, characterization and their electrochemical properties. *Chin. Sci. Bull.* 57 (23): 3045–3050.
42 Nyangiwe, N., Khenfouch, M., Thema, F. et al. (2015). Free-green synthesis and dynamics of reduced graphene sheets via sun light irradiation. *Graphene.* 04 (03): 54–61.
43 Kong, C., Song, W., Meziani, M. et al. (2012). Supercritical fluid conversion of graphene oxides. *J. Supercrit. Fluids.* 61: 206–211.
44 Yeh, M., Lin, L., Chang, L. et al. (2014). Dye-sensitized solar cells with reduced graphene oxide as the counter electrode prepared by a green photothermal reduction process. *ChemPhysChem.* 15 (6): 1175–1181.
45 Hofmann, U. and Frenzel, A. (1934). The reduction of graphite oxide by hydrogen sulfide. *Kolloid-Z.* 68: 149–151.
46 Fernández-Merino, M., Guardia, L., Paredes, J. et al. (2010). Vitamin C is an ideal substitute for hydrazine in the reduction of graphene oxide suspensions. *J. Phys. Chem. C.* 114 (14): 6426–6432.
47 Stankovich, S., Dikin, D., Piner, R. et al. (2007). Synthesis of graphene-based nanosheets via chemical reduction of exfoliated graphite oxide. *Carbon.* 45 (7): 1558–1565.
48 Xu, C., Shi, X., Ji, A. et al. (2015). Fabrication and characteristics of reduced graphene oxide produced with different green reductants. *PLos One.* 10 (12): e0144842.
49 Davies, M., Partridge, D., and Austin, J. (1991). *Vitamin C: It's Chemistry And Biochemistry*, 1e. Cambridge, UK: Royal Society Of Chemistry.
50 Pei, S., Zhao, J., Du, J. et al. (2010). Direct reduction of graphene oxide films into highly conductive and flexible graphene films by hydrohalic acids. *Carbon.* 48 (15): 4466–4474.
51 Zhang, Z., Chen, H., Xing, C. et al. (2011). Sodium citrate: a universal reducing agent for reduction/decoration of graphene oxide with au nanoparticles. *Nano Res.* 4 (6): 599–611.
52 Tian, J. and Zhang, H. (2016). Preparation and characterization of reduced graphene oxide using ascorbic acid and sodium citrate as binary reductant. *Fuller. Nanotub. Carbon Nanostructures* 25 (1): 17–22.
53 Chen, D., Li, L., and Guo, L. (2011). An environment-friendly preparation of reduced graphene oxide nanosheets via amino acid. *Nanotechnology.* 22 (32): 325601.
54 Bose, S., Kuila, T., Mishra, A. et al. (2012). Dual role of glycine as a chemical functionalizer and a reducing agent in the preparation of graphene: an environmentally friendly method. *J. Mater. Chem.* 22 (19): 9696–9703.
55 Liu, J., Fu, S., Yuan, B. et al. (2010). Toward a universal "adhesive nanosheet" for the assembly of multiple nanoparticles based on a protein-induced reduction/decoration of graphene oxide. *J. Am. Chem. Soc.* 132 (21): 7279–7281.
56 Zhu, C., Guo, S., Fang, Y., and Dong, S. (2010). Reducing sugar: new functional molecules for the green synthesis of graphene nanosheets. *ACS Nano.* 4 (4): 2429–2437.

57 Kim, Y., Kim, M., and Min, D. (2011). Biocompatible reduced graphene oxide prepared by using dextran as a multifunctional reducing agent. *Chem. Commun.* 47 (11): 3195–3197.

58 Zhang, J., Yang, H., Shen, G. et al. (2010). Reduction of graphene oxide via L-ascorbic acid. *Chem. Commun.* 46 (7): 1112–1114.

59 Andrijanto, E., Shoelarta, S., Subiyanto, G., and Rifki, S. (2016). Facile synthesis of graphene from graphite using ascorbic acid as reducing agent. *AIP Conf. Proc.* 1725: 020003.

60 Wang, Y., Shi, Z., and Yin, J. (2011). Facile synthesis of soluble graphene via a green reduction of graphene oxide in tea solution and its biocomposites. *ACS Appl. Mater. Interfaces.* 3 (4): 1127–1133.

61 Muthoosamy, K., Geetha Bai, R., Babangida Abubakar, I. et al. (2015). Exceedingly biocompatible and thin-layered reduced graphene oxide nanosheets using an eco-friendly mushroom extract strategy. *Int. J. Nanomed.* 10 (1): 1505–1519.

62 Khan, M., Al-Marri, A., Khan, M. et al. (2015). Green approach for the effective reduction of graphene oxide using Salvadora persica L. Root (Miswak) extract. *Nanoscale Res. Lett.* 10: 281.

63 Kuila, T., Bose, S., Khanra, P. et al. (2012). A green approach for the reduction of graphene oxide by wild carrot root. *Carbon.* 50 (3): 914–921.

64 Thakur, S. and Karak, N. (2012). Green reduction of graphene oxide by aqueous phytoextracts. *Carbon.* 50 (14): 5331–5339.

65 Kartick, B., Srivastava, S., and Srivastava, I. (2013). Green synthesis of graphene. *J. Nanosci. Nanotechnol.* 13 (6): 4320–4324.

66 Tavakoli, F., Salavati-Niasari, M., Qbadiei, A., and Mohandes, F. (2015). Green synthesis and characterization of graphene nanosheets. *Mater. Res. Bull.* 63: 51–57.

67 Haghighi, B. and Tabrizi, M. (2013). Green-synthesis of reduced graphene oxide nanosheets using rose water and a survey on their characteristics and applications. *RSC Adv.* 3 (32): 13365–13371.

68 Lei, Y., Tang, Z., Liao, R., and Guo, B. (2011). Hydrolysable tannin as enviromentally friendly reducer and stablizer for graphene oxide. *Green Chem.* 13 (7): 1655–1658.

69 Akhavan, O. and Ghaderi, E. (2012). Escherichia coli bacteria reduce graphene oxide to bactericidal graphene in a self-limiting manner. *Carbon.* 50 (5): 1853–1860.

70 Salas, E., Sun, Z., Lüttge, A., and Tour, J. (2010). Reduction of graphene oxide via bacterial respiration. *ACS Nano.* 4 (8): 4852–4856.

71 Wang, G., Qian, F., Saltikov, C. et al. (2011). Microbial reduction of graphene oxide by Shewanella. *Nano Res.* 4 (6): 563–570.

72 Habte, A. and Ayele, D. (2019). Synthesis and characterization of reduced graphene oxide (rGO) started from graphene oxide (GO) using the tour method with different parameters. *Adv. Mater. Sci. Eng.* 2019: 1–9.

73 Pei, S. and Cheng, H.M. (2012). The reduction of graphene oxide. *Carbon.* 50 (9): 3210–3228.

74 Fathy, M., Gomaa, A., Taher, F. et al. (2016). Optimizing the preparation parameters of GO and rGO for large-scale production. *J. Mater. Sci.* 51 (12): 5664–5675.

75 Abdolhosseinzadeh, S., Asgharzadeh, H., and Seop Kim, H. (2015). Fast and fully-scalable synthesis of reduced graphene oxide. *Sci. Rep.* 5 (1).

76 Chen, J., Yao, B., Li, C., and Shi, G. (2013). An improved hummers method for eco-friendly synthesis of graphene oxide. *Carbon.* 64: 225–229.

77 Zainuddin, M., Nik Raikhan, N., Othman, N., and Abdullah, W. (2018). Synthesis of reduced graphene oxide (rGO) using different treatments of graphene oxide (GO). *IOP. Conf. Ser. Mater. Sci. Eng.* 358, 012046.

78 Song, J., Wang, X., and Chang, C.T. (2014). Preparation and Characterization of Graphene Oxide. *J. Nanomater.* 276143.

79 Wei, Y., Luo, D., Zhang, C. et al. (2018). Precursor controlled synthesis of graphene oxide supported iron catalysts for Fischer–Tropsch synthesis. *Catal. Sci. Technol.* 8 (11): 2883–2893.

80 Taghavi, S., Asghari, A., and Tavasoli, A. (2017). Enhancement of performance and stability of graphene nano sheets supported cobalt catalyst in Fischer–Tropsch synthesis using Graphene functionalization. *Chem. Eng. Res. Des.* 119: 198–208.

81 Karimi, S., Tavasoli, A., Mortazavi, Y., and Karimi, A. (2015). Cobalt supported on graphene – a promising novel Fischer–Tropsch synthesis catalyst. *Appl. Catal. A: Gen.* 499: 188–196.

82 Torshizi, H., Nakhaei Pour, A., Mohammadi, A. et al. (2020). Fischer-Tropsch synthesis by reduced graphene oxide nanosheets supported cobalt catalysts: role of support and metal nanoparticle size on catalyst activity and products selectivity. *Front. Chem. Sci. Eng.* 15 (2): 299–309.

83 Trzeciak, A., Wojcik, P., Lisiecki, R. et al. (2019). Palladium nanoparticles supported on graphene oxide as catalysts for the synthesis of diaryl ketones. *Catalysts.* 9 (4): 319.

84 Nasser, A., Guo, L., ELnaggar, H. et al. (2018). Mn–Fe nanoparticles on a reduced graphene oxide catalyst for enhanced olefin production from syngas in a slurry reactor. *RSC Adv.* 8 (27): 14854–14863.

85 Guo, X., Jiao, Z., Jin, G., and Guo, X. (2015). Photocatalytic Fischer—Tropsch synthesis on graphene-supported worm-like ruthenium nanostructures. *ACS Catal.* 5 (6): 3836–3840.

86 Hajjar, Z., Doroudian Rad, M., and Soltanali, S. (2016). Novel Co/graphene oxide and Co/ nanoporous graphene a catalysts for Fischer–Tropsch reaction. *Res. Chem. Intermed.* 43 (3): 1341–1353.

87 Li, C., Sayaka, I., Chisato, F., and Fujimoto, K. (2016). Development of high performance graphite-supported iron catalyst for Fischer–Tropsch synthesis. *Appl. Catal. A: Gen.* 509: 123–129.

88 Zafari, R., Abdouss, M., and Zamani, Y. (2020). Effect of Mn and reduced graphene oxide for the Fischer–Tropsch reaction: an efficient catalyst for the production of light olefins from syngas. *React. Kinet. Mech. Catal.* 129 (2): 707–724.

89 Deerattrakul, V., Dittanet, P., Sawangphruk, M., and Kongkachuichay, P. (2016). CO2 hydrogenation to methanol using Cu-Zn catalyst supported on reduced graphene oxide nanosheets. *J. CO2 Util.* 16: 104–113.

11

Catalytic Oxidative Cracking for Light Olefin Production

Huda Sharbini Kamaluddin and Katabathini Narasimharao

Department of Chemistry, Faculty of Science, King Abdulaziz University, Jeddah, Saudi Arabia

11.1 Light Olefins

Light olefins are well-known important intermediate raw materials in the petrochemical industry. Ethylene and propylene are the most notable olefins, with an annual production of approximately 1.5×10^2 million tons (Mt) and 80×10 Mt, respectively [1]. These production requirements are expected to increase owing to the increasing global population and rising living standards. Light olefins are the backbone of modern life, as many daily-use products are manufactured from these building blocks [2, 3]. Olefins are primary building blocks in the production of plastics, fibers, and other polymers [4, 5]. Olefins produce linear alpha olefins (e.g. oligomerization of ethylene or dehydration of alcohols), which are mainly used to manufacture synthetic acids, detergent intermediates, and polymers. As the function of the packaging industry is centered on consumer goods and packaging materials, the demand for olefins production in the international market has increased, and consequently, so has the demand for linear alpha olefins [1]. Linear alpha olefins are the most crucial chemical building blocks in low-density polyethylene (LDPE) and high-density polyethylene (HDPE) [6]. LDPE is primarily used in consumer goods and packaging materials; its global packaging value in 2017 was more than US\$500 billion[1]. In addition, the automobile industry uses internal olefins, which are components of lubricants, to improve the mechanical linkage performance in vehicles [2]. Traditionally, crude oil fractions and natural gas processing products were the primary sources of olefin production. Recently, and for more than half a century, steam cracking (SC) has been the leading technology for light olefin and aromatic production. In this process, hydrocarbons at elevated temperatures are cracked in tubular reactors suspended in a gas-fired furnace [1]. The proportions of different SC feedstocks used worldwide are shown in Figure 11.1 [3].

The primary feedstock for SC is naphtha, accounting for approximately 55% of the total feedstock. However, it is expected that SC technology will switch to utilizing inexpensive feedstock alternatives, such as liquefied petroleum gas (LPG), methanol, ethane, condensates, and associated gas. Asia has already switched to using inexpensive LPG as

Heterogeneous Nanocatalysis for Energy and Environmental Sustainability: Energy Applications, First Edition. Edited by Putla Sudarsanam, Yusuke Yamauchi, and Pankaj Bharali.

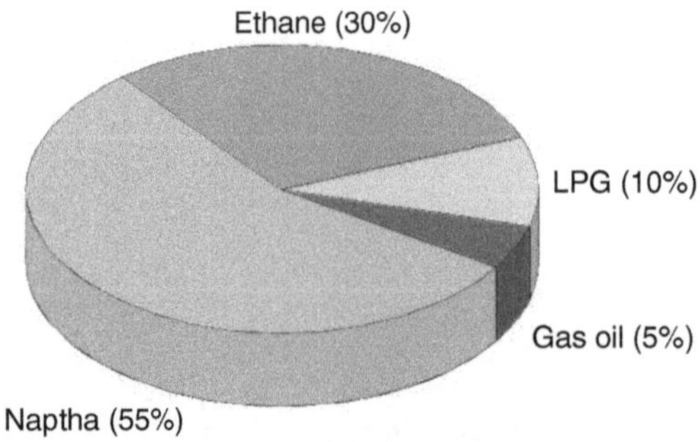

Figure 11.1 Proportion of different feedstocks used in steam cracking [7].

feedstock [8]. Meanwhile, North America and the Middle East use ethane, shale gas, and associated gas as feedstock for crackers, constituting approximately 70–80% of the total feedstock [3]. However, the use of ethane as feedstock threatens the balance of the ethylene-to-propylene ratio because ethane pyrolysis offers high ethylene selectivity (79%) and low propylene selectivity (1%), whereas cracking naphtha leads to unsatisfactory ethylene (30%) and propylene (15%) selectivities [3]. By 2025, the annual demand for ethylene is speculated to be 170 Mt and that for propylene is expected to be 115 Mt, with a propylene:ethylene ratio of 0:68. As propylene constitutes one-third of the total fuel supply, the development of a novel propylene production method is imperative. New processes have been developed to modify the propylene-to-ethylene ratio, such as isomerization, hydro-isomerization, metathesis, interconversion, skeletal isomerization, and catalytic cracking (CC) [3]. In 2019, the worldwide consumption of ethylene and propylene increased to approximately 160 and 115 Mt, respectively [9]. In general, the production of these light olefins depends primarily on several technologies, such as SC, fluid catalytic cracking (FCC), and Fischer–Tropsch to olefins (FTO) [10]. However, the main drawbacks of these processes are the huge amounts of carbon emissions and high energy consumption [9]. Conventional processes, such as FCC and catalytic dehydrogenation (CD), present several drawbacks, such as the insufficient production of light olefins, high energy consumption, and deactivation of the catalyst due to coke formation [10]. The strict in the petrochemical industry during the development of more energy-efficient processes. From this perspective, we will emphasize the most promising alternatives compared with the conventional SC process and the major bottlenecks for each of the competing processes.

11.2 Feedstocks for the Production of Light Olefins

Light olefins are formed via the cracking of ethane, propane, and butane, and complex liquid hydrocarbons (e.g. a mixture of naphtha and gasoil). In the late twentieth century, the increasing demand for light olefin production in the chemical industry diminished this sector's interest in transitioning toward alternative technology and feedstock other than carbon sources [11, 12]. This dilemma led the chemical industry and academia to transition from carbon resources based on crude oil to primary and secondary carbon feedstock alternatives [12] (Figure 11.2a). New alternative technologies could offer new routes that utilize abundant, inexpensive feedstock, such as propane, ethane, and

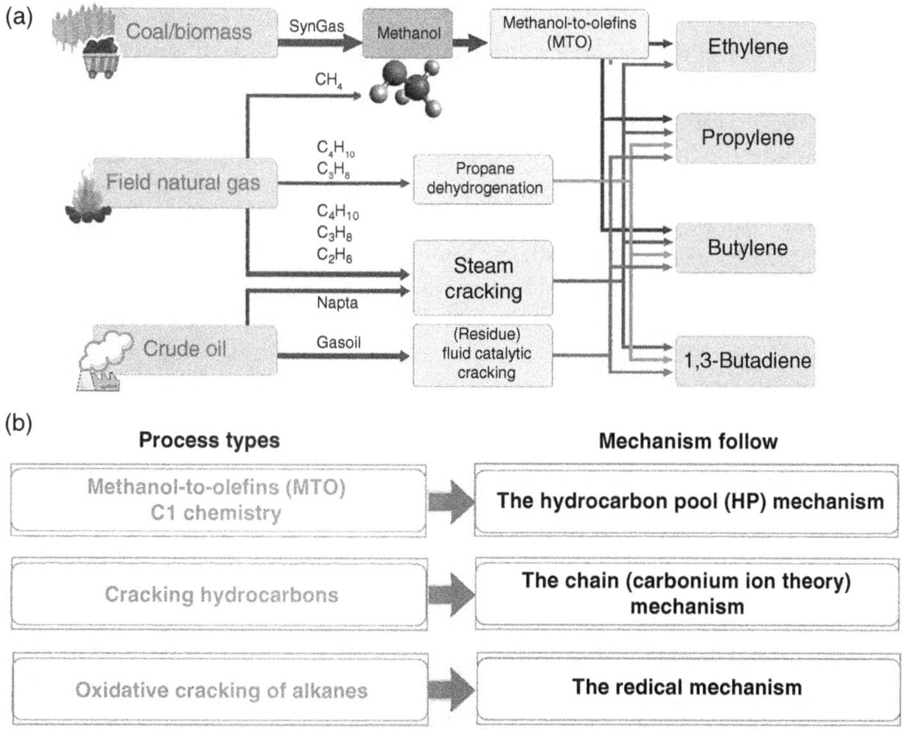

Figure 11.2 (a) Light olefin production using different technologies and feedstocks. *Source:* Adapted from [11]. (b) Mechanisms of existing technologies. *Source:* Adapted from [14–16].

methane formed from shale and standard gases, which utilize the most existing technologies primarily [1]. The use of feedstock produced from heavy oil fractions for light hydrocarbons and methanol also led to new catalytic processes. In addition, by-products from refining and petrochemical plants, such as C_4 and C_5 fractions formed from light hydrocarbon feedstocks, are suitable for cracking into ethylene and propylene [13]. The use of alternative feedstocks such as biomass and waste streams could also be useful for complementing the fossil-based feedstock, which is a better solution than complete replacement in the near future [11].

11.3 Reaction Mechanisms for Olefin Formation

SC is a highly endothermic process that requires external heat input, where hydrocarbon feedstock (ethane to naphtha) is cracked to light olefins, including ethylene and propylene, under steam and high temperatures (700–900 °C) [7, 12], followed by a free-radical mechanism [17–19]. The energy consumption in this process is undesirable; therefore, the use of catalyst materials is an alternative solution to address the energy drawbacks. In general, catalyst materials provide an alternative path with lower energy to increase the reaction rate. In this regard, light olefin formation from alternative carbon feedstocks involves different mechanisms. For example, hydrocarbon cracking with a catalyst follows a carbocation mechanism [20], catalytic oxidation with a selective oxygenated catalyst follows a radical mechanism, and methanol-to-light olefins were proposed to follow the hydrocarbon pool (HP; C_1 chemistry), as shown in Figure 11.2b. C_1 chemistry requires an acidic catalyst (such as zeolites) to convert methanol to olefin (MTO). The process follows the HP mechanism proposed by Dahl and Kolboe [14]. MTO conversion typically involves three stages: initiation, the active stage, and the deactivated stage [15, 21]. The C–C bond of hydrocarbons is a chain reaction that is believed to follow the carbonium ion theory developed by Whitmore [1]. This carbonium mechanism involves three elementary steps: initiation, propagation, and termination. Initiation, the first step, activates the created complex by attacking the active site on the reactant molecule. Whether in the gas phase or liquid super acids, the activation of the complex corresponds to carbocation formation [16]. During propagation, the hybrid ion transfers from a reactant molecule to an adsorbed carbenium ion. Finally, in termination, the initial active sites are restored through the desorption of the adsorbed carbocation and the formation of an olefin [22]. In the oxidative cracking of alkanes over oxygenated catalysts, the system follows a radical mechanism. Hydrogen abstraction from the alkane molecule via an oxygen defect site on the catalyst surface leads to the formation of $[M^+OH^-]$ species during alkane activation. Alkyl radicals are then formed and subsequently leave the catalyst surface. Oxygen enhances the chain propagator radicals $[HO_2{}^*]$ in the gas phase, and the overall system in the gas phase follows a radical chain reaction. Finally, the hydroxyl forms and is released as water, reducing the catalyst [23–25].

11.4 Different Light Olefins

11.4.1 Ethene (C_2H_4)

Ethene (C_2H_4) is the most important and sought light olefin in the global market. The demand for C_2H_4 has incessantly increased because of its application in the plastic industry for the production of intermediate chemicals such as polymers (e.g. polyethylene), fibers, and other chemicals [26, 27]. The global compound annual growth rate (CAGR) for ethylene is 6.78%. The demand for ethylene reached US$391.12 million in 2018, and is expected to grow to US$619.12 million by the end of 2025 [28]. Globally, the production capacity of C_2H_4 was approximately 1.48×10^2 Mt in 2014, representing a 32% increase over the past decade [19]. Further, 57% of C_2H_4 production primarily depends on the use of naphtha and gas oil as feedstocks in SC (Figure 11.3a), while 38% is derived from ethane and LPG. The ratio of the origin feedstock plays a critical role in determining the optimum results. For example, naphtha contains a variable amount of paraffin, aromatic, and olefinic compounds. At high temperatures (650–750 °C), the C_2H_4 yields were approximately 30 and 25 wt% using naphtha and gas oil, respectively. When ethane was mixed with naphtha in the feed stream, the C_2H_4 yield reached 80 wt% [29].

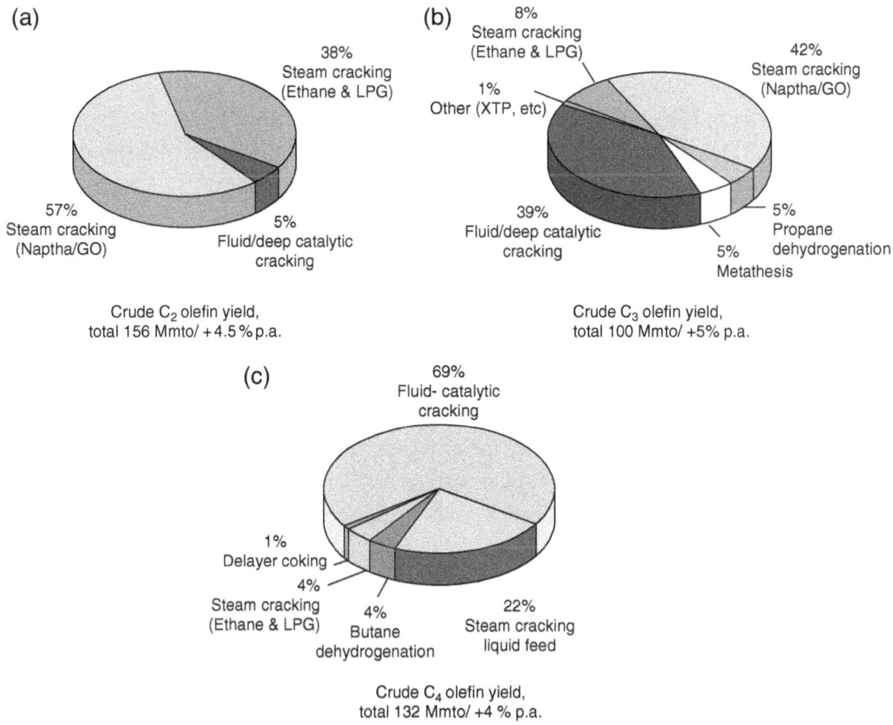

Figure 11.3 Hydrocarbon feedstock for (a) C_2 olefin production, (b) C_3 olefin production, and (c) C_4 olefin production. *Source:* Adapted from [30].

11.4.2 Propene (C_3H_6)

C_3H_6 is the second most crucial light olefin gas in the global market. Its production value reached nearly US$83.7 billion in 2019. In 2015, its CAGR reached 2.95%, which is expected to decrease to 1.8%, i.e. to approximately US$77.8 billion, by 2023 [31]. C_3H_6 is a building block for various industries, such as automotive, packaging, and construction. In addition, C_3H_6 is used as a polymer for poly(propene), as an epoxy of propane, and as a precursor for propanol. Conventionally, C_3H_6 is produced via SC as a by-product or through FCC processes using several feedstocks (Figure 11.3b). The C_3H_6 market is expected to increase globally because plastic is crucial in the automobile industry. In 2014, the global vehicle market was valued at US$1477.2 billion, and in 2018, at US$2085.1 billion. The reason for the increasing C_3H_6 demand is its role as a building block and a monomer for manufacturing polypropylene, a plastic used in the automobile industry. Polypropylene is also used for manufacturing battery cases, trays, bumpers, instrument panels, interior trims, fender liners, door trims, and other components [32]. C_3H_4 production is severely affected when the industry produces C_2H_4 and gasoline, because the steam cracker will change and depend on ethane as feedstock. This leads to reduced C_3H_6 production [33, 34], as well as lesser yields of C_3H_6 as a derivative or co-product [35]. Accordingly, petrochemical companies have been compelled to produce C_3H_6 as the main product, not as an alternative or by-product. Hence, there have been many attempts to modify the traditional FCC unit to increase the C_3H_6 yield by 20% [35]. Recently, other production methods to increase the C_3H_6 production capacity, such as propane dehydrogenation (PDH) [36], olefin metathesis [37], MTO/methanol-to-propene (MTO/MTP) conversion [38, 39], high-severity fluidized catalytic cracking (HS-FCC), coal-to-olefin/coal-to-propene (CTO/CTP) conversion, and gas-to-olefin (GTO) conversion [35], have been implemented. Unconventional feedstocks, such as bio-alcohols and vegetable oils, have also been used [40].

11.4.3 Butene (C_4H_8)

C_4H_8 is an important petrochemical product in the global market. The global market for butene reached US$56.7 M in 2019 and is expected to reach US$80.1 M by the end of 2025, with a CAGR of 5.98% [41]. The most common butenes are 1-butene, *trans*-2-butene, and isobutylene. Butene is a building block for polymer production (i.e. rubbers). It is used in tires in the automobile industry, where the production of 1,3-butadiene (1,3-BD), a monomer for rubber, reached 95% in 2018. It is also used as a blending component and octane enhancer in the fuel industry [26]. *n*-Butene is used as a co-monomer of polyethylene in organic synthesis, and as a monomer for the synthesis of higher olefins [26, 42]. Conventionally, butene is formed from the cracking of naphtha (as a by-product), the dehydrogenation of butane, and the gas oil (Figure 11.3c); however, the yield of butene from these processes is low [26]. Therefore, ethanol was used as feedstock for the formation of butadiene, which could be achieved through two processes: the one-step Lebedev process and the two-step Ostromislensky process. The demand for butene will increase in the future, as it is expected to be used as bio-based feedstock for greener rubbers that have already been proposed with bio-ethanol as feedstock [43].

11.5 Processes for Producing Light Olefins

Owing to environmental regulations and the need for more energy-efficient processes, the petrochemical industry currently faces several challenges. Light olefins such as C_2H_4 and C_3H_6 are building blocks for various applications in the chemical industry. These light olefins are in demand, and their global market is expected to grow by 4% in the upcoming years [7]. Several methods for light olefin production have been reported. These involve either C_1 chemistry or the C-H/C-C bond cleavage of alkanes. This chapter focuses on the latter method, which can be further classified as follows: (i) thermal cracking and (ii) oxidative methods. Thermal cracking involves several processes, i.e. SC, CC, and CD. Meanwhile, the oxidative method involves oxidative dehydrogenation (ODH), oxidative coupling, and oxidative cracking (Figure 11.4).

11.5.1 Thermal Cracking Methods

Cracking involves breaking largely saturated alkanes into smaller, more valuable hydrocarbons, including alkanes and alkenes. For example, the cracking of hexane into butane and C_2H_4 is expressed in Eq. (11.1):

$$C_6H_{14} \rightarrow C_4H_{10} + C_2H_4 \tag{11.1}$$

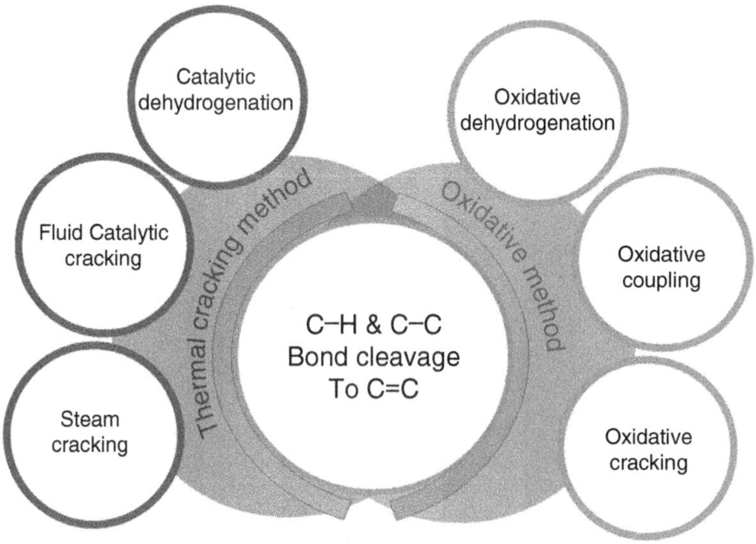

Figure 11.4 Different processes involved in C-H/C-C cleavage to produce C=C components.

The thermal cracking of hydrocarbons can occur in either of two ways: (i) using non-catalysts with high temperature and pressure (i.e. thermal SC processes) or (ii) using a catalyst with low temperature and pressure (i.e. CD and FCC processes). Cracking processes are important for the commercial production of gasoline, olefinic gases, diesel fuel, and other products [44].

11.5.1.1 Thermal Steam Cracking (SC)

Thermal cracking is also known as pyrolysis, and it is an endothermic process wherein complex polymeric materials of high molecular weight are thermally decomposed in an inert atmosphere to produce hydrocarbon molecules of low molecular weight after thermal treatment [45]. The cracking mechanism occurs differently in this process and is dependent on the characteristics of the polymer chain (e.g. the presence of defects, halogens, and other heteroatoms, the aromatic degree, and bond dissociation energy).

The SC of hydrocarbon feedstock (i.e. naphtha, gas oil, and gas condensates) is a leading technology for the production of light olefins (e.g. C_2H_4, C_3H_6) and limited amounts of by-products (higher olefins). However, the drawback of SC is that it is energy-intensive (the dissociation of C-H bonds requires 413 kJ/mol) and is accompanied by high CO_2 emission due to fuel combustion in the chemical industry (Table 11.1) [7]. Among the aforementioned feedstocks, naphtha is the most widely used because of its availability, low cost, and potential for producing olefins (C_2H_4 and C_3H_6). It is exclusively based on thermal cracking, which involves using steam and breaking down macromolecules at high temperatures. The SC unit is divided into three main sections: (i) the hot section, which contains the furnace, primary fractionator, and quench; (ii) the compression section, which comprises the compressors, caustic scrubbers, and dryers; and (iii) the cold section, where the

Table 11.1 Comparison between catalytic oxidative cracking and conventional steam cracking.

Conventional steam cracking	Catalytic oxidative cracking
The reaction is endothermic under atmospheric conditions. The possibility of undesired reactions results in low light olefin selectivity [93].	The reaction uses oxygen as the co-feed and is exothermic. The heat generated in situ reduces the energy efficiency [74].
The reaction under atmospheric conditions accelerates the side reaction that forms coke at high temperatures [89].	The oxygen co-feed minimizes coke formation on the catalyst material [74].
The formation of heavy coke is the precursor to tar and soot formation [74].	The possibility of tar and soot formation is lower due to the limited formation of heavy coke over catalyst [74].
The reaction is associated with significant energy consumption, producing large amounts of CO_2 up to 2 tons/ton $2H_2$ [93].	The presence of the catalyst allows the system to be activated at low temperatures, limiting the formation of CO_x. However, the use of O_2 as the co-feed enhances the formation of CO_x [74].
Steam cracking offers high C_2H_4 production [90].	COC offers different light olefins (C_2H_4, C_3H_6, and C_4H_8) production [90].

pyrolysis products are separated – this section includes the demethanizer, de-ethanizer, depropanizer, and debutanizer. The cracking furnace is the "heart" of any SC plant, and is a key component of a chemical reactor. It generally contains two sections based on the type of heat transfer: the convection section and the radiation section (Figure 11.5a) [46].

11.5.1.2 Fluid Catalytic Cracking (FCC)

The formation of alkenes through FCC is a major process in petroleum refining, enabling the conversion of vacuum gas oil, atmospheric residue, and crude oils. FCC converts hydrocarbons with high molecular weight and high boiling point in crude oils into more valuable products such as gasoline, olefinic gases, and other products. The process requires high temperatures and moderate pressure with a powder catalyst when treated with feedstock. Hence, the long-chain molecules of high-boiling-point hydrocarbon liquids are broken into much shorter molecules that are collected as vapor [3]. There are two approaches for light olefin production, depending on the type of feedstock (heavy or light hydrocarbons). When the feedstock is a heavy hydrocarbon, the primary cracking process involves the formation of light naphtha olefins, followed by further cracking into light olefins. Meanwhile, a feedstock of light hydrocarbons is often composed of by-products from refining and petrochemical plant processes, which is suitable for further cracking to C_2H_4 and C_3H_6 [13].

In FCC, the cracking reactor and catalyst regenerator exhibit the highest thermal efficiency (Figure 11.5b), and the oil gas feed with the injected steam at the entrance (called the "riser") of the reactor interacts with the catalyst particles in the fluidized beds. In the regenerator unit, the feed gas oil evaporates through hot catalyst particles upon contact with the riser. Subsequently, the gas oil vapor and catalyst particles move upward in the reactor, and cracking begins. During the upward movement within the regenerator unit

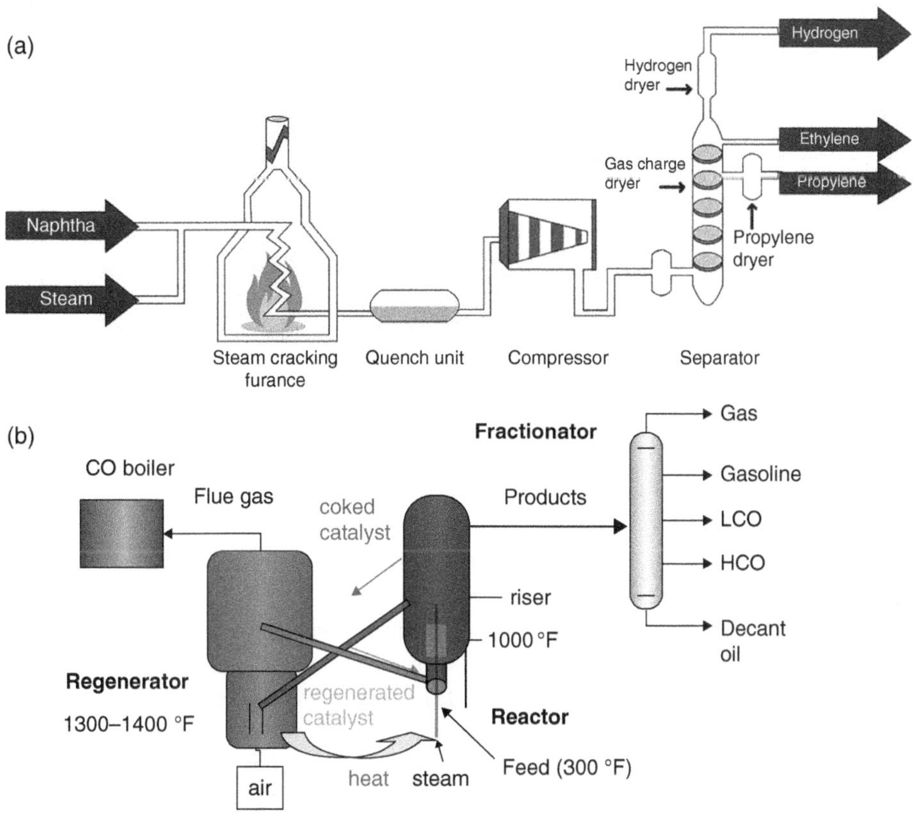

Figure 11.5 Schematic diagram for (a) thermal steam cracking (SC). *Source:* From [46], and (b) fluid catalytic cracking (FCC). *Source:* From [47] / with permission of Semih Eser.

and the endothermic cracking reaction, the evaporation rate of the gas oil and the temperature of the catalyst particles decrease. The cracking reaction generates a large amount of coke in the catalyst, deactivating it. Subsequently, the hydrocarbon adsorbed on the catalyst is removed via steam stripping, and the catalyst is forwarded to the regenerator unit to burn the coke under air. During burning for the removal of coke, the catalyst particles are heated and return to the riser to complete the cycle. In the regenerator, the burning-off of the rejected carbon (coke) provides the necessary energy for cracking without much loss of energy, leading to increased thermal efficiency. The cracking products are sent to a fractionator for recovery after separation from the catalyst particles in the upper section of the reactor [47, 48].

CC often occurs at high temperatures in the presence of a cracking catalyst. The first catalyst used for this process was low alumina, which contains an amorphous solid composed of approximately 87% SiO_2 and 13% Al_2O_3. High-alumina catalysts were then used, which are composed of 25% alumina and 75% silica. However, they were later replaced by a catalyst containing crystalline aluminosilicate (zeolite materials) or molecular sieves. FCC can also be conducted using different essential catalysts such as CaO-SrO-Al_2O_3, WO_3-K_2O-Al_2O_3, and KVO_3/corundum [49].

11.5.1.3 Catalytic Dehydrogenation (CD)

The CD of alkanes is an endothermic reaction. This kind of reaction occurs when the number of moles increases. The reaction is expressed as:

$$\text{Alkane} \Leftrightarrow \text{Olefin} + \text{Hydrogen} \tag{11.2}$$

This reaction is thermodynamically unfeasible because it involves breaking of a C–H bond, which requires relatively high energy (approximately 363 kJ/mol). In comparison, hydrocarbon cracking which also involves C–C bond breakage requires a lesser amount of energy (about 246 kJ/mol). Therefore, a suitable catalyst is required to process C-H cleavage in CD process. Many types of metal oxides (MO_x) and metal catalysts (Ms) have been utilized for CD. MO_x can activate the C–H bond and abstract the H atom, because of the formation of MO_x–H bond which has a higher bond strength than the C–H bond. In the other hand, metal catalysts (Ms) may not be able to complete a cleave of C-H bonds because the formation of Ms–H bond is weaker than the C–H bond. However, the process is thermodynamically feasible if the net strength of the Ms–H and Ms–C bonds exceeds the C–H bond strength [50].

CD is mainly employed for *n*-propane and isobutane dehydrogenation. Recently, C_3H_6 production via CD has increased owing to the demand for C_3H_6 in polymerization. Conventionally, alkane dehydrogenation, using ethane as the alkane feedstock and noble- or heavy-metal catalysts such as platinum (Pt) or chromium (Cr), offers poor ethylene yields. The dehydrogenation of propane to C_3H_6 is the most favorable route in CD, as it is fast and involves exothermic reactions. Intrinsically the CD of *n*-propane reaction could occur on the individual active sites (Scheme 11.1). However, the large particle size of the catalyst could affect the reaction adversely, as it could enhance the side reactions, such as deep dehydrogenation, cracking, etc. [51]. Moreover, the presence of a portion of the steam crackers using inexpensive ethane as a feedstock would complement the dehydrogenation of the propane unit [48].

Propane dehydrogenation (PDH) is anticipated to be increasingly adopted in the near future to fill the C_3H_6 shortage due to the abundance of wet shale gas worldwide. PDH is expected to produce 5 Mt from propane for the next five years. However, CD has some drawbacks, in that

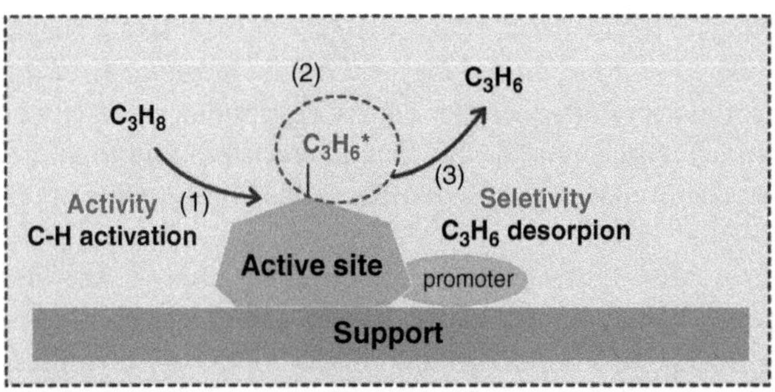

Scheme 11.1 Illustration of the *n*-propane dehydrogenation reaction: (1) C–H activation of propane molecules; (2) C–C cleavage; and (3) desorption of propylene and adsorption of propyl intermediate, which can be regulated via the promoter and support effect. *Source:* From [51] / with permission of American Chemical Society.

it requires high temperatures (>600 °C) and lower pressures (˂30 psig) to be economically feasible. Because this process has a high enthalpy (30 Kcal/mol) and thermodynamic equilibrium limitations, the composition of the catalyst for CD is crucial to ensure commercial viability and stability. Furthermore, because of the required high temperature and low pressure, the catalyst should inhibit side reactions and carbon deposition on the catalyst [47]. Noble (Pt-based catalyst) [52–54] or non-noble metals, such as Cu [55], Zn [56, 57], Fe [58], and VO_x catalysts, have been used in this regard [59]. Although Pt-based catalysts show excellent activity, their use has drawbacks such as coke formation and Pt cluster sintering [60]. Therefore, dedicated efforts have been invested to modify Pt catalysts with secondary elements (e.g. Ce, Cu, Zn, and Sn) as a promoter [61], which improve the selectivity and stability of heavy carbonaceous coke [62]. In addition, the activity of the catalyst is enhanced when the noble metal is dispersed on catalyst supports with a large surface area, such as alumina, silica-alumina, zeolites, titania, magnesium oxide, mixed oxides, spinels, and perovskites, which are suitable for CD [15, 63, 64].

11.5.2 Oxidative Methods

Oxidative routes for light olefin production are promising alternatives because the oxygen present in the feedstock is advantageous in thermodynamic-equilibrium-limited processes and limits the formation of coke. Several approaches have been reported in the literature, including (i) oxidative dehydrogenation (ODH), (ii) oxidative coupling, and (iii) oxidative cracking.

11.5.2.1 Oxidative Dehydrogenation (ODH) of Alkanes

The ODH of alkanes is an exothermic process that does not involve coke formation on the catalyst surface. It is a promising alternative route because it is an exothermic system that requires a low reaction temperature. ODH has a complicated reaction mechanism that involves various reaction channels under aerobic conditions [9]. Non-catalytic ODH is limited by its low selectivity for light olefins; when the reaction occurs under partial oxidation conditions, oxygenated products are formed, and a deep oxidation reaction leads to the formation of CO_x. Hence, it is necessary to use a catalyst to produce the required products with high selectivity for olefins under an oxidative atmosphere. Accordingly, there have been extensive efforts to improve the catalytic activity in ODH, and also to improve the selectivity for light olefin products, most of which are metal-based catalysts.

Metal-based catalysts show high reactivity in the oxidative conversion of alkanes; however, CO_x unfavorably forms due to deep oxidation, leading to CO_2 emission due to a strong reaction between the electrophilic defects in metal sites and the electrons that are abundant in olefin products. A study was conducted to improve the output characteristics of ODH using hydrogen, wherein hydrogen was oxidized into water, and there was no limitation in equilibrium [9].

$$C_3H_8 + O_2 \rightarrow C_3H_6 + H_2O \qquad \Delta H = -86\,kJ/mol \tag{11.3}$$

However, the selectivity for olefin formation remains a crucial issue. For example, the maximum yield for propane ODH is 30%, which is not promising for commercialization on a larger scale [4]. To date, the only commercialized ODH process involves the conversion of ethylbenzene to styrene [65]. ODH requires a suitable catalyst for the hydrocarbon–oxygen mixture; therefore, three types of catalytic materials were investigated: redox catalysts, non-redox catalysts, and noble-metal-coated monolith catalysts [66].

Propane oxidative dehydrogenation (PODH) is an exothermic process. During adsorption and desorption, the chemical species interact with each other and the catalyst, generating heat within the reactor. As PODH is an exothermic reaction, efficient heat transfer during reactor operation is the primary concern. The simple design for carrying out ODH in a fixed-bed reactor leads to a gradual increase in the reaction temperature, leading to challenging conditions for PODH to occur. Unless the catalyst bed is diluted with inert particles with a suitable ratio of large inert particles/catalyst particles (which is a very expensive process for commercialization), it will be difficult to produce C_3H_6 using the traditional ODH reactor.

Many researchers have attempted to solve this issue using a multi-tubular reactor with periodic air injection, membrane reactors, and a wall-cooled catalytic reactor to control the produced heat from the exothermic PODH process. In addition, they used a microchannel reactor to overcome the exothermicity and lower costs. The microchannel reactor is believed to enable excellent heat transfer and heat removal. The catalyst used in the PODH reaction within a small-sized fixed-bed and microchannel reactor was pure $MoVNbTeO_x$ catalyst under the same conditions as ODH. This study aimed to demonstrate the advantages of heat management in the microchannel reactor. X-ray diffraction (XRD), scanning electron microscope (SEM), and inductively coupled plasma mass spectroscopy (ICP-MS) characterization results indicated that the catalyst possessed high stability in the microchannel system. The circulating fluidized bed reactor and downer reactor for ODH (Figure 11.6a) have received more attention owing to several shortcomings of fixed-bed reactors. If operated using fine particles sized 60–100 μm, the downer reactor enables controlled thermal change, as well as adequate particle concentration and residence times, resulting in ODH with high olefin selectivity [67].

11.5.2.2 Oxidative Coupling of Alkanes

In the oxidative coupling process, methane is used as feedstock to produce hydrocarbons, mostly C_2H_4. Breaking the C–H bond in methane is difficult, and harsh conditions are required. A one-step process is not feasible on an industrial scale because of the parallel and consecutive undesired reactions that limit C_2H_4 formation [69]. The primary reactions in the oxidative coupling of methane (OCM) are as follows:

$$2CH_4 + O_2 \rightarrow C_2H_4 + 2H_2O \tag{11.4}$$

$$CH_4 + 2O_2 \rightarrow CO_2 + 2H_2O \tag{11.5}$$

$$CH_4 + O_2 \rightarrow CO + H_2O + H_2 \tag{11.6}$$

Conventional techniques, in which oxygen and methane are mixed, produce low C_{2+} components due to combustion and other consecutive reactions that limit the formation of higher C_{2+} hydrocarbons. The maximum yield of a stable catalyst is always below 25% [70].

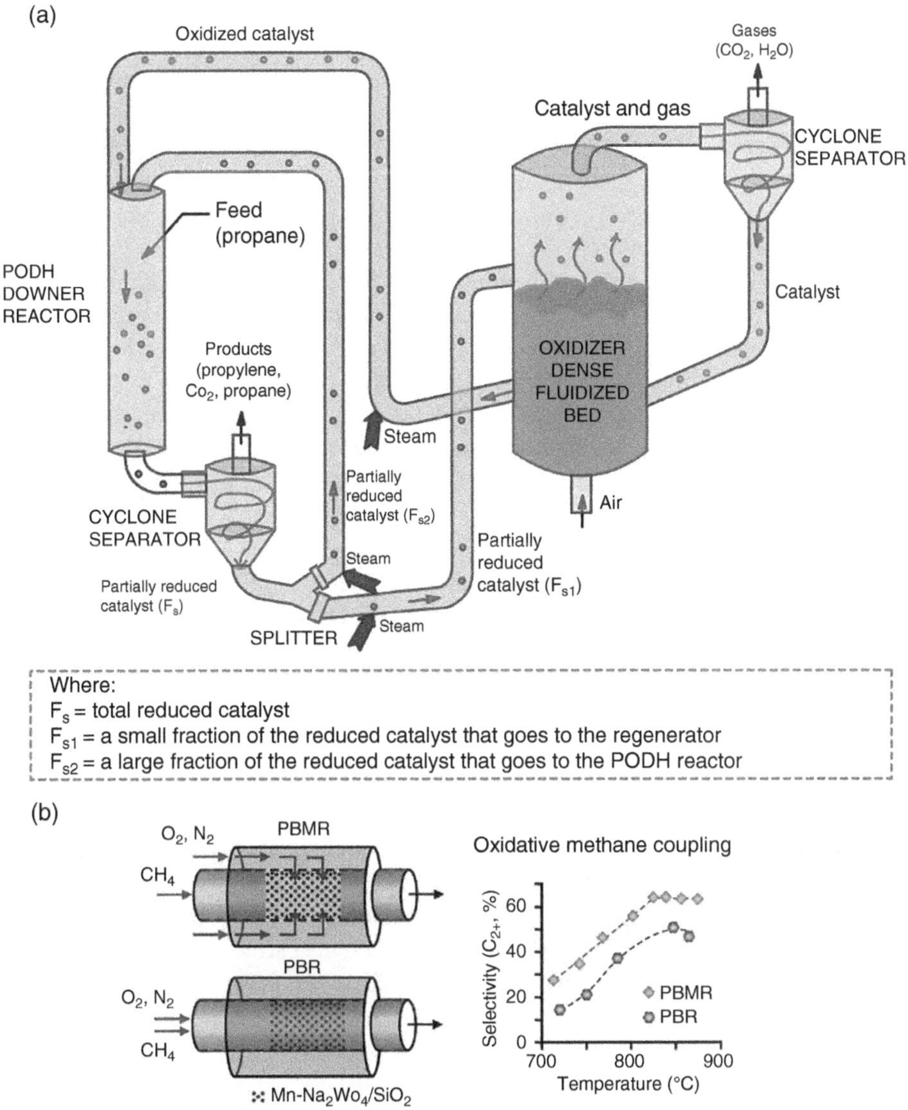

Figure 11.6 (a) Flowchart of oxidative dehydrogenation (ODH) process displaying the expected industrial requirements for PODH. *Source:* From [67] / MDPI / CC BY 4.0 (b) Systematic comparison between a packed-bed membrane reactor (PBMR) and packed-bed reactor (PBR) for the oxidative coupling of methane (OMC) on Na_2WO_4-Mn/SiO_2 catalyst, revealing the importance of distributing the oxygen feed along the catalyst bed length. *Source:* From [68]/with permission of Elsevier.

Therefore, the yield should be improved if OCM is used on an industrial scale [69, 71]. In general, methane is activated by a catalyst to form methyl radicals that are released to the gas phase. The formed methyl radical reacts with other CH_4 molecule or methyl radical to form ethane and a subsequent ethane dehydrogenation results the formation of C_2H_4. The possibility of the methyl radical undergoing a side reaction to form carbon dioxide leads to low C_2H_4 selectivity. The chemical looping methodology could be an alternative strategy for oxidative naphtha cracking, as this process does not depend on crude oil [72]. The essential factor in the methane coupling mechanism is the desorption of the radical from the catalyst surface into the gas phase. For example, a non-redox catalyst is used in methane oxidative coupling. Oxygen in the catalyst $[Li^+O^-]$ is activated, and it subsequently abstracts hydrogen from the methane molecule to form a surface hydroxyl $[Li^+OH^-]$ species [66]. The efficiency of catalyst was characterized by using several techniques such as mass spectrometry, matrix-isolation IR, and matrix-isolation EPR [45, 46].

The OCM enables the catalytic and homogeneous conversion of methane into valuable products, mostly olefins such as C_2H_4, and it can be implemented using different methods. The technology presented here involves a new feeding policy that increases the efficiency of packed-bed membrane reactors (PBMRs), especially when used for the oxidative coupling of alkanes, such as methane. In addition to the problems mentioned above, the conversion of methane into higher hydrocarbons such as ethane and ethylene poses an additional challenge, as the activation of methane is difficult due to its thermodynamic stability. The feeding policy of this technology can be used in single reactors and reactor networks. In conventional PBMRs, a large amount of oxygen is fed to the outer shell, and only a small amount of methane is fed to the inner reactor core. With the feeding policy proposed by this invention, it is possible to introduce a large amount of methane and nitrogen into the outer shell while introducing only a small amount of oxygen into the core. This is achieved by separating the feed before it enters the reactor. Thus, it is possible to ensure a higher oxygen concentration in the feed without negatively influencing the conversion or yield [73]. The PBMR allows for the controlled distribution of O_2 along the length of the catalyst bed, resulting in a nearly constant, high local hydrocarbon-to-oxygen ratio. For this reason, distributed feed membrane reactors have been used in selected partial oxidation reactions, including the ODH of alkanes to olefins and methanol to formaldehyde. The two most common types of membranes used in distributed feed reactors are permselective dense membranes and non-selective porous membranes. While dense membranes generally have higher perm selectivity than porous membranes, the permeation flux is low, limiting the reactor productivity. In contrast, a porous ceramic membrane has a much higher flux, but its selectivity is lower (Figure 11.6b) [68].

11.5.2.3 Oxidative Cracking of Alkanes

Oxidative cracking is an alternative route to produce light olefins, e.g. C_2H_4 and C_3H_6. Three types of processes occur in this category: (i) gas-phase oxidative cracking (GOC, non-catalytic); (ii) catalytic oxidative cracking (COC); and (iii) redox oxidative cracking (ROC, using a catalyst without O_2 feed) (Scheme 11.2).

11.6 Mechanism for the Catalytic Oxidative Cracking (COC) of Alkanes

Extensive spectroscopic studies by Knözinger et al. [63, 64] and other recent literature reports reveal that the activity of MgO catalysts depends on the defect sites and not on the MgO phase. The MgO material exhibits poor catalytic activity due to the lack of electron

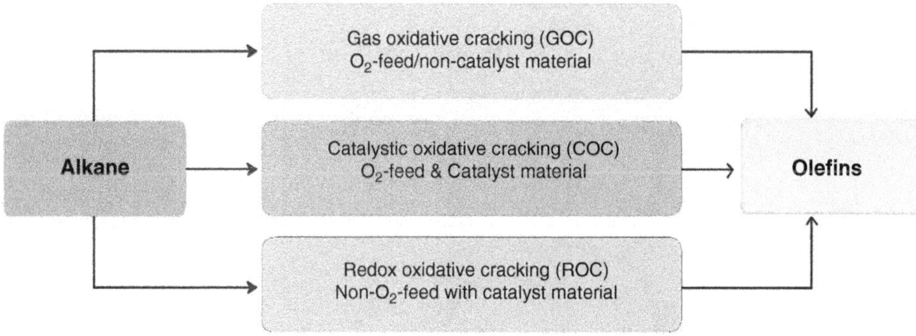

Scheme 11.2 Illustration of different catalytic and non-catalytic processes using O_2 feed.

mobility. The presence of low coordinated [$Mg\delta^{+}$-O^{2-}] sites in a catalyst abstracts the hydrogen from alkanes during the oxidative cracking reaction [74]. Lithium (Li) metal is the best promoter for COC activity, because Li^+ and Mg^{2+} has no typical redox behavior and inherent strong Brønsted basicity *via* forming more defect sites in MgO compared to the formal acidic or redox-type catalysts, e.g., Al_2O_3 or V_2O_5. Therefore, Li/MgO catalyst shows less activity in undesired processes such as re-adsorption and sequential combustion of formed alkenes. Boyadjian et al. [74] provided more details regarding the nature of Li/MgO active sites. Most oxidative cracking reactions follow a reaction mechanism similar as the OCM process. First, alkane (for instance CH_4) reacts with the active sites [Li^+O^-] to form [Li^+OH^-] and $CH_3^{\bullet}$ radicals:

$$CH_4 + Li^+O^- \rightarrow Li^+OH^- + CH_3^{\bullet} \tag{11.7}$$

The methyl radical is released into the gas phase, and there are two possible paths: (i) the methyl radical reacts with another methyl radical Eq. (11.8):

$$2CH_3^{\bullet} \rightarrow C_2H_6 \tag{11.8}$$

Alternatively, (ii) methyl radical reacts with oxygen in the gas phase to form a methyl peroxy radical, which acts as a chain propagator and accelerates the gas chemistry reaction:

$$CH_3^{\bullet} + O_2 \rightarrow CH_3O_2^{\bullet} \tag{11.9}$$

In Eq. (11.8), C_2H_6 can react with oxygen to form ethyl radicals, which then react with the gas phase and form C_2H_4.

$$C_2H_6 + O_s^- \rightarrow C_2H_5 + OH^- \tag{11.10}$$

$$C_2H_5 + O^{2-} \rightarrow OC_2H_5^- + e^- \tag{11.11}$$

$$OC_2H_5^- \rightarrow C_2H_4 + OH^- \tag{11.12}$$

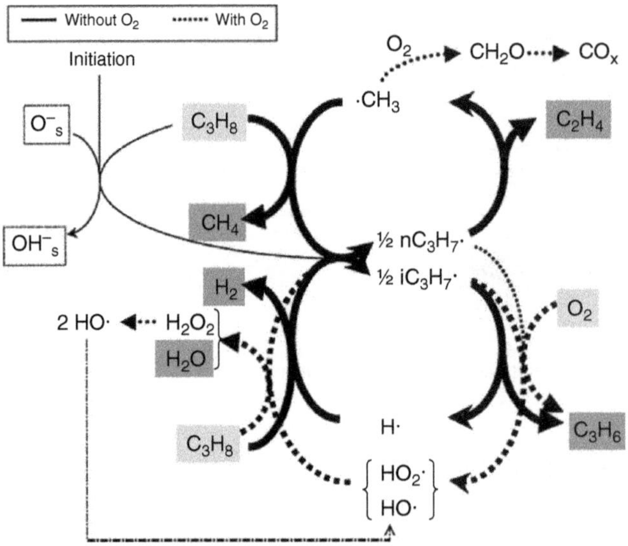

Figure 11.7 Reaction mechanism for the oxidative cracking of *n*-propane with and without oxygen in the feed. *Source:* From [74]/with permission of John Wiley & Sons.

Therefore, alkanes such as propane, *n*-butane, and *n*-hexane with defects at the active sites on the catalyst lead to hydrogen abstraction from the alkane to form alkyl radicals. Hydrogen abstraction from alkanes is the rate-determining step. The alkyl radicals react with the oxygen co-feed and form light olefins, alkanes, CO_x, and H_2O [74]:

$$C_nH_{2n+2} + O_s^- \rightarrow i - C_nH_{2n+1} + OH_s^- \quad (11.13)$$

$$C_nH_{2n+1} + O_2 \rightarrow C_1 - C_5\left[\text{alkanes}\right] + C_2 - C_4\left[\text{olefins}\right] + CO_x + H_2O \quad (11.14)$$

The length of the carbon chain number affects the product. For example, the oxidative cracking of *n*-butane results in the formation of C_4H_8, C_3H_6, and C_2H_4, while the oxidative cracking of ethane produces only C_2H_4 [74]. Interestingly, the MCO process over Li/MgO catalyst could proceed in with or without oxygen gas (Figure 11.7). The MCO process in presence of oxygen, the conversion of the alkane is improved due to generate more active sites on the catalyst surface and enhance radical formation of the $HO_2^·$ and $OH^·$ radicals, which act as a chain propagator of the gas phase chemistry [74].

11.7 Types of Oxidative Cracking of Alkanes

11.7.1 Gas-Phase Oxidative Cracking (GOC)

GOC is an endothermic process because of presence of oxygen in the reactor feed, see reactor of GOC (Figure 11.9a), which leads to an accelerated reaction and rapid alkane conversion [74]. Oxygen as the co-feed offers the advantages of the pure air operation, i.e. increased selectivity and product yield, lower pressure that saves the resources for unit operation, increased possible capacity, low product processing cost, operational flexibility, and unit compactness. For example, when cracking naphtha in the presence of O_2 as the co-feed, the endothermic system requires intensive heat. This causes O_2 gas to combust with H_2 and other hydrocarbons via a non-selective process, leading to heat production [75]. However, O_2 gases with light alkanes are less reactive than liquid hydrocarbons, requiring high temperatures [76]. The gas-phase oxidation of light alkanes (C_2–C_4) mainly depends on C–H bond breakage [77, 78]. Meanwhile, with O_2 gases and heavy hydrocarbons (i.e. *n*-hexane, cyclohexane, *i*-octane, and *n*-decane), the main cracking focuses on C–C bond breakage, which produces majorly light alkenes such as ethylene and propylene. [79]. GOC is similar to the pyrolysis process, and the differences are based on the use of O_2 gases in the GOC. When the reaction is performed at 750 °C, the light alkene yields for ethylene, propylene, butylene, and butadiene (over 50–85%) are higher with GOC than with pyrolysis (44–64%) (Figure 11.8) [79]. O_2 is an excellent free radical that reduces the activation energy of the initial step, thus changes the thermodynamics and kinetics of the reaction

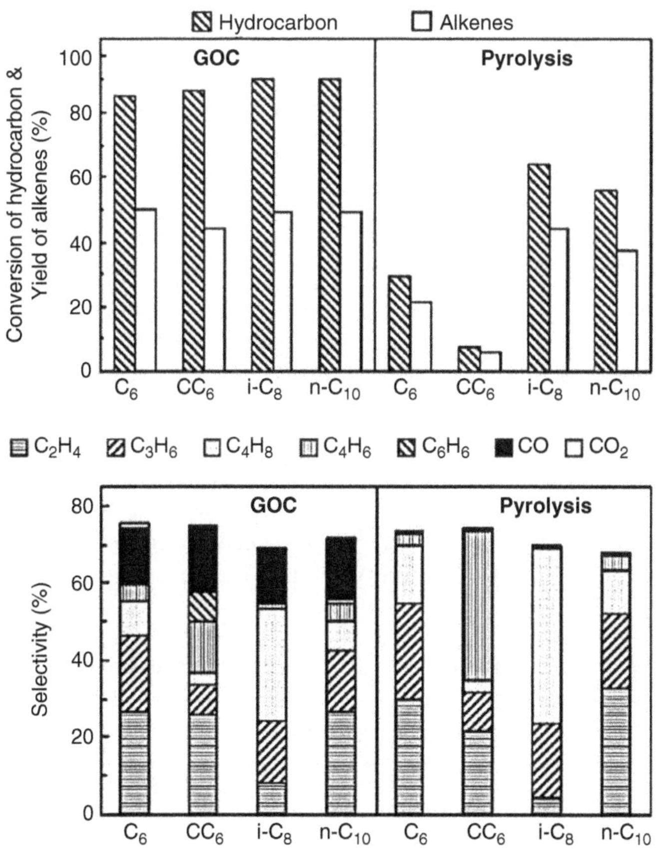

Figure 11.8 Comparing various hydrocarbon yields between pyrolysis and GOC at 750 °C, at a contact time of 0.7 seconds. GOC: atom mole ratio of C/O = 2.4, no dilution. Pyrolysis: nitrogen gas is used. *Source:* From [79] / with permission of Springer Nature.

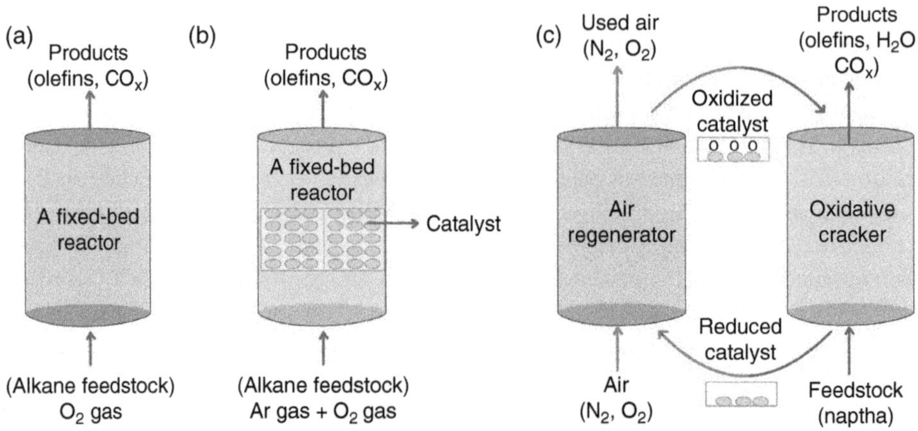

Figure 11.9 Schematic of (a) a fixed-bed reactor for the non-catalytic oxidative cracking or gas-phase oxidative cracking (GOC). *Source:* Adapted from [85]. (b) catalytic oxidative cracking, and (c) redox oxidative cracking (ROC) of naphtha redox catalyst. *Source:* From [105] / with permission of Royal Society of Chemistry.

mechanism [79]. In addition, light alkanes can be mixed with diluted O_2 gas in a specific ratio such as $[O_2]_0$:$[C_2H_6]_0 = 0.2–1$ at temperatures of up to 750 °C. The results reveal reaction times of up to 4 seconds [76].

11.7.2 Catalytic Oxidative Cracking (COC)

COC is a potential alternative to SC, and it involves the use of two main components, a catalyst and continuous flow of oxygen as the co-feed, see Figure 11.9b. This process is a combination of a homogenous reaction (combustion of O_2 gas with the reactant) and a heterogeneous reaction (on the catalyst surface, followed by gas-phase cracking) [74]. COC has some advantages over conventional SC (Table 11.1). Liu et al. [79, 80] compared COC and GOC processes (Table 11.2) and found that the conversion of *n*-hexane by the latter process at 750 °C reaches to 85% with 50% of yield of light alkenes [80]. Meanwhile, the conversion of *n*-hexane using by COC over 0.25% Li/MgO reached 64% conversion, and the selectivity yield of olefins was 67% at 700 °C. However, both processes require O_2 gases,

Table 11.2 Differences between catalytic oxidative cracking and gas-phase oxidative cracking.

Gas-phase oxidative cracking	Catalytic oxidative cracking
The reaction depends on the gas-phase (homogenous) reaction and does not require a catalyst [74].	The reaction is a combination of heterogeneous (catalyst surface) and homogenous (oxidative in the gas phase) reactions [74].
The absence of a catalyst necessitates high reaction temperatures, leading to the formation of higher hydrocarbons [74].	The catalyst assists the formation of light olefins [74].

thereby leading to non-selective combustion and the production of CO_x compounds (14.2–15.4% CO_x selectivity) [79]. Alkanes used as feedstock in past studies include naphtha [81], 1-methylnaphthalene [82], *n*-propane [83–87], *n*-butane [88, 89], and *n*-hexane [20, 90–93]. Furthermore, conventional catalysts used in the COC of hydrocarbons to light olefins include Li-doped MOs [94–97], metal oxychlorides [98], noble-metal-coated monoliths [89, 99–102], supported Co catalysts [20, 103, 104], and supported Au catalysts [85, 87]. However, these catalysts are rapidly deactivated and require the optimum reaction conditions of GOC, with the most serious drawback being the formation of CO_2 [20, 88].

11.8 Different Catalytic Systems for the Catalytic Oxidative Cracking (COC) of *n*-Alkanes

11.8.1 *n*-Hexane to Light Olefins

The conversion of *n*-hexane to light olefins using different catalysts has been investigated. Among all the reported catalysts, Li/MgO catalysts found to be promising for the COC process. Boyadjian et al. [20] synthesized Li/MgO catalysts using several methods such as (i) sol-gel and (ii) conventional impregnation method. The authors observed that the catalyst prepared using sol-gel route exhibited better performance in COC, due to formation more defect [Li^+O^-] active sites, which could be used for cleavage of C-H bond. These catalysts exhibited *n*-hexane conversion of 28 mol% and selectivity of light olefins of 60 mol% in the temperature range of 475–600 °C. The authors proposed that Li^+O^- defect sites abstract hydrogen from a secondary atom, which activated *via* O^- active sites. In another report, Boyadjian et al. [90] used Li/MgO catalyst to compare its performance in the COC and SC processes. The catalytic tests were carried out under atmospheric pressure and isothermal conditions in a conventional fixed-bed reactor. It was observed that the COC was more advantageous than the SC process for the following reasons; (i) the reaction proceeded at 575 °C, which is much lower than that required for SC (800 °C), and (ii) the distribution of light olefin production could be controlled by increasing the ratio of $(C_4^= + C_3^=)/C_2^=$. However, the Li/MgO catalyst suffers from low catalytic activity and undergoes slight deactivation. To avoid these disadvantages, Boyadjian et al. [106] modified the Li/MgO catalyst with MoO_3 (loadings in the range of 0.23–7.1 wt%) (Table 11.3). The catalyst with 0.5 wt% MoO_3 loading on Li/MgO gave the best yield of C_2–C_4 olefins at 575 °C, with a hexane conversion rate of 40% and light olefin (C_2–C_4) yield of approximately 60%. The loading of MoO_3 led to a larger surface area and improved the catalyst's stability by minimizing carbonate formation. Meanwhile, another study focused on the reaction conditions and used integrated plasma-Li/MgO, plasma, and Li/MgO systems (Figure 11.10) for the COC at low temperatures (500–600 °C). The light olefin $(C_2^= + C_5^=)$ yield improved, especially with the integrated plasma-Li/MgO-catalyst reactor due to the synergy between the plasma and the catalyst. Figure 11.11d elucidates the significant influence of both the Li/MgO catalyst and the plasma on hexane activation [107]. The influence of pressure on the oxidative cracking of light alkanes C_2–C_4 was investigated by Nikitin et al. [108], who observed that the elevated pressure in the system decreased the temperature for the

Table 11.3 Performance of different catalysts for oxidative catalytic cracking (COC) of *n*-hexane at different reaction conditions.

Catalyst	Synthesis methods	Total number of acid sites[a] (mmol/g)	BET surface area (m^2/g)	Reaction conditions: Temperature (°C)	GHSV (h)[b]	Feed alkane	Concentration of alkane (mol%)	Light olefins $C_2^= - C_4^=$ (mol%): $C_2^=$	$C_3^=$	$C_4^=$	CO_x[f]	Ref.
Li/MgO	Sol-gel method	–	–	475	15.4[c]	*n*-hexane	4.19	–	–	–	–	[20]
		–	–	525	15.4[c]	*n*-hexane	11.17	–	–	–	–	[20]
		–	–	575	15.4[c]	*n*-hexane	28.35	–	–	–	–	[20]
Li/MgO	Conventional method	–	–	475	15.4[c]	*n*-hexane	3.82	1.89	4.27	3.79	–	[20]
		–	–	525	15.4[c]	*n*-hexane	9.92	5.34	10.81	10.35	–	[20]
		–	–	575	15.4[c]	*n*-hexane	24.49	18.36	20.04	12.23	–	[20]
Li/MgO	Sol-gel method	–	106	475	100[d]	*n*-hexane	4.26		14.77		38.80	[106]
				525	100[d]	*n*-hexane	20.29		40.62		52.07	[106]
				575	100[d]	*n*-hexane	40.01		60.66		24.55	[106]
0.23Mo-Li/MgO	Wet impregnation of Sol-gel synthesized Li/MgO using an aqueous solution of the ammonium molybdate	–	36	475	100[d]	*n*-hexane	6.36		14.78		82.77	[106]
				525	100[d]	*n*-hexane	19.70		38.74		54.91	[106]
				575	100[d]	*n*-hexane	40.67		58.90		28.21	[106]
0.5Mo-Li/MgO		–	70	475	100[d]	*n*-hexane	10.72		30.40		64.17	[106]
				525	100[d]	*n*-hexane	23.21		47.12		76.29	[106]
				575	100[d]	*n*-hexane	40.40		60.22		28.09	[106]
3.6Mo-Li/MgO		–	76	475	100[d]	*n*-hexane	12.12		11.45		84.38	[106]
				525	100[d]	*n*-hexane	16.64		32.86		63.79	[106]
				575	100[d]	*n*-hexane	24.03		45.74		47.29	[106]
7.1Mo-Li/MgO		–	82	475	100[d]	*n*-hexane	12.51		5.42		96.42	[106]
				525	100[d]	*n*-hexane	13.45		11.69		61.22	[106]
				575	100[d]	*n*-hexane	20.38		24.60		48.43	[106]
VO/SrO-γAl_2O_3	Surfactant-free	–	60	525	–	*n*-hexane	57.1[e]	7.5	71.3	2.7	–	[92]
				600	–	*n*-hexane	81.4[e]	5.3	53.9	12.9	–	[92]
Li/MgO	Sol-gel	–	176	575	15.4	*n*-hexane	28.0[e]		60		–	[20]
$Na_2WO_4/CaMnO_3$	–	–	–	725	4500	*n*-hexane	–		55–58		–	[90]

[a] Total number of acid sites obtained from TPD-NH_3.
[b] Gas hourly space velocity (h^{-1}).
[c] Weight hourly space velocity (WHSV) (h^{-1}).
[d] feed (ml/min).
[e] Conversion of alkanes (%).
[f] $CO_x = CO + CO_2$.

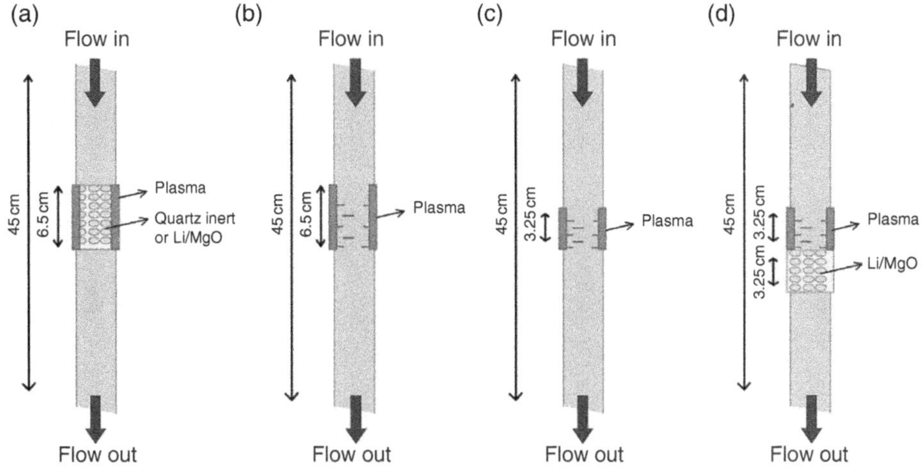

Figure 11.10 Reactor configurations used: (a) Plasma applied to Li/MgO (P-Li/MgO) or quartz (P-Quartz); (b, c) plasma applied to an empty reactor (P-ER); and (d) Li/MgO downstream to plasma zone (PB-Li/MgO). *Source:* From [107] / with permission of Springer Nature.

oxy-cracking of light alkanes and increased the pressure for the selective oxy-cracking of light olefins C_2–C_4 associated gas from 2 to 4 atm. However, as the pressure increased further, the above-mentioned effect weakened.

11.8.2 *n*-Butane to Light Olefins

The conversion of *n*-butane using several catalysts via the COC has been reported (see Table 11.4). Otsuka et al., Campbell et al., and Tong et al. studied *n*-butane conversion via the COC using rare-earth oxides, such as La_2O_3, Nd_2O_3, Sm_2O_3, Eu_2O_3, and Gd_2O_3, at a reaction temperature of 600 °C. The results suggest that the selectivity toward (C_2–C_3) olefins was 20%–28%. In contrast, CeO_2, Pr_6O_{11}, and Tb_4O_7 catalysts exhibited poor light olefin selectivity, and the dominant product was carbon dioxide [109–111]. Kijima et al. [89] performed the COC of *n*-butane using BiOCl, SmOCl, and BiOCl catalysts at two reaction temperatures, i.e. 625 and 650 °C (Table 11.3). When the BiOCl catalyst was used, the light olefin (C_2–C_4) selectivity at 625 °C was 52%, and the C_2–C_3 yield at 650 °C was 34%. McDermott et al. [88] studied the feed of reactants in the COC of *n*-butane using hexagonal boron nitride (hBN) catalysts. The selectivity of the products was controlled by controlling the ratio of *n*-butane to O_2 within the reactant feed. They observed that the adjustment of reactant feed from O_2-rich to O_2-lean changed the ethylene and propylene selectivity. In addition, the use of a nitrogen-containing cobalt catalyst (Co-N) on an alumina catalyst in the COC of *n*-butane to light olefins at 600 °C resulted in an *n*-butane conversion of 82% and total light olefin yield of 47.7 wt%; the ethylene and propylene selectivity was 31 and 13%, respectively [98]. Co-N on an alumina catalyst was prepared by depositing the precursor, sodium salt of (4,5-carboxy)-cobalt phthalocyanine, on γ-Al_2O_3 or SiO_2, followed by its thermal decomposition in a stream of He gas at 700 °C [112]. Nakamura et al. [103] studied the COC of *n*-butane with rare-earth MOs and Li^+-containing Y_2O_3 catalysts at 750 °C

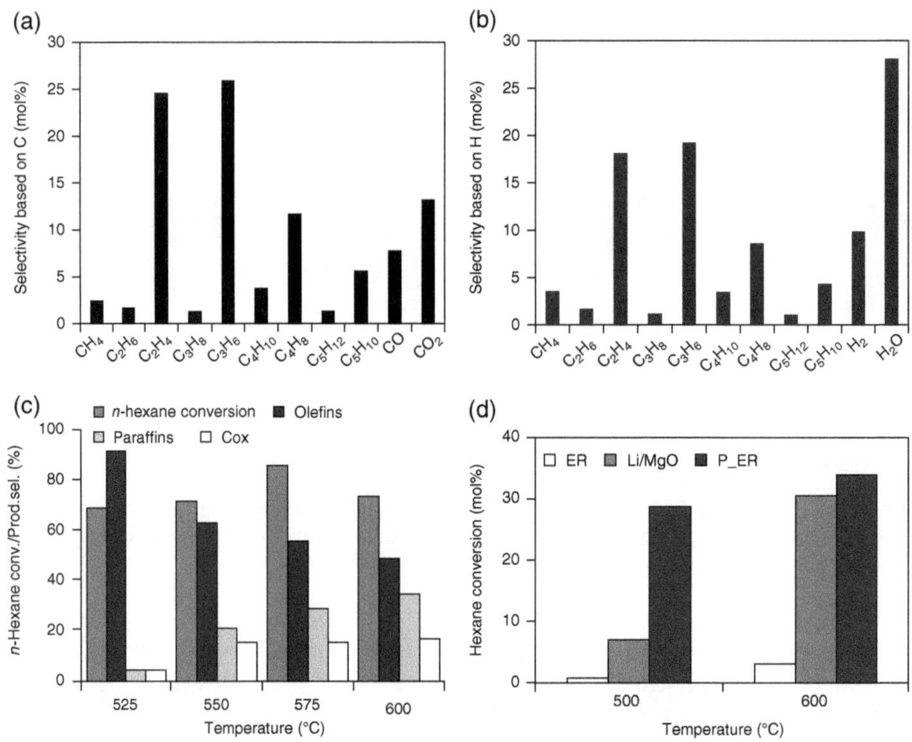

Figure 11.11 *n*-Hexane conversion via catalytic oxidative cracking (COC) to light olefins: (a) product distribution based on carbon (C atom) and (b) (H atom) during the oxidative cracking of *n*-hexane using Li/MgO (feed: 100 ml/min, 10% hexane, 8% oxygen, helium balance; catalyst wt.: 300 mg; temperature: 575 °C; *n*-hexane conversion 45 mol%). *Source:* From [90]/with permission of American Chemical Society. (c) conversion of *n*-hexane to olefins using VO_x/γ-Al_2O_3 (*n*-hexane feed: 0.2 ml; catalyst wt.: 0.425 g; time: 25 second) at different reaction temperatures. *Source:* From [92]/ with permission of Elsevier. (d) *n*-hexane conversions using different reactors: (ER) in an empty reactor, (Li/MgO) reactor filled with Li/MgO, and (P-ER) in plasma reactor. Oxygen conversions (mol%): At 500 °C; 4 mol% (ER), 56 mol% (Li/MgO), 71 mol% (P-ER) and at 600 °C; 10 mol% (ER), 100 mol% (Li/MgO), 100 mol% (P-ER). Reaction conditions: 100 ml/min, 10% *n*-hexane, 8% oxygen, and helium balance. Weight hourly space velocity (WHSV) = 3.08 h^{-1}. *Source:* From [107] / with permission of Springer Nature.

(Table 11.3). They observed that the selectivity toward light olefins over Li^+-containing Y_2O_3 was 58%.

11.8.3 *n*-Propane to Light Olefins

The COC of propane using several catalysts has been reported. Choudhary et al. [113] studied the performance of the COC of *n*-propane using SrO/La_2O_3/SA5205 catalysts at reaction temperatures of 700–850 °C in the presence of steam and limited oxygen. They observed that the highest *n*-propane conversion was 90% at 850 °C, with light olefin selectivity values of 50 and 25% for C_2 and C_3, respectively. Narasimharao et al. [83] synthesized bulk Y_2O_3, loaded the support with gold (Au) (0.5, 1, 1.5, and 2 wt%), and studied the COC performance of *n*-propane at 600 °C (Table 11.5). The bulk yttrium oxide (Y_2O_3) nanorod

Table 11.4 Performance of different catalysts for catalytic oxidative cracking (COC) of *n*-butane at different reaction conditions.

Type of catalysts	Synthesis methods	Total number of acid sites[a] (mmol/g)	BET Surface area (m^2/g)	Reaction conditions: Temperature (°C)	Reaction conditions: GHSV (h^{-1})[b]	Reaction conditions: Feed alkane	Concentration of alkane (mol%)	Light olefins $C_2^= - C_4^=$ (mol%): $C_2^=$	$C_3^=$	$C_4^=$	CO_x[e]	Ref.
LaOCl	Heat treatment	–	2.5	500	–	*n*-butane	6.4[d]	2.9[d]	4.0[d]	10.3[d]	82.5	[89]
		–		550	–	*n*-butane	17.4[d]	6.9[d]	5.9[d]	9.4[d]	77.4	[89]
		–		600	–	*n*-butane	38.3[d]	13.7[d]	8.3[d]	6.5[d]	70.2	[89]
SmOCl		–	6.8	500	–	*n*-butane	5.0[d]	4.9	11.3	0.0	64.2	[89]
		–		550	–	*n*-butane	14.0[d]	10.0	11.3	0.6	61.6	[89]
		–		600	–	*n*-butane	27.4[d]	19.2	13.6	1.1	52.5	[89]
BiOCl		–	6.3	500	–	*n*-butane	3.6	6.2	9.5	20.9	30.4	[89]
		–		550	–	*n*-butane	15.5	16.8	11.3	15.8	22.4	[89]
		–		600	–	*n*-butane	50.7	28.3	10.1	15.1	25.6	[89]
0.7 Co (wt.%)/ Υ-Al_2O_3	–	–	–	550	5[c]	*n*-butane	51.2	15.2	7.3	7.1	–	[98]
0.8Co-N (wt.%)/ Υ-Al_2O_3	–	–	–	550	5[c]	*n*-butane	73.7	26.5	11.7	4.0	–	[98]
0.8Co-N[b] (wt.%)/ Υ-Al_2O_3	–	–	–	550	5[c]	*n*-butane	82.0	30.8	13.2	3.7	–	[98]
2.3 Co (wt.%)/ SiO_2	–	–	–	550	5[c]	*n*-butane	23.2	1.9	1.8	13.8	–	[98]

(Continued)

Table 11.4 (Continued)

Type of catalysts	Synthesis methods	Total number of acid sites[a] (mmol/g)	BET Surface area (m^2/g)	Reaction conditions: Temperature (°C)	GHSV (h^{-1})[b]	Feed alkane	Concentration of alkane (mol%)	Light olefins $C_2^{=}-C_4^{=}$ (mol%): $C_2^{=}$	$C_3^{=}$	$C_4^{=}$	CO_x[e]	Ref.
0.7 Co-N (wt.%)/SiO_2	–	–	–	550	5[c]	*n*-butane	27.6	6.0	5.8	7.8	–	[98]
0.7 V (wt.%)/ Y-Al_2O_3	–	–	–	550	5[c]	*n*-butane	13.5	0.4	0.4	8.6	–	[98]
30 V (wt.%)/ MgO	–	–	–	550	5[c]	*n*-butane	35.6	5.0	2.6	20.1	–	[98]
1.3 Co (wt.%)/ θ-Al_2O_3	–	–	–	750	1.2[c]	*n*-butane	70.6	15.5	18.0	4.8	–	[98]
2.9 P (wt.%)/ θ-Al_2O_3	–	–	–	750	1.2[c]	*n*-butane	65.7	17.8	19.0	5.6	–	[98]
5 La (wt.%)/ θ-Al_2O_3	–	–	–	750	1.2[c]	*n*-butane	53.5	15.2	17.0	3.6	–	[98]
1 K (wt.%)/ ZSM-5	–	–	–	750	1.2[c]	*n*-butane	56.4	17.7	15.4	4.0	–	[98]
HY-12	–	–	–	750	1.2[c]	*n*-butane	67.5	20.0	16.2	5.7	–	[98]
3.3 K/HY-12	–	–	–	750	1.2[c]	*n*-butane	58.0	17.1	14.8	4.2	–	[98]
$CaAlO_2$	–	–	–	750	1.2[c]	*n*-butane	49.3	15.1	16.2	2.1	–	[98]
Y_2O_3	–	–	–	700	–	*n*-butane	65.9	36	26	4	25	[103]
				750	–	*n*-butane	80.6	33	21	3	22	[103]
Sm_2O_3	–	–	–	700	–	*n*-butane	66.5	35	23	4	26	[103]
				750	–	*n*-butane	74.1	36	23	4	25	[103]

Gd_2O_3	–	–	–	700	–	*n*-butane	66.3	34	24	4	24	[103]
				750	–	*n*-butane	77.7	34	22	4	25	[103]
La_2O_3	–	–	–	700	–	*n*-butane	66.1	34	20	3	25	[103]
				750	–	*n*-butane	96.6	34	19	4	24	[103]
CeO_2	–	–	–	700	–	*n*-butane	98.2	29	12	2	43	[103]
				750	–	*n*-butane	54.3	34	15	1	31	[103]
quartz	–	–	–	700	–	*n*-butane	10.1	34	45	6	1	[103]
				750	–	*n*-butane	42.3	37	41	5	1	[103]
hBN	–	–	–	500	55[c]	*n*-butane	8.4	21.7	35.5	28.81	1.6	[88]
				525	27.5[c]	*n*-butane	8.4	23	40.3	21	1.1	[88]
				550	15.7[c]	*n*-butane	7.8	23.8	43.7	15.2	0.9	[88]

[a] Total number of acid sites obtained from TPD-NH_3.
[b] Gas hourly space velocity (h^{-1}).
[c] Weight hourly space velocity (WHSV) (h^{-1}).
[d] Conversion of alkanes and selectivity of light olefins (%).
[e] $CO_x = CO + CO_2$.

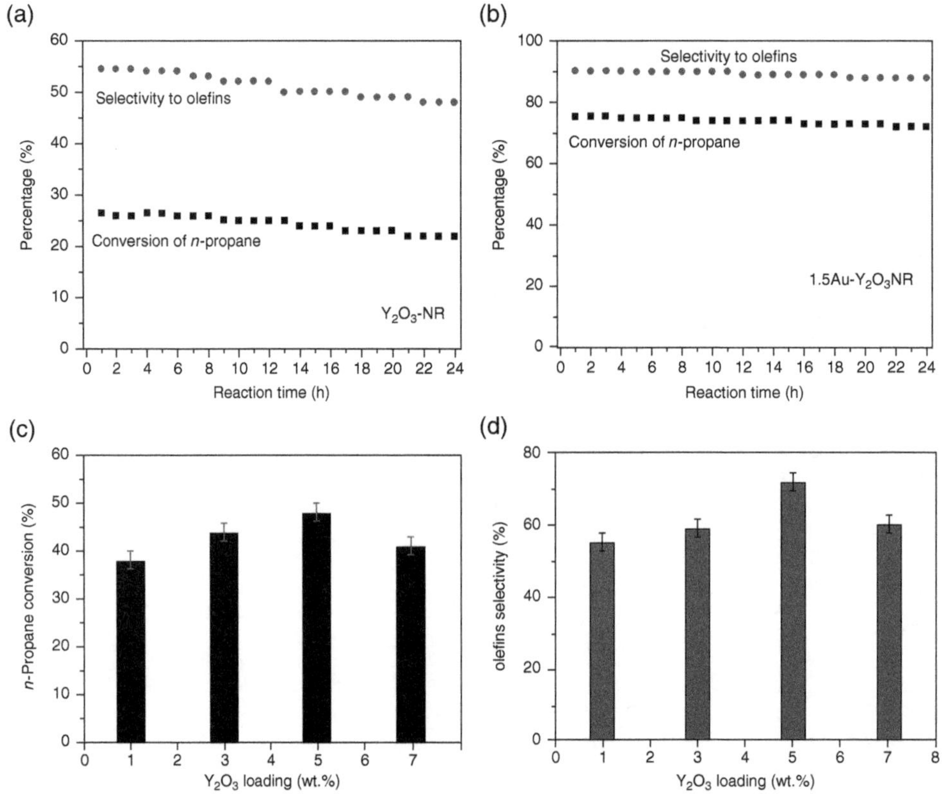

Figure 11.12 (a) Time on stream analysis of Y_2O_3-NR and (b) 1.5 Au-Y_2O_3-NR catalysts (reaction temperature: 600 °C, GHSV: 48000 h^{-1}). *Source:* From [83]/with permission of Elsevier. Influence of Y_2O_3 loading on (c) *n*-propane conversion and (d) olefin selectivity (reaction conditions: temperature: 500 °C; GHSV = 48000 h^{-1}). *Source:* From [86]/with permission of Elsevier.

showed *n*-propane conversion of 20%, with light olefin (C_2–C_3) selectivity of 25%. With 2.5 wt% gold-supported Y_2O_3 nanorods (1.5 Au-Y_2O_3-NR), *n*-propane conversion of 75.5% and light olefin selectivity of 90.3% were achieved (Figure 11.12a and b). They indicated that the superior catalytic activity of the 1.5 Au-Y_2O_3-NR sample was due to the presence of greater amounts of mobile oxygen species, Lewis acidic sites, and $Au\delta^+$-O-Y species in the catalyst. In addition, La_2O_3 and 0.2Au-La_2O_3 nanorod catalysts were synthesized using a hydrothermal method and modified with Y_2O_3 (1, 3, 5, and 7 wt%) using the impregnation method applied for the COC of *n*-propane at 600 °C (Table 11.5). They observed that the 5Y-0.2Au-La_2O_3 catalyst exhibited an *n*-propane conversion of 78% and light olefin selectivity of 72% (Figure 11.12c and d). They suggested that the samples possessed a greater number of Lewis acid sites and mobile oxygen species [86]. In another study, the COC of *n*-propane using La_2O_3, 0.2 wt% Au/La_2O_3, and 10 wt% SO_4^{2-}/La_2O_3 catalysts at 500 °C was reported. The La_2O_3 catalyst showed an *n*-propane conversion of 10% and olefin selectivity of 15%, whereas 0.2 wt% Au/La_2O_3 catalyst exhibited *n*-propane conversion of 25% and olefin selectivity of 39%. The most significant improvement was observed when 10 wt%

Table 11.5 Performance of different catalysts for catalytic oxidative cracking (COC) of *n*-propane at different reaction conditions.

Type of catalysts	Synthesis methods	Total number of acid sites (mmol/g)[a]	BET surface area (m^2/g)	Reaction conditions: Temperature (°C)	GHSV (h)[b]	Feed alkane	Concentration of alkane (mol%)	Light olefins $C_2^= - C_4^=$ (mol%): $C_2^=$	$C_3^=$	$C_4^=$	CO_x[e]	Ref.
La_2O_3-nanoroad (NR)	Hydrothermal method	14.5[c]	59	650	48 000	*n*-propane	23.1		22.3		67.0	[87]
1 wt% Y/La_2O_3 (NR)		14.6[c]	47	650	48 000	*n*-propane	30.1		30.1		73.6	[87]
3 wt% Y/La_2O_3 (NR)		15.5[c]	44	650	48 000	*n*-propane	35.3		35.3		69.0	[87]
5 wt% Y/La_2O_3 (NR)		15.7[c]	42	650	48 000	*n*-propane	29.3		29.3		71.7	[87]
7 wt% Y/La_2O_3 (NR)		16.2[c]	39	650	48 000	*n*-propane	29.2		29.2		71.1	[87]
S-La_2O_3 (NR)		20.8[c]	71	650	48 000	*n*-propane	29.2		29.2		73.6	[87]
1 wt% Y-S/La_2O_3 (NR)		22.7[c]	62	650	48 000	*n*-propane	32.3		32.3		74.6	[87]
3 wt% Y-S/La_2O_3 (NR)		24.1[c]	59	650	48 000	*n*-propane	38.4		38.4		65.0	[87]
5 wt% Y-S/La_2O_3 (NR)		23.5[c]	48	650	48 000	*n*-propane	34.1		34.1		70.4	[87]
7 wt% Y-S/La_2O_3 (NR)		22.2[c]	45	650	48 000	*n*-propane	33.2		33.2		70.8	[87]
La_2O_3	–	9.5	30	650	48 000	*n*-propane	7.5		15.0		71.3	[85]
5 SO_4^{2-}/La_2O_3	Impregnation technique	21.5	–	650	48 000	*n*-propane	15.8		18.4		78.8	[85]
10 SO_4^{2-}/La_2O_3		30.1	46	650	48 000	*n*-propane	18.2		19.5		78.9	[85]
0.05 Au/La_2O_3		–	55	650	48 000	*n*-propane	20.5		26.0		71.9	[85]
0.2 Au/La_2O_3		–	60	650	48 000	*n*-propane	25.0		39.0		66.7	[85]
0.05 Au/5 SO_4^{2-}/La_2O_3		–	42	650	48 000	*n*-propane	27.0		28.0		69.9	[85]
0.05 Au/10 SO_4^{2-}/La_2O_3		–	–	650	48 000	*n*-propane	32.0		28.0		69.0	[85]
0.2 Au/5 SO_4^{2-}/La_2O_3		22.6	65	650	48 000	*n*-propane	35.0		47.0		48.5	[85]
0.2 Au/10 SO_4^{2-}/La_2O_3		–	–	650	48 000	*n*-propane	40.6		54.2		41.6	[85]
La_2O_3	–	–	–	550	48 000	*n*-propane	23		22		67.0	[86]
0.2 Au/La_2O_3	–	14.2[c]	52	550	48 000	*n*-propane	32		48		48.9	[86]
1Y-0.2Au-La_2O_3	–	15.5[c]	46	550	48 000	*n*-propane	38		55		41.3	[86]

(Continued)

Table 11.5 (Continued)

Type of catalysts	Synthesis methods	Total number of acid sites (mmol/g)[a]	BET surface area (m^2/g)	Reaction conditions: Temperature (°C)	Reaction conditions: GHSV (h)[b]	Reaction conditions: Feed alkane	Concentration of alkane (mol%)	Light olefins $C_2^=-C_4^=$ (mol%): $C_2^=$	$C_3^=$	$C_4^=$	CO_x[e]	Ref.
3Y-0.2Au-La_2O_3	–	17.1[c]	40	550	48 000	*n*-propane	44		59		36.6	[86]
5Y-0.2Au-La_2O_3	–	21.6[c]	36	550	48 000	*n*-propane	58		72		22.6	[86]
7Y-0.2Au-La_2O_3	–	17.8[c]	30	550	48 000	*n*-propane	41		60		34.0	[86]
Y_2O_3	–	–	–	550	48 000	*n*-propane	18		20		72.0	[86]
0.5 wt% Au/Y_2O_3	Hydrothermal method	11.7	48	600	48 000	*n*-propane	49.6	23.4	52.5	–	–	[83]
1.0 wt% Au/Y_2O_3	Hydrothermal method	17	34	600	48 000	*n*-propane	56.8	25.1	56.2	–	–	[83]
1.5 wt% Au/Y_2O_3	Hydrothermal method	23	22	600	48 000	*n*-propane	75.5	26.2	64.1	–	–	[83]
2.0 wt% Au/Y_2O_3	Hydrothermal method	17.5	14	600	48 000	*n*-propane	59.5	24.3	55.8	–	–	[83]
Pt/$Ce_{0.5}Zr_{0.5}O_2$	–	–	–	500	36[d]	*n*-propane	18.0	0.5	0.1	0.8	16.5	[84]
		–	–	550	36[d]	*n*-propane	23.0	0.5	0.1	0.8	16.4	[84]
		–	–	600	36[d]	*n*-propane	28.0	0.5	0.1	0.8	16.7	[84]
PtS/$Ce_{0.5}Zr_{0.5}O_2$	–	–	–	500	36[d]	*n*-propane	25.0	0.0	0.0	0.2	16.0	[84]
		–	–	550	36[d]	*n*-propane	28.0	0.0	0.0	0.0	15.1	[84]
		–	–	600	36[d]	*n*-propane	32.0	0.0	0.0	0.0	15.0	[84]
Au/$Ce_{0.5}Zr_{0.5}O_2$	–	52.7[c]	–	500	36[d]	*n*-propane	35.2	10.8	4.5	18.2	12.0	[84]
		52.7[c]	–	550	36[d]	*n*-propane	39.8	12.9	5.3	19.8	12.9	[84]
		52.7[c]	–	600	36[d]	*n*-propane	44.6	14.0	6.4	23.0	13.5	[84]
Au/$SCe_{0.5}Zr_{0.5}O_2$	–	70.5[c]	103	500	36[d]	*n*-propane	41.5	16.3	7.8	26.1	10.0	[84]
		70.5[c]	103	550	36[d]	*n*-propane	49.5	17.7	8.4	27.5	11.4	[84]
		70.5[c]	103	600	36[d]	*n*-propane	55.8	18.7	9.8	28.5	12.3	[84]

[a] Total number of acid sites obtained from TPD-NH_3.
[b] Gas hourly space velocity (h^{-1}).
[c] Measurements by DRIFT pyridine adsorption.
[d] Weight hourly space velocity (WHSV) (h^{-1}).
[e] $CO_x = CO + CO_2$.

SO_4^{2-}/La_2O_3 catalyst was used, which had an *n*-propane conversion of 46% and light olefin selectivity of 53%, and it showed stable activity even after 20 hours. They explained that the partially present oxide gold species and sulfate ions resulted in the enhancement of the Lewis and Bronsted acidity of the catalysts [85]. Furthermore, the use of Y_2O_3-supported La_2O_3 nanorods for the COC of *n*-propane conversion at 550 °C has also attracted attention. The Y_2O_3-supported (1, 3, 5, and 7 wt%) La_2O_3 was incorporated using the impregnation method, which showed conversion of 15% and light olefin (C_2–C_3) selectivity of 22%. The COC of *n*-propane using 3 wt% Y_2O_3-supported on La_2O_3 nanorods showed conversion of 25% and light olefin selectivity of 36% [87] (Table 11.5).

11.9 Redox Oxidative Cracking (ROC) Catalyst via Chemical Looping

The latest anaerobic process is also known as the redox oxidative cracking or chemical looping process, which is an alternative strategy for conversion of alkanes which do not use O_2 gas as a co-feed [72]. The oxygen atoms in crystal lattice of the catalyst provides the oxygen needed and plays a crucial role in alkane oxidation. The reduced catalyst material is subsequently regenerated by the gaseous oxygen or air and recycled to complete the redox catalyst, see (Figure 11.9c). The ROC process has three advantages over the traditional SC and GOC/COC processes:

1) Avoiding the premixing of O_2 with the alkane which also eliminates the oxygen separation from products because of the utilization of oxygen from the catalyst lattice.
2) The side reaction that leads to the consumption of light olefins because of non-selective combustion with oxygen is avoided.
3) The energy efficiency of alkane conversion is improved by eliminating the oxygen separation unit and combustion unit in the plant [72].

However, oxidative cracking is disadvantageous because it forms CO_x gases [20]. There has been an attempt to minimize the CO_x formation. The study considers the oxidative cracking of *n*-hexane using the oxygen lattice of $VO_x/Ce\text{-}Al_2O_3$ catalysts and the operation of a circulating fluidized bed, as shown in Figure 11.9c. The circulating interconnected fluidized beds contain two separate chambers: (1) oxidative crackers (reduces the catalyst) and (2) catalyst regenerator (re-oxidizes the catalyst). The solid-supported catalysts are circulated between the two reactor chambers. In the oxidative cracking unit, the oxidative cracking reaction of hexane using a solid oxide catalyst in the presence of O_2 gas occurs, and the catalyst is reduced to the initial catalyst. The reduced catalyst is sent to the regenerator under air, and the catalyst is then regenerated via re-oxidation with air, which is recycled back to the oxidative cracker reactor. The circulating fluidized bed has the advantages of achieving higher olefin selectivity owing to its controllable reaction temperature, stable catalyst, and kinetically favorable system. The performance of the process is affected by several factors: (1) the amount of lattice oxygen readily released by the catalyst, which reacts with the hydrocarbon substrate, and (2) the susceptibility of the reduced catalyst to re-oxidization by gas-phase oxygen (air) [114]. The possible mechanism of oxidative cracking of alkanes is presented in Eqs. (11.14) through (11.16). Equation (11.17) shows the re-oxidation process for the catalyst inside the catalyst regenerator. The chemical looping process may have an undesired reaction, as shown in Eq. (11.15) and (11.16), which should be controlled to enhance the light olefin selectivity [92, 115].

The following reactions between the metal oxide (MO_x) catalyst and alkane (C_nH_{2n+2}) generally occur in the reductive oxidative cracking unit.

Desired reaction:

$$C_nH_{2n+2} + MO_x \rightarrow C_nH_{2n} + H_2O + MO_{x-1} \tag{11.15}$$

Undesired reaction:

$$C_nH_{2n+2} + MO_x \rightarrow CO_2 + CH_4 + C_2H_6 \ldots + MO_{x-1} \tag{11.16}$$

$$C_nH_{2n} + MO_x \rightarrow CO_2 + CH_4 + C_2H_6 \ldots + MO_{x-1} \tag{11.17}$$

Reactions in the catalyst regenerator:

$$MO_{x-1} + \frac{1}{2}O_2 \rightarrow MO_x \tag{11.18}$$

The conversion of alkanes into light olefins via chemical looping using several catalysts has been reported. Amusa et al. [92] studied novel $VO_x/SrO\text{-}\gamma\text{-}Al_2O_3$ catalysts with chemical looping at 525–600 °C. They synthesized $SrO/\gamma\text{-}Al_2O_3$ with different ratios through a surfactant-free method, while VO_x was incorporated into $SrO\text{-}\gamma\text{-}Al_2O_3$ through the impregnation method. Chemical looping was performed using a fluidized riser reactor in the absence of gas-phase oxygen. The best performing catalyst was 1 : 1 $VO_x/SrO\text{-}\gamma\text{-}Al_2O_3$, which exhibited an *n*-hexane conversion of 84% and light olefin selectivity of 88% (Table 11.6). This is due to its balanced acidity and moderate VO_x support interactions. Moreover, perovskite oxides have also been used for the ROC of *n*-hexane. Dudek et al. converted *n*-hexane into 20 wt% Na_2WO_4, $SrMnO_3$, and $CaMnO_3$ through oxy-cracking using the chemical loop process at 725 °C. The catalytic test was conducted in a 1/8 in ID fixed-bed quartz U-tube reactor that was heated externally using a furnace controlled with a K-type thermocouple. Cycling between the reduction and oxidation steps was achieved using an automated valve-switching system. The 20 wt% $Na_2WO_4/CaMnO_3$ exhibited a light olefin selectivity of 55–58%, while $Na_2WO_4/SrMnO_3$ achieved 41% light olefin selectivity, as shown in Figure 11.13 and Table 11.6 [93]. Elbadawi et al. [114] studied the performance of the chemical loop process using $VO_x/Ce\text{-}Al_2O_3$ at 525–600 °C (Table 11.6). The conversion of *n*-hexane using $VO_x/Ce\text{-}Al_2O_3$ at 550 °C resulted in an *n*-hexane conversion of 30% and light olefin selectivity (C_2–C_4) of 27%.

11.10 Conclusions

In this chapter, we presented a comprehensive review of the scientific literature on the production of light olefins from alkanes. The primary aim was to compare different processes for the production of olefins using ethane and hexane under different reaction

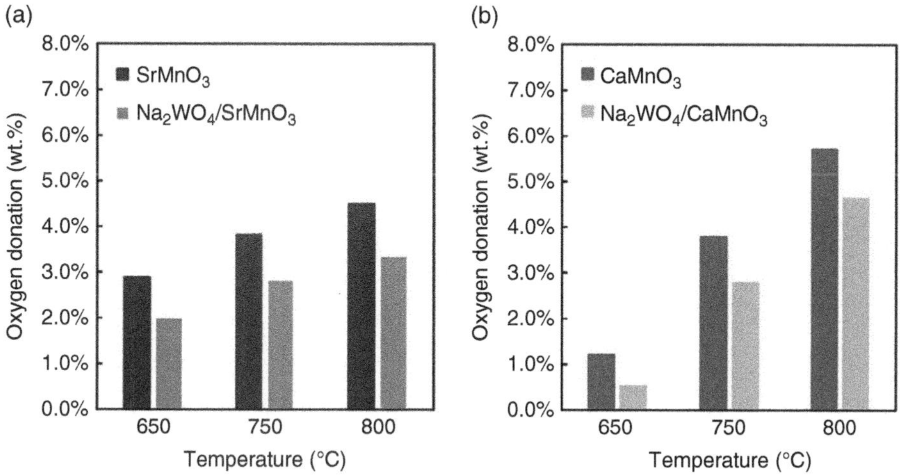

Figure 11.13 Oxygen donation by (a) $SrMnO_3$, $Na_2WO_4/SrMnO_3$, and (b) $CaMnO_3$, $Na_2WO_4/CaMnO_3$, as determined by the material balanced on H_2/C_2H_4 co-feed experiments in a U-tube reactor (feed: 100 ml/min; $H_2/C_2H_4/Ar = 0.4/0.4/0.2$; catalyst wt: 100 mg; step duration = 10 seconds) at different reaction temperatures. *Source:* From [93]/with permission of Elsevier.

conditions. The COC of short-chain alkanes to olefins is an effective route to the conventional SC of naphtha. An effective oxidative cracking catalyst offers active and selective catalytic sites for alkane activation and olefin formation and reduces the extent of CO_x formation. A comparison study of different catalysts reported in the literature shows that high yields of olefins were achieved with oxidative cracking at short contact times using Pt-Al_2O_3-based catalysts. However, these catalyst systems operate at high temperatures.

This chapter also discusses various catalysts for oxidative cracking of different alkanes, comparing different catalysts, such as mixed oxide catalysts, noble, rare-earth, alkali-metal-supported catalysts, metal oxychlorides, and metal-nitride-based catalysts. The reported studies suggested that Co-N-supported Al_2O_3 and gold-supported sulfated $Ce_{0.5}Zr_{0.5}O_2$ catalysts offer a significant activity for oxidative cracking at lower reaction temperatures. Gold-supported catalysts appear to be efficient catalysts. However, the minimization of CO_2 formation remains a challenge for the technical and economic feasibility of the process.

Recently, an alternative strategy, known as the ROC of alkanes without O_2 as a co-feed, was reported. The oxygen lattice of the catalyst provides oxygen and plays a crucial role in oxidizing alkanes. The reduced catalyst material is subsequently recovered by the oxygen from the carrier air and recycled to complete the redox catalyst. It was observed that a VO_x/SrO-γ-Al_2O_3 catalyst exhibited superior performance in the ROC process. However, future work should focus on improving the catalyst systems toward commercially acceptable yields of light alkenes in oxidative cracking processes. The catalyst should possess meso- to macropores to enhance the gas-phase propagation of intermediate radicals. The combination of an optimum catalyst and reactor system is important to achieve adequate olefin yields.

Table 11.6 Performances of different catalysts in the redox oxidative cracking (ROC) of *n*-hexane at different reaction temperatures.

Catalyst types	Synthesis methods	Total number of acid sites[a] (mmol/g)	BET surface area (m^2/g)	Reaction conditions: Temperature (°C)	Gas flow	Pressure vacuum (kPa)	Feed alkane	Concentration of alkane (%)	Light olefins $C_2^= - C_4^=$ (%): $C_2^=$	$C_3^=$	$C_4^=$	CO_x[b]	Ref.
$VO_x/\gamma\text{-}Al_2O_3$	Sol-gel method	0.985	60	525	Ar	20.7	*n*-hexane	68.6	14.6	73.4	3.02	4.4[c]	[92]
				600	Ar	20.7	*n*-hexane	73.6	3.1	6.7	38.9	16.6[c]	[92]
VO_x/ $SrO\text{-}\gamma Al_2O_3$		0.561	76	525	Ar	20.7	*n*-hexane	57.1	7.5	71.3	2.7	4.1[c]	[92]
				600	Ar	20.7	*n*-hexane	81.4	5.3	53.9	12.9	12.2[c]	[92]
VO_x/ $SrO\text{-}\gamma\text{-}Al_2O_3$				550	Ar	20.7	*n*-hexane	84		88		1.7	[92]
Na_2WO_4/ $CaMnO_3$	–	–	–	725	–	–	*n*-hexane	49.4	25	36.8	9.1	0	[93]
	–	–	–	750	–	–	*n*-hexane	65.7	24	40.3	10	0	[93]
	–	–	–	775	–	–	*n*-hexane	81.7	21.9	43.2	10.5	0	[93]
Na_2WO_4/ $SrMnO_3$	–	–	–	725	–	–	*n*-hexane	53		43.5		0	[93]
VO_x/Ce-Li/ MgO	–	–	–	550	–	–	*n*-hexane	30		60		30	[116]
Plasma-Li/ MgO	–	–	–	600	–	–	*n*-hexane	64		19		29.7	[116]
$VO_x/\gamma\text{-}Al_2O_3$	–	–	–	525	–	–	*n*-hexane	68.6	14.6	73.4	3.02	4.4	[92]
	–	–	–	600	–	–	*n*-hexane	73.6	3.1	6.7	38.9	16.6	[92]

VO_x/ SrO/γ-Al_2O_3	–	–	–	525	–	–	*n*-hexane	57.1	7.5	71.3	2.7	4.1	[92]
	–	–	–	600	–	–	*n*-hexane	81.4	5.3	53.9	12.9	12.2	[92]
VO_x/ Ce-Al_2O_3	–	–	–	525	–	–	*n*-hexane	21	14.3	10.4	8.5	27	[114]
	–	–	–	550	–	–	*n*-hexane	30	15.2	6.14	5.7	30	[114]
	–	–	–	575	–	–	*n*-hexane	31	11.4	3.76	4.4	33	[114]
	–	–	–	600	–	–	*n*-hexane	33	9.4	2.0	3.29	37	[114]

[a] Total number of acid sites obtained from TPD-NH_3.
[b] $CO_x = CO + CO_2$.
[c] $CO_x = CO/CO_2$.

Acknowledgments

The authors (H.S.K. and K.N.) gratefully acknowledge the Research and Development Office (RDO) of King Abdulaziz University, Jeddah, Saudi Arabia for the Postdoctoral Fellowship to H.S.K.

References

1 Amghizar, I., Vandewalle, L.A., Van Geem, K.M., and Marin, G.B. (2017). New trends in olefin production. *Engineering* 3 (2): 171–178.

2 Mascal, M. (2012). Chemicals from biobutanol: technologies and markets, Biofuels. *Bioprod. Bioref.* 6: 483–493.

3 Zhu, G., Xie, C., Li, Z., and Wang, X. (2017). Catalytic processes for light olefin production BT. In: *Handbook of Petroleum Technology* (ed. C.S. Hsu and P.R. Robinson), 1063–1079. Cham: Springer International Publishing.

4 Manjula, B., Reddy, A.B., Sadiku, E.R. et al. (2017). 18 - Use of polyolefins in hygienic applications. In: *The Textile Institute Book Series*, 2e (ed. S.C.O.B.T.-P.F. Ugbolue), 539–560. Woodhead Publishing.

5 Kissin, Y.V. (2005). Olefin polymers, introduction. *Kirk-Othmer Encycl. Chem. Technol.* 3: 1–11.

6 Benvenuto, M. (2015). Linear alpha olefins (LAO). In: *Industrial Chemistry* (ed. M.A. Benvenuto), 19–22. De Gruyter.

7 Ren, T., Patel, M., and Blok, K. (2006). Olefins from conventional and heavy feedstocks: Energy use in steam cracking and alternative processes. *Energy* 31 (4): 425–451.

8 Al-Douri, A., Sengupta, D., and El-Halwagi, M.M. (2017). *J. Nat. Gas Sci. Eng.* 45: 436–455.

9 Sheng, J., Yan, B., Lu, W.-D. et al. (2021). Oxidative dehydrogenation of light alkanes to olefins on metal-free catalysts. *Chem. Soc. Rev.* 50 (2): 1438–1468.

10 Corma, A., Melo, F.V., Sauvanaud, L., and Ortega, F. (2005). Light cracked naphtha processing: controlling chemistry for maximum propylene production. *Catal. Today* **107–108**: 699–706.

11 Fakhroleslam, M. and Sadrameli, S.M. (2020). Thermal cracking of hydrocarbons for the production of light olefins: a review on optimal process design, operation, and control. *Ind. Eng. Chem. Res.* 59 (27): 12288–12303.

12 Lee, R.P. (2019). Alternative carbon feedstock for the chemical industry? – assessing the challenges posed by the human dimension in the carbon transition. *J. Clean. Prod.* 219: 786–796.

13 Xieqing, W., Chaogang, X., Zaiting, L., and Genquan, Z. (2006). Catalytic processes for light olefin production. In: *Practical Advances in Petroleum Processing* (ed. C. Hsu and P. Robinson), 149–168. New York: Springer.

14 Dahl, I.M. and Kolboe, S. (1993). On the reaction mechanism for propene formation in the MTO reaction over SAPO-34. *Catal. Letters* 20 (3): 329–336.

15 Borodina, E., Sharbini Harun Kamaluddin, H., Meirer, F. et al. (2017). Influence of the reaction temperature on the nature of the active and deactivating species during methanol-to-olefins conversion over H-SAPO-34. *ACS Catal.* 7 (4): 5268–5281.

16 Kazansky, V.B., Frash, M.V., and van Santen, R.A. (1996). Quantumchemical study of the isobutane cracking on zeolites. *Appl. Catal. A Gen.* 146 (1): 225–247.

17 Masoumi, M., Shahrokhi, M., Sadrameli, M., and Towfighi, J. (2006). Modeling and control of a naphtha thermal cracking pilot plant. *Ind. Eng. Chem. Res.* 45 (10): 3574–3582.

18 Rice, F.O. (1931). The thermal decomposition of organic compounds from the standpoint of free radicals. I. saturated hydrocarbons. *J. Am. Chem. Soc.* 53 (5): 1959–1972.

19 Gholami, Z., Gholami, F., Tišler, Z., and Vakili, M. (2021). *Energies* 14: 8190.

20 Boyadjian, C., Lefferts, L., and Seshan, K. (2010). Catalytic oxidative cracking of hexane as a route to olefins. *Appl. Catal. A Gen.* 372 (2): 167–174.

21 Yarulina, I., Chowdhury, A.D., Meirer, F. et al. (2018). Recent trends and fundamental insights in the methanol-to-hydrocarbons process. *Nat. Catal.* 1 (6): 398–411.

22 Corma, A. and Orchillés, A.V. (2000). Current views on the mechanism of catalytic cracking. *Microporous Mesoporous Mater.* **35–36**: 21–30.

23 Nguyen, K.T. and Kung, H.H. (1990). Generation of gaseous radicals by a V-Mg-O catalyst during oxidative dehydrogenation of propane. *J. Catal.* 122 (2): 415–428.

24 Cavani, F., Ballarini, N., and Cericola, A. (2007). Oxidative dehydrogenation of ethane and propane: how far from commercial implementation? *Catal. Today* 127 (1): 113–131.

25 Leveles, L., Seshan, K., Lercher, J.A., and Lefferts, L. (2003). Oxidative conversion of propane over lithium-promoted magnesia catalyst: I. Kinetics and mechanism. *J. Catal.* 218 (2): 296–306.

26 Torres Galvis, H.M. and de Jong, K.P. (2013). Catalysts for production of lower olefins from synthesis gas: a review. *ACS Catal.* 3 (9): 2130–2149.

27 Zimmermann, H. and Walzl, R. (2009). Ethylene. In: *Ullmann's Encyclopedia of Industrial Chemistry*, vol. 13, 465–529.

28 Ramezani, S.F., Karimi, M., Panahi, M., and Rafiee, A. (2020). Sustainable dimethyl carbonate production from ethylene oxide and methanol. *Chem. Eng. Technol.* 43: 2484–2492.

29 Matar, S. and Hatch, L. (2001). *Chemistry of Petrochemical Processes*. Butterworth-Heinemann: Gulf Professional Publishing.

30 Bender, M. (2014). An overview of industrial processes for the production of olefins – C_4 hydrocarbons. *ChemBioEng Rev.* 1 (4): 136–147.

31 (2021). Propylene oxide market research report by production process, by application - global forecast to 2025 – cumulative impact of COVID-19. *ReportLinker*. https://www.reportlinker.com/p05911310/Propylene-Oxide-Market-Research-Report-by-Production-Process-by-Application-Global-Forecast-to-Cumulative-Impact-of-COVID-19.html?utm_source=GNW.

32 The Business Research Company (2020). Propylene global market report 2020. 150. https://www.thebusinessresearchcompany.com/report/propylene-global-market-report

33 Ding, J. and Hua, W. (2013). Game changers of the C_3 value chain: gas, coal, and biotechnologies. *Chem. Eng. Technol.* 36 (1): 83–90.

34 Eisele, P. and Killpack, R. (2012). Propene. In: *Ullmann's Encyclopedia of Industrial Chemistry*, vol. 30, 281–293.

35 The Business Research Company (2020). 2020 global ethylene market and propylene industry research analysis by TBRC. https://www.prnewswire.com/news-releases/2020-global-ethylene-market-and-propylene-industry-research-analysis-by-tbrc-301014325.html

36 Sahebdelfar, S. and Zangeneh Tahriri, F. (2010). Dehydrogenation of propane to propylene over Pt-Sn/Al_2O_3 catalysts: the influence of operating conditions on product selectivity. *Iran. J. Chem. Eng.* 7 (2): 51–57.

37 Lwin, S. and Wachs, I.E. (2014). Olefin metathesis by supported metal oxide catalysts. *ACS Catal.* 4 (8): 2505–2520.

38 Chen, D., Moljord, K., and Holmen, A. (2012). A methanol to olefins review: diffusion, coke formation and deactivation on SAPO type catalysts. *Microporous Mesoporous Mater.* 164: 239–250.

39 Tian, P., Wei, Y., Ye, M., and Liu, Z. (2015). Methanol to Olefins (MTO): from fundamentals to commercialization. *ACS Catal.* 5 (3): 1922–1938.

40 Werpy, T., and Petersen, G. (2004) Top value added chemicals from biomass: volume I – results of screening for potential candidates from sugars and synthesis gas.

41 (2020). The global butane market is expected to grow from USD 56 565.18 million in 2019 to USD 80 171.63 million by the end of 2025 at a compound annual growth rate (CAGR) of 5.98%. *ReportLinker*. http://www.globenewswire.com/fr/news-release/2020/07/01/2056054/0/en/The-Global-Butane-Market-is-expected-to-grow-from-USD-56-565-18-Million-in-2019-to-USD-80-171-63-Million-by-the-end-of-2025-at-a-Compound-Annual-Growth-Rate-CAGR-of-5-98.html

42 Kotov, S.V. and Kankaeva, I.N. (1994). Commercial production and principal trends in the use of 1-butene. *Chem. Technol. Fuels Oils* 30 (5): 240–245.

43 James Matthew, K. (2020) Review of established and emergent methods for the production of C_4 olefins.

44 Symoens, S.H., Olahova, N., Muñoz Gandarillas, A.E. et al. (2018). State-of-the-art of coke formation during steam cracking: anti-coking surface technologies. *Ind. Eng. Chem. Res.* 57 (48): 16117–16136.

45 Speight, J. (2020). *handbook of Industrial Hydrocarbon Processes*. Cambridge: Elsevier Inc.

46 Lousdad, S. (2020). What is steam cracking / pyrolysis? *AONG website*, 4.

47 Dutton, J.A. (2020). Fluid Catalytic Cracking (FCC). *e-Education Institute, PennState*.

48 Eser, S. (2017). Fluid catalytic cracking (FCC).

49 Yoshimura, Y., Kijima, N., Hayakawa, T. et al. (2001). Catalytic cracking of naphtha to light olefins. *Catal. Surv. from Japan* 4 (2): 157–167.

50 Resasco, D.E. (2000). Dehydrogenation by heterogeneous catalysts. *Encycl. Catal.* 52: 49–52. https://doi.org/10.1002/0471227617.eoc074.

51 Chen, S., Pei, C., Sun, G. et al. (2020). Nanostructured catalysts toward efficient propane dehydrogenation. *Accounts Mater. Res.* 1 (1): 30–40.

52 Sattler, J.J.H.B., Ruiz-Martinez, J., Santillan-Jimenez, E., and Weckhuysen, B.M. (2014). Catalytic dehydrogenation of light alkanes on metals and metal oxides. *Chem. Rev.* 114 (20): 10613–10653.

53 Marcinkowski, M.D., Darby, M.T., Liu, J. et al. (2018). Pt/Cu single-atom alloys as coke-resistant catalysts for efficient C–H activation. *Nat. Chem.* 10 (3): 325–332.

54 Sun, X., Han, P., Li, B., and Zhao, Z. (2018). Tunable catalytic performance of single Pt atom on doped graphene in direct dehydrogenation of propane by rational doping: a density functional theory study. *J. Phys. Chem. C* 122 (3): 1570–1576.

55 Schäferhans, J., Gómez-Quero, S., Andreeva, D.V., and Rothenberg, G. (2011). Novel and effective copper–aluminum propane dehydrogenation catalysts. *Chem. – A Eur. J.* 17 (44): 12254–12256.

56 Schweitzer, N.M., Hu, B., Das, U. et al. (2014). Propylene hydrogenation and propane dehydrogenation by a single-site Zn^{2+} on silica catalyst. *ACS Catal.* 4 (4): 1091–1098.
57 Liu, G., Zeng, L., Zhao, Z.-J. et al. (2016). Platinum-modified ZnO/Al_2O_3 for propane dehydrogenation: minimized platinum usage and improved catalytic stability. *ACS Catal.* 6 (4): 2158–2162.
58 Tan, S., Hu, B., Kim, W.-G. et al. (2016). Propane dehydrogenation over alumina-supported iron/phosphorus catalysts: structural evolution of iron species leading to high activity and propylene selectivity. *ACS Catal.* 6 (9): 5673–5683.
59 Hu, P., Lang, W.-Z., Yan, X. et al. (2018). Influence of gelation and calcination temperature on the structure-performance of porous VO_X-SiO_2 solids in non-oxidative propane dehydrogenation. *J. Catal.* 358: 108–117.
60 Iglesias-Juez, A., Beale, A.M., Maaijen, K. et al. (2010). A combined in situ time-resolved UV–Vis, Raman and high-energy resolution X-ray absorption spectroscopy study on the deactivation behavior of Pt and PtSn propane dehydrogenation catalysts under industrial reaction conditions. *J. Catal.* 276 (2): 268–279.
61 Zhang, Y., Zhou, Y., Huang, L. et al. (2011). Sn-modified ZSM-5 as support for platinum catalyst in propane dehydrogenation. *Ind. Eng. Chem. Res.* 50 (13): 7896–7902.
62 de Miguel, S., Castro, A., Scelza, O. et al. (1996). FTIR and XPS study of supported PtSn catalysts used for light paraffins dehydrogenation. *Catal. Letters* 36 (3): 201–206.
63 Cheung, T.-K., Lange, F.C., and Gates, B.C. (1996). Propane conversion catalyzed by sulfated zirconia, iron- and manganese-promoted sulfated zirconia, and USY zeolite. *J. Catal.* 159 (1): 99–106.
64 Bricker, J.C. (2012). Advanced catalytic dehydrogenation technologies for production of olefins. *Top. Catal.* 55 (19): 1309–1314.
65 Castro, A.J.R., Soares, J.M., Filho, J.M. et al. (2013). Oxidative dehydrogenation of ethylbenzene with CO_2 for styrene production over porous iron-based catalysts. *Fuel* 108: 740–748.
66 Ito, T., Wang, J., Lin, C.H., and Lunsford, J.H. (1985). Oxidative dimerization of methane over a lithium-promoted magnesium oxide catalyst. *J. Am. Chem. Soc.* 107 (18): 5062–5068.
67 Rostom, S. and de Lasa, H. (2020). Propane oxidative dehydrogenation on vanadium-based catalysts under oxygen-free atmospheres. *Catalysts* 10 (4): 418–441.
68 Aseem, A. and Harold, M.P. (2018). C_2 yield enhancement during oxidative coupling of methane in a nonpermselective porous membrane reactor. *Chem. Eng. Sci.* 175: 199–207.
69 Cruellas, A., Melchiori, T., Gallucci, F., and van Sint Annaland, M. (2020). Oxidative coupling of methane: a comparison of different reactor configurations. *Energy Technol.* 8 (8): 1900148–1900163.
70 Cruellas, A., Melchiori, T., Gallucci, F., and van Sint Annaland, M. (2017). Advanced reactor concepts for oxidative coupling of methane. *Catal. Rev.* 59 (3): 234–294.
71 Spallina, V., Velarde, I.C., Jimenez, J.A.M. et al. (2017). Techno-economic assessment of different routes for olefins production through the oxidative coupling of methane (OCM): advances in benchmark technologies. *Energy Convers. Manag.* 154: 244–261.
72 Haribal, V.P., Chen, Y., Neal, L., and Li, F. (2018). Intensification of ethylene production from naphtha via a redox oxy-cracking scheme: process simulations and analysis. *Engineering* 4 (5): 714–721.
73 Luo, L., You, R., Liu, Y. et al. (2019). Gas-Phase reaction network of Li/MgO-Catalyzed oxidative coupling of methane and oxidative dehydrogenation of ethane. *ACS Catalysis* 9: 2514–2520.

74 Boyadjian, C. and Lefferts, L. (2018). Catalytic oxidative cracking of light alkanes to alkenes. *Eur. J. Inorg. Chem.* **2018** (19): 1956–1968.

75 Arutyunov, V.S. and Magomedov, R.N. (2012). Gas-phase oxypyrolysis of light alkanes. *Russ. Chem. Rev.* 81 (9): 790–822.

76 Magomedov, R.N., Proshina, A.Y., and Arutyunov, V.S. (2013). Gas-phase oxidative cracking of ethane in a nitrogen atmosphere. *Kinet. Catal.* 54 (4): 383–393.

77 Liu, X., Li, W., Xu, H., and Chen, Y. (2004). A comparative study of non-oxidative pyrolysis and oxidative cracking of cyclohexane to light alkenes. *Fuel Process. Technol.* 86 (2): 151–167.

78 Zhu, H., Liu, X., Ge, Q. et al. (2006). Production of lower alkenes and light fuels by gas phase oxidative cracking of heavy hydrocarbons. *Fuel Process. Technol.* 87 (7): 649–657.

79 Liu, X., Li, W., Zhu, H. et al. (2004). Light alkenes preparation by the gas phase oxidative cracking or catalytic oxidative cracking of high hydrocarbons. *Catal. Letters* 94 (1): 31–36.

80 Liu, X., Li, W., Xu, H., and Chen, Y. (2004). Production of light alkenes with low CO_2 emission from gas phase oxidative cracking (GOC) of hexane. *React. Kinet. Catal. Lett.* 81 (2): 203–209.

81 Mohiuddin, E., Mdleleni, M.M., and Key, D. (2018). Catalytic cracking of naphtha: the effect of Fe and Cr impregnated ZSM-5 on olefin selectivity. *Appl. Petrochemical Res.* 8 (2): 119–129.

82 Do, L.T. and Shin, E.W. (2019). Reinforcement of steam oxidative cracking of 1-methyl naphthalene by the synergistic combination of hydrogenation and oxidation over nickel-containing mixed oxide catalysts. *J. Catal.* 371: 245–254.

83 Narasimharao, K. and Alshehri, A. (2020). Gold supported yttrium oxide nanorods for catalytic oxidative cracking of n-propane to light olefins. *Fuel* 278: 118375–118387.

84 Narasimharao, K. and Ali, T.T. (2013). Catalytic oxidative cracking of propane over nanosized gold supported $Ce_{0.5}Zr_{0.5}O_2$ Catalysts. *Catal. Letters* 143 (10): 1074–1084.

85 Al-Sultan, F.S., Basahel, S.N., and Narasimharao, K. (2018). Catalytic oxidative cracking of n-propane over nanosized gold supported La_2O_3 catalysts. *Fuel* 233: 796–804.

86 Narasimharao, K. and Al-Sultan, F.S. (2020). Y_2O_3 modified Au-La_2O_3 nanorod catalysts for oxidative cracking of n-propane. *Fuel* 280: 118599–118610.

87 Al-Sultan, F.S., Basahel, S.N., and Narasimharao, K. (2020). Yttrium oxide supported La_2O_3 nanomaterials for catalytic oxidative cracking of n-propane to olefins. *Catal. Letters* 150 (1): 185–195.

88 McDermott, W.P., Venegas, J., and Hermans, I. (2020). Selective oxidative cracking of n-butane to light olefins over hexagonal boron nitride with limited formation of CO_x. *ChemSusChem* 13 (1): 152–158.

89 Kijima, N., Matano, K., Saito, M. et al. (2001). Oxidative catalytic cracking of n-butane to lower alkenes over layered BiOCl catalyst. *Appl. Catal. A Gen.* 206 (2): 237–244.

90 Boyadjian, C., Seshan, K., Lefferts, L. et al. (2011). Production of C_3/C_4 olefins from n-hexane: conceptual design of a catalytic oxidative cracking process and comparison to steam cracking. *Ind. Eng. Chem. Res.* 50 (1): 342–351.

91 Hao, F., Gao, Y., Neal, L. et al. (2020). Sodium tungstate-promoted $CaMnO_3$ as an effective, phase-transition redox catalyst for redox oxidative cracking of cyclohexane. *J. Catal.* 385: 213–223.

92 Amusa, H.K., Adamu, S., Bakare, I.A. et al. (2020). High-performance VOx on SrO-γAl_2O_3 catalyst for oxidative cracking of n-hexane to light olefins under anaerobic environment. *J. Ind. Eng. Chem.* 89: 339–350.

93 Dudek, R.B., Tian, X., Blivin, M. et al. (2019). Perovskite oxides for redox oxidative cracking of n-hexane under a cyclic redox scheme. *Appl. Catal. B Environ.* 246: 30–40.

94 Leveles, L., Fuchs, S., Seshan, K. et al. (2002). Oxidative conversion of light alkanes to olefins over alkali promoted oxide catalysts. *Appl. Catal. A Gen.* 227 (1): 287–297.

95 Lange, J.-P., Schoonebeek, R.J., Mercera, P.D.L., and van Breukelen, F.W. (2005). Oxycracking of hydrocarbons: chemistry, technology and economic potential. *Appl. Catal. A Gen.* 283 (1): 243–253.

96 Subramanian, R., Panuccio, G.J., Krummenacher, J.J. et al. (2004). Catalytic partial oxidation of higher hydrocarbons: reactivities and selectivities of mixtures. *Chem. Eng. Sci.* 59 (22): 5501–5507.

97 Buyevskaya, O.V. and Baerns, M. (1998). Catalytic selective oxidation of propane. *Catal. Today* 42 (3): 315–323.

98 Kaliya, M.L. and Kogan, S.B. (2005). Highly effective oxidative cracking of n-butane in light olefins on a novel cobalt catalyst. *Catal. Today* 106 (1): 95–98.

99 Basini, L., Cimino, S., Guarinoni, A. et al. (2012). Olefins via catalytic partial oxidation of light alkanes over Pt/$LaMnO_3$ monoliths. *Chem. Eng. J.* **207–208**: 473–480.

100 Buyevskaya, O.Va., Müller, D., Pitsch, I., and Baerns, M. (1998). Selective oxidative conversion of propane to olefins and oxygenates on boria-containing catalysts. In: *Natural Gas Conversion V*, vol. 119 (ed. A. Parmaliana, D. Sanfilippo, F. Frusteri, et al.), 671–676. Elsevier.

101 Sui, Z., Zhou, J., Dai, Y., and Yuan, W. (2005). Oxidative dehydrogenation of propane over catalysts based on carbon nanofibers. *Catal. Today* 106 (1): 90–94.

102 Jiang, G., Zhang, L., Zhao, Z. et al. (2008). Highly effective P-modified HZSM-5 catalyst for the cracking of C_4 alkanes to produce light olefins. *Appl. Catal. A Gen.* 340 (2): 176–182.

103 Nakamura, M., Takenaka, S., Yamanaka, I., and Otsuka, K. (2000). Production of alkenes through oxidative cracking of n-butane over OCM catalysts. In: *12th International Congress on Catalysis*, vol. 130 (ed. A. Corma, F.V. Melo, S. Mendioroz and J.L.G. Fierro), 1781–1786. Elsevier.

104 Sá, J., Ace, M., Delgado, J.J. et al. (2011). Activation of alkanes by gold-modified lanthanum oxide. *ChemCatChem* 3 (2): 394–398.

105 Tian, X., Dudek, R.B., Gao, Y. et al. (2019). Redox oxidative cracking of n-hexane with Fe-substituted barium hexaaluminates as redox catalysts. *Catal. Sci. Technol.* 9 (9): 2211–2220.

106 Boyadjian, C., van der Veer, B., Babich, I.V. et al. (2010). Catalytic oxidative cracking as a route to olefins: oxidative conversion of hexane over MoO_3-Li/MgO. *Catal. Today* 157 (1): 345–350.

107 Boyadjian, C., Ağıral, A., Gardeniers, J.G.E. et al. (2011). Oxidative conversion of hexane to olefins-influence of plasma and catalyst on reaction pathways. *Plasma Chem. Plasma Process.* 31 (2): 291–306.

108 Nikitin, A.V., Dmitruk, A.S., and Arutyunov, V.S. (2016). Effect of pressure on the oxidative cracking of C_2–C_4 alkanes. *Russ. Chem. Bull.* 65 (10): 2405–2410.

109 Otsuka, K., Jinno, K., and Morikawa, A. (1986). Active and selective catalysts for the synthesis of C_2H_4 and C_2H_6 via oxidative coupling of methane. *J. Catal.* 100 (2): 353–359.

110 Campbell, K.D., Zhang, H., and Lunsford, J.H. (1988). Methane activation by the lanthanide oxides. *J. Phys. Chem.* 92 (3): 750–753.

111 Tong, Y., Rosynek, M.P., and Lunsford, J.H. (1989). Secondary reactions of methyl radicals with lanthanide oxides: their role in the selective oxidation of methane. *J. Phys. Chem.* 93 (8): 2896–2898.

112 Kaliya, M.L., Kogan, S.B., Froumin, N., and Herskowitz, M. (2002). Novel nitrogen containing heterogeneous catalysts for oxidative dehydrogenation of light paraffins. *Catal. Commun.* 3 (8): 327–333.

113 Choudhary, V.R., Rane, V.H., and Rajput, A.M. (2000). High-temperature catalytic oxidative conversion of propane to propylene and ethylene involving coupling of exothermic and endothermic reactions. *Ind. Eng. Chem. Res.* 39 (4): 904–908.

114 Elbadawi, A.H., Khan, M.Y., Quddus, M.R. et al. (2017). Kinetics of oxidative cracking of n-hexane to olefins over $VO_x/Ce\text{-}Al_2O_3$ under gas phase oxygen-free environment. *AIChE J.* 63 (1): 130–138.

115 Elbadawi, A.H., Ba-Shammakh, M.S., Al-Ghamdi, S. et al. (2017). Phenomenologically based kinetics of ODH of ethane to ethylene using lattice oxygen of $VO_x/Al_2O_3\text{–}ZrO_2$ catalyst. *Chem. Eng. Res. Des.* 117: 733–745.

116 Rahman, F., Loughlin, K.F., Al-Saleh, M.A. et al. (2010). Kinetics and mechanism of partial oxidation of ethane to ethylene and acetic acid over MoV type catalysts. *Appl. Catal. A Gen.* 375 (1): 17–25.

12

The Magic of Heterogeneous Nanocatalysis in VLPC for Sustainable Organic Synthesis

Suresh Iyer[1], *Arumugam Sudalai*[1], *and Alwar Ramani*[2]

[1] *CSIR-National Chemical Laboratory, Pune, Maharashtra, India*
[2] *School of Energy, Geoscience, Infrastructure and Society, Heriot-Watt University, Edinburg, Scotland*

12.1 Introduction

The sun is an endless source of sustainable energy. Artificial light in light-emitting diode (LED) and other sources is universally available in abundance, making solar energy and visible light harvesting a fruitful endeavor for human civilization. Sustainable manufacturing with photon energy has metamorphosed globally with the rapid escalation of visible light activation covering chemical science. Sensitized semiconductors in organic synthesis are a new class of heterogeneous and recyclable catalysts initiating single electron transfer (SET) processes with low to medium power light irradiation. Visible light and LED illumination is a powerful technology with fine-tunable duration and intensity of illumination. Selective chemical processes can be initiated by accurate modulation to match the absorption corresponding to the reactive molecules' electronic transition [1].

The burgeoning of catalytic systems that can harvest photons proficiently and stimulate selective chemical transformation is a universal research effort. Photocatalysis using nanostructured materials has opened new and efficient operational windows. Small molecule photocatalysts have combined with heterogeneous materials to achieve harmonious effects. Semiconductors have been used for over 30 years in photovoltaics and represent an encouraging category of materials that harvest light energy enabling chemical reactions and synthesis by initiating a charge separation process. The visible light activation of semiconductors opens up new operational windows hitherto restricted to ultraviolet (UV) irradiation and simple synthetic schemes.

The quest for benign technology has gifted an incentive to arrive at visible light photocatalysis (VLPC) by harvesting sunlight energy. Abundant sunlight energy is available from a natural source; from an environmental perspective, sunlight energy to carry out the organic reaction is clean, inexpensive, environmentally compatible, and sustainable. With the emergence of VLPC technology in the early twentieth century, this area of research is well past the infancy stage with the aid of organometallic compounds that can

Heterogeneous Nanocatalysis for Energy and Environmental Sustainability: Energy Applications, First Edition. Edited by Putla Sudarsanam, Yusuke Yamauchi, and Pankaj Bharali.

absorb visible light. Photostimulation with various photocatalysts, semiconductors, organic photoconductors, plasmonic metal nanoparticles (NPs), and novel heterogeneous photoelectronic materials generated active, reactive intermediates in the reaction medium; with the progression of this technology, the photoredox catalyst coupled with Lewis acids, organocatalysts, and chiral catalysts designed asymmetric catalytic synthesis of enantiomerically pure complex organic compounds.

Interestingly, a combination of two or more catalysts together exhibiting higher order of synergism, furthermore, assisted chemists in achieving an astonishing variety of molecular construction. UV lamps operating at a specific set of wavelengths carried out photochemical reactions. However, VLPC technology uses readily available sunlight or artificial light; consequently, it is feasible to scale up to an industrial level.

In a nutshell, typical photocatalysis refers to any chemical reaction initiated by a solid catalyst that uses photonic energy (photons) instead of heat (thermal) energy as the driving force toward the chemical transformations. Semiconductors with a filled valence band (VB) and an empty conductance band (CB) are the catalysts for this purpose and undergo absorption of photons by the action of the sun or LED lights. An electron hole is formed by the electrons migrating from the VB to the CB because of the energy in the semiconductor. Therefore, photoinduced electron-hole couples possess high redox potential, and they transfer an electron to organic substrates, activating toward subsequent transformations. Ru and Ir complexes were the most generally employed, although Cu and Au complexes and other compounds were also explored [1].

12.2 Heterogeneous Photocatalysis: Theory, Mechanism, and Reactions

The advent of Ru and Ir photosensitizers has made forays in organic synthesis mediated by visible light illumination. Visible light, though abundant, cannot directly activate organic molecules and requires photocatalysts acting as a bridge for the transfer of energy between light and organic substrates. Heterogeneous photocatalysis focuses on the narrow window of operation of UV light activation of suitable semiconductors. Plasmonic metal NPs and several other materials are heterogeneous photocatalysts. Absorption of light photons by the semiconductors, CdS, $BiVO_4$, and mpg-C_3N_4 (mesoporous graphitic carbon nitride), with equal to or more tremendous energy than their bandgap causes the transition of the VB electrons to the CB, leaving VB holes. Electron pairs and holes with various redox potentials are generated by different semiconductors displaying disparate bandgap widths and positions. The photocatalyst efficiency is lowered by the hole electron pair migration to and recombination on the catalyst surface. Semiconductors with significant bandgaps still perform only with specific substrates under visible light irradiation, e.g. benzyl alcohol and benzylamines. The interaction of surface complexes formed by the active metal sites and benzylic alcohols or amines explains the photoresponse of TiO_2 or Nb_2O_5. The different mechanisms of action from CdS and mpg-C_3N_4 with small bandgaps are the visible light excitation of electrons to the CB of TiO_2. TEMPO ([2,2,6,6-tertamethylpiperidinyl-1-yl]oxyl) quenches the electron directly injected by the dye-sensitized semiconductor into the TiO_2 CB. A pivotal role in the oxidation of the substrate is played by TEMPO-oxidizing species [2, 3].

Metal oxides, metal sulfides, organic photoconductors, plasmonic metal NPs, and heterogeneous photoelectronic materials are novel heterogeneous photocatalysts. They catalyze the sustainable photo-organic reactions, oxidation and reduction, C—C and C-hetero atom bond formation, cycloaddition, annulation, and enamine reactions. Heterogeneous photocatalysis depends on the UV activation of appropriate semiconductors. For example, TiO_2 acts as a UV sensitizer. The high-energy UV light leads to side reactions or decompositions due to the incompatibility of many organic and highly functionalized target molecules. The shift of light excitation range from UV to the visible spectrum leads to an abundant light source from the naturally available sunlight to artificial light, which is also harmonious with most organic substrates and derived products with high functional group tolerance. In recent years, TiO_2 was modified with NPs, antenna molecules, and in situ surface tempering.

12.2.1 Oxidations

The photocatalytic application of dye-sensitized semiconductors in oxidation reactions is widely studied. A common objective is substituting expensive or hazardous stoichiometric oxidizing agents with the more benign and ubiquitous oxygen or air. The photooxidation of a wide range of organic molecules like hydrocarbons, alcohols, and other aromatic compounds by TiO_2 is studied widely. Alcohols acting as hole scavengers in the photocatalytic systems have less positive redox potentials than the VB boundary of photocatalytic metal oxides. The overoxidation of alcohols to carboxylic acids and mineralization to CO_2 is a drawback in their oxidation to carbonyl compounds. Several parameters like solvent, photocatalyst, presence of O_2, and structure of the alcohol affect the reaction. Heterogeneous photocatalysts achieve high oxidative conversions and selectivity with alcohol. Citronellol, geraniol, and trans-2-penten-1-ol undergo well-explored photocatalytic oxidations. The aerobic oxidation of glycerol in water and piperonol to piperonal is catalyzed by the sol–gel-encapsulated photocatalytic species in silica-based matrices. The magnetically separable TiO_2/maghemite-silica nanocomposite was superior to other transition metal catalysts in the catalytic photooxidation of benzyl alcohol to benzaldehyde (90% selectivity) [3].

The photooxidation of aromatics is another industrially important reaction. The TiO_2-H_2O system photocatalyzed the direct hydroxylation of several aromatics, benzene, toluene, and acetophenone. Examples of such widely used chemicals, precursors of resins, and pharmaceuticals are phenols, hydroquinone, and catechol.

Ortho and para isomer selectivities were determined by the substituents like electron-withdrawing group (EWG) (NO_2, CN, CO_2H, and Ar-$COCH_3$) and electron-donating group (EDG) (OH, NH_2, and $NHCOCH_3$) or combinations. Naphthalenes were photooxidized to 1,4-naphthoquinones, (E,Z)-2-formyl cinnamaldehydes, and naphthols (traces) [4, 5].

The photooxidation of aromatic alcohols to aldehydes was carried out by Higashimoto using nitrobenzene instead of O_2 as electron acceptor and CdS-TiO_2 nanocomposite catalyst (CdS – 15%) (Scheme 12.1). Simultaneously, nitrobenzene was photoreduced to aniline [6].

The Au/TiO_2 nanocomposites' efficiency was increased by Ir, Cu, Ag, Rh, Ru, Pd, and Pt cocatalysts. Isopropyl alcohol is readily oxidized to acetone. Simultaneously, phenethyl alcohol was photooxidized to acetophenone by Au-Pd/ZrO_2. Similarly, the plasmonic Au/CeO_2 oxidized 4-amino benzyl alcohol to 4-amino benzaldehyde (Scheme 12.2).

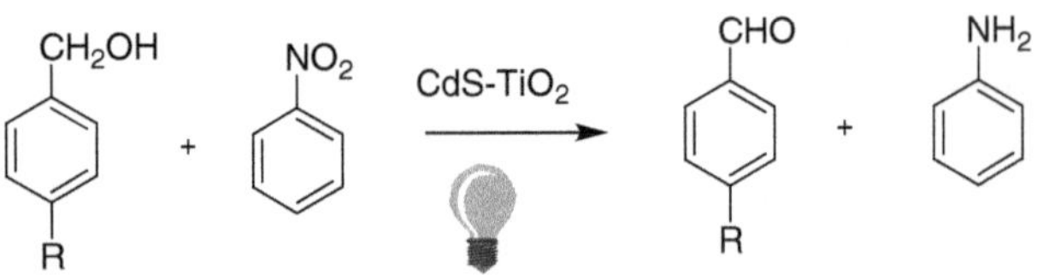

Scheme 12.1 CdS-TiO_2-catalyzed photooxidation of benzyl alcohol.

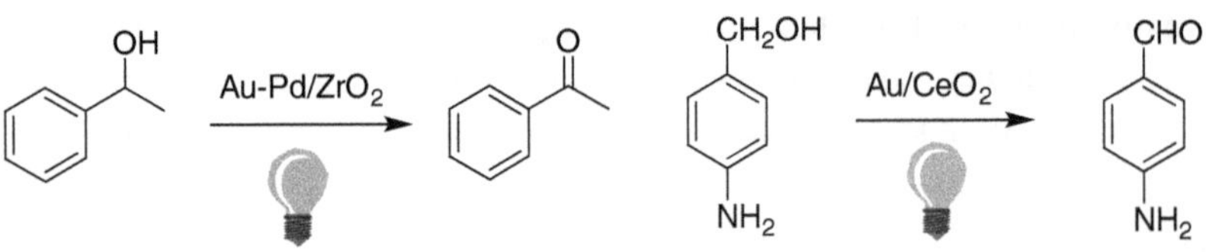

Scheme 12.2 Plasmonic photocatalytic oxidation.

Higashimoto reported nano TiO_2 dispersion in CH_3CN at room temperature (RT) in the presence of O_2-photooxidized primary benzylic alcohols under light LED (480 nm) irradiation. Conversion to the corresponding aldehydes was more efficient with the (EDG–meta-orientation or ortho- and para-orientation of EWG) less delocalized benzylic radical [7].

Nb_2O_5 (Tanaka and coworkers), ZnTl-layered double hydroxide (Wu et al.), and a semiconductor titanate monolayer nanosheet all demonstrated a similar reactivity [8].

Au/SiO_2 nanocomposite catalyzed the tandem oxidation/amidation reaction of benzyl alcohol with morpholine under laser irradiation with wavelength $\lambda = 532$ nm (Scheme 12.3). Initial oxidation to benzaldehyde, in H_2O_2 and KOH, followed by the reaction with morpholine, produced the amide [2, 5]. Alizarin red sodium-dye-doped TiO_2 photocatalyst gave sulfoxides in the presence of O_2 with TEMPO as a cocatalyst (Scheme 12.4).

TiO_2 P25 (Evonik-Degussa Aeroxide P25) catalyzed the photooxidation of thioether and benzylamine to give sulfoxide and *N*-formyl benzylamine (Scheme 12.5).

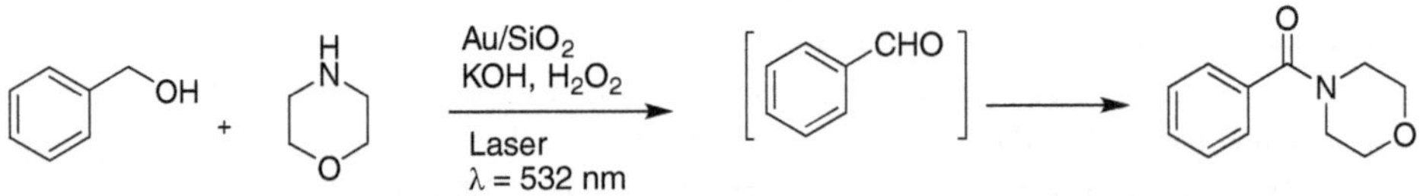

Scheme 12.3 Au/SiO_2-photocatalyzed synthesis of amide from aldehyde and morpholine.

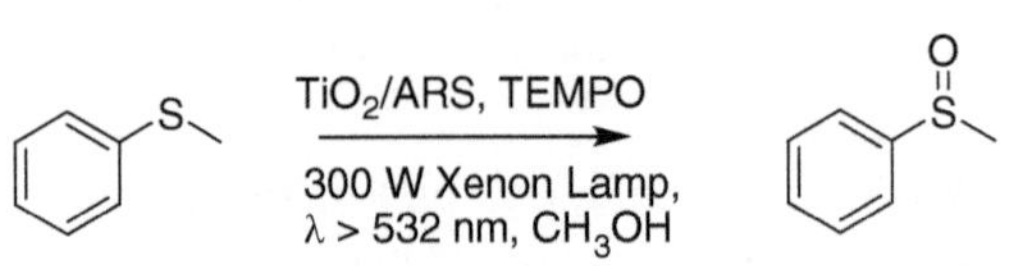

Scheme 12.4 Visible light-mediated sulfide oxidation to sulfoxides.

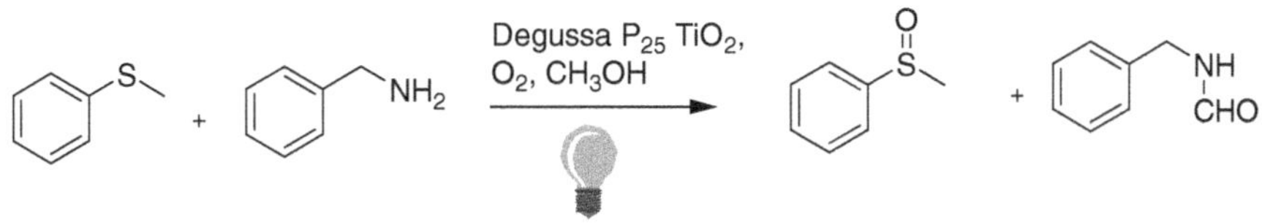

Scheme 12.5 Visible light oxidation of thioanisole.

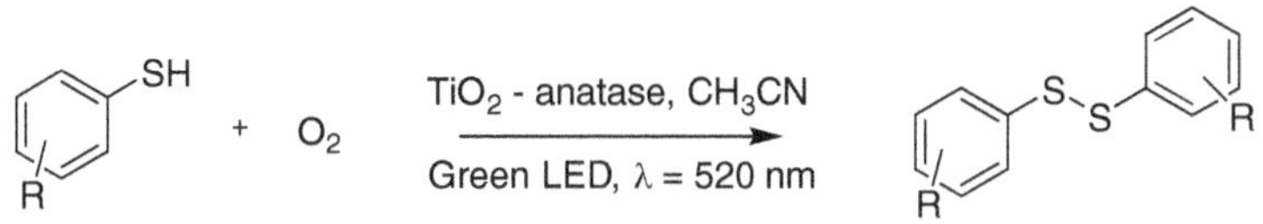

Scheme 12.6 Visible light oxidation of thiols.

Thiols gave disulfides in high yields in the selective photooxidation catalyzed by anatase TiO_2 in molecular O_2 by irradiation with green LED at $\lambda = 520\,nm$ (Scheme 12.6). $BiVO_4$ or Pt/$BiVO_4$ catalyzes the photocatalytic oxidation of sulfides to sulfoxides and sulfones in water as the favorable solvent for the oxidation giving a percentage selectivity of 98 : 2 in favor of sulfoxide to sulfone (Scheme 12.7). Solvents are affected by the interaction between reactants and the catalyst surface.

The oxidation of cyclohexane further exemplifies this. The weak degree of adsorption on TiO_2 led to the oxidation of cyclohexane to cyclohexanol in the polar solvent (acetone). In contrast, a nonpolar solvent like CH_2Cl_2 yielded cyclohexanone with a substantial adsorption extent and then mineralized to CO_2 [2].

Ag/ZnO/graphene photocatalyzed the synthesis of vitamin B3 from pyridine-3-methanol under visible light and tert-butyldimethyl silyl chloride (Scheme 12.8). The reagent enhanced the active oxidizing species (OH radicals) formation in the oxidation process.

Benzyl amine was oxidized efficiently catalyzed by the dye-sensitized nano TiO_2 to the corresponding imine (Scheme 12.9).

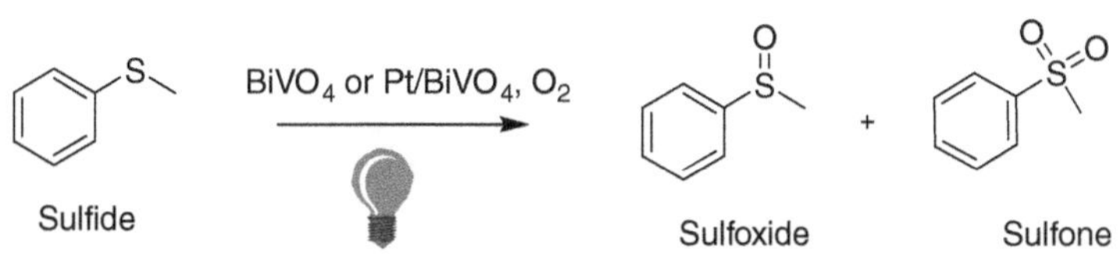

Scheme 12.7 $BiVO_4$- or Pt/$BiVO_4$-photocatalyzed oxidation of sulfides.

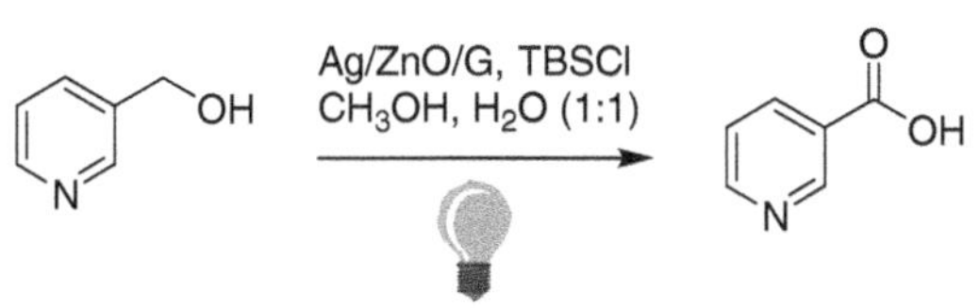

Scheme 12.8 Oxidation of pyridine-3-methanol under visible light.

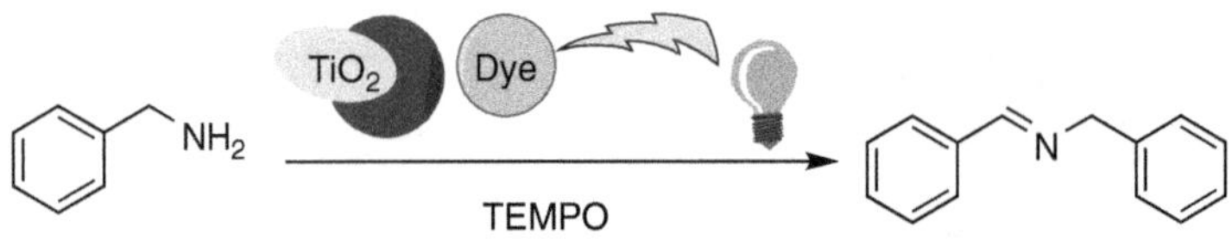

Scheme 12.9 TiO_2-dye-sensitized photooxidation of benzylamine.

The mechanism postulated was the visible light excited the dye followed by the electron transfer from the VB to the CB of TiO_2, and subsequently quenched by TEMPO, acting as a redox mediator agent.

12.2.2 Reductions

Several workers have reported the photocatalytic reduction of nitroarenes. The Ag cocatalyst-reformed plasmonic Au/TiO_2, more than Ag-free catalyst, increased the nitrobenzene reduction to aniline and the simultaneous quantitative oxidation of isopropanol to acetone (Scheme 12.10) [2]. Alcohol solvents serve as hole-scavenging agents in these reactions acting as sacrificial reagents. Au-Pd/TiO_2 catalyzed the photoreductive dehalogenation of chlorobenzene in the presence of isopropanol to benzene. TiO_2 NPs and the conjugated poly-p-phenylene affected the photo-dehalogenation of simple haloarenes. The high surface area of TiO_2 increased the reaction rate.

The photocatalytic transfer of electrons from the VB to the CB is enhanced by the wavelength of the incident light, leaving positive holes. A chemical reaction favorable for organic transformation promotes the interaction of the electron with the reacting species producing thermodynamic products. The Au/CeO_2 nanocatalyst demonstrates selectivity in reductions by changing the illumination wavelength. Azo, carbonyl, and epoxide groups readily react efficiently (Scheme 12.11).

The surface plasmon resonance of the gold surface played a remarkable role in reducing Au/CeO_2-adsorbed visible light at a wavelength below 520 nm, showing increased catalytic

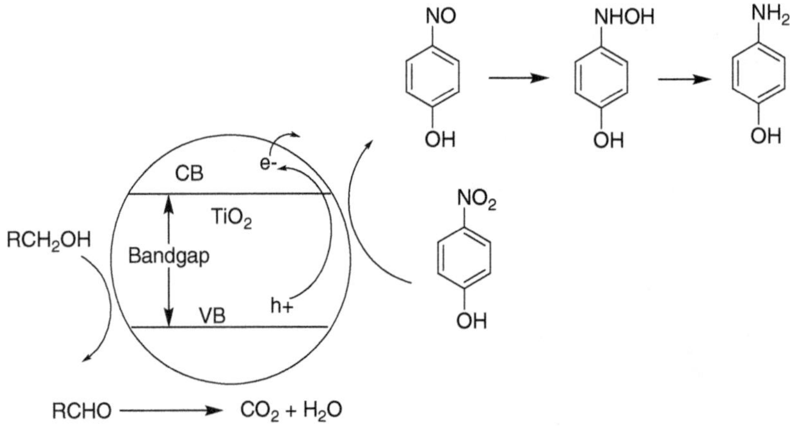

Scheme 12.10 Photoreduction of 4-nitrophenol.

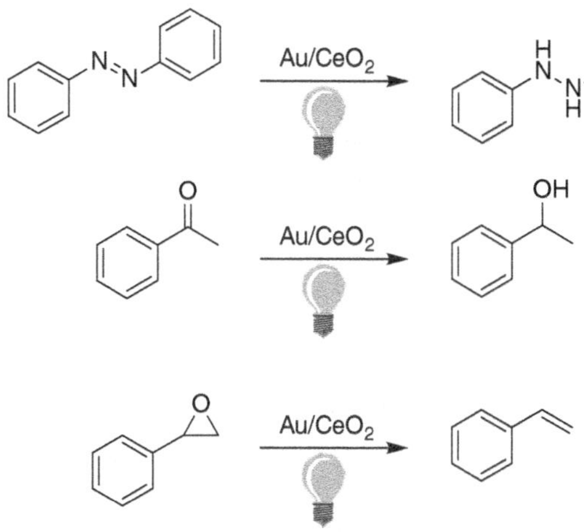

Scheme 12.11 Photocatalytic reductions catalyzed by Au/CeO_2 photo nanocatalyst.

efficiency. The activation of Ag/AgX/reduced graphene oxide (Ag/AgX/r-GO) with the addition of trimethylsilyl chloride (TMSCl) greatly enhanced the photochemical transformation of 4-nitrophenol to 4-aminophenol and paracetamol. The addition of catalytic Ag/AgBr/r-GO and 1% TMSCl to the reaction mixture reduced the reaction time from 15 to 8 minutes producing the highest yields of paracetamol (Scheme 12.12) [9].

In another study, the addition of oxalic acid as the sacrificial agent improved the TiO_2 photocatalytic reduction of nitroarenes (Scheme 12.13). The photoreduction of nitrobenzene to aniline was catalyzed by TiO_2 NPs reformed with amino acids, asparagine, serine, phenylalanine, or tyrosine. The amino acids incarcerated the hole on the TiO_2 surface and confined electron-hole (e^-h^+) recombination. Compared to bare TiO_2, the Pt-Ag/TiO_2 enhanced the nitrobenzene reduction. The Pt-Ag/TiO_2 nanocomposite catalyzed the higher aniline production with high conversion by increased adsorption of nitrobenzene and suppressing the reabsorption of aniline.

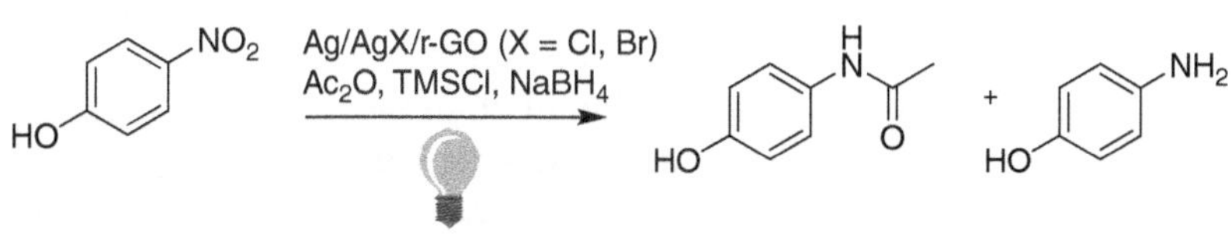

Scheme 12.12 Ag/AgX/r-GO-TMSCl-photocatalyzed synthesis of *para*-aminophenol (PAP) to acetyl *para*-aminophenol (APAP).

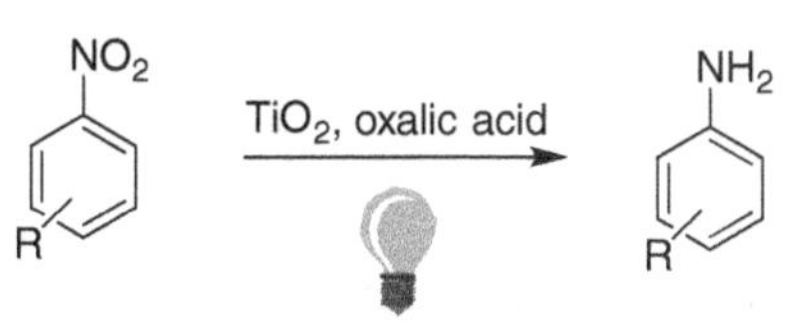

Scheme 12.13 The photoreduction of nitroarenes using oxalic acid.

The C=C, C=O, N=O, and C-X bonds undergo mild and facile transfer hydrogenation catalyzed by TiO_2 with various low boiling amines and alcohols acting as a hydrogen source. TiO_2-photoinduced holes oxidatively remove two hydrogen atoms from the transfer hydrogenation agents at RT upon photoexcitation. The simultaneous photoinduced combination of these two H atoms with the conduction band electrons results in smooth hydrogenation of multiple bonds and C-X bonds [10].

12.2.3 Photocatalytic C=C Bond Formation

Photo cross-coupling and other visible light C=C bond formation reactions:

Graphitic C_3N_4 is one of the most promising artificial polymeric materials finding boundless applications in industry and science. The earth-abundant visible light-sensitive g-C_3N_4 has a tunable electronic and unique 2D arrangement, robust stability, an enticing electronic band constitution with multiple utilizations in catalysis, and several material science and technology. The anchoring of stabilized Pd NPs (~3 nm) inside the mesoporous g-C_3N_4 or on its surface provided effective Mott–Schottky junction catalyst (PdNPs@g-C_3N_4) [11, 12]. The RT photo Suzuki–Miyaura coupling of aryl iodides and bromides with various aryl boronic acids gave good to excellent yields. The activation of the two substrates into electron-rich by the location of the energetic electrons at the noble metal NPs and electron-deficient by the holes situated at the g-C_3N_4 plane expedited the photo Suzuki coupling.

β-SiC cubic silicon carbide (bandgap ~2.4 eV) can absorb visible light and transfer electrons to Pd NP in a Mott–Schottky junction, leaving holes in the SiC. A high TOF of 1053 h^{-1} for the Suzuki coupling between iodobenzene and phenylboronic acid was catalyzed by the visible light-energized Pd NPs at 30 °C (Scheme 12.14) [13].

The green and visible light-mediated Suzuki–Miyaura, Sonogashira, Stille, Hiyama, Ullmann, and Buchwald–Hartwig cross-couplings were catalyzed by Pd—Au alloys supported on ZrO_2 at ambient RTs. The conduction electrons of the nanoalloys gain energy by absorbing visible light at the surface Pd sites [14–16]. The localized surface plasmon resonance (LSPR) effect of Au NPs predominantly causes the crucial intense visible light absorption nature of Pd—Au nanoalloys, making them useful in photocatalysis. The incident wavelength significantly affects the reaction rate. The UV light propels the reaction of chlorobenzene and substantially improves the reactions of bromobenzene substrates, while visible light irradiation failed. This finding reveals the fact that chemical transformations can be driven by low energy and density sources, such as sunlight.

The activation of chloroarenes is a challenging problem in the Suzuki–Miyaura coupling. The Pd/Au/porous CeO_2 (PN-CeO_2) catalyzed the first C—Cl bond activation

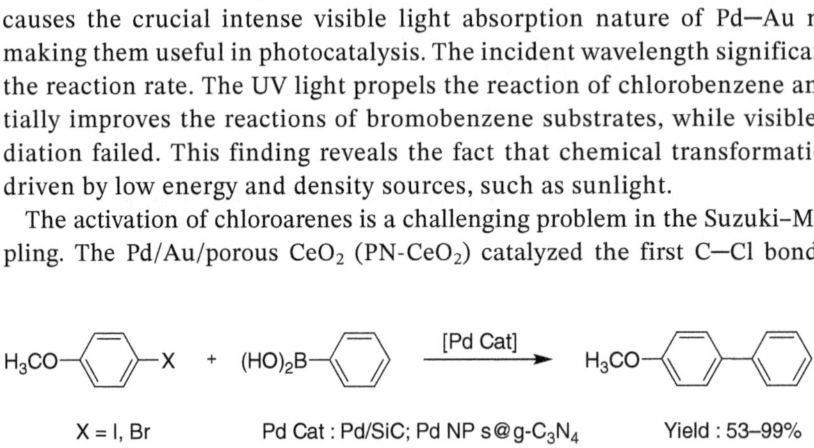

Scheme 12.14 Photocatalyzed Suzuki coupling.

under visible light irradiation (> 400 nm). Aryl boronic acids are activated by the enhanced electron density and holes on the metal NPs due to the photogenerated electrons by the strongly basic PN-CeO_2. The visible light-excited hot electrons due to the strong local plasma resonance (LPR) of Au are injected into the spatially connected Pd nanocatalysts, enriching their electron density, which activates chloroarenes proficiently for cross-coupling [17].

A field of exploration in photocatalysis is the enhancement of chemical reactions under visible light and the phenomenon of surface plasmon resonance (SPR). The ultrafast decay of hot electrons generated by SPR remains a challenging task in their efficient utility. The optimum utilization of hot electrons generated in SPR-enhanced reactions is carried out by well-designed Pd-graphene nanorod nanocomposites. Due to its exceptionally high electron mobility, SPR-induced electrons can be swiftly transported to catalytically active metals by graphene. Iodobenzene and phenylboronic acid undergo the photo Suzuki–Miyaura coupling accelerated by Au nanorods coated with homogenously r-GO having highly dispersed Pd nanoclusters. The electroconductive r-GO layer plays a crucial rate-accelerating role by transporting the SPR-generated hot electrons from Au to Pd [18, 19].

The visible light-mediated combination and coupling of two C—H bonds in two substrate molecules are catalyzed by TiO_2 by functionalization of two saturated sp^3 C—H bonds. Efficient and recyclable photoredox metal oxides (TiO_2 and ZnO) catalyze the C—C and C—P bond formation reactions by the light-mediated heterogeneous cross-dehydrogenative coupling [20].

The *aza*-Henry reaction mediated by visible light (Scheme 12.15), the oxidative Mannich reaction (Scheme 12.16), and the oxidative cyanation reactions (Scheme 12.17) proceeded with high yields when catalyzed by the commercially available TiO_2 P25 NPs.

ZnO catalyzed the oxidative phosphorylation reaction (Scheme 12.18). Both catalysts were efficiently recycled without loss in reactivity and selectivity in consecutive catalytic cycles.

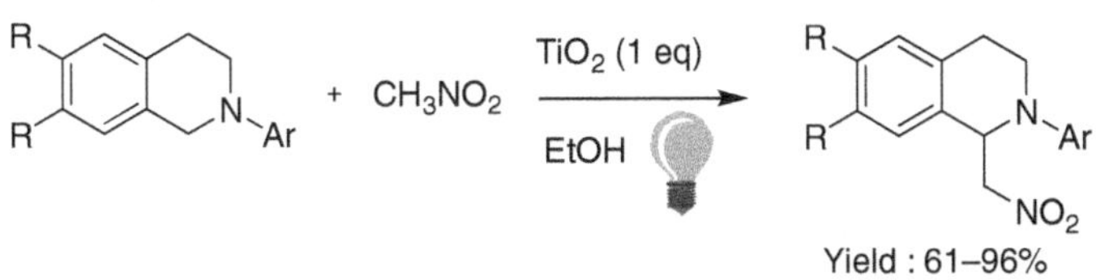

Scheme 12.15 TiO_2-catalyzed visible light-mediated oxidative *aza*-Henry reaction.

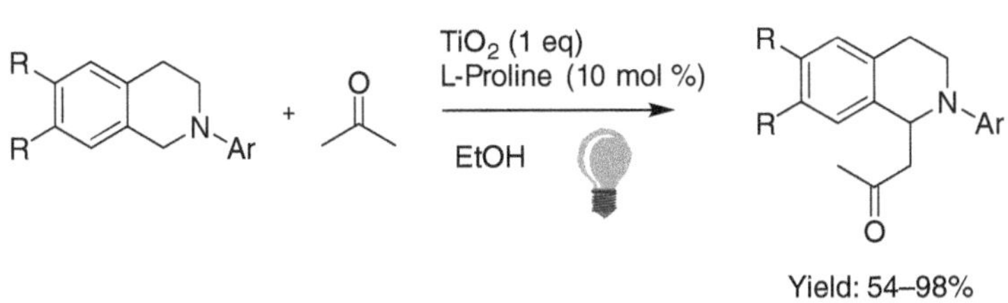

Scheme 12.16 TiO_2-mediated visible light-oxidative Mannich reaction.

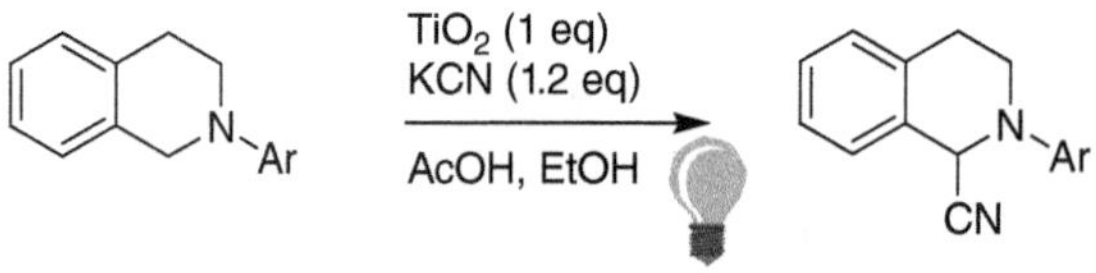

Scheme 12.17 TiO_2-catalyzed oxidative cyanation.

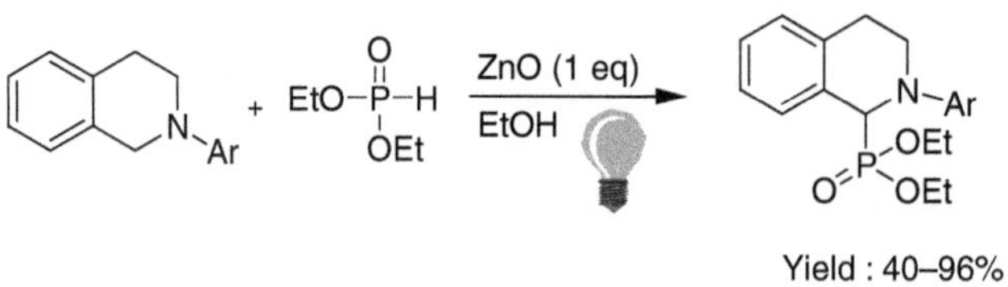

Scheme 12.18 ZnO-mediated oxidative phosphorylation reaction.

Another visible light-mediated C—C bond reaction with oxygen as the terminal oxidizing agent was catalyzed by a heterogeneous metal-free photosensitizer, the mpg-C_3N_4 semiconductor [21]. A dual Mannich-type photo coupling reaction of ketones with iminium ions was catalyzed by combining proline organocatalysis with the mpg-C_3N_4 photocatalytic oxidation (Scheme 12.19).

Tertiary amines produced iminium ions that could undergo further transformations with carbon nucleophiles under photoredox conditions. The photoreaction of nitroalkanes and malonates with tetrahydroisoquinoline (THIQ) proceeded in good to excellent yields (Scheme 12.20).

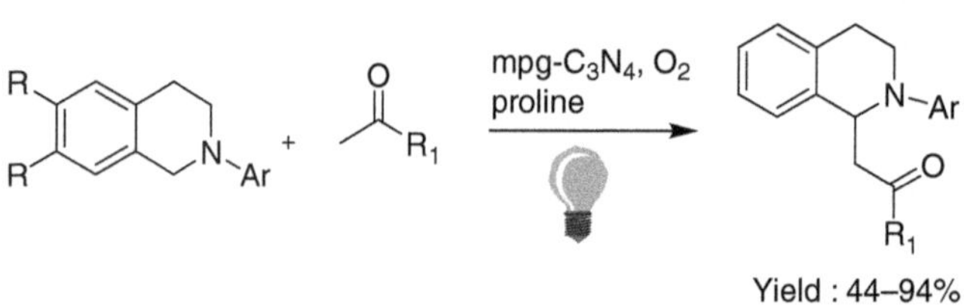

Scheme 12.19 Visible light dual-catalytic Mannich reaction.

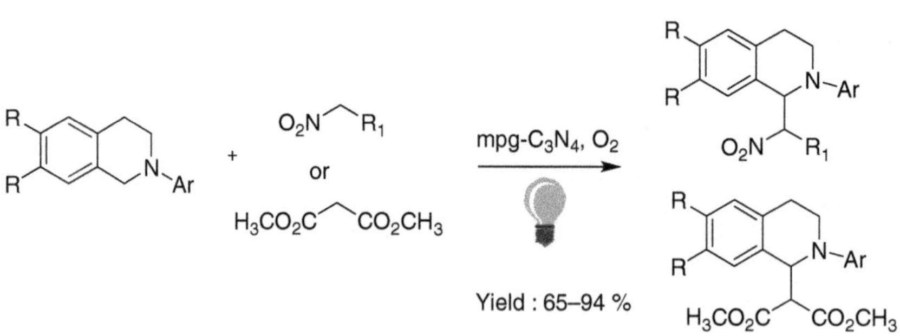

Scheme 12.20 Visible light-mediated oxidative coupling reactions of tetrahydroisoquinolines.

12.2.4 Visible Light C—H Activation

The photo C—C and C-heteroatom bond formation from inert C—H bonds are triggered by TiO_2 under UV light, sunlight, and even visible light at RT. The TiO_2 photocatalysis-based photoinduced hole reactivity can utilize cheap and abundant raw materials without pre-functionalization for highly selective synthesis of halides, boronates, aldehydes, ketones, enamines, and silanes. It does not require the harsh reaction conditions of transition metal catalysis. Aryl diazonium salts reacted with various heteroarenes catalyzed by commercially available and inexpensive TiO_2 P25. Dantrolene is a specific and essential muscle relaxant. It was derived from the arylation product of furfural and nitrophenyl diazonium salt (Scheme 12.21). A robust and recyclable TiO_2 catalyzed the photoreaction of electron-rich and electron-poor heteroarenes in moderate to high yields [22].

Wang and coworkers developed the PANI-g-C_3N_4-TiO_2, a user-friendly composite catalyst for photolytic transformations. It catalyzed the C—H functionalization of several heteroarenes, enol acetate, and benzoquinone using various diazonium salts under visible light (Scheme 12.22).

Aniline was converted in situ to aryl diazonium salts using t-BuONO and $MeSO_3H$ and transformed into radicals on visible light irradiation and catalyzed by the ternary PANI-g-C_3N_4-TiO_2 composite. Several functional groups were tolerated in this visible light-mediated transformation. The synergistic effect of the three components of the PANI-g-C_3N_4-TiO_2 catalyst contributes to enhanced photocatalysis and recyclability [23]. The incident photons of the visible light were absorbed by a surface complex of diazoether-TiO_2 in the

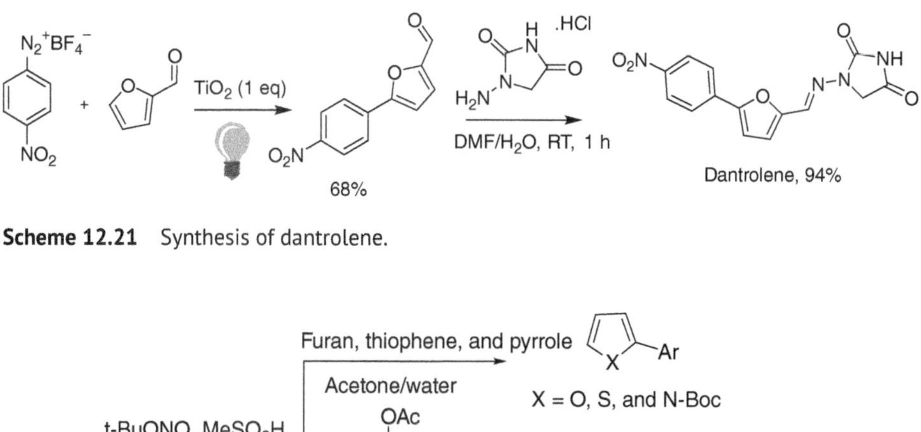

Scheme 12.21 Synthesis of dantrolene.

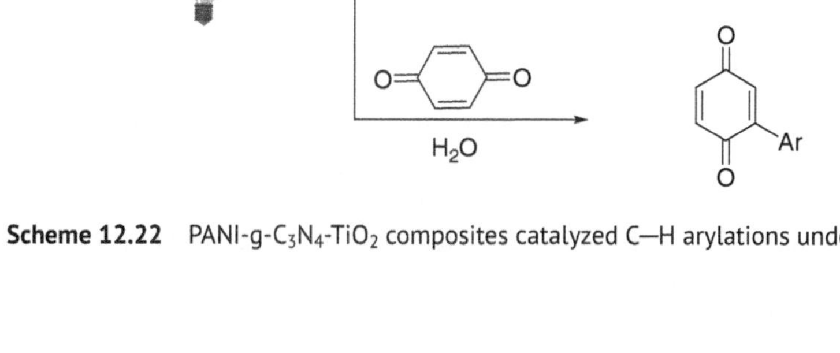

Scheme 12.22 PANI-g-C_3N_4-TiO_2 composites catalyzed C—H arylations under visible light.

Rueping system compared to the narrow gap semiconductor – PANI in Wang's system. While in the earlier system, the catalysis was limited to polar aromatic compounds containing a heteroatom such as O, N, or S, containing functional groups; surface coordination of the substrates to TiO_2 was essential. The latter Wang system claimed a broader substrate scope because visible light absorption was by PANI itself and did not require any heteroatom substrates. Good functional group tolerance and high chemoselectivity and regioselectivity were achieved with excellent yields.

The 2,5-diaryl-1,3,4-oxadiazoles having wide pharmaceutical application were conveniently synthesized under visible light illumination catalyzed by the synergistic ternary PANI-supported catalyst system. The RT visible light illumination mediated the decarboxylative cyclization of keto acids with acyl hydrazines (Scheme 12.23). Several substrates survived the reaction conditions with recyclability of the composite photocatalysts without much loss of activity for six consecutive runs [24].

12.2.5 Dye-Sensitized SET

Dye-sensitized TiO_2 P25 served as efficient catalysts for C—H functionalization. The benzylic $C(sp^3)$-H bonds of *N*-aryl THIQ undergo a decarboxylative alkylation under visible light illumination (Scheme 12.24). The precursors for the unactivated alkyl radicals using dye-sensitized TiO_2 for the dye-sensitized SET were the *N*-hydroxyphthalimide esters [25].

The TiO_2 P25 efficiently sensitized by organic dyes and decorated with transition metal oxides NiO served as an effective sensitizer for organic synthesis.

12.2.6 Photo Annulations

The $Ni(acac)_2{\cdot}2H_2O$ and TiO_2 P25 by a simple chemisorption and calcination protocol was reformed into NiO/TiO_2 nanocomposite. The *N*-methyl C—H bond of dimethyl anilines was derivatized catalytically by the NiO/TiO_2 nanocomposite to construct the C—C-coupling product of tetrahydroquinoline. Irradiation with a 365-nm UV LED lamp in the presence of monodispersed TiO_2 compared with commercial P25 nanomaterials demonstrated parallel

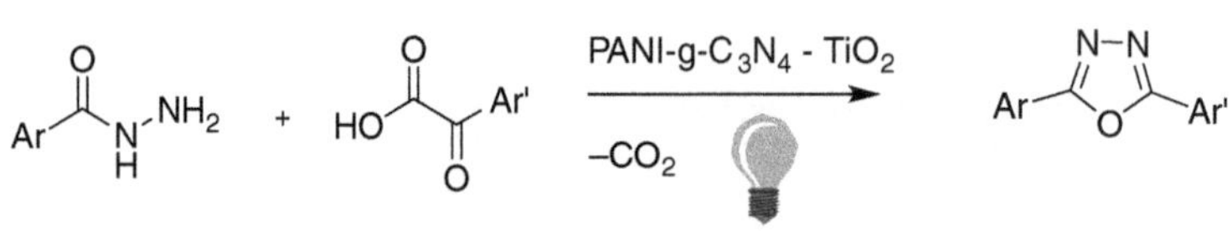

Scheme 12.23 The synthesis of 2,5-diaryl-1,3,4-oxadiazoles catalyzed by the PANI-g-C_3N_4-TiO_2 composite.

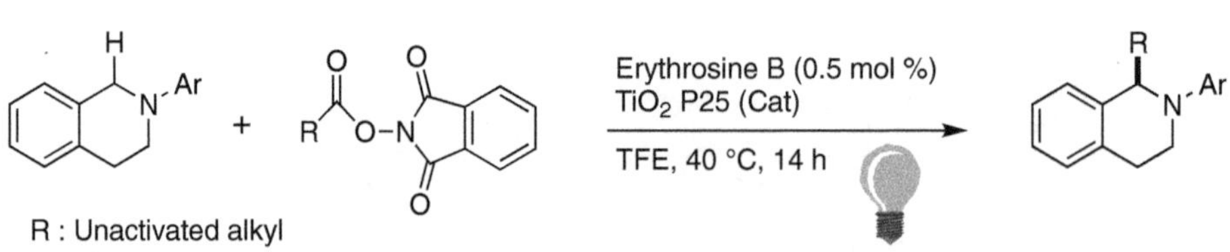

Scheme 12.24 Dye-sensitized SET, decarboxylative alkylation of tetrahydroisoquinoline.

activity toward the C—H functionalization of *N*-methyl aniline and *N,N*-dimethylaniline *N*-methyl bonds of maleic imide olefinic C—H bond [26].

The absorption boundary was expanded to the visible light region by the elevated TiO_2 VB of the NiO-decorated TiO_2. The TiO_2 bandgap decreased significantly by the modification but with enough oxidation ability for *N*-methyl C—H functionalization (Scheme 12.25). Undesired transformations could be prevented by stepping up the h^+ VB redox potential while attenuating photoinduced h^+ VB oxidizability by NiO [27].

12.2.7 McMillan Chiral Catalysis Using Enamines

TiO_2 was also applied to the visible light C—H bond alkylation transformations. Konig and coworkers reported the highly chemoselective and stereoselective aldehyde α-alkylation with electron-deficient alkyl bromides catalyzed by a combination of secondary amine organocatalysts with organic dye-modified TiO_2 photocatalyst (Scheme 12.26) [27, 28].

Presumably, the proposed pathway for photoredox catalysis is followed by the alkylation reaction. An enamine is formed by condensation of the octanal with a chiral amine organocatalyst. An α-carbonyl radical spawned by electron transfer from a semiconductor to the halogenated carbonyl adds to this enamine. The VB hole oxidizes the α-amino radical, producing the iminium ion, releasing catalyst and product (Figure 12.1).

12.2.8 The Photo Mannich Reactions

l-Proline on CdS photocatalyzes the Mannich reaction of *N*-aryltetrahydroisoquinolines with ketones. The excited chromophore is quenched in the crucial step reductively, producing radical amine cations subsequently converted into an iminium ion. Holes in the

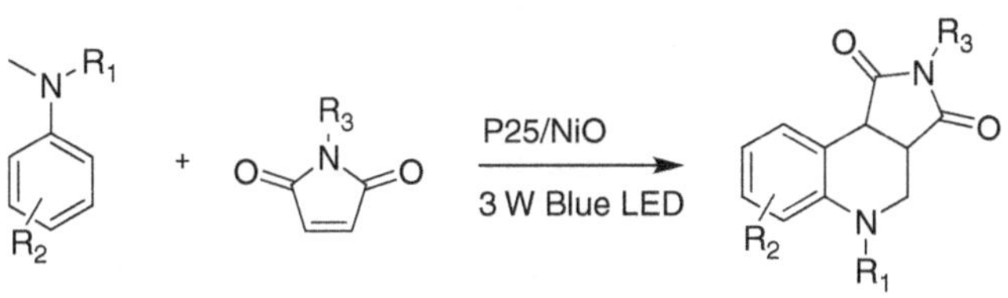

Scheme 12.25 TiO_2-photocatalyzed tandem addition–cyclization reaction.

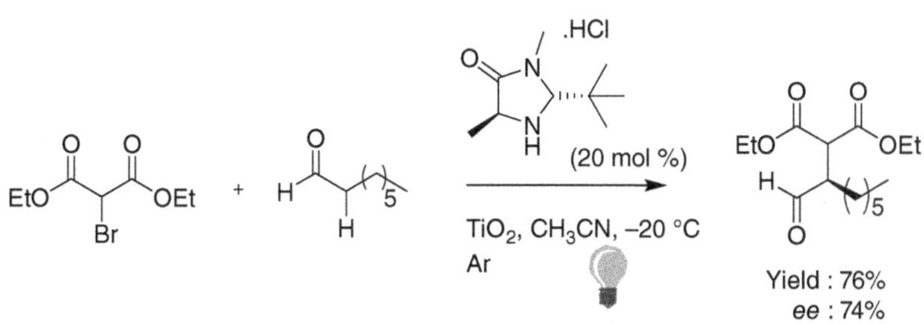

Scheme 12.26 TiO_2 photocatalytic enantioselective chiral enamine alkylations.

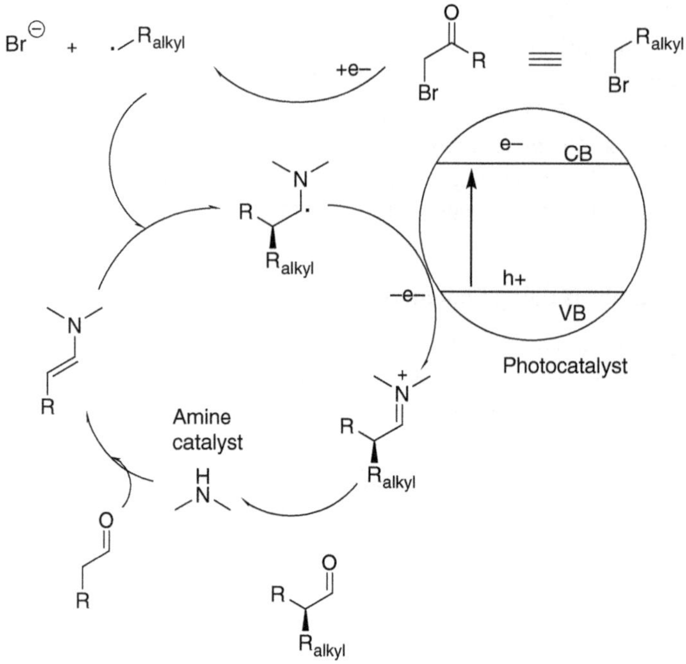

Figure 12.1 Proposed mechanism of the semiconductor photoredox catalysis.

VB in such reactions were photogenerated by the inorganic semiconductors. CdS should be an appropriate heterogeneous photocatalyst based on the bandgap and redox potential for the oxidative creation of the desired iminium cations. The synergy of inorganic photocatalyst CdS and organo-catalyst l-proline promotes the oxidative photo Mannich reaction of aryltetrahydroisoquinoline with ketones under blue light irradiation (Scheme 12.27).

There should be a noncovalent interaction between the organocatalyst proline and CdS, which otherwise retards or quenches the reaction by oxidative decomposition of the organocatalyst. Several available inorganic semiconductors with different bandgaps and redox potentials are ideal partners for chiral organocatalysis in stereoselective photoreactions with simple recovery from the reaction mixtures' recyclability [27, 28].

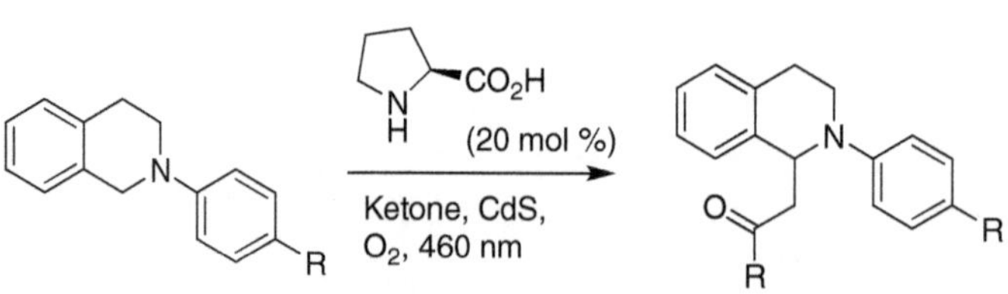

Scheme 12.27 Photocatalytic Mannich reaction catalyzed by l-proline on CdS.

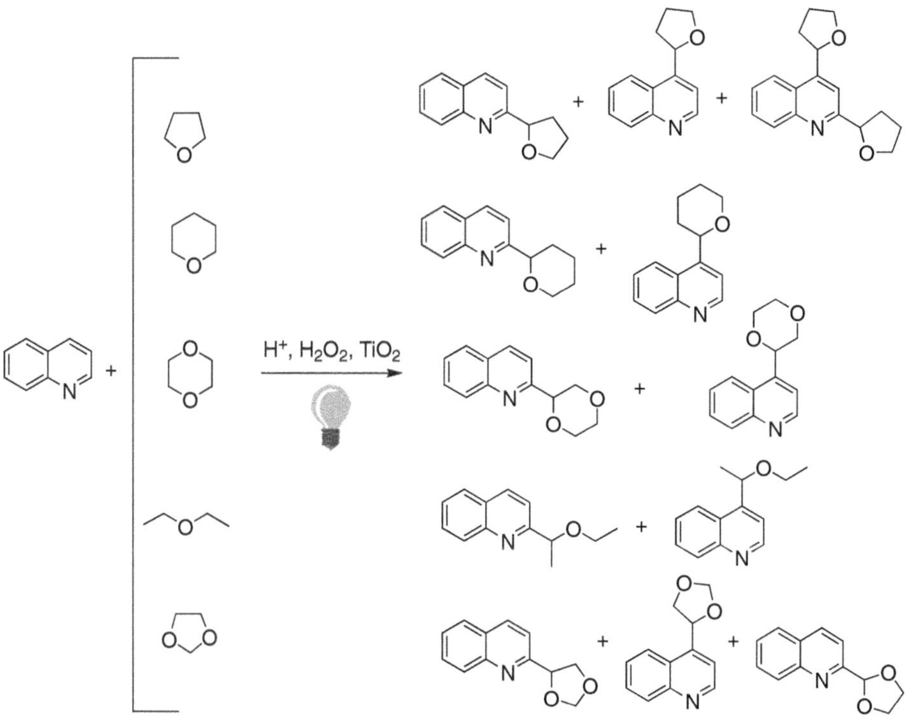

Scheme 12.28 TiO_2-photocatalyzed reaction of quinoline with ethers.

12.2.9 Activation of Ethers via Photocatalysis

The Pt/F-TiO_2-P25, with surface OH modification by F, photocatalyzed the formation of 2,3-butanediol via anaerobic C—C coupling of ethanol. The reaction of quinoline with different ethers, tetrahydrofuran (THF), dioxane, diethyl ether, tetrahydropyran, and dioxolane was catalyzed by TiO_2 on irradiation with visible light (Scheme 12.28).

Various heterocyclic bases like quinoline, pyridine, quinoxaline, isoquinoline, and 4-cyanopyridine were also functionalized with trioxane leading to masked aldehydes [27].

12.2.10 Visible Light-Mediated Hydroamination of Terminal Alkynes

A new variety of Cu catalysts was heterogenized on activated carbon (Cu/AC) by the impregnation-adsorption method. Cu^{2+} was chemically adsorbed on large surface-activated carbon surfaces by abundant oxygenated functional groups. The highly stable and inexpensive AC is a widely used support system and has a large surface area. The well-developed porous structure and oxygenated surface groups of the AC heterogenized the Cu catalysts. Cu/AC catalyzed the hydroamination of terminal alkynes to form imines by the irradiation with visible light selectively. Internal alkynes showed no reaction. The mechanisms of hydroamination are shown in Figure 12.2. During this photocatalytic process, the Cu-active site was immobilized by AC support enriched by adsorbing the reactants, contributing to the alkynes' activation [29].

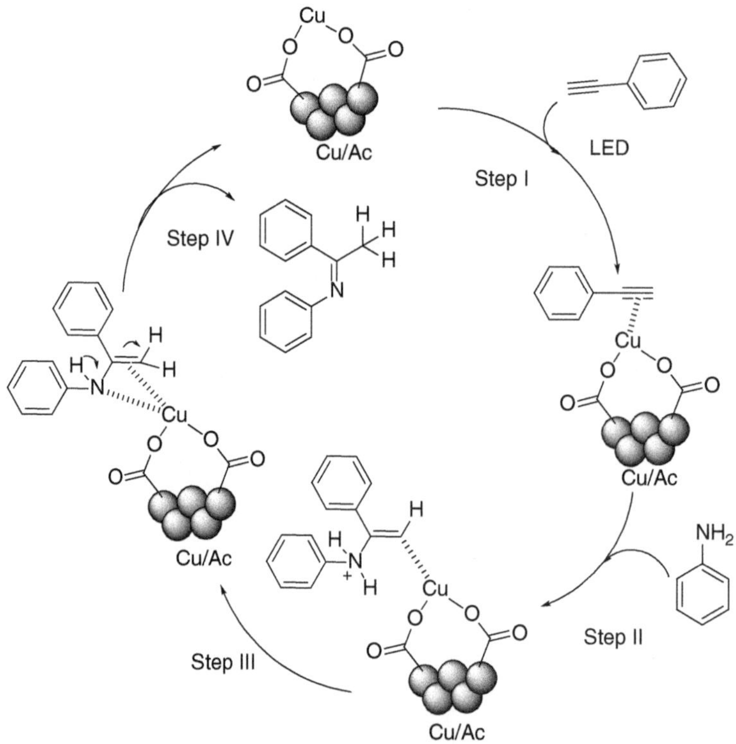

Figure 12.2 Mechanism of Cu/AC-catalyzed hydroamination of terminal alkynes.

12.2.11 Photo C—H Functionalization – Weinreb Olefination

Rueping and coworkers proposed the C—H functionalization reaction to be catalyzed by heterogeneous semiconductors. Sustainable C—H functionalization is achieved by photoactive semiconductors with appropriate bandgap and reduction potentials and expedites the active metal catalyst regeneration. Equivalent measures of tertiary butyl hydroperoxide (TBHP), MnO_2, Cu, or Ag salts are required for the C—H activation catalyzed by transition metal complexes. Visible light-active semiconductors catalyze efficient and sustainable C—H functionalization with appropriate bandgap and reduction potential facilitating the regeneration of the active metal catalysts. $BiVO_4$ (bandgap energy – 2.4 eV) and WO_3 (2.6–3.0 eV) seemed to be the excellent starting points for these metal-catalyzed C—H functionalization investigations (Scheme 12.29).

Mechanistically, these semiconductors are a suitable replacement for metal salts acting as visible light photoredox catalysts [30].

The different visible light-absorbing heterogeneous semiconductors were evaluated for the Ru-catalyzed *ortho*-olefination of protected phenols. Initially, high yields of the expected product were achieved using stoichiometric quantities of $BiVO_4$. Later, the use of heterogeneous $BiVO_4$ catalyst provided the olefinated product (71%). The concept of photoredox and metal catalysis expanded to C-H functionalization catalyzed by Pd, envisioning the intramolecular reaction of enamides to indole core structures (Scheme 12.30).

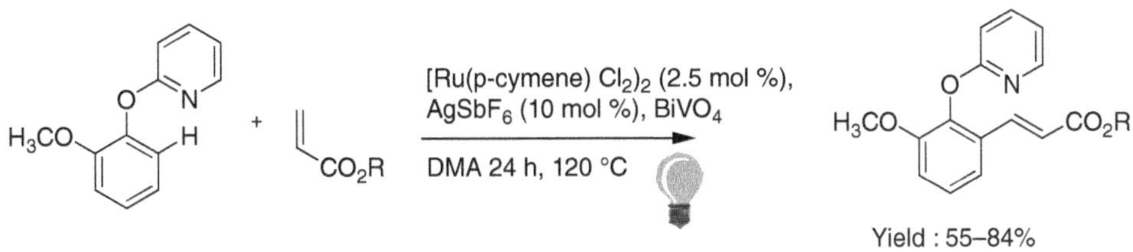

Scheme 12.29 *Ortho*-olefination of phenol ethers photocatalyzed by Ru/$BiVO_4$.

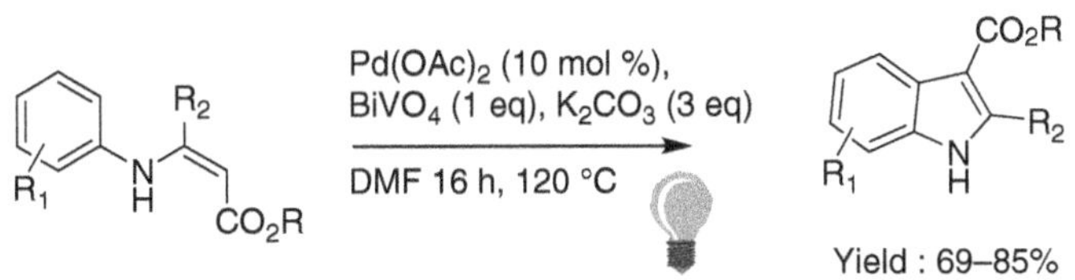

Scheme 12.30 Heterogeneous photoredox-catalyzed indole synthesis.

$Pd(OAc)_2$ and bismuth vanadate under visible light irradiation catalyzed the cyclization of several enamides.

Rh catalysts were used for the C—H functionalization, having proved the concept with Ru and Pd. In this study, WO_3 provided the highest yields of olefination. Various Weinreb and alkyl amides were comfortably *ortho*-olefinated using catalytic amount of $[RhCp^*Cl_2]_2$ (Scheme 12.31).

12.2.12 Quantum Dot (QD) – Photocatalysis

Quantum dots (QDs) are very tiny photostable semiconductors with particles ranging from 2 to 5 nm leading to various applications. They have a high absorption of photons per molecule, are inherently photostable, and have vast absorption spectra. QDs can be fine-tuned and customized to specific photochemical reactions. They are synthesized from various elements (Zn, Cd, Se, and S) with a modified surface structure and unchanged electronic constitution. QDs can enhance the charge transfer rates while interacting with several substrates. Single-sized CdSe QDs (2–5 nm) catalyzed the photoredox C—C bond construction in the β-alkylation of aldehydes with acceptor alkenes. Ketones, when irradiated with blue LED, reacted with imines to give γ-amino alkylation product (Scheme 12.32). CdSe QDs

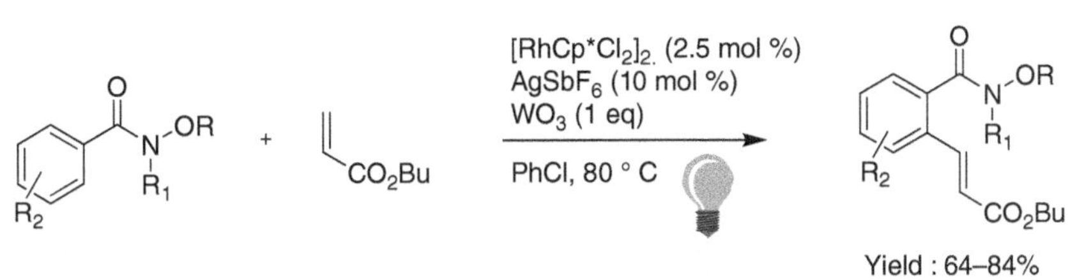

Scheme 12.31 Heterogeneous photoredox-catalyzed Weinreb olefination.

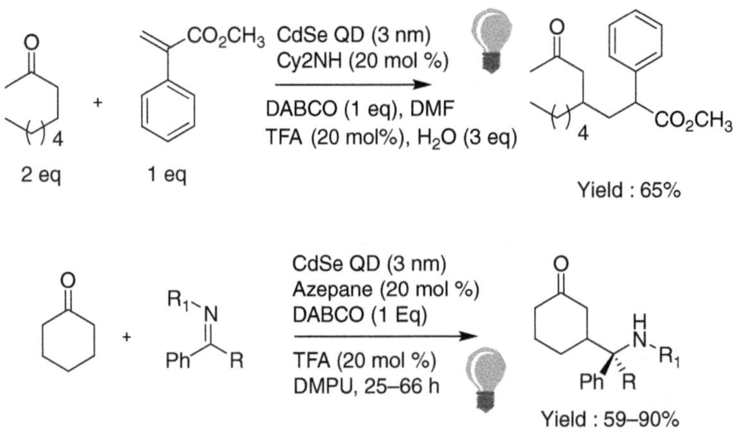

Scheme 12.32 CdSe QD-catalyzed β-alkylation and γ-amino alkylation.

were suitable for Ru and Ir photosensitizers with comparative turn over numbers (TONs) and turn over frequencies (TOFs) [31].

12.2.13 Visible Light-Mediated C-Heteroatom Bond Formation

The coupling of aryl halides with amines and alcohols was efficiently catalyzed by the dual photoredox catalyst system of $NiCl_2$ and CdS (Scheme 12.33).

These two transformations, combined with the use of robust CdS and inexpensive Ni salts, are highly attractive. Aryl bromides underwent a highly fruitful nucleophilic reaction with water catalyzed by a heterogeneous catalyst system. Only electron-poor aryl halides undergo the reaction. Aniline, phenols, and analogs also do not participate in these coupling reactions [32].

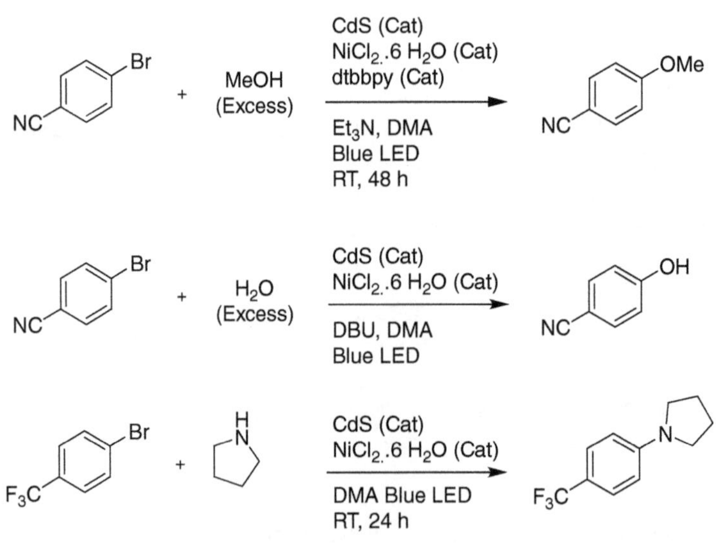

Scheme 12.33 Dual Ni—Cd-catalyzed photo heterocoupling reactions.

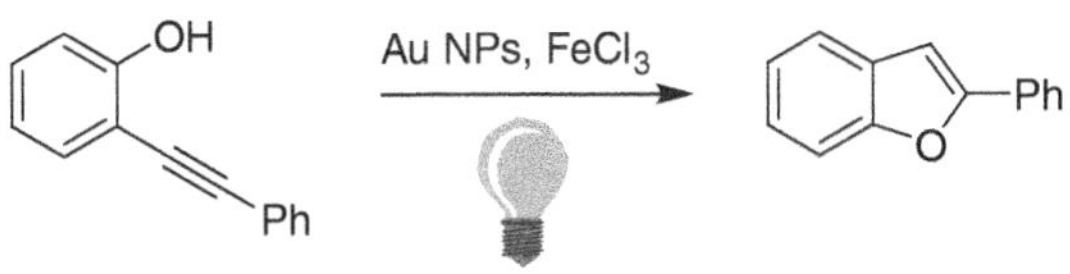

Scheme 12.34 Light-driven photocyclization of 2-(phenethynyl)phenol.

12.2.14 Photocyclization Reaction

Sub-nanosized Au clusters were prepared by an oxidative etching process using $FeCl_3$. These were active catalysts for the photocyclization of 2-(phenethynyl)phenol and afforded 2-phenylbenzofuran in a high 99% yield (Scheme 12.34).

Irradiating with incident light in the near-UV region ranging between 390 and 420 nm gave quantitative yields of benzofuran. The results indicate that hot carriers generated efficiently by interband transitions compared to those produced solely by LSPR of Au, Ag, and Cu exhibited higher reactivity [33].

12.2.15 Photocycloaddition Reaction

The earth-abundant TiO_2 photocatalysis integrates two C—C couplings and the formation of one C=O to synthesize tetralones. Cyclic tetralones are formed from tethered alkenes' intramolecular reaction, providing tricyclic tetralones with excellent diastereoselectivity (Scheme 12.35) [34].

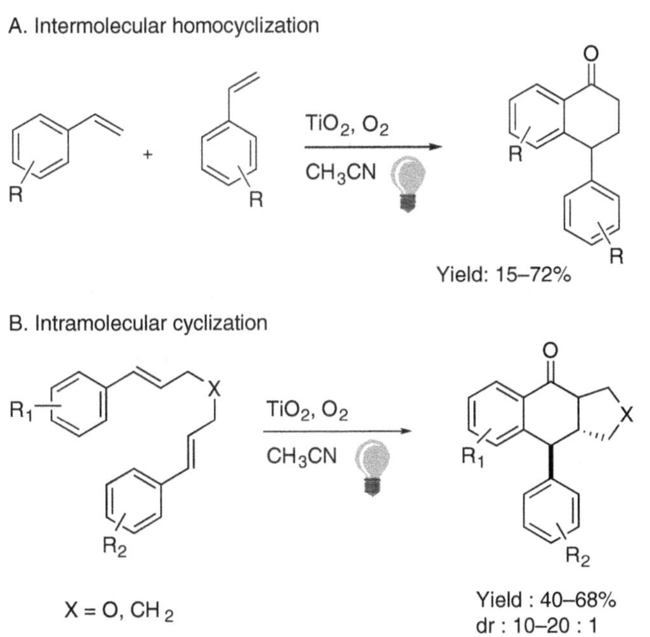

Scheme 12.35 TiO_2-catalyzed photocycloadditions to aryl tetralones.

12.3 Conclusion

When every research and phenomenon takes time to mature to higher levels, VLPC as a new domain made rapid progress and attained greater heights, a boon to chemists and technologists. The history of organic synthesis has given birth to various strategies. However, these methodologies took decades to reach the stereoselective synthetic routes. On the other hand, exponential heights have been reached rapidly by the nano heterogeneous school of photocatalysis. It is fascinating to note that the application of photocatalysis has rendered escalated access to pharmaceutical compounds and their intermediates, fine chemicals, commodity chemicals, and agrochemicals. Eventually, the manufacturing sector industrial process will employ the VLPC regime's strategies to enforce green and cost-effective protocols. On the elementary level, every synthetic scheme focuses on minimizing the synthetic operations for numerous benefits, with the proven track record by nano heterogeneous photocatalysis. Chemists will soon rewrite the textbooks of the synthetic disconnection approach. The rapid access of structurally complex compounds by nano heterogeneous photocatalysis will become a powerful instrument to synthesize analogs of molecules of biological and pharmacological significance (bioactive complex scaffold) and correspondingly to elucidate the structural activity relationship. The growth so far indicates that nano heterogeneous photocatalysis-mediated synthetic protocols from the immediate to distant future would render more exciting shortcut routes to complex agro and pharmaceutical molecules; most notably, the pharmaceutical-manufacturing sector will function with time and cost-effective benign sustainable methodologies.

This new technology complements the existing UV photochemistry with added advantages; conventional photochemistry excites the most polarizable functional groups and reduces the selectivity, thereby reducing the yield of the product. Usually, fluorescent lamps, LEDs, and the sun are used as light sources in photocatalysis. Since the energy is less, it does not disrupt the molecular bonds, unlike UV photochemistry.

While the present report focuses on the synthetic utility of heterogeneous nanocatalysts in VLPC, many other aspects need investigation. They are: (i) the light cannot be uniform throughout the reaction medium; the intensity of the light at different parts of the reactor is to be established. (ii) The quantum yield is reported as high. However, the influence of light absorbing and scattering due to solvents, substrates, and reagents need to be estimated. (iii) The nature of the spent catalyst should be examined to understand its properties after the reaction. (iv) The kinetic studies are not reported; this domain will soon end up in the industrial processes, and more thermodynamic and kinetic properties are needed.

Nano heterogeneous photocatalysis has reached its heights rapidly. It has portrayed its competence in synthesizing fine, specialty, agrochemicals, complex organic compounds, and total synthesis in one pot. Consequently, in the industry, it will bring employment to chemists and chemical engineers.

References

1 Lang, X., Chen, X., and Zhao, J. (2014). Heterogeneous visible-light photocatalysis for selective organic transformations. *Chem. Soc. Rev.* 43 (1): 473–486.

2 Mahmoud, Y., Attia, Y., El Nazer, H., and Solum, E. (2021). An overview of recent development in visible light-mediated organic synthesis over heterogeneous photo-nanocatalysts. *Curr. Org. Synth.* 18 (1): 23–36.

3 Riente Paiva, P. and Noel, T. (2019). Application of metal oxide semiconductors in light-driven organic transformations. *Catal. Sci. Technol.* 9: 5186–5232.

4 Friedmann, D., Hakki, A., Kim, H. et al. (2016). Heterogeneous photocatalytic organic synthesis: state-of-the-art and future perspectives. *Green Chem.* 18 (20): 5391–5411.

5 Munir, S., Dionysiou, D.D., Khan, S.B. et al. (2015). Development of photocatalysts for selective and efficient organic transformations. *J. Photochem. Photobiol. B* 148: 209–222.

6 Chen, J., Cen, J., Xu, X., and Li, X. (2016). The application of heterogeneous visible light photocatalysts in organic synthesis. *Catal. Sci. Technol.* 6 (2): 349–362.

7 Higashimoto, S., Nakai, Y., Azuma, M. et al. (2014). One-pot synthesis of imine from benzyl alcohol and nitrobenzene on visible-light responsive CdS–TiO_2 photocatalysts. *RSC Adv.* 4 (71): 37662–37668.

8 Franchi, D. and Amara, Z. (2020). Applications of sensitized-semiconductors as heterogeneous visible-vight photocatalysts in organic synthesis. *ACS Sustain. Chem. Eng.* 8: 15405–15429.

9 Attia, Y.A. and Mohamed, Y.M.A. (2018). Silicon-grafted Ag/AgX/rGO nanomaterials (X = Cl or Br) as dip-photocatalysts for highly efficient p-nitrophenol reduction and paracetamol production. *Appl. Organomet. Chem.* 33 (3): e4757.

10 Ma, D., Zhai, S., Wang, Y. et al. (2019). TiO_2 photocatalysis for transfer hydrogenation. *Molecules* 24 (2): 330.

11 Li, X.-H., Baar, M., Blechert, S., and Antonietti, M. (2013). Facilitating room-temperature Suzuki coupling reaction with light: Mott-Schottky photocatalyst for C-C-coupling. *Sci. Rep.* 3 (1): 1743.

12 Wang, N., Ma, L., Wang, J. et al. (2019). Graphitic carbon nitride (g-C_3N_4) supported Pd species: an efficient heterogeneous photocatalyst surpassing homogeneous thermal heating systems for Suzuki Coupling. *ChemPlusChem.* 84 (8): 1164–1168.

13 Jiao, Z., Zhai, Z., Guo, X., and Guo, X.-Y. (2015). Visible-light-driven photocatalytic Suzuki–Miyaura coupling reaction on Mott–Schottky-type Pd/SiC catalyst. *J. Phy. Chem. C* 119 (6): 3238–3243.

14 Xiao, Q., Sarina, S., Bo, A. et al. (2014). Visible light-driven cross-coupling reactions at lower temperatures using a photocatalyst of palladium and gold alloy nanoparticles. *ACS Catal.* 4 (6): 1725–1734.

15 Gu, Q., Jia, Q., Long, J., and Gao, Z. (2018). Heterogeneous photocatalyzed C—C cross- coupling reactions under visible-light and near-infrared light irradiation. *ChemCatChem.* 11 (2): 669–683.

16 Xiao, Q., Sarina, S., Jaatinen, E. et al. (2014). Efficient photocatalytic Suzuki cross-coupling reactions on Au—Pd alloy nanoparticles under visible light irradiation. *Green Chem.* 16 (9): 4272–4285.

17 Zhang, S., Chang, C., Huang, Z. et al. (2015). Visible- light-activated Suzuki–Miyaura coupling reactions of aryl chlorides over the multifunctional Pd/Au/Porous nanorods of CeO_2 catalysts. *ACS Catal.* 5 (11): 6481–6488.

18 Wang, P., Huang, B., Dai, Y., and Whangbo, M.-H. (2012). Plasmonic photocatalysts: harvesting visible light with noble metal nanoparticles. *Phy. Chem. Chem. Phy.* 14 (28): 9813–9825.

19 Yoshii, T., Kuwahara, Y., Mori, K., and Yamashita, H. (2019). Design of Pd – Graphene–Au nanorod nanocomposite catalyst for boosting Suzuki–Miyaura coupling reaction by assistance of surface plasmon resonance. *J. Phys. Chem. C.* 123: 24575–24583.

20 Rueping, M., Zoller, J., Fabry, D.C. et al. (2012). Light-mediated heterogeneous cross dehydrogenative coupling reactions: metal oxides as efficient, recyclable, photoredox catalysts in C—C bond- forming reactions. *Chem. Eur. J.* 18 (12): 3478–3481.

21 Möhlmann, L., Baar, M., Rieß, J. et al. (2012). Carbon nitride-catalyzed photoredox C—C bond formation with N- aryltetra- hydroisoquinolines. *Adv. Synth. Catal.* 354 (10): 1909–1913.

22 Zoller, J., Fabry, D.C., and Rueping, M. (2015). Unexpected dual role of titanium dioxide in the visible light heterogeneous catalyzed C—H arylation of heteroarenes. *ACS Catal.* 5 (6): 3900–3904.

23 Wang, L., Shen, J., Yang, S. et al. (2018). C—H arylation reactions through aniline activation catalyzed by a PANI-g-C_3N_4-TiO_2 composite under visible light in aqueous medium. *Green Chem.* 20 (6): 1290–1296.

24 Wang, L., Wang, Y., Chen, Q., and He, M. (2018). Photocatalyzed facile synthesis of 2,5- diaryl 1,3,4-oxadiazoles with polyaniline- g-C_3N_4 -TiO_2 composite under visible light. *Tetrahedron Lett.* 59 (15): 1489–1492.

25 Ren, L. and Cong, H. (2018). Visible-light-driven decarboxylative alkylation of C—H bond catalyzed by dye-sensitized semiconductor. *Org. Lett.* 20 (11): 3225–3228.

26 Tang, J., Grampp, G., Liu, Y. et al. (2015). Visible light mediated cyclization of tertiary anilines with maleimides using nickel(II) oxide surface-modified titanium dioxide catalyst. *J. Org. Chem.* 80 (5): 2724–2732.

27 Wang, Y., Liu, A., Ma, D. et al. (2018). TiO_2 photocatalyzed C—H bond transformation for C—C coupling reactions. *Catalysts* 8 (9): 355.

28 Cherevatskaya, M., Neumann, M., Füldner, S. et al. (2012). Visible-light-promoted stereoselective alkylation by combining heterogeneous photocatalysis with organocatalysis. *Angew. Chem. Int. Ed.* 51 (17): 4062–4066.

29 Gu, X., Ma, P., Liu, P. et al. (2020). Visible-light-driven hydroamination of alkynes over a new type of activated carbon immobilized Cu^{2+} photocatalyst. *Chem. Res. Chin. Univ.* 36: 1039–1044.

30 Fabry, D.C., Zoller, J., and Rueping, M. (2019). Semiconductors as heterogeneous visible light photoredox catalysis in combined dual metal-catalyzed C—H functionalizations. *Org. Chem. Front.* 6: 2635–2639.

31 Caputo, J.A., Frenette, L.C., Zhao, N. et al. (2017). General and efficient C—C bond forming photoredox catalysis with semiconductor quantum dots. *J. Am. Chem. Soc.* 139 (12): 4250–4253.

32 Lu, L.-Q., Liu, Y.-Y., Liang, D., and Xiao, W.-J. (2019). Practical heterogeneous photoredox/nickel dual catalysis for C—N and C—O coupling reactions. *Chem. Commun.* 55: 4853–4856.

33 Zhao, J., Nguyen, S.C., Ye, R. et al. (2017). A comparison of photocatalytic activities of gold nanoparticles following plasmonic and interband excitation and a strategy for harnessing interband hot carriers for solution phase photocatalysis. *ACS Cent. Sci.* 3 (5): 482–488.

34 Liu, Y., Zhang, M., Tung, C.-H., and Wang, Y. (2016). TiO_2 photocatalytic cyclization reactions for the syntheses of aryltetralones. *ACS Catal.* 6 (12): 8389–8394.

13

Metal Nanoparticles-Catalyzed Hydrogen Generation from Ammonia Borane

Xiugang Li[1,2] *and Zhang-Hui Lu*[1]

[1]*Key Laboratory of Functional Small Molecules for Ministry of Education, College of Chemistry and Chemical Engineering, Jiangxi Normal University, Nanchang, China*
[2]*College of Material and Chemical Engineering, Tongren University, Tongren, China*

13.1 Introduction

With the rapid development of population growth and industrial production, the increasing demand for energy has provoked the search for effective, renewable, and eco-friendly alternative energy. Renewable energy, such as solar, wind, tidal, and geothermal energy, has shown many advantages. However, these kinds of energy are relatively scattered, due to the restriction of geographical and climatic conditions. Hydrogen is extensively considered as an ideal clean energy carrier because of its high energy density and near-zero emission [1].

Secure and efficient hydrogen storage is crucial to make hydrogen as a viable energy carrier [2]. Numerous hydrogen storage methods have been developed. Among them, solid materials or chemical hydrides are considered as one of the most effective ways to store hydrogen, which provide higher energy density and higher stability than that of gas or liquid H_2 storage systems [3]. According to the latest hydrogen storage targets set by the U.S. Department of Energy (DOE) (Table 13.1) for the year 2020, the minimum gravimetric hydrogen storage capacity for onboard vehicles and portable power is 4.5 and 4.0 wt%, respectively [4]. The B–N adducts, with safe storability, high content of hydrogen [5], and easy release under mild conditions, have been identified to be the promising chemical hydrogen storage material [6]. As the simplest B–N compound, ammonia borane (AB) has become an ideal candidate, because of its nontoxicity, long-term stability in solvents, as well as high gravimetric hydrogen capacity (19.6 wt%) [7].

AB was first reported to generate H_2 through thermal decomposition upon melting in 1987 [8]. However, high dehydrogenation temperatures and undesired by-products severely limited the practical application of AB [9]. Another way to release H_2 from AB is the solvolysis in protic solvents, for instance, in water (hydrolysis, Eq. 13.1) or in CH_3OH

Heterogeneous Nanocatalysis for Energy and Environmental Sustainability: Energy Applications, First Edition. Edited by Putla Sudarsanam, Yusuke Yamauchi, and Pankaj Bharali.

Table 13.1 Hydrogen storage targets for onboard vehicle and portable power set by DOE.

	Vehicle (>5 kg H_2)	Portable (1~50 g H_2)
Gravimetric hydrogen storage capacity	4.5%	4.0%
Volumetric hydrogen storage capacity g H_2/L	30	50
System costs US$/g H_2	0.3	3.3
The cost of hydrogen US$/kg H_2	4	N.A.

(methanolysis, Eq. 13.2). When proper catalysts are used, hydrogen will be released from the hydrolysis or methanolysis of AB under mild conditions [10].

$$NH_3BH_3 + 2H_2O \rightarrow NH_4BO_2 + 3H_2 \tag{13.1}$$

$$NH_3BH_3 + 4CH_3OH \rightarrow NH_4B(CH_3O)_4 + 3H_2 \tag{13.2}$$

Up to now, tremendous attempts have been made to fabricate highly active and highly stable heterogeneous catalysts for AB hydrolysis or methanolysis, such as metal phosphides [11], metal nanoparticles (NPs) [12], metal borides [13], and polymetallic nanoalloys [14]. The significant advances have not only been made to increase the understanding of these fields but also lead to the establishment of a complete and practical system toward AB solvolysis for hydrogen production. In this chapter, we chiefly review the progress of developing highly efficient metal NPs catalysts for AB solvolysis in the last several years, such as monometallic and heterometallic nanocatalysts. We also discuss the mechanism of solvolysis of AB and its regeneration method, which provides a reference for the future research in these fields.

13.2 Hydrolytic Dehydrogenation of AB

In 2006, Xu et al. first reported the stoichiometric hydrogen generation from AB hydrolysis using noble metal catalysts [15]. However, noble metals are expensive and scarce. In order to reduce their consumption, two strategies have been developed: one is to increase the efficiency of precious metal utilization by reducing the size and/or increasing the specific surface area of the metal NPs; the other is to develop non-noble metals or alloys NPs catalysts. In view of this purpose, numerous monometallic and heterometallic catalysts have been fabricated toward the hydrolysis of AB [16].

13.2.1 Monometallic Catalysts

13.2.1.1 Noble Metal Catalysts

Noble metal catalysts, including Rh, Ru, Pt, Pd, Au, and Ag, were firstly used for AB hydrolysis to hydrogen production [15]. A variety of solid porous materials, including γ-Al_2O_3,

SiO_2, covalent organic frameworks (COFs), metal-organic frameworks (MOFs), and porous carbons, have been employed as the carrier to prepare well-dispersed noble metal NPs catalysts. Xu et al. first reported the supported catalysts using SiO_2, γ-Al_2O_3, and carbon as supports [16b]. Pt, Rh, and Ru were proved to be highly efficient, and Pd had moderate catalytic activity, while Au and Ag were inactive. Özkar et al. developed monodisperse Rh NPs catalysts for the hydrolysis of AB. The well-dispersed Rh NPs immobilized on polyaminoborane displayed a high catalytic performance, giving a turnover frequency (TOF) of 130 min^{-1} [17]. Xu's group developed a double solvent method (DSM) method to encapsulate ultrasmall Pt NPs in the cavities of materials of institute Lavoisior (MIL)-101 to avoid the aggregation of metal NPs [18]. The obtained ultrafine Pt NPs (1.8 $\pm$ 0.2 nm) demonstrated a high catalytic performance (TOF = 414 min^{-1}) toward AB hydrolysis. Using the same method, Jiang's group synthesized a core-shell Pd@Co nanocomposite encapsulated in the cavities of MIL-101(Figure 13.1) [19]. Co^{2+} and Pd^{2+} precursor can easily diffuse into the cavities of MIL-101 using DSM and form core-shell structure via in situ reduction of AB based on different reduction potentials. The confined Pd@Co@MIL-101 nanocatalyst demonstrated a superior catalytic performance (TOF = 51 min^{-1}) toward the hydrolysis of AB, compared to their monometallic and alloy counterparts.

Our group synthesized a supported Ru catalyst by employing CNT as the support for AB hydrolysis [20]. The obtained Ru/CNT catalyst with ultrasmall Ru NPs (2.3 nm) provided an ultrahigh TOF of 706 min^{-1} at room temperature. Recently, Yu's group fabricated a single-atom Rh catalyst encapsulated in Mobil fifth (MFI) aluminosilicate ZSM-5 molecular sieve with $[Rh(NH_2CH_2CH_2NH_2)_3]Cl_3$ complex as the precursor via a ligand protection and hydrogen direct reduction method (Figure 13.2). The obtained Rh@S-1-H catalyst can efficiently catalyze AB hydrolysis, and the TOF value can reach up to 699 min^{-1} [21]. The catalytic activities of all these noble metal-based catalysts for AB hydrolysis are illustrated in Table 13.2.

13.2.1.2 Non-noble Metal Catalysts

Despite the remarkable catalytic activities of noble metal catalysts, attention to the practical applications of noble metal has promoted the search for a low-cost catalyst toward AB hydrolysis. The fabrication of non-noble metal NPs catalysts with high catalytic performance is obviously an attractive target. For this purpose, the first-row transition metals, including Cu, Ni, Co, and Fe, were employed for AB hydrolysis.

Sun et al. prepared a monodisperse Ni NPs with a mean size of about 3.2 nm by the reduction of $Ni(acac)_2$ with borane tributylamine complex (BTB) in oleic acid (OA) and oleylamine (OAm) [47]. The obtained catalyst displayed a high catalytic activity (TOF = 8.8 min^{-1}) for AB hydrolysis. Yu's group investigated the influence of NPs size of the same Ni NPs on the catalytic performance. A typical volcano curve for catalytic activities was obtained and 8.9 nm Ni NPs presented the highest catalytic activity for AB hydrolysis [48].

Xu's group prepared a variety of non-noble nanocatalysts, including Cu, Ni, Co, as well as Fe, using SiO_2, γ-Al_2O_3, and C as carriers [49]. Among them, Co NPs supported on C displayed the highest catalytic performance toward AB hydrolysis. However, the supported Cu catalyst had low activity, and the supported Fe catalyst had almost no catalytic activity. Subsequently, the same group prepared amorphous Fe NPs by reducing $FeSO_4$

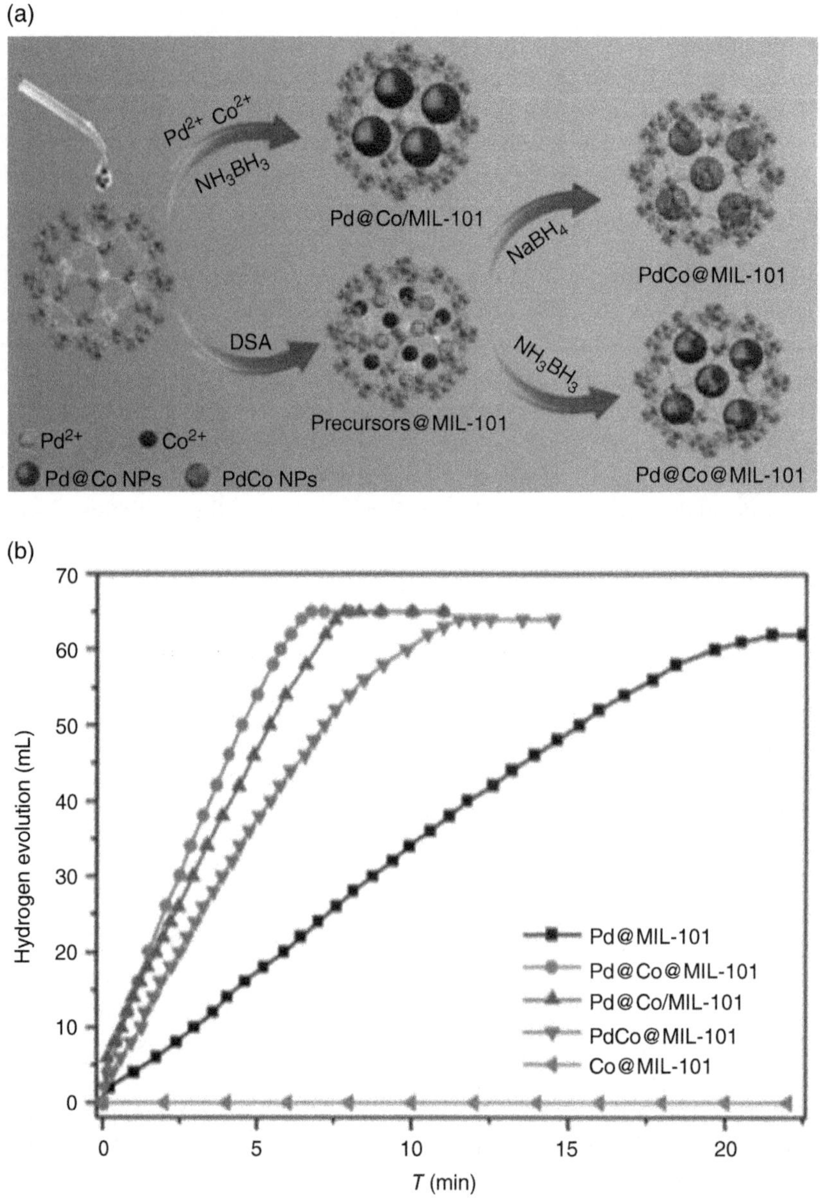

Figure 13.1 (a) Synthetic schematic illustration of Pd@Co/MIL-101, PdCo@MIL-101, and Pd@Co@MIL-101. (b) Time course plots of H_2 generation from AB hydrolysis over different catalysts at 303 K. *Source:* From [19] / with permission of John Wiley & Sons.

with AB and $NaBH_4$ as co-reducing agents [50]. The as-synthesized Fe nanocatalyst provided high catalytic performance for AB hydrolysis due to its amorphous structure.

Later on, various supported non-noble metal NPs were developed for hydrolysis of AB. Xu's group developed an encapsulated Co NPs catalyst by reducing the complex of Co and dendrimers [51]. The obtained Co NPs exhibited high activity for AB hydrolysis. Gu

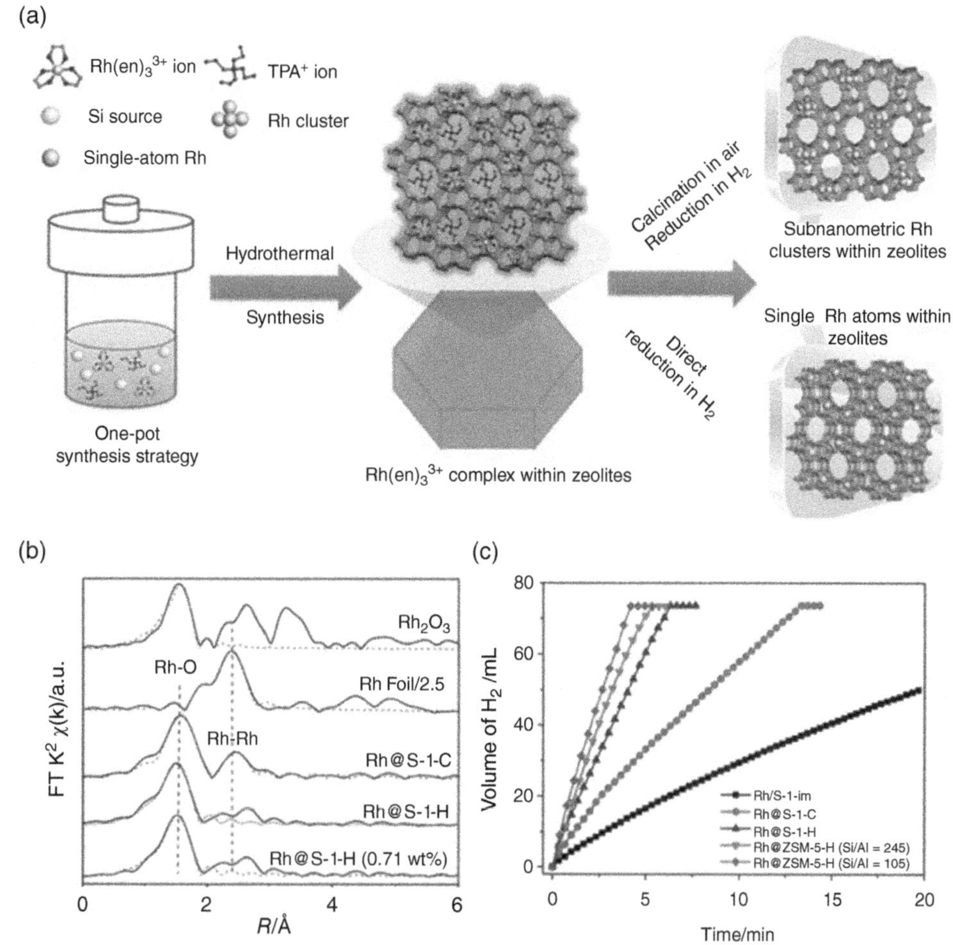

Figure 13.2 (a) Synthetic schematic illustration of Rh@S-1 catalysts; (b) Fourier transform of k^2-weighted extended X-ray absorption fine structure (EXAFS) spectra of Rh foil and various zeolite-encaged Rh catalysts; (c) Time course plots of H_2 generation from the hydrolysis of AB over different catalysts at 298 K. *Source:* From [21] / with permission of John Wiley & Sons.

et al. used ultrasound-assisted reduction method to immobilize Co NPs on an MOF. The as-synthesized Co/MIL-101-1-U catalyst can efficiently catalyze AB hydrolysis and give a high TOF value of 51.4 min^{-1} [52]. Tao et al. prepared nitrogen-doped porous nanowire using Co-MOF as the precursor by high-temperature calcination. The loaded Co NPs catalyst (Co/NPCNW) displayed excellent activity toward AB hydrolysis, and the activity was still maintained 94.6% after recycled 10 times [53]. He et al. loaded Co NPs on covalent triazine framework materials (CTF) and the optimal catalyst Co/CTF displayed a high catalytic performance (TOF = 42.3 min^{-1}) at 298 K.

Recently, our group synthesized a porous silica nanosphere confined Cu NPs catalyst via a reverse micelle method (Figure 13.3) [54]. The aggregation of Cu NPs could be effectively prevented by an outer shell of silica. The core-shell catalyst Cu@SiO_2 displayed a higher

Table 13.2 Catalytic activities of different noble metal-based catalysts for AB hydrolysis.

Catalysts	Temperature (K)	n_{metal}/n_{AB}	TOF (min^{-1})	E_a (kJ/mol)	Reference
Commercial Pt/C	298	0.018	83.3	–	[15]
Pt/γ-Al_2O_3	298	0.018	222.2	21	[16b]
Pt@MIL-101	298	0.0029	413.8	–	[18]
Pt/CNTs	298	0.004	407.2	38	[22]
Rh@S-1	298	0.0011	699	75.5	[21]
Pt@CeO_2 nanonecklace	298	0.018	133.3	–	[23]
Pt-CeO_2/rGO	298	0.0079	93.8	64.7	[24]
Pt_2/graphene	300	0.0011	2800	–	[25]
Pt@SiO_2	303	0.00245	158.6	53.9	[26]
Pt@PC-POP	303	0.001	104.33	56.42	[27]
Pt25@TiO_2	298	0.0016	311	–	[28]
Pt@h-mNSiO_2	298	0.0013	371.7	49.4	[29]
Rh/γ-Al_2O_3	298	0.018	128	21	[16b]
Rh/NaY	298	0.002	92	66.9	[30]
Rh/CeO_2	298	0.0008	2010	42.6	[31]
Rh/AC	298	0.0038	188	39.9	[32]
Rh/graphene	298	0.004	325	19.7	[33]
Rh/CNTs	298	0.0025	706	32	[20]
In situ Rh/C	298	0.0114	1246	40.9	[34]
Rh@UiO-66	298	0.0056	219.8	38.4	[35]
Ru/γ-Al_2O_3	298	0.018	55.5	23	[16b]
Ru/carbon black	298	0.0085	429.5	34.81	[36]
Ru/C-300	298	0.0034	643	38.7	[37]
Ru@SBA-15	298	0.0025	231	34.8	[16f]
Ru@SiO_2	298	0.002	200	38.2	[10b]
Ru/MWCNT	298	0.00094	329	33	[38]
Metastable Ru	298	0.0025	21.8	27.5	[39]
Ru/g-C_3N_4	298	0.0013	384.6	37.4	[40]
Ru/graphene	298	0.002	600	12.7	[41]
Pd/MCN	298	0.03	125	57	[42]
Pd/RGO	298	0.004	6.25	51	[43]
Ag/SBA-15	298	0.093	12	–	[44]
Pd/SiO_2-$CoFe_2O_4$	298	0.0019	254	52	[45]
Ag/SiO_2-$CoFe_2O_4$	298	0.0028	264	53.4	[46]

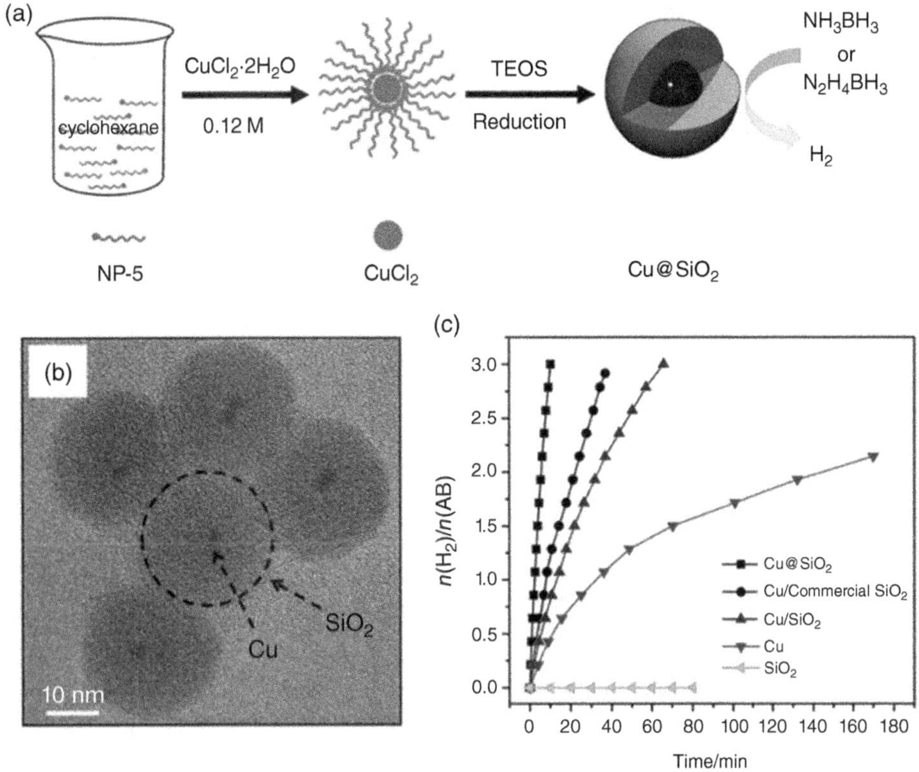

Figure 13.3 (a) Synthetic schematic illustration of Cu@SiO$_2$ catalyst; (b) TEM image of the Cu@SiO$_2$ catalyst; (c) Time course plots of H_2 release from AB hydrolysis (0.2 M, 5 ml) over different catalysts at 25 °C. *Source:* From [54] / with permission of Springer Nature.

catalytic performance than that of Cu/commercial SiO$_2$, Cu/SiO$_2$, as well as support-free Cu NPs toward the hydrolysis of AB. Moreover, Cu@SiO$_2$ displayed long-term stability and retained about 90% of its initial catalytic activity even after recycled 10 times. The catalytic activities of all these non-noble metal nanocatalysts for AB hydrolysis are shown in Table 13.3.

13.2.2 Heterometallic Catalysts

Compared to monometallic catalysts, heterometallic catalysts have drawn more attention because of their synergistic effect between components which improves the activity and stability of the catalysts [73]. The surface metal sites and electronic structure can be controlled by adjusting different compositions of metals, thus affecting the catalytic activities of the catalysts [74]. In particular, Xu's research group synthesized spherical PdPt and cubic PdPt catalysts [75]. Cubic PdPt displayed a higher catalytic performance (TOF = 50.02 min^{-1}) than spherical PdPt toward AB hydrolysis. Although the size of spherical PdPt NPs (3–4 nm) is smaller, cubic PdPt particles (15–18 nm) can expose more (100)

Table 13.3 Catalytic activities of different non-noble base catalysts toward AB hydrolysis.

Catalyst	Temperature (K)	n_{metal}/n_{AB}	TOF (min^{-1})	E_a (kJ/mol)	Reference
In situ Fe NPs	RT	0.12	3.12	–	[50]
Co/γ-Al_2O_3	RT	0.018	2.27	62	[49]
In situ Co NPs	RT	0.04	44.1	–	[55]
Co/PEI-GO	298	0.11	39.9	28.2	[56]
Co/MIL-101-1-U	298	0.02	51.4	31.3	[52]
Co/NPCNW	298	0.075	7.29	25.4	[53]
Co@N-C-700	298	0.057	5.6	31	[57]
Co/HPC	323	0.11	2.94	32.8	[58]
Co/CTF	298	0.05	42.3	42.7	[59]
Co NCs@PCC-2a	298	0.07	90.1	–	[60]
Co@C-N@SiO_2-800	298	–	8.4	36.1	[60]
G6-OH(Co_{60})	298	0.013	10	50.2	[51]
Co-(CeO_x)0.91	298	0.04	79.5	31.82	[61]
Ni/γ-Al_2O_3	RT	0.018	2.5	-	[49]
Ni/C	298	0.0425	8.8	28	[47]
Ni/SiO_2	298	0.0225	13.2	34	[62]
Ni/ZIF-8	RT	0.016	14.2	–	[63]
Ni@MSC-30	RT	0.016	30.7	–	[64]
Ni-CeO_x/graphene	298	0.08	68.2	28.9	[10c]
Ni@3D-(N)GFs	RT	0.009	41.7	–	[65]
Ni/CNT	298	–	26.2	32.3	[66]
Ni/PDA-$CoFe_2O_4$	298	0.017	7.6	50.8	[67]
Ni/Ketjen black	298	0.13	7.5	66.6	[48]
Cu/γ-Al_2O_3	RT	0.018	0.23	–	[49]
Cu@Cu_2O	293	0.15	0.32	–	[68]
Zeolite-confined Cu	298	0.013	1.25	51.8	[69]
Cu/$CoFe_2O_4$@SiO_2	298	0.0031	40	–	[70]
Cu@SiO_2	298	0.08	3.24	36	[54]
Cu/RGO	298	0.1	3.61	38.2	[71]
p(AMPS)-Cu	303	0.069	0.72	48.8	[72]

crystal planes, which makes it exhibit superior performance for AB hydrolysis. Zhou et al. prepared PtRu alloy NPs with an uniform structure by the dealloying method [76]. Compared with single metal Pt and Ru, PtRu NPs displayed a better catalytic performance (TOF = 59.6 min^{-1}) for AB hydrolysis.

To reduce the cost of the catalysts, non-noble metals are introduced into the heterometallic systems. The obtained cost-efficient heterometallic catalysts display superior catalytic performance for AB hydrolysis, which makes the dehydrogenation of AB easier to industrialize. Different kinds of heterometallic alloy catalysts containing non-noble metal, such as Pt-Co [77], Pt-Ni [78], Ru-Ni [79], Ru-Co [80], Ru-Cu [80a, 81], Au-Ni [82], Au-Co [16e, 83], Ag-Ni [19, 84], Ag-Co [85], Pd-Ni [74c, 74g, 86], and Pd-Co [19, 47, 87], have been exploited for H_2 generation from AB hydrolysis with high activities.

Pt and non-noble metal alloy NPs were firstly reported for AB hydrolysis because of the outstanding catalytic activities. Lu et al. prepared 2.3 nm bimetallic PtCo NPs using polyethyleneimine (PEI)-decorated GO as the support [88]. The synergistic effect was obviously dependent on the ratio of Pt/Co, and $Pt_{0.17}Co_{0.83}$ was found as the most efficient catalyst (TOF = 377.8 min^{-1}). Xu' group prepared an alloy Au-M (M = Ni or Co) nanocatalysts immobilized by MOF toward the hydrolysis of AB [93]. The location and particle size of Au-M NPs could be efficiently controlled through a liquid-phase concentration-controlled reduction (CCR) strategy. The confined Au-M NPs catalyst encapsulated in the pores of MOF displayed an ultrahigh catalytic performance toward AB hydrolysis because of the confined effect as well as the bimetallic synergistic effect.

As the research moves along, non-noble metal heterometallic alloy nanocatalysts have been considerably fabricated, because of the numerous advantages, such as abundant reserves, low-cost, and high catalytic activity. Numerous noble metal-free heterometallic catalysts, for example Fe–Co [89], Fe–Ni [90], Cu–Fe [91], Cu–Co [16a, 92], Ni–Co [93], and Cu–Ni [94], have been synthesized for AB hydrolysis. Typically, our group fabricated magnetically recyclable $Cu_{1-x}Fe_x$ alloy NPs via in situ reduction strategy for the hydrolysis of AB. The resulting $Cu_{0.33}Fe_{0.67}$ alloy NPs catalyst displayed the highest catalytic performance (13.95 min^{-1}) for AB hydrolysis [95]. We also doped a series of inactive metals (Mo, Cr, and W) into the active metal NPs (Ni, Co, and Cu) to prepare composite metal NPs catalysts [96]. The dispersion and activity of these catalysts can be improved by introducing large specific surface area support, such as graphene, r-GO, and porous carbon. The optimized catalyst Ni-Mo/graphene exhibits an ultrahigh Pt-like catalytic performance for AB hydrolysis at 298 K (Figure 13.4) [97].

As is known to all, Fe, Ni, and Co NPs are magnetic and tend to aggregate easily. In order to improve their stability, numerous porous materials, for example MOFs, silica spheres, as well as porous carbon materials, have been employed to prepare non-noble bimetallic NPs catalysts with high activity and stability [16a, 92, 95, 98]. Yan's group synthesized Co-decorated Cu NPs supported on graphene via a self-catalytic method [92a]. Xu et al. developed a bimetallic CuCo nanocatalyst supported on a diamine-functionalized rGO. The obtained catalyst CuCo/PDA-rGO displayed a higher catalytic activity (41 min^{-1}) than that of CuCo/rGO (20.6 min^{-1}) stabilized on unfunctionalized rGO [92b]. The encapsulation of ultrafine metal NPs into specific porous materials has been considered as a striking strategy to enhance the stability and activity of the catalysts. The confinement effect can effectively control the growth of metal NPs, diminish NPs aggregation, and enhance their stability. To this end, our group successfully prepared a confined bimetallic nanocatalyst by encapsulating ultrasmall Cu-Co NPs into SiO_2 nanospheres. The obtained catalyst $Cu_{0.5}Co_{0.5}@SiO_2$ displayed an enhanced catalytic activity and stability for AB hydrolysis, and preserved about 93% of the initial activity even after recycled for 10 times [98a].

Trimetallic non-noble nanocatalysts were also studied toward the hydrolysis of AB. Our group fabricated CuCoMo NPs via a co-reduction strategy [99]. The resulting trimetallic

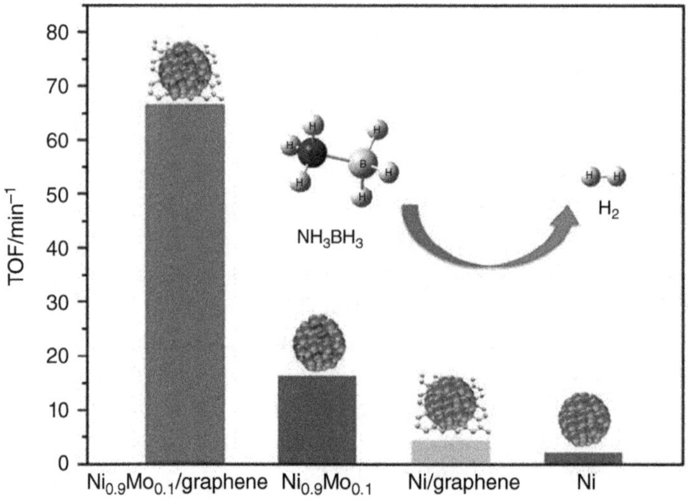

Figure 13.4 TOF values of the different catalysts for hydrogen release from AB hydrolysis. *Source:* From [97] / with permission of Royal Society of Chemistry.

catalyst $Cu_{0.72}Co_{0.18}Mo_{0.1}$ displayed a higher catalytic performance (46.0 min^{-1}) than that of their bimetallic or monometallic NPs catalysts toward the hydrolysis of AB. The catalytic performances of some representative heterometallic catalysts are illustrated in Table 13.4.

13.3 Methanolytic Dehydrogenation of AB

AB hydrolysis has been widely studied; however, some inherent problems have not yet been solved, such as the limitation of operating temperature (>273 K), the inevitable NH_3 production, and the unrecyclable by-products [106]. Methanolysis of AB provides an effective way to solve these problems. Interestingly, AB methanolysis has even more superiorities than hydrolysis, for instance, higher dissolvability in CH_3OH, higher hydrogen purity, efficient regeneration of by-product, and possible applications at lower temperatures (≤0 °C) [10b, d, f, 106b, 107].

Precatalyst metal salts ($RhCl_3$, $PdCl_2$, $RuCl_3$, $NiCl_2$, and $CoCl_2$) and supported metal NPs catalysts have been reported for AB methanolysis [106c, 108]. Ramachandran et al. firstly reported H_2 generation for AB methanolysis using transition metal as the catalysts [108]. They demonstrated a fast generation of pure hydrogen gas without NH_3 contamination from the methanolysis of AB over $CoCl_2$, Raney Ni, $RuCl_3$, and $RhCl_3$ at 298 K. However, metal NPs without supports are unstable and prone to aggregate, resulting in a decrease in catalytic performance. Supported metal NPs catalysts have demonstrated a higher catalytic activity and stability as compared to the supported-free metal NPs catalysts. Based on the abovementioned analysis, numerous metal NPs catalysts stabilized by specific supports have been fabricated for the methanolysis of AB.

Table 13.4 Catalytic activities of different heterometallic catalysts for AB hydrolysis.

Catalyst	Temperature (K)	n_{metal}/n_{AB}	TOF (min^{-1})	E_a (kJ/mol)	Reference
PtPd NPs	298	0.002	50.02	57.3	[75]
$Pt_{70}Ru_{30}$-NP	298	0.001	59.6	38.9	[76]
PdRh-PVP	298	0.003	1333	46.1	[100]
Ag_1Pd_4@UIO-66-NH_2	298	0.0125	90	51.77	[101]
PtNi@SiO_2	303	0.036	5.54	54.76	[102]
Ni_2Pt@ZiF-8	293	0.01	361.4	23.3	[78g]
Pt-Co/dendrimer	293	0.01	303	28.8	[77b]
$PtCo_{20}$/CNTs	298	–	675.1[a]	42.5	[77c]
RuNi/TiO_2	298	0.001	914	28.1	[79d]
RuCo/γ-Al_2O_3	338	–	32.9	47	[80a]
RuCu/γ-Al_2O_3	338	–	16.4	52	[80a]
AuNi@MIL-101	RT	0.017	66.2	–	[82b]
AuCo@MIL-101	RT	0.017	23.5	–	[83a]
Ag-doped Ni/MIL-101	298	0.017	20.2	25	[103]
AgCo/PAMAM	298	0.033	15.84	35.66	[85]
$Pd_{67}Ni_{33}$/MCN	298	0.03	222.2	54.1	[42]
$Ni_{30}Pd_{70}$/rGO	298	0.01	28.7	45	[74c]
Ni_3Pd_7/CS	298	0.024	187.5	35.32	[86]
$Co_{35}Pd_{65}$/C	298	0.024	22.7	27.5	[104]
Pd@Co@MIL-101	303	0.011	51	22	[19]
$Fe_{0.3}Co_{0.7}$ alloy	293	0.12	13.9	16.3	[89b]
$Fe_{0.5}Ni_{0.5}$ alloy	293	0.12	11.4	–	[90a]
CuCo/graphene	293	0.02	9.18	–	[92a]
$Cu_{0.2}Co_{0.8}$/PDA-rGO	303	0.05	51.5	54.9	[92b]
CuNi/CMK-1	298	0.072	54.8	–	[94a]
NiMo/graphene	298	0.05	66.7	21.8	[97]
$Cu_{0.5}Co_{0.5}$@SiO_2	298	0.08	4.26	24	[98a]
$Cu_{0.3}Co_{0.7}$@MIL-101	RT	0.034	19.6	–	[98b]
$Cu_{0.72}Co_{0.18}Mo_{0.1}$	298	0.04	119.0[b]	45	[99]
$Cu_{0.72}Co_{0.18}Mo_{0.1}$	298	0.04	46.0	–	[99]
CuCo@MIL-101-1-U	298	0.02	51.7	30.5	[52]
$Cu_{0.33}Fe_{0.67}$	298	0.04	13.95	43.2	[91]
$Cu_{0.8}Co_{0.2}$O-GO	298	0.024	70.0	45.5	[105]
$Cu_{0.2}Ni_{0.8}$/MCM-41	298	0.05	10.7	38	[94e]

[a] TOF is calculated based on Pt.
[b] TOF was calculated based on the addition of promoter NaOH (0.1 M).
RT, room temperature.

13.3.1 Monometallic Catalysts

13.3.1.1 Noble Metal Catalysts

For AB methanolysis, noble metal-based catalysts, for instance Rh, Ru, Pt, as well as Pd, have been proved to be highly active catalysts. Lara et al. synthesized a terphenylphosphane-stabilized Pt NPs catalyst by reducing [Pt(dba)$_2$] precursor [109]. The obtained dispersed Pt NPs exhibited a highly catalytic activity (TOF = 284 min^{-1}) toward AB methanolysis. Özkar's group synthesized a series of Rh NPs catalysts supported on different oxide nano-powders: (i) hydroxyapatite [110], (ii) silica [111], (iii) alumina [112], and (iv) ceria [113]. The highest catalytic activity and minimum mean particle size of Rh NPs were found as 218 min^{-1} and 3.6 nm in nanoalumina. Confined Rh NPs catalysts are also studied due to their higher stability and activity. Özkar et al. also reported zeolite-confined Rh NPs to catalyze hydrogen releasing from AB methanolysis [114]. The confined Rh NPs were highly active, stable, and reusable in AB methanolysis. Very recently, Xu's group synthesized an organic molecular cage (CC3-R) encapsulated Rh NPs catalyst (Figure 13.5) [115]. Compared to heterogeneous catalyst Rh@CC3-R-hetero, homogeneous catalyst Rh@CC3-R-homo displayed a much higher catalytic performance (TOF = 215.3 min^{-1}) due to the combined merits of homogeneous and heterogeneous catalysts. Wang et al. prepared highly porous rhodium nanospheres (HPRhS) through an iodide-mediated templating strategy [116]. The resulting HPRhS displayed a predominant catalytic performance (TOF = 296.1 min^{-1}) due to the singularly high surface area (100 m^2/g) and controllable nanoporous structure. Fan's group prepared an L-proline (PRO)-stabilized Rh NPs and the resulting Rh NPs deposited on carbon (Rh-PRO/C) exhibited an ultrahigh catalytic activity

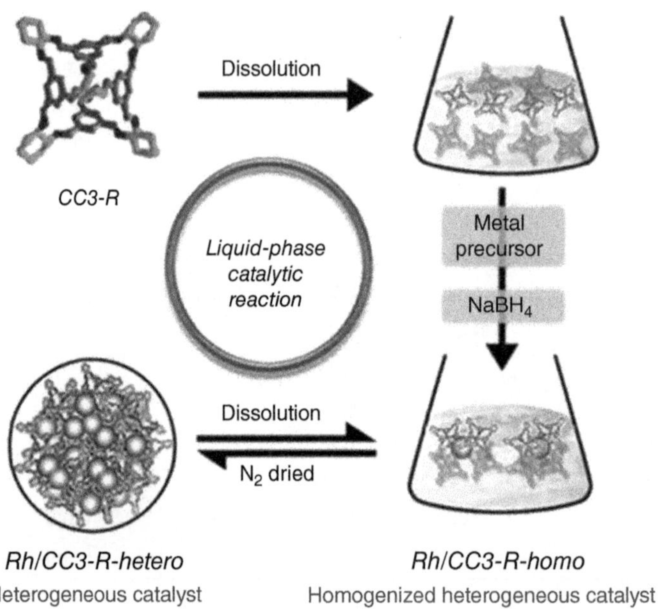

Figure 13.5 Synthetic schematic illustration of the catalyst Rh/CC3-R-homo and Rh@CC3-R-hetero. *Source:* From [115] / with permission of American Chemical Society.

(TOF = $1035\,min^{-1}$) under base conditions [117]. We synthesized a confined Rh NPs catalyst using a nitrogen-rich COF as support through an M–N coordination reduction strategy [118]. The optimal catalyst displayed an ultrasmall size of Rh NPs (1.4–2.6 nm), providing an ultrahigh catalytic performance (TOF = $505\,min^{-1}$) for AB methanolysis at room temperature.

Ru is also used as active metal NPs to catalyze the methanolysis of AB. Özkar et al. reported a poly (*N*-vinyl-2-pyrrolidone) (PVP)-stabilized Ru NPs catalyst. The resulting catalyst displayed a small size of Ru NPs (2.4 nm) and high catalytic activity (TOF = $67\,min^{-1}$) at room temperature [119]. Wang's group fabricated montmorillonite immobilized Ru nanocomposite, in which Ru NPs were embedded into the interlayers of silica and montmorillonite (MMT) [120]. The resulting catalyst Ru/MMT displayed remarkable durability and activity (TOF = $118.1\,min^{-1}$) due to the anchoring effect of MMT. Zhou's group reported an electrostatic attraction strategy to encapsulate fcc Ru NPs into the octahedral cavities of PCC-2 (Figure 13.6) [121]. Due to the confined effect, these Ru NPs exhibited high activity (TOF = $304\,min^{-1}$), selectivity, and stability for AB methanolysis.

Palladium is the fourth most active noble metal reported for AB methanolysis. Özkar group reported an in situ generated Pd NPs stabilized by PVP [122]. The resulting catalyst Pd@PVP displayed high activity and provided a TOF of $23.2\,min^{-1}$. Erturk et al. synthesized a confined Pd NPs catalyst by encapsulating Pd NPs in poly (amidoamine) (PAMAM) dendrimer. The effect of terminal groups, such as -NH_2, -COOH, and -OH, and core type on the catalytic performance were examined. The highest catalytic performance was obtained with -NH_2 as Jeffamine core and terminal groups of Pd NPs, giving a TOF of $55.8\,min^{-1}$ at 298 K and apparent activation energy (E_a) of 48 kJ/mol. Xu's group developed a reverse double-solvents approach to encapsulate ultrafine Pd NPs into organic molecular cage RCC3 [123]. The resulting catalyst Pd@RCC3 with a precisely controllable size of 0.72 nm exhibited high dispersibility and solubility, as well as excellent accessibility in solution, giving an ultrahigh catalytic performance (TOF = $176\,min^{-1}$) for AB methanolysis at 303 K.

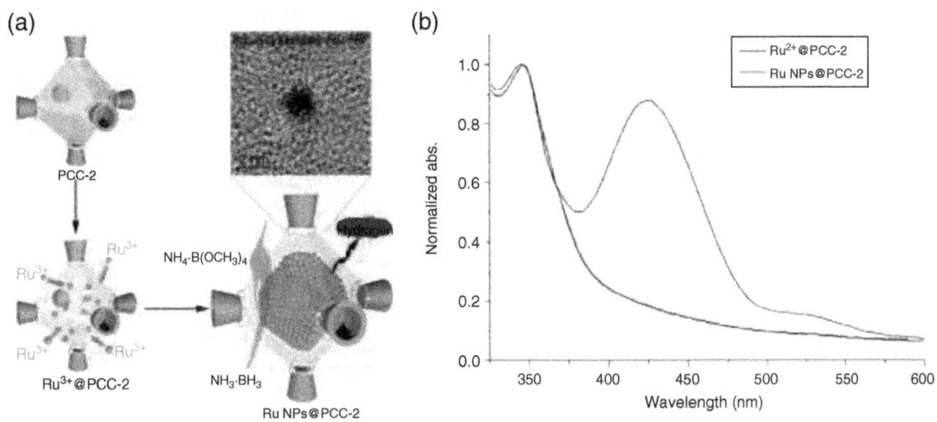

Figure 13.6 (a) Synthesis of Ru NPs@PCC-2 catalyst; (b) Time course plots of H_2 generation from AB methanolysis over Ru NPs@PCC-2 and Ru NPs@PVP. *Source:* From [121] / with permission of Elsevier.

13.3.1.2 Non-noble Metal Catalysts

Non-noble metal NPs catalysts are more attractive for AB methanolysis than noble metals like AB hydrolysis, because of the low cost and wide range of sources [124]. Distinguish from the widespread studies on the hydrolysis of AB over non-noble metal catalysts, there are few reports on the methanolysis of AB over non-noble metal catalysts (Table 13.3) [107b, 108, 125]. Among them, highly active non-noble monometallic NPs, such as Co, Ni, and Cu, have been widely used for the methanolysis of AB.

Luo et al. synthesized amorphous Co NPs in situ [126]. The resulting catalyst displayed a TOF of 515 h^{-1} toward AB methanolysis. Through the ^{11}B NMR, trimethylborate was proved to be the only product. Özkar's group reported a PVP-stabilized Ni NPs catalyst and the optimal catalyst displayed a TOF value of 12.1 min^{-1} [125a]. Later on, our group synthesized numerous mesoporous Cu NPs catalysts with different morphologies, such as nanoflower, nanosheet, nano-bundle, and dandelion-like nanostructures via a facile wet-chemical method [107b]. The flower-like Cu NPs displayed the highest catalytic activity (TOF = 2.41 min^{-1}) toward AB methanolysis as compared with the other nanostructures of Cu NPs. Zahmakiran's group synthesized Cu-Cu_2O-CuO/C nanocomposites. The obtained catalyst exhibited high activity (24 min^{-1}) at room temperature [127]. Thereafter, Liu's group developed a bunch-like copper oxide nanowire arrays on copper foam through a simple oxidation method and the obtained catalyst b-CuO NA/CF that showed a high catalytic performance (13.3 min^{-1}) toward AB methanolysis at 298 K [128].

13.3.2 Heterometallic Catalysts

Heterometallic NPs can be also used to catalyze AB methanolysis due to the synergy effect between different metals as similar to the hydrolysis of AB. Sun et al. prepared monodisperse CoPd NPs (about 7 nm) by reducing the precursors of Co and Pd in the system of trioctylphosphine and OAm (Figure 13.7) [124a]. The resulting catalyst $Co_{48}Pd_{52}/C$ deposited on Ketjen carbon showed the highest catalytic activity (TOF = 27.7 min^{-1}) for AB methanolysis. Subsequently, the same group replaced Co with Cu and Ag to prepare alloy CuPd and AgPd NPs catalysts, and the obtained CuPd and AgPd alloy NPs exhibited an enhanced catalytic activity than CoPd NPs. Additionally, $Cu_{48}Pd_{52}/C$ and $Ag_{30}Pd_{70}/C$ catalysts demonstrated the highest catalytic performance, providing a TOF of 53.2 and 366.4 min^{-1}, respectively [129].

Jagirdar et al. fabricated a series of Ni-Ni_3B and Co-Co_2B nanocomposites by reducing Ni^{2+} and Co^{2+} precursors with AB in CH_3OH [125d]. They also synthesized Co-Ni-B nanocomposite via a coreduction strategy. The obtained catalyst Co-Ni-B displayed a higher catalytic performance than individual metal–metal boride nanocomposite with a TOF of 10 min^{-1}. After that, Sun et al. synthesized a graphene-supported CuNi alloy NPs and the optimal CuNi/G catalyst exhibited remarkable recyclability and activity (49.1 min^{-1}) for AB methanolysis [130]. Chen et al. recently fabricated a Cu/$Co(OH)_2$ hybrid catalyst by controlling the reactant–surface electrostatic interactions [131]. The optimal catalyst demonstrated an ultrahigh catalytic activity (61.63 min^{-1}) toward AB methanolysis at 298 K. The catalytic performances of the typical heterometallic catalysts are shown in Table 13.5.

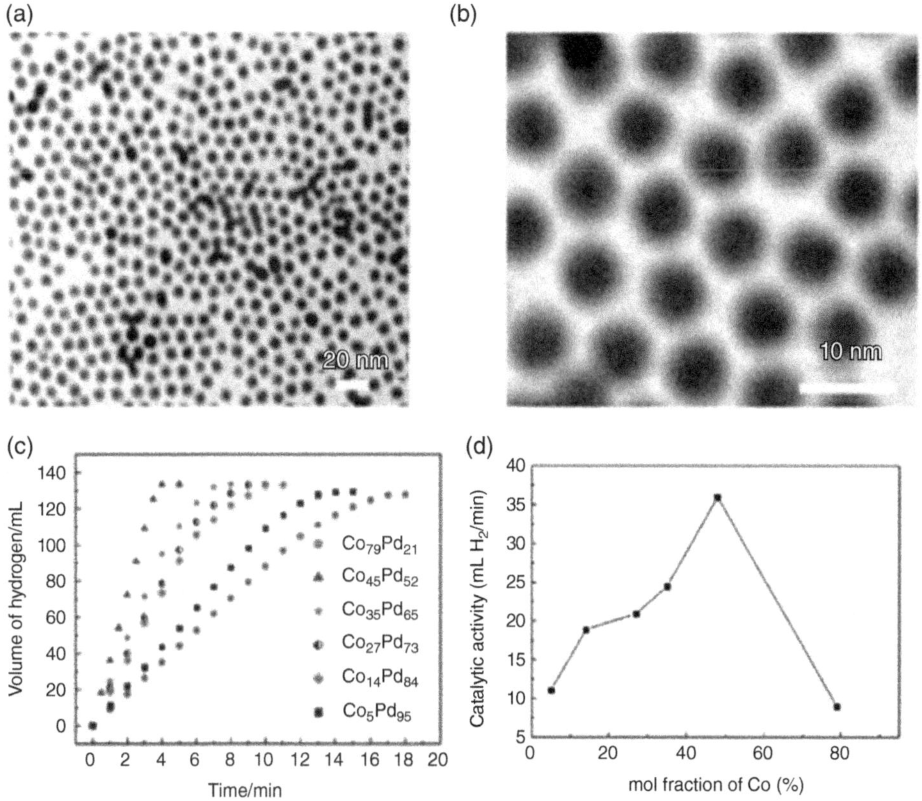

Figure 13.7 (a, b) TEM images of monodisperse $Co_{48}Pd_{52}$ alloy NPs; (c) Time course plots of H_2 generation from AB methanolysis over different CoPd/C catalysts; (d) Plot of catalytic activities versus mole fractions of Co. From [124a] / with permission of American Chemical Society.

13.4 The Dehydrogenation Mechanism and Regeneration of AB

13.4.1 The Dehydrogenation Mechanism of AB

A particular understanding of the mechanism for AB hydrolysis has been established via the thorough investigations of the hydrolysis reaction [136]. Three plausible mechanisms of AB hydrolysis catalyzed by metal NPs catalysts have been proposed. (i) Based on the analogous mechanism of the hydrolysis of $NaBH_4$, the formation of M–H intermediate was considered as the key step of AB hydrolysis [49, 68, 137]. AB was first adsorbed on metal NPs surface and an activated intermediate M–H was formed by the binding of H in B–H bond with the metal. Then, the H was produced by the attack of H_2O molecules to the M–H intermediate. As a result, the formation of the M–H activated complex was considered as the rate-determining step (RDS) (Figure 13.8) [49]. (ii) The attack of the H_2O molecule to the B–N bond of AB was proposed to be an important step [10a, 56, 105]. After the adsorption of the reactant AB on the catalyst surface, a transition state B···O and N···H_2 was

Table 13.5 Catalytic activities of different catalysts for AB methanolysis.

Catalyst	Temperature (K)	n_{metal}/n_{AB}	TOF (min^{-1})	E_a (kJ/mol)	Reference
$RhCl_3$	298	0.02	100	–	[108]
$RuCl_3$	298	0.02	150	–	[108]
$PdCl_2$	298	0.02	1.5	–	[108]
Pd/C	298	0.02	1.9	–	[108]
Rh(0)/nanoHAP	298	0.00203	147	56	[110]
Rh(0)/nanoAl_2O_3	298	0.00245	218	62	[112]
Rh(0)/nanoCeO_2	298	0.00245	144	75	[113]
Zeolite-confined Rh	298	0.005	6.5	40	[114]
Rh/CC3-R-homo	298	0.0199	215.3	–	[115]
Rh/CC3-R-hetero	298	0.0199	65.5	–	[115]
HPRhS	298	0.00486	296.1	–	[116]
Rh-PRO/C	298	0.008	1035	38.2	[117]
Rh/PC-COF	298	0.0125	505	31.8	[118]
Rh(0)/nanoSiO_2	298	0.00245	168	62	[111]
Rh(0)/nanoceria	298	0.00245	144	75	[113]
PVP-stabilized Ru	298	0.0006	67	58	[119]
Ru/MMT	298	0.02	131	23.8	[120]
Ru NPs@PCC-2	298	0.0035	304	–	[121]
Ru/graphene	298	0.003	99.4	54	[132]
PVP-stabilized Pd	298	0.005	22.3	35	[122]
Pd@RCC3	303	0.01	176	–	[123]
Pd/GNS	298	0.01	101	46	[133]
Colloidal Cu NPs	298	0.15	0.1	–	[68]
Cu_2O NPs	298	0.15	0.2	–	[68]
Cu@Cu_2O	298	0.15	0.16	–	[68]
$CoCl_2$	298	0.02	3.7	–	[108]
$NiCl_2$	298	0.02	2.9	–	[108]
PVP-stabilized Ni	298	0.02	12.1	62	[125a]
Ni-Ni_3B	RT	0.2	5.0	–	[125d]
Co-Co_2B	RT	0.2	7.5	–	[125d]
Co-Ni-B	RT	0.2	10.0	–	[125d]
Cu-Cu_2O-CuO/C	298	0.04	24.0	67.9	[127]
b-CuO NA/CF	298	0.018	13.3	34.7	[128]
$Cu_{36}Ni_{64}$/graphene	298	0.072	49.1	24.4	[130]
Mesoporous Cu	298	0.15	2.41	34.2	[107b]
$Cu_{48}Pd_{52}$/C	298	0.072	53.2	–	[134]
$Co_{48}Pd_{52}$/C	298	0.027	27.7	25.5	[104]
AgPd alloy	298	0.05	366.4	37.5	[135]

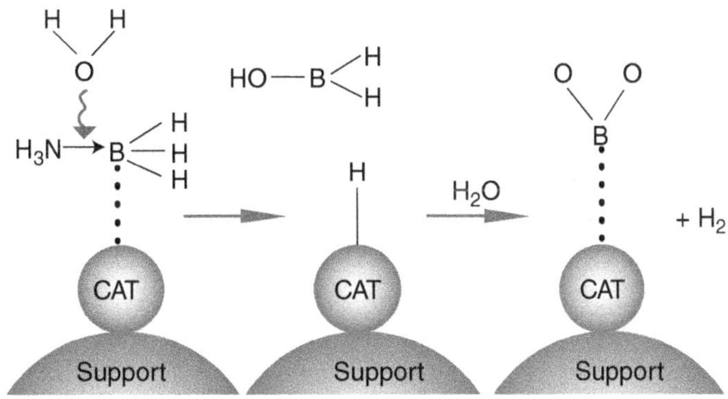

Figure 13.8 The schematic illustration of AB hydrolysis catalyzed by metal NPs catalysts. *Source:* Adapted from [49].

formed by the attack of O and H atoms in H_2O to B and N atoms in AB. Then, the following attack of another H_2O molecule resulted in the concerted decomposition of the H in the B–H bond and H_2O molecule. As a result, H_2 was formed and escaped from the reaction system. (iii) The breaking of the O–H bond of H_2O molecule was regarded as the RDS of AB hydrolysis [136b, 138]. On the basis of the experiments, Astruc's research group proposed that the cleavage of the O–H bond in H_2O was the RDS in the hydrolysis reaction of AB using kinetic isotope effects (KIE) [136b]. An activated BH_3OH intermediate could be formed by the attack of the O–H bond of the H_2O molecule to the BH_3 group, and then H_2 was produced by combining with the H in the H_2O molecule. As mentioned above, most research can only focus on a particular step or intermediate during AB hydrolysis by experimental and theoretical studies, and there is a lack of comprehensive understanding of the entire hydrolysis process. As a result, a complete investigation on the hydrolysis mechanism of AB should be required.

Similar to AB hydrolysis, the mechanism of AB methanolysis has also been studied. Unlike the widespread studies on the mechanism of AB hydrolysis, there are few studies on AB methanolysis [117, 118]. Fan's group proposed a mechanism to explain the formation of H_2 from AB methanolysis catalyzed by Rh-PRO/C catalyst [117]. In the presence of Rh NPs, H_2 molecule was formed by the attack of CH_3OH to the $H_3B(OCH_3)$ unit. Our group also explored the mechanism of AB methanolysis (Figure 13.9) [118]. The formation of the final methanolysis product $(CH_3O)_4B^-$ was explained by us and proposed the entire cycle of the H_2 production reaction for AB methanolysis.

Based on the abovementioned mechanisms of hydrolysis and methanolysis of AB, the whole reaction process of AB solvolysis can be divided into several subprocesses: adsorption of the reactants, activation of B–H bonds, attack of active molecules (H_2O or MeOH), breaking of B–H and O–H bonds, and the formation and desorption of hydrogen. As a result, developing efficient catalysts for AB hydrolysis or methanolysis should target the orderly and efficient improvement of each subprocess, as well as the RDS, which is regarded as the most challenging step of the whole process.

13.4.2 Regeneration of AB

Looking for ways for efficient regeneration of AB is needed to reduce the costs, which is a key to the industrial applications of AB as a chemical hydrogen storage material [139]. As described above, borate species are the main by-products of AB solvolysis. These borate

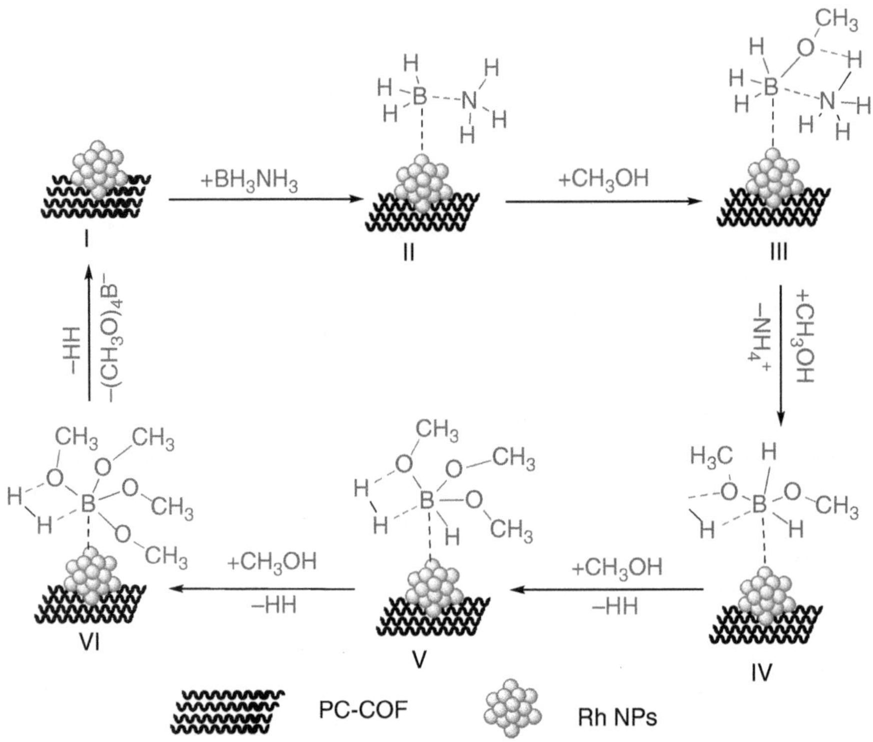

Figure 13.9 A plausible mechanism of hydrogen generation from AB methanolysis over Rh/PC-COF. *Source:* From [118] / with permission of Royal Society of Chemistry.

species are so stable that a tremendous amount of energy should be required for the regeneration of AB from them. Ramachandran et al. reported an efficient regeneration method of AB from the by-product of methanolysis ($NH_4B(CH_3O)_4$) by reacting with $LiAlH_4$ and NH_4Cl under mild conditions (Figure 13.10a) [139e, f]. Nevertheless, the recovery of $LiAlH_4$ remains a huge challenge for practical applications. Later on, Yu et al. demonstrated the regeneration process of the by-product from AB hydrolysis. First, the by-product NH_4BO_2 reacted with H_2O and CH_3OH to produce trimethoxyborane ($B(OMe)_3$) (Figure 13.10b). Then, AB can be regenerated from trimethoxyborane through a similar strategy as described in Figure 13.10a [139c, e]. Accordingly, more energy is required to regenerate AB from hydrolysis by-product than that from methanolysis.

13.5 Conclusions and Future Perspectives

For the past few years, significant progress has been made for hydrogen production via AB solvolysis process. Many monometallic and heterometallic catalysts have been developed, which exhibited a superior catalytic performance toward the hydrolysis or methanolysis of AB to produce hydrogen. Nevertheless, several issues related to the solvolysis of AB should be solved. First of all, the high cost of AB severely limits its industrial application due to the

Figure 13.10 The regeneration cycle of AB from the by-products of (a) methanolysis and (b) hydrolysis. *Source:* From [139f] / with permission of American Chemical Society.

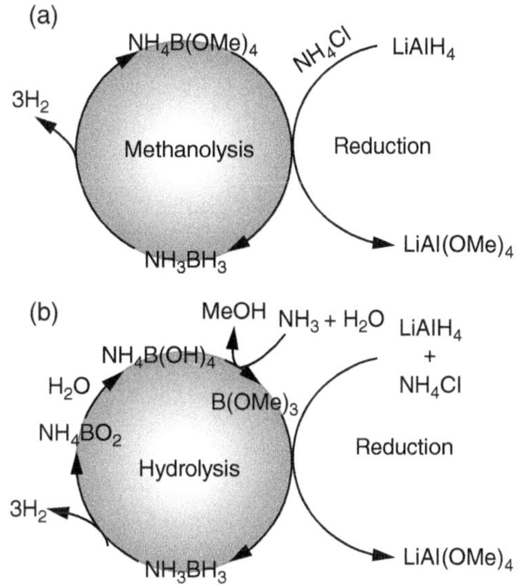

irreversibility of consumption. Regenerating AB from the by-products of solvolysis, especially hydrolysis, is cost-inefficient because of the undesired by-products and multistep transformations. Another one is the catalyst itself, the high cost of noble metal and the easy oxidation of the non-noble metal. Additionally, the undesired by-product NH_3 generated from AB hydrolysis is also one of the shortcomings limiting the applications of AB, due to the noxiousness to fuel cells. In spite of these problems, AB has also been a promising hydrogen storage material, due to its intrinsic superiority, for instance high stability in solvents, high capacity of hydrogen, and the fast hydrogen production rate under ambient conditions.

Acknowledgement

This work was financially supported by the National Natural Science Foundation of China (No. 21763012). We are grateful to the editors for the kind invitation.

References

1 (a) Schlapbach, L. and Züttel, A. (2011). *Materials for Mobile Applications, Materials for Sustainable Energy: A Collection of Peer-Reviewed Research and Review*, 265–270. Nature Publishing Group, World Scientific. (b) Eberle, U., Felderhoff, M., and Schuth, F. (2009). Chemical and physical solutions for hydrogen storage. *Angew. Chem. Int. Ed.* 48: 6608–6630. (c) Shi, J.H., Qiu, F., Yuan, W.B. et al. (2020). Novel electrocatalyst of nanoporous FeP cubes prepared by fast electrodeposition coupling with acid-etching for efficient hydrogen evolution. *Electrochim. Acta* 329: 9. (d) Shi, J., Qiu, F., Yuan, W. et al. (2021). Nitrogen-doped carbon-decorated yolk-shell CoP@FeCoP micro-polyhedra derived from MOF for efficient overall water splitting. *Chem. Eng. J.* 403: 126312.

2 (a) Mandal, T.K. and Gregory, D.H. (2010). ChemInform abstract: hydrogen storage materials: present scenarios and future directions. *Cheminform* 41: 21–54. (b) Armaroli, N. and Balzani, V. (2011). The hydrogen issue. *Chemsuschem* 4: 21–36.

3 (a) Rosi, N.L., Eckert, J., Eddaoudi, M. et al. (2003). Hydrogen storage in microporous metal-organic frameworks. *Science* 300: 1127–1129. (b) Chen, P., Xiong, Z.T., Luo, J.Z. et al. (2002). Interaction of hydrogen with metal nitrides and imides. *Nature* 420: 302–304.

4 Das, L.M. (1996). On-board hydrogen storage systems for automotive application. *Int. J. Hydrogen Energ.* 21: 789–800.

5 Hamilton, C.W., Tombaker, R., Staubitz, A., and Manners, I. (2009). B–N compounds for chemical hydrogen storage. *Chem. Soc. Rev.* 38: 279–293.

6 Züttel, A. (2003). Materials for hydrogen storage. *Mater. Today* 6: 24–33.

7 (a) Jiang, H.L., Singh, S.K., Yan, J.M. et al. (2010). Liquid-phase chemical hydrogen storage: catalytic hydrogen generation under ambient conditions. *ChemSusChem* 3: 541–549. (b) Xu, Q. and Chandra, M. (2007). A portable hydrogen generation system: catalytic hydrolysis of ammonia–borane. *J. Alloy. Compd.* 446: 729–732.

8 (a) Wendlandt, V.S.A.G.W. (1987). The thermal dissociation of NH_3BH_3. *Thermochim. Acta* 113: 379–382. (b) Gutowska, A., Li, L., Shin, Y. et al. (2005). Nanoscaffold mediates hydrogen release and the reactivity of ammonia borane. *Angew. Chem. Int. Ed.* 44: 3578–3582.

9 Zhongyue, L., Guangshan, Z., Gaoqing, L. et al. (2010). Ammonia borane confined by a metal–organic framework for chemical hydrogen storage: enhancing kinetics and eliminating ammonia. *J. Am. Chem. Soc.* 132: 1490–1491.

10 (a) Peng, C.Y., Kang, L., Cao, S. et al. (2015). Nanostructured Ni_2P as a robust catalyst for the hydrolytic dehydrogenation of Ammonia–Borane. *Angew. Chem. Int. Ed.* 54: 15725–15729. (b) Yao, Q., Shi, W., Feng, G. et al. (2014). Ultrafine Ru nanoparticles embedded in SiO_2 nanospheres: highly efficient catalysts for hydrolytic dehydrogenation of ammonia borane. *J. Power Sources* 257: 293–299. (c) Yao, Q., Lu, Z.H., Yang, Y. et al. (2018). Facile synthesis of graphene-supported Ni-CeO_x nano-composites as highly efficient catalysts for hydrolytic dehydrogenation of ammonia borane. *Nano. Res.* 11: 4412–4422. (d) Jiang, H.L., Akita, T., Ishida, T. et al. (2011). Synergistic catalysis of Au@Ag core-shell nanoparticles stabilized on metal-organic framework. *J. Am. Chem. Soc.* 133: 1304–1306. (e) Liu, J., Chen, L., Cui, H. et al. (2014). Applications of metal-organic frameworks in heterogeneous supramolecular catalysis. *Chem. Soc. Rev.* (f) Sun, J.K., Kochovski, Z., Zhang, W.Y. et al. (2017). A general synthetic route towards highly dispersed metal clusters enabled by poly(ionic liquid)s. *J. Am. Chem. Soc.* 139: 8971.

11 Song, H., Cheng, Y., Li, B. et al. (2020). Carbon dots and RuP_2 nanohybrid as an efficient bifunctional catalyst for electrochemical hydrogen evolution reaction and hydrolysis of ammonia borane. *ACS Sustain. Chem. Eng.* 8: 3995–4002.

12 Yan, J.M., Zhang, X.B., Han, S. et al. (2008). Iron-nanoparticle-catalyzed hydrolytic dehydrogenation of ammonia borane for chemical hydrogen storage. *Angew. Chem. Int. Ed.* 120: 2319–2321.

13 Fernandes, R., Patel, N., Miotello, A. et al. (2012). Dehydrogenation of Ammonia Borane with transition metal-doped Co–B alloy catalysts. *Int. J. Hydrogen Energ.* 37: 2397.

14 Karatas, Y., Bulut, A., Yurderi, M. et al. (2016). PdAu-MnO_x nanoparticles supported on amine-functionalized SiO_2 for the room temperature dehydrogenation of formic acid in the absence of additives. *Appl. Catal. B-Environ.* 180: 586–595.

15 Chandra, M. and Xu, Q. (2006). A high-performance hydrogen generation system: Transition metal-catalyzed dissociation and hydrolysis of ammonia–borane. *J. Power Sources* 156: 190–194.

16 (a) Bulut, A., Yurderi, M., Ertas, İ.E. et al. (2016). Carbon dispersed copper-cobalt alloy nanoparticles: A cost-effective heterogeneous catalyst with exceptional performance in the hydrolytic dehydrogenation of ammonia-borane. *Appl. Catal. B-Environ.* 180: 121–129. (b) Chandra, M. and Xu, Q. (2007). Room temperature hydrogen generation from aqueous ammonia-borane using noble metal nano-clusters as highly active catalysts. *J. Power Sources* 168: 135–142. (c) Yan, J.M., Zhang, X.B., Han, S. et al. (2009). Synthesis of longtime water/air-stable Ni nanoparticles and their high catalytic activity for hydrolysis of ammonia-borane for hydrogen generation. *Inorg. Chem.* 48: 7389–7393. (d) Umegaki, T., Yan, J.M., Zhang, X.B. et al. (2009). Boron- and nitrogen-based chemical hydrogen storage materials. *Int. J. Hydrogen Energ.* 34: 2303–2311. (e) Lu, Z.H., Jiang, H.L., Yadav, M. et al. (2012). Synergistic catalysis of Au-Co@SiO_2 nanospheres in hydrolytic dehydrogenation of ammonia borane for chemical hydrogen storage. *J. Mater. Chem.* 22: 5065. (f) Yao, Q., Lu, Z.H., Yang, K. et al. (2015). Ruthenium nanoparticles confined in SBA-15 as highly efficient catalyst for hydrolytic dehydrogenation of ammonia borane and hydrazine borane. *Sci. Rep.* 5: 15186.

17 Karahan, S., Zahmakiran, M., and Ozkar, S. (2012). A facile one-step synthesis of polymer supported rhodium nanoparticles in organic medium and their catalytic performance in the dehydrogenation of ammonia-borane. *Chem. Commun.* 48: 1180–1182.

18 Aijaz, A., Karkamkar, A., Choi, Y.J. et al. (2012). Immobilizing highly catalytically active Pt nanoparticles inside the pores of metal-organic framework: a double solvents approach. *J. Am. Chem. Soc.* 134: 13926–13929.

19 Chen, Y.-Z., Xu, Q., Yu, S.-H., and Jiang, H.-L. (2015). Tiny Pd@Co core–shell nanoparticles confined inside a metal–organic framework for highly efficient catalysis. *Small* 11: 71–76.

20 Yao, Q., Lu, Z.-H., Jia, Y. et al. (2015). In situ facile synthesis of Rh nanoparticles supported on carbon nanotubes as highly active catalysts for H_2 generation from NH_3BH_3 hydrolysis. *Int. J. Hydrogen Energ.* 40: 2207–2215.

21 Sun, Q., Wang, N., Zhang, T. et al. (2019). Zeolite-encaged single-atom rhodium catalysts: highly-efficient hydrogen generation and shape-selective tandem hydrogenation of nitroarenes. *Angew. Chem. Int. Ed.* 58: 18570–18576.

22 Chen, W., Ji, J., Feng, X. et al. (2014). Mechanistic insight into size-dependent activity and durability in Pt/CNT catalyzed hydrolytic dehydrogenation of ammonia borane. *J. Am. Chem. Soc.* 136: 16736–16739.

23 Xiao, W., Liu, D., Song, S., and Zhang, H. (2012). Synthesis of highly active Pt-CeO2 hybrids with tunable secondary nanostructures for the catalytic hydrolysis of ammonia borane. *Chem. Commun.* 48: 10207.

24 Yao, Q., Shi, Y., Zhang, X. et al. (2016). Facile synthesis of platinum–cerium(IV) oxide hybrids arched on reduced graphene oxide catalyst in reverse micelles with high activity and durability for hydrolysis of ammonia borane. *Chem.– Asian J.* 11: 3251–3257.

25 Yan, H., Lin, Y., Wu, H. et al. (2017). Bottom-up precise synthesis of stable platinum dimers on graphene. *Nat. Commun.* 8: 1070.

26 Hu, Y., Wang, Y., Lu, Z.-H. et al. (2015). Core–shell nanospheres Pt@SiO_2 for catalytic hydrogen production. *Appl. Surf. Sci.* 341: 185–189.

27 Zhao, H., Yu, G., Yuan, M. et al. (2018). Ultrafine and highly dispersed platinum nanoparticles confined in a triazinyl-containing porous organic polymer for catalytic applications. *Nanoscale* 10: 21466–21474.
28 Khalily, M., Aref, E., Hamit, A. et al. (2016). Facile synthesis of three-dimensional Pt-TiO_2 nano-networks: a highly active catalyst for the hydrolytic dehydrogenation of ammonia-borane. *Angew. Chem. Int. Ed.* 55: 12257–12261.
29 Xu, D., Wang, W.D., Tian, M., and Dong, Z. (2018). Immobilization of Pt nanoparticles in hollow mesoporous silica nanocapsules: an aggregation- and leaching-resistant catalyst. *J. Colloid Interface Sci.* 516: 407–415.
30 Zahmakıran, M. and Özkar, S. (2009). Zeolite framework stabilized rhodium(0) nanoclusters catalyst for the hydrolysis of ammonia-borane in air: outstanding catalytic activity, reusability and lifetime. *Appl. Catal. B-Environ.* 89: 104–110.
31 Akbayrak, S., Tonbul, Y., and Özkar, S. (2016). Ceria supported rhodium nanoparticles: superb catalytic activity in hydrogen generation from the hydrolysis of ammonia borane. *Appl. Catal. B-Environ.* 198: 162–170.
32 Akbayrak, S., Özçifçi, Z., and Tabak, A. (2019). Noble metal nanoparticles supported on activated carbon: highly recyclable catalysts in hydrogen generation from the hydrolysis of ammonia borane. *J. Colloid Interface Sci.* 546: 324–332.
33 Shen, J., Yang, L., Hu, K. et al. (2015). Noble metal nanoparticles supported on activated carbon: Highly recyclable catalysts in hydrogen generation from the hydrolysis of ammonia borane. *Int. J. Hydrogen Energ.* 40: 1062–1070.
34 Hu, M., Ming, M., Xu, C. et al. (2018). Towards high-efficiency hydrogen production through in situ formation of well-dispersed rhodium nanoclusters. *ChemSusChem* 11: 3253–3258.
35 Zhang, H., Huang, M., Wen, J. et al. (2019). Sub-3 nm Rh nanoclusters confined within a metal–organic framework for enhanced hydrogen generation. *Chem. Commun.* 55: 4699–4702.
36 Liang, H., Chen, G., Desinan, S. et al. (2012). In situ facile synthesis of ruthenium nanocluster catalyst supported on carbon black for hydrogen generation from the hydrolysis of ammonia-borane. *Int. J. Hydrogen Energ.* 37: 17921–17927.
37 Xu, C., Ming, M., Wang, Q. et al. (2018). Facile synthesis of effective Ru nanoparticles on carbon by adsorption-low temperature pyrolysis strategy for hydrogen evolution. *J. Mater. Chem. A* 6: 14380–14386.
38 Akbayrak, S. and Özkar, S. (2012). Ruthenium(0) nanoparticles supported on multiwalled carbon nanotube as highly active catalyst for hydrogen generation from ammonia–borane. *ACS Appl. Mater. Interfaces* 4: 6302–6310.
39 Abo-Hamed, E.K., Pennycook, T., Vaynzof, Y. et al. (2014). Highly active metastable ruthenium nanoparticles for hydrogen production through the catalytic hydrolysis of ammonia borane. *Small* 10: 3145–3152.
40 Fan, Y., Li, X., He, X. et al. (2014). Effective hydrolysis of ammonia borane catalyzed by ruthenium nanoparticles immobilized on graphic carbon nitride. *Int. J. Hydrogen Energ.* 39: 19982–19989.
41 Du, C., Ao, Q., Cao, N. et al. (2015). Facile synthesis of monodisperse ruthenium nanoparticles supported on graphene for hydrogen generation from hydrolysis of ammonia borane. *Int. J. Hydrogen Energ.* 40: 6180–6187.

42 Wang, W., Lu, Z.-H., Luo, Y. et al. (2018). Mesoporous carbon nitride supported Pd and Pd-Ni nanoparticles as highly efficient catalyst for catalytic hydrolysis of NH_3BH_3. *ChemCatChem* 10: 1620–1626.

43 Xi, P., Chen, F., Xie, G. et al. (2012). Surfactant free RGO/Pd nanocomposites as highly active heterogeneous catalysts for the hydrolytic dehydrogenation of ammonia borane for chemical hydrogen storage. *Nanoscale* 4: 5597–5601.

44 Fuku, K., Hayashi, R., Takakura, S. et al. (2013). The synthesis of size- and color-controlled silver nanoparticles by using microwave heating and their enhanced catalytic activity by localized surface plasmon resonance. *Angew. Chem. Int. Ed.* 125: 7594–7598.

45 Akbayrak, S., Kaya, M., Volkan, M., and Özkar, S. (2014). *Appl. Catal. B-Environ.* 147: 387–393.

46 Chen, J., Lu, Z.-H., Wang, Y. et al. (2015). Magnetically recyclable Ag/SiO_2–$CoFe_2O_4$ nanocomposite as a highly active and reusable catalyst for H_2 production. *Int. J. Hydrogen Energ.* 40: 4777–4785.

47 Metin, O., Mazumder, V., Özkar, S., and Sun, S. (2010). Monodisperse nickel nanoparticles and their catalysis in hydrolytic dehydrogenation of ammonia borane. *J. Am. Chem. Soc.* 132: 1468–1469.

48 Guo, K., Li, H., and Yu, Z. (2018). Size-dependent catalytic activity of monodispersed nickel nanoparticles for the hydrolytic dehydrogenation of ammonia borane. *ACS Appl. Mater. Interfaces* 10: 517–525.

49 Qiang, X. and Chandra, M. (2006). Catalytic activities of non-noble metals for hydrogen generation from aqueous ammonia–borane at room temperature. *J. Power Sources* 163: 364–370.

50 Yan, J.-M., Zhang, X.-B., Han, S. et al. (2008). Iron-nanoparticle-catalyzed hydrolytic dehydrogenation of ammonia borane for chemical hydrogen storage. *Angew. Chem. Int. Ed.* 47: 2287–2289.

51 Aranishi, K., Zhu, Q.-L., and Xu, Q. (2014). Dendrimer-encapsulated cobalt nanoparticles as high-performance catalysts for the hydrolysis of ammonia borane. *ChemCatChem* 6: 1375–1379.

52 Liu, P., Gu, X., Kang, K. et al. (2017). Highly efficient catalytic hydrogen evolution from ammonia borane using the synergistic effect of crystallinity and size of noble-metal-free nanoparticles supported by porous metal–organic frameworks. *ACS Appl. Mater. Interfaces* 9: 10759–10767.

53 Zhou, L., Meng, J., Li, P. et al. (2017). Ultrasmall cobalt nanoparticles supported on nitrogen-doped porous carbon nanowires for hydrogen evolution from ammonia borane. *Mater. Horiz.* 4: 268–273.

54 Yao, Q., Lu, Z.H., Zhang, Z. et al. (2014). One-pot synthesis of core-shell $Cu@SiO_2$ nanospheres and their catalysis for hydrolytic dehydrogenation of ammonia borane and hydrazine borane. *Sci. Rep.* 4: 7597.

55 Yan, J.-M., Zhang, X.-B., Shioyama, H., and Xu, Q. (2010). Room temperature hydrolytic dehydrogenation of ammonia borane catalyzed by Co nanoparticles. *J. Power Sources* 195: 1091–1094.

56 Hu, J., Chen, Z., Li, M. et al. (2014). Amine-capped Co nanoparticles for highly efficient dehydrogenation of ammonia borane. *ACS Appl. Mater. Interfaces* 6: 13191–13200.

57 Wang, H., Zhao, Y., Cheng, F. et al. (2016). Cobalt nanoparticles embedded in porous N-doped carbon as long-life catalysts for hydrolysis of ammonia borane. *Catal. Sci. Technol.* 6: 3443–3448.

58 Zhang, X.-L., Zhang, D.-X., Chang, G.-G. et al. (2019). Bimetallic (Zn/Co) MOFs-derived highly dispersed metallic Co/HPC for completely hydrolytic dehydrogenation of ammonia–borane. *Ind. Eng. Chem. Res.* 58: 7209–7216.

59 Fang, Y., Xiao, Z., Li, J. et al. (2018). Formation of a highly reactive cobalt nanocluster crystal within a highly negatively charged porous coordination cage. *Angew. Chem. Int. Ed.* 57: 5283–5287.

60 Chen, M., Xiong, R., Cui, X. et al. (2019). SiO_2-encompassed Co@N-doped porous carbon assemblies as recyclable catalysts for efficient hydrolysis of ammonia borane. *Langmuir* 35: 671–677.

61 Men, Y., Su, J., Huang, C. et al. (2018). Three-dimensional nitrogen-doped graphene hydrogel supported Co-CeO_x nanoclusters as efficient catalysts for hydrogen generation from hydrolysis of ammonia borane. *Chinese Chem. Lett.* 29: 1671–1674.

62 Metin, O., Özkar, S., and Sun, S. (2010). Monodisperse nickel nanoparticles supported on SiO_2 as an effective catalyst for the hydrolysis of ammonia-borane. *Nano Res.* 3: 676–684.

63 Li, P.Z., Aranishi, K., and Xu, Q. (2012). ZIF-8 immobilized nickel nanoparticles: highly effective catalysts for hydrogen generation from hydrolysis of ammonia borane. *Chem. Commun.* 48: 3173–3175.

64 Pei-Zhou, L., Arshad, A., and Qiang, X. (2012). Highly dispersed surfactant-free nickel nanoparticles and their remarkable catalytic activity in the hydrolysis of ammonia borane for hydrogen generation. *Angew. Chem. Int. Ed.* 124: 6753–6756.

65 Mahyari, M. and Shaabani, A. (2014). Nickel nanoparticles immobilized on three-dimensional nitrogen-doped graphene as a superb catalyst for the generation of hydrogen from the hydrolysis of ammonia borane. *J. Mater. Chem. A* 2: 16652–16659.

66 Zhang, J., Chen, C., Yan, W. et al. (2016). Ni nanoparticles supported on CNTs with excellent activity produced by atomic layer deposition for hydrogen generation from the hydrolysis of ammonia borane. *Catal. Sci. Technol.* 6: 2112–2119.

67 Manna, J., Akbayrak, S., and Özkar, S. (2017). Nickel(0) nanoparticles supported on bare or coated cobalt ferrite as highly active, magnetically isolable and reusable catalyst for hydrolytic dehydrogenation of ammonia borane. *J. Colloid Interface Sci.* 508: 359–368.

68 Kalidindi, S.B., Sanyal, U., and Jagirdar, B.R. (2008). Nanostructured Cu and Cu@Cu_2O core shell catalysts for hydrogen generation from ammonia–borane. *Phys. Chem. Chem. Phys.* 10: 5870–5874.

69 Zahmakıran, M., Durap, F., and Özkar, S. (2010). Zeolite confined copper(0) nanoclusters as cost-effective and reusable catalyst in hydrogen generation from the hydrolysis of ammonia-borane. *Int. J. Hydrogen Energ.* 35: 187–197.

70 Kaya, M., Zahmakiran, M., Özkar, S., and Volkan, M. (2012). Copper(0) nanoparticles supported on silica-coated cobalt ferrite magnetic particles: cost effective catalyst in the hydrolysis of ammonia-borane with an exceptional reusability performance. *ACS Appl. Mater. Interfaces* 4: 3866–3873.

71 Yang, Y., Lu, Z.-H., Hu, Y. et al. (2014). Facile in situ synthesis of copper nanoparticles supported on reduced graphene oxide for hydrolytic dehydrogenation of ammonia borane. *RSC Adv.* 4: 13749–13752.

72 Ozay, O., Inger, E., Aktas, N., and Sahiner, N. (2011). Hydrogen production from ammonia borane via hydrogel template synthesized Cu, Ni, Co composites. *Int. J. Hydrogen Energ.* 36: 8209–8216.

73 Singh, A.K. and Xu, Q. (2013). Synergistic catalysis over bimetallic alloy nanoparticles. *ChemCatChem* 5: 652–676.

74 (a) Tong, Y., Lu, X., Sun, W. et al. (2014). Electrospun polyacrylonitrile nanofibers supported Ag/Pd nanoparticles for hydrogen generation from the hydrolysis of ammonia borane. *J. Power Sources* 261: 221–226. (b) Jiao, W., Hu, X., Ren, H. et al. (2014). Magnetic Ni and Ni/Pt hollow nanospheres and their catalytic activities for hydrolysis of ammonia borane. *J. Mater. Chem. A* 2: 18171–18176. (c) Ciftci, N.S. and Metin, O. (2014). Monodisperse nickel–palladium alloy nanoparticles supported on reduced graphene oxide as highly efficient catalysts for the hydrolytic dehydrogenation of ammonia borane. *Int. J. Hydrogen Energ.* 39: 18863–18870. (d) Rakap, M. (2015). Poly(N-vinyl-2-pyrrolidone)-stabilized palladium–platinum nanoparticles-catalyzed hydrolysis of ammonia borane for hydrogen generation. *J. Power Sources* 276: 320–327. (e) Wang, H., Zhou, L., Han, M. et al. (2015). CuCo nanoparticles supported on hierarchically porous carbon as catalysts for hydrolysis of ammonia borane. *J. Alloy. Comp.* 651: 382–388. (f) Park, J.W., Lai, S.W., and Cho, S.O. (2015). Catalytic hydrogen generation from hydrolysis of ammonia borane using octahedral Au@Pt nanoparticles. *Int. J. Hydrogen Energ.* 40: 16316–16322. (g) Zhang, L., Zhou, L., Yang, K. et al. (2016). PdNi nanoparticles supported on MIL-101 as high-performance catalysts for hydrogen generation from ammonia borane. *J. Alloy. Comp.* 677: 87–95.

75 Amali, A.J., Aranishi, K., Uchida, T., and Xu, Q. (2013). PdPt nanocubes: a high-performance catalyst for hydrolytic dehydrogenation of ammonia borane. *Part. Part. Syst. Char.* 30: 888–892.

76 Zhou, Q. and Xu, C. (2016). Nanoporous PtRu alloys with unique catalytic activity toward hydrolytic dehydrogenation of ammonia borane. *Chem.–Asian J.* 11: 705–712.

77 (a) Yang, X., Cheng, F., Tao, Z., and Chen, J. (2011). Hydrolytic dehydrogenation of ammonia borane catalyzed by carbon supported Co core–Pt shell nanoparticles. *J. Power Sources* 196: 2785–2789. (b) Wang, Q., Fu, F., Yang, S. et al. (2019). Dramatic synergy in CoPt nanocatalysts stabilized by "click" dendrimers for evolution of hydrogen from hydrolysis of ammonia borane. *ACS Catal.* 9: 1110–1119. (c) Zhang, J., Chen, W., Ge, H. et al. (2018). Synergistic effects in atomic-layer-deposited $PtCo_x$/CNTs catalysts enhancing hydrolytic dehydrogenation of ammonia borane. *Appl. Catal. B-Environ.* 235: 256–263.

78 (a) Cheng, F., Ma, H., Li, Y., and Chen, J. (2007). $Ni_{1-x}Pt_x$ (x = 0-0.12) hollow spheres as catalysts for hydrogen generation from ammonia borane. *Inorg. Chem.* 46: 788–794. (b) Yang, X., Cheng, F., Jing, L. et al. (2009). PtxNi1-x nanoparticles as catalysts for hydrogen generation from hydrolysis of ammonia borane. *Int. J. Hydrogen Energ.* 34: 8785–8791. (c) Yi, R., Shi, R., Gao, G. et al. (2009). Microspheres: fabrication and characterization. *J. Phys. Chem. C* 113: 1222–1226. (d) Du, J., Cheng, F., Si, M. et al. (2013). Nanoporous Ni-based catalysts for hydrogen generation from hydrolysis of ammonia borane. *Int. J. Hydrogen Energ.* 38: 5768–5774. (e) Yang, X., Cheng, F., Liang, J. et al. (2011). Carbon-supported $Ni_{1-x}@Pt_x$ (x = 0.32, 0.43, 0.60, 0.67, and 0.80) core–shell nanoparticles as catalysts for hydrogen generation from hydrolysis of ammonia borane. *Int. J. Hydrogen Energ.* 36: 1984–1990. (f) Ge, Y., Ye, W., Shah, Z.H. et al. (2017). PtNi/NiO clusters coated by hollow sillica: novel design for highly efficient hydrogen production from ammonia-borane. *ACS Appl. Mater. Interfaces* 9: 3749–3756. (g) Fu, F., Wang, C., Wang, Q. et al. (2018). Highly selective and sharp volcano-type synergistic Ni_2Pt@ZIF-8-catalyzed hydrogen evolution from ammonia borane hydrolysis. *J. Am. Chem. Soc.* 140: 10034–10042.

79 (a) Chen, G., Desinan, S., Rosei, R. et al. (2012). Hollow ruthenium nanoparticles with small dimensions derived from Ni@Ru core@shell structure: synthesis and enhanced catalytic dehydrogenation of ammonia borane. *Chem. Commun.* 48: 8009–8011. (b) Chen, G., Desinan, S., Nechache, R. et al. (2011). Bifunctional catalytic/magnetic Ni@Ru core-shell nanoparticles. *Chem. Commun.* 47: 6308–6310. (c) Li, X., Zeng, C., and Fan, G. (2015). Ultrafast hydrogen generation from the hydrolysis of ammonia borane catalyzed by highly efficient bimetallic RuNi nanoparticles stabilized on $Ti_3C_2X_2$ (X = OH and/or F). *Int. J. Hydrogen Energ.* 40: 3883–3891. (d) Mori, K., Miyawaki, K., and Yamashita, H. (2016). Ru and Ru–Ni nanoparticles on TiO_2 support as extremely active catalysts for hydrogen production from ammonia–borane. *ACS Catal.* 6: 3128–3135. (e) Roy, S., Pachfule, P., and Xu, Q. (2016). Inside cover: high catalytic performance of MIL-101-immobilized NiRu alloy nanoparticles towards the hydrolytic dehydrogenation of ammonia borane. *Ber. Dtsch. Chem. Ges.* 2016: 4530.

80 (a) Rachiero, G.P., Demirci, U.B., and Miele, P. (2011). Bimetallic RuCo and RuCu catalysts supported on γ-Al_2O_3. A comparative study of their activity in hydrolysis of ammonia-borane. *Int. J. Hydrogen Energ.* 36: 7051–7065. (b) Lu, D., Yu, G., Li, Y. et al. (2017). RuCo NPs supported on MIL-96(Al) as highly active catalysts for the hydrolysis of ammonia borane. *J. Alloy. Comp.* 694: 662–671. (c) Chen, M., Zhou, L., Di, L. et al. (2017). RuCo bimetallic alloy nanoparticles immobilized on multi-porous MIL-53(Al) as a highly efficient catalyst for the hydrolytic reaction of ammonia borane. *Int. J. Hydrogen Energ.* 43: 1439–1450.

81 Pachfule, P., Yang, X., Zhu, Q.L. et al. (2017). From Ru nanoparticle-encapsulated metal–organic frameworks to highly catalytically active Cu/Ru nanoparticle-embedded porous carbon. *J. Mater. Chem. A* 5: 4835–4841.

82 (a) Jiang, H.L., Umegaki, T., Akita, T. et al. (2010). Bimetallic Au-Ni nanoparticles embedded in SiO_2 nanospheres: synergetic catalysis in hydrolytic dehydrogenation of ammonia borane. *Chem.-Eur. J.* 16: 3132–3137. (b) Zhu, Q.L., Li, J., and Xu, Q. (2013). Immobilizing metal nanoparticles to metal-organic frameworks with size and location control for optimizing catalytic performance. *J. Am. Chem. Soc.* 135: 10210–10213. (c) Guo, L., Gu, X., Kang, K. et al. (2015). Porous nitrogen-doped carbon-immobilized bimetallic nanoparticles as highly efficient catalysts for hydrogen generation from hydrolysis of ammonia borane. *J. Mater. Chem. A* 3: 22807–22815. (d) Kang, K., Gu, X., Guo, L. et al. (2015). Efficient catalytic hydrolytic dehydrogenation of ammonia borane over surfactant-free bimetallic nanoparticles immobilized on amine-functionalized carbon nanotubes. *Int. J. Hydrogen Energ.* 40: 12315–12324.

83 (a) Li, J., Zhu, Q.L., and Xu, Q. (2014). Highly active AuCo alloy nanoparticles encapsulated in the pores of metal–organic frameworks for hydrolytic dehydrogenation of ammonia borane. *Chem. Commun.* 50: 5899–5901. (b) Yan, J.-M., Zhang, X.-B., Akita, T. et al. (2010). One-step seeding growth of magnetically recyclable Au@Co core–shell nanoparticles: highly efficient catalyst for hydrolytic dehydrogenation of ammonia borane. *J. Am. Chem. Soc.* 132: 5326–5327.

84 Wen, M., Sun, B., Zhou, B. et al. (2012). Controllable assembly of Ag/C/Ni magnetic nanocables and its low activation energy dehydrogenation catalysis. *J. Mater. Chem.* 22: 11988–11993.

85 Ke, D., Li, Y., Wang, J. et al. (2016). Facile fabrication of poly(amidoamine) (PAMAM) dendrimers-encapsulated Ag–Co bimetallic nanoparticles for highly efficient dehydrogenation of ammonia borane. *Int. J. Hydrogen Energ.* 41: 2564–2574.

86 Shang, N., Zhou, X., Feng, C. et al. (2017). Synergetic catalysis of NiPd nanoparticles supported on biomass-derived carbon spheres for hydrogen production from ammonia borane at room temperature. *Int. J. Hydrogen Energ.* 42: 5733–5740.

87 Wang, J., Qin, Y.-L., Liu, X., and Zhang, X.-B. (2012). In situ synthesis of magnetically recyclable graphene-supported Pd@Co core–shell nanoparticles as efficient catalysts for hydrolytic dehydrogenation of ammonia borane. *J. Mater. Chem.* 22: 12468–12470.

88 Li, M., Hu, J., Chen, Z., and Lu, H. (2014). A high-performance Pt–Co bimetallic catalyst with polyethyleneimine decorated graphene oxide as support for hydrolysis of ammonia borane. *RSC Adv.* 4: 41152–41158.

89 (a) Qiu, F., Li, L., Liu, G. et al. (2013). Synthesis of $Fe_{0.3}Co_{0.7}$/rGO nanoparticles as a high performance catalyst for the hydrolytic dehydrogenation of ammonia borane. *Int. J. Hydrogen Energ.* 38: 7291–7297. (b) Qiu, F.Y., Wang, Y.J., Wang, Y.P. et al. (2011). Dehydrogenation of ammonia borane catalyzed by in situ synthesized Fe–Co nano-alloy in aqueous solution – ScienceDirect. *Catal. Today* 170: 64–68. (c) Qiu, F., Li, L., Liu, G. et al. (2013). In situ synthesized Fe–Co/C nano-alloys as catalysts for the hydrolysis of ammonia borane. *Int. J. Hydrogen Energ.* 38: 3241–3249.

90 (a) Yan, J.M., Zhang, X.B., Han, S. et al. (2009). Magnetically recyclable Fe–Ni alloy catalyzed dehydrogenation of ammonia borane in aqueous solution under ambient atmosphere. *J. Power Sources* 194: 478–481. (b) Zhou, X., Chen, Z., Yan, D., and Lu, H. (2012). Deposition of Fe–Ni nanoparticles on polyethyleneimine-decorated graphene oxide and application in catalytic dehydrogenation of ammonia borane. *J. Mater. Chem.* 22: 13506. (c) Lai, S.W., Lin, H.L., Lin, Y.P., and Yu, T.L. (2013). Hydrolysis of ammonia–borane catalyzed by an iron–nickel alloy on an SBA-15 support. *Int. J. Hydrogen Energ.* 38: 4636–4647. (d) Cui, C., Liu, Y., Mehdi, S. et al. (2020). Enhancing effect of Fe-doping on the activity of nano Ni catalyst towards hydrogen evolution from NH_3BH_3. *Appl. Catal. B-Environ.* 265: 118612.

91 Lu, Z.H., Li, J., Zhu, A. et al. (2013). Catalytic hydrolysis of ammonia borane via magnetically recyclable copper iron nanoparticles for chemical hydrogen storage. *Int. J. Hydrogen Energ.* 38: 5330–5337.

92 (a) Yan, J.M., Wang, Z.L., Wang, H.L., and Jiang, Q. (2012). Rapid and energy-efficient synthesis of a graphene–CuCo hybrid as a high performance catalyst. *J. Mater. Chem.* 22: 10990–10993. (b) Song, F.-Z., Zhu, Q.-L., Yang, X.-C., and Xu, Q. (2016). Monodispersed CuCo nanoparticles supported on diamine-functionalized graphene as a non-noble metal catalyst for hydrolytic dehydrogenation of ammonia borane. *ChemNanoMat* 2: 942–945. (c) Liu, Y., Zhang, J., Guan, H. et al. (2018). Preparation of bimetallic Cu-Co nanocatalysts on poly (diallyldimethylammonium chloride) functionalized halloysite nanotubes for hydrolytic dehydrogenation of ammonia borane. *Appl. Surf. Sci.* 427: 106–113.

93 (a) Zhou, S., Wen, M., Wang, N. et al. (2012). Highly active NiCo alloy hexagonal nanoplates with crystal plane selective dehydrogenation and visible-light photocatalysis.

J. Mater. Chem. 22: 16858–16864. (b) Wang, Q., Zhang, Z., Liu, J. et al. (2017). Bimetallic non-noble CoNi nanoparticles monodispersed on multiwall carbon nanotubes: highly efficient hydrolysis of ammonia borane. *Mater. Chem. Phys.* 204: 58–61.

94 (a) Yen, H., Seo, Y., Kaliaguine, S., and Kleitz, F. (2015). Role of metal–support interactions, particle size, and metal–metal synergy in CuNi nanocatalysts for H_2 generation. *ACS Catal.* 5: 5505–5511. (b) Meng, X., Li, S., Xia, B. et al. (2014). Decoration of graphene with tetrametallic Cu@FeCoNi core–shell nanoparticles for catalytic hydrolysis of amine boranes. *RSC Adv.* 4: 32817–32825. (c) Gao, D., Zhang, Y., Zhou, L., and Yang, K. (2017). CuNi NPs supported on MIL-101 as highly active catalysts for the hydrolysis of ammonia borane. *Appl. Surf. Sci.* 427: 114–122. (d) Li, S.-J., Wang, H.-L., Yan, J.-M., and Jiang, Q. (2017). Oleylamine-stabilized Cu0.9Ni0.1 nanoparticles as efficient catalyst for ammonia borane dehydrogenation. *Int. J. Hydrogen Energ.* 42: 25251–25257. (e) Lu, Z.H., Li, J., Feng, G. et al. (2014). Synergistic catalysis of MCM-41 immobilized Cu–Ni nanoparticles in hydrolytic dehydrogeneration of ammonia borane. *Int. J. Hydrogen Energ.* 39: 13389–13395.

95 Guo, K., Ding, Y., Luo, J. et al. (2019). NiCu bimetallic nanoparticles on silica support for catalytic hydrolysis of ammonia borane: composition-dependent activity and support size effect. *Mater. Interfaces* 2: 5851–5861.

96 Yang, K., Yao, Q., Huang, W. et al. (2016). Enhanced catalytic activity of NiM (M = Cr, Mo, W) nanoparticles for hydrogen evolution from ammonia borane and hydrazine borane. *Int. J. Hydrogen Energ.* 42: 6840–6850.

97 Yao, Q., Lu, Z.H., Huang, W. et al. (2016). Highly Pt-like activity of Ni-Mo/graphene catalyst for hydrogen evolution from hydrolysis of ammonia borane. *J. Mater. Chem. A* 4: 8579–8583.

98 (a) Yao, Q., Lu, Z.-H., Wang, Y. et al. (2015). Synergetic catalysis of non-noble bimetallic Cu–Co nanoparticles embedded in SiO_2 nanospheres in hydrolytic dehydrogenation of ammonia borane. *J. Phys. Chem. C* 119: 14167–14174. (b) Li, J., Zhu, Q.L., and Xu, Q. (2014). Non-noble bimetallic CuCo nanoparticles encapsulated in the pores of metal–organic frameworks: synergetic catalysis in the hydrolysis of ammonia borane for hydrogen generation. *Catal. Sci. Technol.* 5: 525–530.

99 Yao, Q., Yang, K., Hong, X. et al. (2018). Base-promoted hydrolytic dehydrogenation of ammonia borane catalyzed by noble-metal-free nanoparticles. *Catal. Sci. Technol.* 8: 870–877.

100 Rakap, M. (2015). The highest catalytic activity in the hydrolysis of ammonia borane by poly(N-vinyl-2-pyrrolidone)-protected palladium–rhodium nanoparticles for hydrogen generation. *Appl. Catal. B-Environ.* 163: 129–134.

101 Shang, N.-Z., Feng, C., Gao, S.-T., and Wang, C. (2016). Ag/Pd nanoparticles supported on amine-functionalized metal–organic framework for catalytic hydrolysis of ammonia borane. *Int. J. Hydrogen Energ.* 41: 944–950.

102 Qi, X., Li, X., Chen, B. et al. (2016). Highly active nanoreactors: patchlike or thick Ni coating on Pt nanoparticles based on confined catalysis. *ACS Appl. Mater. Interfaces* 8: 1922–1928.

103 Chen, Y.-Z., Liang, L., Yang, Q. et al. (2015). A seed-mediated approach to the general and mild synthesis of non-noble metal nanoparticles stabilized by a metal–organic framework for highly efficient catalysis. *Mater. Horiz.* 2: 606–612.

104 Sun, D., Mazumder, V., Metin, Ö., and Sun, S. (2012). Methanolysis of ammonia borane by CoPd nanoparticles. *ACS Catal.* 2: 1290–1295.

105 Feng, K., Zhong, J., Zhao, B. et al. (2016). $Cu_xCo_{1-x}O$ nanoparticles on graphene oxide as a synergistic catalyst for high-efficiency hydrolysis of ammonia–borane. *Angew. Chem. Int. Ed.* 55: 11950–11954.

106 (a) Lo, C.T.F., Karan, K., and Davis, B.R. (2007). Kinetic studies of reaction between sodium borohydride and methanol, water, and their mixtures. *Ind. Eng. Chem. Res.* 46: 5478–5484. (b) Brockman, A., Yuan, Z., and Gore, J. (2010). A study of catalytic hydrolysis of concentrated ammonia borane solutions. *Int. J. Hydrogen Energ.* 35: 7350–7356. (c) Ramachandran, P.V., Mistry, H., Kulkarni, A.S., and Gagare, P.D. (2014). Ammonia-mediated, large-scale synthesis of ammonia borane. *Dalton Trans.* 43: 16580–16583.

107 (a) Peng, C.Y., Kang, L., Cao, S. et al. (2016). Nanostructured Ni_2P as a robust catalyst for the hydrolytic dehydrogenation of ammonia–borane. *Angew. Chem. Int. Ed.* 127: 15951–15955. (b) Yao, Q., Huang, M., Lu, Z.H. et al. (2015). Methanolysis of ammonia borane by shape-controlled mesoporous copper nanostructures for hydrogen generation. *Dalton T.* 44: 1070–1076. (c) Song, J., Gu, X., Cheng, J. et al. (2018). Remarkably boosting catalytic H_2 evolution from ammonia borane through the visible-light-driven synergistic electron effect of non-plasmonic noble-metal-free nanoparticles and photoactive metal-organic frameworks. *Appl. Catal. B-Environ.* 225: 424–432.(d)Wang, H.E., Hua, C., Liu, C. et al. (2011). Facile synthesis and electrochemical characterization of porous and dense TiO_2 nanospheres for lithium-ion battery applications. *J. Power Sources* 196: 6394–6399.

108 Ramachandran, P.V. and Gagare, P.D. (2007). Preparation of ammonia borane in high yield and purity, methanolysis, and regeneration. *Inorg. Chem.* 46: 7810–7817.

109 Lara, P., Philippot, K., and Suárez, A. (2019). Phosphane-decorated platinum nanoparticles as efficient catalysts for H2 generation from ammonia borane and methanol. *ChemCatChem* 11: 766–771.

110 Özhava, D. and Özkar, S. (2015). Rhodium(0) nanoparticles supported on hydroxyapatite nanospheres and further stabilized by dihydrogen phosphate ion: A highly active catalyst in hydrogen generation from the methanolysis of ammonia borane. *Int. J. Hydrogen Energ.* 40: 10491–10501.

111 Özhava, D. and Özkar, S. (2016). Rhodium(0) nanoparticles supported on nanosilica: highly active and long lived catalyst in hydrogen generation from the methanolysis of ammonia borane. *Appl. Catal. B-Environ.* 181: 716–726.

112 Özhava, D. and Özkar, S. (2017). Nanoalumina-supported rhodium(0) nanoparticles as catalyst in hydrogen generation from the methanolysis of ammonia borane. *Mol. Catal.* 439: 50–59.

113 Özhava, D. and Özkar, S. (2018). Nanoceria supported rhodium (0) nanoparticles as catalyst for hydrogen generation from methanolysis of ammonia borane. *Appl. Catal. B-Environ.* 237: 1012–1020.

114 Çalışkan, S., Zahmakıran, M., and Özkar, S. (2010). Zeolite confined rhodium(0) nanoclusters as highly active, reusable, and long-lived catalyst in the methanolysis of ammonia-borane. *Appl. Catal. B-Environ.* 93: 387–394.

115 Sun, J.-K., Zhan, W.-W., Akita, T., and Xu, Q. (2015). Toward homogenization of heterogeneous metal nanoparticle catalysts with enhanced catalytic performance: soluble porous organic cage as a stabilizer and homogenizer. *J. Am. Chem. Soc.* 137: 7063–7066.

116 Zou, H., Jin, B., Wang, R. et al. (2018). Iodide-mediated templating synthesis of highly porous rhodium nanospheres for enhanced dehydrogenation of ammonia borane. *J. Mater. Chem. A* 6: 24166–24174.

117 Luo, W., Cheng, W., Hu, M. et al. (2019). Ultrahigh catalytic activity of l-proline-functionalized Rh nanoparticles for methanolysis of ammonia borane. *ChemSusChem* 12: 535–541.

118 Li, X., Zhang, C., Luo, M. et al. (2020). Ultrafine Rh nanoparticles confined by nitrogen-rich covalent organic frameworks for methanolysis of ammonia borane. *Inorg. Chem. Front.* 7: 1298–1306.

119 Erdoğan, H., Metin, Ö., and Özkar, S. (2011). Hydrogen generation from the methanolysis of ammonia borane catalyzed by in situ generated, polymer stabilized ruthenium(0) nanoclusters. *Catal. Today* 170: 93–98.

120 Dai, H.-B., Kang, X.-D., and Wang, P. (2010). Ruthenium nanoparticles immobilized in montmorillonite used as catalyst for methanolysis of ammonia borane. *Int. J. Hydrogen Energ.* 35: 10317–10323.

121 Fang, Y., Li, J., Togo, T. et al. (2018). Ultra-small face-centered-cubic Ru nanoparticles confined within a porous coordination cage for dehydrogenation. *Chem* 4: 555–563.

122 Erdoğan, H., Metin, Ö., and Özkar, S. (2009). In situ-generated PVP-stabilized palladium(0) nanocluster catalyst in hydrogen generation from the methanolysis of ammonia–borane. *Phys. Chem. Chem. Phys.* 11: 10519–10525.

123 Yang, X., Sun, J.-K., Kitta, M. et al. (2018). Encapsulating highly catalytically active metal nanoclusters inside porous organic cages. *Nat. Catal.* 1: 214–220.

124 (a) Sun., D., Mazumder, V., Metin, Ö., and Sun, S. (2012). Methanolysis of Ammonia Borane by CoPd Nanoparticles. *ACS Catal.* 2: 1290–1295. (b) Byun, S., Song, Y., and Kim, B.M. (2016). Heterogenized bimetallic Pd–Pt–Fe_3O_4 nanoflakes as extremely robust, magnetically recyclable catalysts for chemoselective nitroarene reduction. *ACS Appl. Mater. Interfaces* 8: 14637–14647.

125 (a) Özhava, D., Kılıçaslan, N.Z., and Özkar, S. (2015). PVP-stabilized nickel(0) nanoparticles as catalyst in hydrogen generation from the methanolysis of hydrazine borane or ammonia borane. *Appl. Catal. B-Environ.* 162: 573–582. (b) Amoo, K.O., Onyeozili, E.N., Kalu, E.E. et al. (2016). Activity of varying compositions of Co–Ni–P catalysts for the methanolysis of ammonia borane. *Int. J. Hydrogen Energ.* 41: 21221–21235. (c) Zhang, J., Duan, Y., Zhu, Y. et al. (2017). Evenly dispersed microspherical amorphous alloy Co_xB_{1-x}: robust and magnetically recyclable catalyst for alcoholyzing ammonia borane to release H_2. *Mater. Chem. Phys.* 201: 297–301. (d) Kalidindi, S.B., Vernekar, A.A., and Jagirdar, B.R. (2010). Co-Co_2B, Ni-Ni_3B and Co-Ni-B nanocomposites catalyzed ammonia-borane methanolysis for hydrogen generation. *Phys. Chem. Chem. Phys.* 40: 770–775.

126 Chen, H., Yu, Z.J., Xu, D.D. et al. (2019). In-situ formed amorphous co nanoparticles for efficiently catalytic hydrogen production from the methanolysis of ammonia borane. *Chin. J. Inorg. Chem.* 35: 141–148.

127 Yurderi, M., Bulut, A., Ertas, İ.E. et al. (2015). Supported copper–copper oxide nanoparticles as active, stable and low-cost catalyst in the methanolysis of ammonia–borane for chemical hydrogen storage. *Appl. Catal. B-Environ.* 165: 169–175.

128 Cui, L., Cao, X., Sun, X. et al. (2018). A bunch-like copper oxide nanowire array as an efficient, durable, and economical catalyst for the methanolysis of ammonia borane. *ChemCatChem* 10: 710–715.

129 Li, P.Y., Xiao, Z.L., Liu, Z.Y. et al. (2015). Highly efficient hydrogen generation from methanolysis of ammonia borane on CuPd alloy nanoparticles. *Nanotechnology* 26: 9.

130 Yu, C., Fu, J., Muzzio, M. et al. (2017). CuNi nanoparticles assembled on graphene for catalytic methanolysis of ammonia borane and hydrogenation of nitro/nitrile compounds. *Chem. Mat.* 29: 1413–1418.

131 Chen, Q.Q., Li, Q., Hou, C.C. et al. (2019). Enhancing electrostatic interactions to activate polar molecules: ammonia borane methanolysis on a $Cu/Co(OH)_2$ nanohybrid. *Catal. Sci. Technol.* 9: 2828–2835.

132 Peng, S., Liu, J., Zhang, J., and Wang, F. (2015). An improved preparation of graphene supported ultrafine ruthenium (0) NPs: very active and durable catalysts for H_2 generation from methanolysis of ammonia borane. *Int. J. Hydrogen Energ.* 40: 10856–10866.

133 Karataş, Y., Gülcan, M., Çelebi, M., and Zahmakiran, M. (2017). Pd(0) nanoparticles decorated on graphene nanosheets (GNS): synthesis, definition and testing of the catalytic performance in the methanolysis of ammonia borane at room conditions. *ChemistrySelect* 2: 9628–9635.

134 Li, P., Xiao, Z., Liu, Z. et al. (2014). Highly efficient hydrogen generation from methanolysis of ammonia borane on CuPd alloy nanoparticles. *Nanotechnology* 26: 025401.

135 Sun, D., Li, P., Yang, B. et al. (2016). Monodisperse AgPd alloy nanoparticles as a highly active catalyst towards the methanolysis of ammonia borane for hydrogen generation. *RSC Adv.* 6: 105940–105947.

136 (a) Lin, Y., Yang, L., Jiang, H. et al. (2019). Boosted reactivity of ammonia borane dehydrogenation over Ni/Ni_2P heterostructure. *J. Phys. Chem. Lett.* 10: 1048–1054. (b) Wang, C., Tuninetti, J., Wang, Z. et al. (2017). Hydrolysis of ammonia-borane over Ni/ZIF-8 nanocatalyst: high efficiency, mechanism, and controlled hydrogen release. *J. Am. Chem. Soc.* 139: 11610–11615.

137 Ma, H. and Na, C. (2015). Isokinetic temperature and size-controlled activation of ruthenium-catalyzed ammonia borane hydrolysis. *ACS Catal.* 5: 1726–1735.

138 (a) Li, Z., He, T., Liu, L. et al. (2017). Covalent triazine framework supported non-noble metal nanoparticles with superior activity for catalytic hydrolysis of ammonia borane: from mechanistic study to catalyst design. *Chem. Sci.* 8: 781–788. (b) Hou, C.-C., Li, Q., Wang, C.-J. et al. (2017). Ternary Ni–Co–P nanoparticles as noble-metal-free catalysts to boost the hydrolytic dehydrogenation of ammonia-borane. *Energy Environ. Sci.* 10: 1770–1776.

139 (a) Peng, B. and Chen, J. (2008). Ammonia borane as an efficient and lightweight hydrogen storage medium. *Energy Environ. Sci.* 1: 479–483. (b) Summerscales, O.T. and Gordon, J.C. (2013). Regeneration of ammonia borane from spent fuel materials. *Dalton T.* 42: 10075–10084. (c) Tan, Y. and Yu, X. (2013). Chemical regeneration of hydrogen storage materials. *RSC Adv.* 3: 23879–23894. (d) Veeraraghavan Ramachandran, P.D.G.P., Sakavuyi, K., and Clark, P. (2010). Reductive amination using ammonia borane. *Tetrahedron Lett.* 51: 3167–3169. (e) Veeraraghavan Ramachandran, P., Raju, B.C., and Gagare, P.D. (2012). One-pot synthesis of ammonia–borane and trialkylamine–boranes from trimethyl borate. *Org. Lett.* 14: 6119–6121. (f) Zhan, W.-W., Zhu, Q.-L., and Xu, Q. (2016). Dehydrogenation of ammonia borane by metal nanoparticle catalysts. *ACS Catal.* 6: 6892–6905. (g) Sutton, A.D., Burrell, A.K., Dixon, D.A. et al. (2011). Regeneration of ammonia borane spent fuel by direct reaction with hydrazine and liquid ammonia. *Science* 331: 1426–1429.

Author Index

Heterogeneous Nanocatalysis for Energy and Environmental Sustainability: Energy Applications, First Edition. Edited by Putla Sudarsanam, Yusuke Yamauchi, and Pankaj Bharali.

u

x

y

z

Subject Index

Heterogeneous Nanocatalysis for Energy and Environmental Sustainability: Energy Applications,
First Edition. Edited by Putla Sudarsanam, Yusuke Yamauchi, and Pankaj Bharali.